Materials Horizons: From Nature to Nanomaterials

Series Editor

Vijay Kumar Thakur, School of Aerospace, Transport and Manufacturing, Cranfield University, Cranfield, UK

Materials are an indispensable part of human civilization since the inception of life on earth. With the passage of time, innumerable new materials have been explored as well as developed and the search for new innovative materials continues briskly. Keeping in mind the immense perspectives of various classes of materials, this series aims at providing a comprehensive collection of works across the breadth of materials research at cutting-edge interface of materials science with physics, chemistry, biology and engineering.

This series covers a galaxy of materials ranging from natural materials to nanomaterials. Some of the topics include but not limited to: biological materials, biomimetic materials, ceramics, composites, coatings, functional materials, glasses, inorganic materials, inorganic-organic hybrids, metals, membranes, magnetic materials, manufacturing of materials, nanomaterials, organic materials and pigments to name a few. The series provides most timely and comprehensive information on advanced synthesis, processing, characterization, manufacturing and applications in a broad range of interdisciplinary fields in science, engineering and technology.

This series accepts both authored and edited works, including textbooks, monographs, reference works, and professional books. The books in this series will provide a deep insight into the state-of-art of Materials Horizons and serve students, academic, government and industrial scientists involved in all aspects of materials research.

Review Process

The proposal for each volume is reviewed by the following:

1. Responsible (in-house) editor
2. One external subject expert
3. One of the editorial board members.

The chapters in each volume are individually reviewed single blind by expert reviewers and the volume editor.

Sibdas Bandyopadhyay
Editor

Application of Low Cost Ceramic Membranes in Wastewater Treatment

Issues on Reusability of Treated Water

Springer

Editor
Sibdas Bandyopadhyay
Formerly at Ceramic Membrane Division
CSIR—Central Glass and Ceramic
Research Institute (CGCRI)
Kolkata, West Bengal, India

ISSN 2524-5384 ISSN 2524-5392 (electronic)
Materials Horizons: From Nature to Nanomaterials
ISBN 978-981-97-5841-8 ISBN 978-981-97-5842-5 (eBook)
https://doi.org/10.1007/978-981-97-5842-5

This Springer imprint is published by the registered company Springer Nature Singapore Pte Ltd.
The registered company address is: 152 Beach Road, #21-01/04 Gateway East, Singapore 189721, Singapore

In memory of my beloved Father,
who imbibed the sense of waste minimization and utilization during my early days of learning, while struggling for living in a rural environment, which I tried to adopt in my academic pursuits as well.

—Sibdas Bandyopadhyay

Foreword

Wastewater treatment holds significant importance in sustainable development through environmental protection particularly by preventing contamination of the natural water resources. Reclamation and reuse of wastewater is also emerging as a promising solution in addressing the water scarcity. Ceramic membranes possess unique advantages in wastewater treatment because of their greater stability under harsh chemical environments, ease of cleaning and longer lifetime. However, higher cost of the commercial ceramic membranes restricts their widespread application. In this backdrop, this book opens up an avenue in using the low cost ceramic membranes thereby bridging the gap in exploiting their full potential in wastewater treatment. The scope of application of low cost ceramic membranes has been presented for treatment of effluents from various types of sources covering textile, tannery, petrochemical, metallurgical, food processing and pharmaceutical industries. The book has also highlighted the emerging issues, challenges and reusability aspects.

The editor is known to me personally and I had the opportunity to visit his lab in connection with a collaborative work on arsenic removal. It was indeed impressive to see the community model for purification of arsenic and iron contaminated groundwater using low cost clay-alumina based ceramic membranes, which was effectively utilized for point of use application in providing quality drinking water to the arsenic affected people of West Bengal, India. The group has also carried out extensive work on development and application of low cost ceramic membranes particularly in wastewater treatment.

As far as I know, such book on the application of low cost ceramic membranes for wastewater treatment is being published for the first time covering the work carried out globally in this direction. The book looks very well structured and elaborated.

I'm convinced that the book is a valuable contribution in the domain of membrane research and hope it will be immensely useful for the academia and industries as well.

Dr. Alberto Figoli
Director, Institute on Membrane Technology
National Research Council of Italy (CNR-ITM)
Rende (CS), Italy

Preface

Sustainability of the aquatic ecosystem is a priority concern particularly in the context of emerging contaminants. The foremost issue that needs urgent attention appears to safeguard the freshwater resources from contamination of wastewater generated as a result of increasing human activities. Meeting the ever-growing water demand is another significant feature for sustainable development which can be addressed through reuse and recycling of treated wastewater in line with the concept of circular economy. Thus, wastewater treatment serves dual purpose, viz. freshwater conservation and renovation for non-potable application. Prioritizing environmental sustainability through wastewater treatment is one of the Sustainable Development Goals (SDG) of the United Nation.

Membrane technology contributes to over 50% of the global water treatment volume in which polymeric membranes play a major role. By virtue of the inherent materials characteristics like chemical and thermal stabilities, ceramic membranes have already demonstrated their potential in challenging separation applications and superior performance compared to the polymer membranes under harsh environmental conditions, thereby showcasing their viability and reliability for wastewater treatment. However, their widespread application in wastewater treatment seems to be not so attractive especially for such low value product like treated wastewater due to the prohibitive cost of commercial ceramic membranes manufactured using synthetic ceramic powders. Therefore, low cost ceramic membranes have been evolved for developing cost effective process for reclamation and recycling of wastewater upholding the waste-to-resource strategy. This book highlights the scope and salient features of wastewater treatment process using low cost ceramic membrane demonstrating their effectiveness in pressure-driven MF and UF processes.

This is the first book on the feasibility of low cost ceramic membranes based process for wastewater treatment containing the comprehensive details about source specific complexities and separation performances which was covered under three sections. In the introduction part, the challenges in wastewater treatment process have been highlighted emphasizing the progressive developments including membrane bioreactors and diverse approaches for addressing the inevitable membrane fouling aspects along with the scope of further investigation. The first section deals with the global scenario on membrane-based wastewater management followed with the multitudes of industry-specific case studies on wastewater treatment covering major industrial sectors such as textile, tannery, petrochemicals and electroplating industry. The issues related to higher loading of organics in the wastewater from food processing and pharmaceutical industries have been discussed in a separate section. Finally, the emerging issues like micro-pollutants, toxicity and reusability aspects, recovery of byproducts from highly alkaline waste streams and prospect of bio-sorption assisted hybrid membrane processes have been addressed in the third section.

Hope the book will serve as a valuable reference for researchers and professionals concerned with cost efficient wastewater treatment process, water pollution and sustainability aspects. It will also be useful for industries and policy makers in addressing the water scarcity issues through reclamation and reuse of wastewater resources.

Kolkata, India Sibdas Bandyopadhyay

Contents

About the Editor

Dr. Sibdas Bandyopadhyay obtained his Ph.D. in applied chemistry from the erstwhile Indian School of Mines (Indian Institute of Technology-ISM), Dhanbad, India, in 1981. He started his professional career as research associate and lecturer at ISM for applied research in materials science specializing on cementing materials using pozzolanic clays along with an exposure in the area of separation technology particularly on fine particle processing aiming at utilization of low-grade ores and minerals. He served CSIR—Central Glass and Ceramic Research Institute, Kolkata, for 23 years in different capacities, out of which last six years as Chief Scientist. As the founder head of ceramic membrane research, Dr. Bandyopadhyay spearheaded the development, demonstration and deployment of the patented technologies on low cost ceramic membrane based small scale plants for point of use application, especially for purification of arsenic and iron contaminated groundwater for providing safe drinking water in the eastern and north-eastern region of India. He was also instrumental in implementing six international collaborations and organizing several national and international seminars, conferences, workshops, etc. After serving CSIR for 30 years (1982–2012), he joined his alma mater as Adjunct Professor in the Department of Environmental Science and Engineering, IIT (ISM), where he was associated as Member of the Research Council. In 2014, he joined as CSIR Emeritus Scientist in the Department of Environmental Studies, Visva-Bharati, Santiniketan, India.

Chapter 1
An Overview on Recent Progress in Wastewater Treatment Using Low Cost Ceramic Membrane

Sibdas Bandyopadhyay

1 Introduction

The rising burden on freshwater resources and increasing requirements for clean water demand a greater attention for wastewater management in moving towards safe and green environment. United Nations Sustainable Development Goal-6 emphasized universal access to clean water and sanitation while the demand for water has outpaced population growth and water scarcity is projected to increase with the rise of global temperatures because of climate change [1]. In this context, valorisation is a new concept in wastewater management promoting the principle of sustainable development. A substantial shift in wastewater management can help to address the dual purpose by minimizing the risk of freshwater contamination and to augment the water supply scenario for non-potable application. However, wastewater production from varying sources containing different types of contaminants and/or pollutants poses a critical challenge for a universal wastewater treatment technique and its reuse. Contamination of the emerging pollutants like pharmaceuticals and personal care products (PPCPs) aggravates the situation further necessitating a more aggressive role in utilizing the advanced technologies for combating the situation [2–4].

Membrane based separation technology has already demonstrated as a highly efficient treatment option for purification and concentration of products in liquid and gas separation applications owing to their convenient operation, space and energy saving process. Of the advanced separation processes, membrane technology has

S. Bandyopadhyay (✉)
Formerly at CSIR-Central Glass and Ceramic Research Institute, Kolkata 700032, India
e-mail: sibdas4@rediffmail.com; sibdas2013@gmail.com

Emeritus Scientist CSIR, Department of Environmental Studies, Siksha-Bhavana, Visva-Bharati, Santiniketan 731235, India

Department of Environmental Science & Engineering, IIT (Indian School of Mines), Dhanbad 826004, India

S. Bandyopadhyay (ed.), *Application of Low Cost Ceramic Membranes in Wastewater Treatment*, Materials Horizons: From Nature to Nanomaterials,
https://doi.org/10.1007/978-981-97-5842-5_1

proved the scope of reducing cotamiation of water resources and controlling environmental pollution producing high-quality effluent. Although ceramic membranes have demonstrated their potential and key performance advantages like longer life and suitability under harsh process conditions as required for wastewater treatment, higher memrane cost limits their greater usage particularly for a low value product like water. During the past two decades, research efforts have therefore been directed towards cost reduction of ceramic membranes using cheaper raw materials like natural minerals and industrial wastes considering the waste as a secondary raw material substituting the costly ceramic oxide powders being used currently. A recent review highlighted the scope of wastewater treatment using low-cost ceramic membranes developed by reusing waste materials and by-products for sustainable development [5]. A synergistic approach in zero waste strategy can uphold the principle of sustainable development and move forward towards the concept of circular economy [6]. As a concerted effort in this direction, the current status on low-cost ceramic membranes has been addressed by the author in a separate book to pursue and propagate the waste to resource strategy more effectively [7]. This book deals with the applications of low-cost ceramic membranes in wastewater treatment addressing the industry-specific case studies. It also highlights the salient features of wastewater treatment for minimizing the prohibitive cost of ceramic membranes upholding the princilpe of circular economy. The emerging issues on membrane-based processes have been discussed while projecting a glimpse on the future scope of work in this direction. In this chapter, current strategies on wastewater utilization and membrane technologies in wastewater treatment have been discussed highlighting the potential of membrane bioreactors (MBR) and the membrane fouling aspects particularly for the beginners.

2 Current Strategies for Wastewater Utilization

Renovation and recycling of urban and industrial wastewater discharges is envisaged as a promising approach to mitigate of the rising requirements for fresh water. In line with the concept of circular economy (CE) for sustainable growth, the waste materials can be considered as a resource and reused as the secondary raw material in the production process as long as possible thereby minimizing generation or even elimination of the waste [8–10]. Following the strategy of circular economy for developing wastewater industry, an approach towards tapping the potential for clean water production and safe reuse of treated wastewater integrated with recovery of water, nutrients, energy and valuable constituents can lead to several advantages like boosting up food production, managing water scarcity, helping land scape development, etc. thereby taking a step forward in achieving the coveted goal of clean and green environment [10–12]. However, implementation of the CE and smart cities concept demands adoption of more efficient technologies for addressing the challenges in the wastewater treatment plants (WWTPs) [8].

2.1 *Wastewater Resources and Classification*

Wastewater is generated from various sources and their characteristics vary depending on the origin as given in Table 1. The criteria for classifying wastewater may be based on the concentration of the contaminants and when these are present in the ppm range, the treatment process is considered in the category of dilute separations. Wastewater can be further classified based on the nature of species to be removed such as ionic species containing salts and metal ions, organic pollutants, presence of high value materials, etc. [13]. The scope of treatment and reuse of grey water, municipal and industrial wastewater is discussed below.

2.1.1 Grey Water

Grey water being the most generated wastewater in urban areas, its recycling could be one of the main options since the effluent is of sufficiently better quality that can be beneficially used instead of being discharged. There is greater interest in the reuse of grey water, given that its pollutant load is lower than that of other types of wastewater, it can be treated for reuse before its discharge into the waste network. Application of porous membrane filters would help to recirculate a significant amount of water and reduce demand for fresh water within the same community [14–16]. Wastewaters can be classified as weak (<400 and <200 mg/L), medium (400 and 350 mg/L), strong (1000 and 500 mg/L) and very strong (>1500 and >750 mg/L) based on COD and BOD concentrations [17]. Domestic wastewater is defined as the water produced from community use containing human body wastes including toilet flushing and sullage. Earlier literature covering treatment of domestic wastewater based on application of ceramic membranes is available in the literature on 'Environmental Protection, Pollution and Treatment' [18]. Jeong et al. reported higher COD removal and specific water flux using ceramic membrane compared to that of polymeric membrane in anaerobic membrane bioreactor treatment of domestic wastewater [19].

2.1.2 Municipal Wastewater

Reclamation of municipal wastewater of relatively consistent quality is a potential resource for water reuse, particularly due to the availability of a substantial volume in the centralized collection system. Utilization of treated municipal wastewater is also an attractive proposition due to its scope of reuse for diverse applications. Depending on the treatment process and output quality, its utilisation has been contemplated in several areas like agricultural and landscape irrigation, environmental enhancements, industrial uses, non-potable urban reuse and groundwater recharge. Arab countries have reclaimed 21% of municipal wastewater primarily for agricultural irrigation and groundwater replenishment [20]. In a recent publication, Hamedi et al. have dealt

Table 1 Category of wastewater and their characteristics

Category	Sewage		Non-sewage		Others
Activity	Human activities		Industrial operations	Agriculture	Natural sources/ precipitation
Sources	Residential buildings, hotels, restaurants, schools, shopping centres, etc.		Process industries, food processing, drugs and pharmaceuticals, energy and petrochemicals	Surface irrigation processes	Storm water, rainwater or flood collected in rivers, lakes, or other water bodies
Subcategories and source	Grey water from kitchen, bathrooms, washing machines, etc.	Black water (and Yellow water) from toilets	Toxic and nontoxic	Agricultural runoff & irrigation	Surface runoff
Broad composition	Detergents, liquid soaps, oils and fats, etc.	Organic matter, pathogens, pharmaceutical and personal care products, etc.	Inorganic and organic contaminants	Fertilizers, pesticides, herbicides, etc.	Primarily suspends solids, pathogens
Quality and scope of treatment. Scope of recovery/ reuse	Less polluted and not regarded as pathogenic. Generally less complex treatment. Potential source for recovery	Major source of municipal wastewater. Scope of reuse in industry.	More polluted requiring complex treatment approaches. Needs more aggressive efforts for reuse and recycling.	Significant water pollution. More complex treatment due to emerging contaminants.	Less polluted. Modest treatment techniques. Generally used for drinking water supply.

with the relevant issues on wastewater irrigation for future sustainable agriculture and food security in the UAE [21].

Tomczak and Gryta summarized the physico-chemical characteristics of real municipal wastewater treated in AnMBRs [22]. It has been found that the studies regarding the application of AnMBRs for treating municipal wastewater have been performed with wide range of COD and BOD varying from about 373 to 1729 mg/L and 160 to 694 mg/L, respectively. AnMBR studies based on a weak, medium as well as strong types of municipal wastewater have been reported in the literature [22]. Presence of organic micropollutants in municipal wastewater is an important criteria for its reuse after treatment [23]. Literature study revealed the current trends on treatment of municipal wastewater covering various aspects of membrane based process viz. application of anaerobic membrane bioreactor (AnMBR), MBR, NF membranes, ceramic membranes [24–27].

2.1.3 Industrial Wastewater

Industrial wastewater contains wide range of pollutants depending on the nature of starting materials and treatment conditions used as well as processing stages adopted in manufacturing the products requiring use of water primarily for removing undesirable constituents. The discharge from most of the industrial categories is laden with pollutants like suspended solids, pathogens, inorganic salts/heavy metals, synthetic organic chemicals, microplastics, emerging pollutants, etc. which are generally classified as hazardous, toxic, biodegradable, non-degradable/persistent types based on impacts on human health and ecological perspectives. Ranade et al. provided an overview of industrial wastewater treatment, recycling, and reuse and also dealt with the advanced physico-chemical methods of treatment for industrial wastewater [13].

Based on physicochemical aspects, advanced industrial wastewater treatment systems have been developed following discharge norms and to meet the criteria for reuse utilizing tertiary treatment techniques especially using membrane technology and advanced oxidation processes (AOPs) particularly for certain organic pollutants [28]. In a recent review, Singh et al. observed that efficient management of wastewater disposal in minimizing the environmental impact using advanced water-efficient treatment technologies is one of the key challenges for sustainable industrial development. And the future of wastewater management relies on evolving decentralized wastewater treatment systems in addressing complex contaminants and recycling wastewater [29]. Considering the importance and issues involved in the treatment of various types wastewater, this book deals with several case studies primarily classified in two sections, viz. industrial wastewater treatment covering some typical technological processes like textile, leather, paper, metallurgical/electroplating, oil containing wastewater primarily aiming at petroleum industries while food processing and pharmaceutical wastewater treatment having organic and emerging contaminants have been discussed separately with a view to project the future direction in using low cost membrane based separation solutions.

3 Membrane Technologies in Wastewater Treatment

UN sustainable development goals envisage improvement in water quality by reducing pollution, eliminating dumping and minimising release of hazardous chemicals and materials, halving the proportion of untreated wastewater and substantially increasing recycling and safe reuse globally by 2030 [30]. Traditional methods like activated sludge process and anaerobic digestion were not designed for this purpose to address the current challenges of wastewater management, like recovery of treated water, removal of emerging contaminants and recalcitrant pollutants. Recent advances on research in wastewater treatment and technological innovation on more efficient processes for removal of organics and micropollutants needs to be deployed for fostering a sustainable and resilient wastewater management future.

The membrane acts as a semipermeable barrier which allows certain components to pass through it selectively according to size of the particles or molecules present in the system based on the principle of sieving. In membrane separation processes, separation occurs due to variation in the rate of transport of the components in the feed stream under the action of a driving force or potential difference (such as pressure, temperature, concentration or electrical potential) across the membrane interphase. As a rapidly evolving area, membrane technology has revolutionized the separation and purification processes in water and wastewater treatment including various industries like food and beverage, pharmaceuticals, biotechnology, etc. Global scenario and sustainable solutions for wastewater management using membrane technology has been elaborated in the subsequent chapters.

3.1 Ceramic Membranes in Wastewater Treatment

Membranes are generally classified depending on the materials of construction (e.g. polymeric, ceramic, metallic, etc.) and their efficacy varies based on the morphological structure like pore size, porosity, etc., surface charge and chemical affinity of the barrier layer [31]. Based on the decrease in pore size of the membrane used, the separation process is termed as microfiltration (MF), ultrafiltration/(UF) and nanofiltration (NF). Ceramic membranes are being increasingly used for wastewater treatment due to their excellent performance under harsh conditions showcasing a steady yearly expansion of 12% for their inherent advantages [32]. Lin et al. provided a brief overview on application of ceramic membrane in municipal water and industrial wastewater treatment [27].

In comparison with the polymeric membranes, ceramic membranes are characterized by higher hydrophilicity and lower surface charge which attribute to reduction in susceptibility towards organic fouling while their surface roughness plays a positive role in minimizing irreversible fouling and superior chemical stability contributes to more effective cleaning leading to improved flux recovery [33].

The prospects of ceramic membranes in large scale wastewater treatment applications have been reviewed by several authors particularly with respect to two major issues viz. reducing cost and increasing efficiency [34–41]. Samaei et al. reviewed the application of pressure-driven ceramic membrane technology for the treatment of industrial wastewaters [42]. Issues on mitigating ceramic membranes bottlenecks have also been reviewed along with the improvement strategies in water purification processes [43]. Gisi et al. provided an overview of technologies wastewater treatment used in major industrial sectors with case studies and salient aspects like wastewater characteristics, main process parameters, and performance in terms of pollutant removal based on start-of-pipe and end-of-pipe approaches [44]. Asif et al.

presented a review covering full-scale applications of ceramic membrane technology wastewater treatment addressing associated issues [38].

Ceramic membrane based technologies have established their foothold in industrial wastewater treatment to overcome the shortcomings of polymeric membranes primarily for separation applications under pressing conditions. However, despite the technological advantages of ceramic membranes, their potential are yet to be fully realized in industrial wastewater treatment, primarily due to their higher cost compared to the polymeric counterpart. Therefore, this books aims at exploring the scope of applications of low cost ceramic membranes produced from natural minerals and industrial waste for treatment of wastewater generated from some typical industries.

3.2 *Challenges for Industrial Wastewater Treatment*

This section deals with the recent literature about complexities envisaged in treating the effluents from major industrial sectors such as textile, leather, petrochemicals, food and pharmaceuticals, etc. with a view to realize the advantages, challenges and status of ceramic membrane technology in industrial wastewater treatment. Ultrafiltration is considered as a viable, environmentally friendly alternative for industrial effluent treatment in the new perspectives on reclamation of wastewater [45]. Application of membrane based separation technologies gained momentum primarily due to better effluent quality following discharge norms. The scope of recycling wastewater after treatment opened up the prospect of using the microfiltration/ultrafiltration (MF/UF) process for removal of suspended solids and organic pollutants [46].

Textile industry, which plays a crucial role in the economies of developing countries, generates a large amount of wastewater during multiple washing at different stages of processing. It has been estimated that dyeing and finishing operations in textile production causes about 20% clean water pollution with serious environmental problems. The challenges in the treatment of textile wastewater are primarily for removal of colour which are difficult to degrade and treatment of microplastics fibres [47].

In the tannery industry, hide processing generates effluents containing several non-biodegradable compounds and to realize the concept of "Zero Liquid Discharge (ZLD)", membrane technology could be utilized for treatment of such complex tannery wastewater [48]. However, based on a review of the global research trends, Yang et al. remarked about the possibility of achieving "ZLD" standard adopting a single-process in addressing the problems of tannery wastewater with high salt concentration and organic matter using membrane technologies [49].

In the petroleum industry, it has been estimated that 3–8 barrels of oily wastewater is generated for the production of every barrel of crude oil and as per discharge norms, the oil content should be restricted to 10–15 mg/L in the effluent. There is a wide scope of utilizing oilfield produced water, after treatment in agricultural irrigation and industrial operations. Various types of treatments such as gravity-based separation, oil skimming, flocculation etc. have been used for oily wastewater containing free, dispersed and emulsified oil with droplets size in the range of >150 to <20 μm. Xiaodong et al. used ceramic membranes (0.1 and 0.2 μm pore size) to treat the water produced in Baolang Oilfield with better quality (oil content generally less than 30 mg/L, median particle size <1.5 μm), and the average suspended solids content in production water was less than 0.6 mg/L with oil content below the detection limit [50]. Several authors reported use of low-cost mullite based membranes for treatment of oily water with encouraging results [51–53]. Visible light-driven photocatalysts modified ceramic membrane exhibited high performance during ultrafiltration of oily wastewater [54]. The prospect of using integrated membrane systems for recovery of the valuable components from the by-product of the oil and gas industry was explored in a recent review [55]. Abuhantash et al. reviewed the prospect of using hydrophilic, oleophilic and mixed matrix membranes (MMMs) for effective treatment of oily wastewater [56].

The metal-working industries discharge large volumes of oily wastewater containing high content of spent cutting-oils. Model solution treated using ceramic (Al_2O_3/ZrO_2) and polymeric (PVC) membranes showed that ceramic membrane operated effectively achieving higher fluxes with reversible nature of fouling while it was irreversible with the polymeric one [57].

Although the food industry contributes significantly to the economic growth, this sector is associated with various environmental problems, particularly due to the production of large volumes of wastewater. Barbera and Gurnari discussed the key parameters and various methodologies, viz. separation or concentration systems, centrifugation, evaporation, filtration, flotation, gravity separation, membrane techniques, aerobic and anaerobic biological treatments, as well as hybrid systems highlighting specific methods, specifically a new adsorption method for the removal of metal ions [58]. Generally, food industry wastewaters are easily biodegradable and largely non-toxic, and are characterized by high concentrations of BOD and COD, suspended and dissolved solids (oils, fats, grease, etc.) along with high concentration of nitrogen (N), phosphorous (P) and minerals (salts). Tekerlekopoulou et al. presented an overview on the characteristics of wastewater originated from different agri-food industrial activities such as olive oil production, dairies and wineries, including the technologies for the treatment of such types of wastewater, case studies with emphasis on biological and electrochemical processes, the quality of treated wastewater and their potential for reuse [59]. Costa et al. reviewed the most recent approaches for the valorization of food industry wastewater having different origins (dairy, brewery, oil mill, fish, meat, among others) [60]. Gbadeyan et al. reviewed various technologies for obtaining energy from sugarcane wastewater, including anaerobic digestion for biogas, fermentation for bioethanol, and the direct production of electricity using microbial fuel cells [61]. Low cost ceramic membranes fabricated using kaolin and alumina with a single step firing was used to evaluate the performance for treatment of the dairy industry effluent. Microfiltration membrane process, with a permeate flux of 897 L/h m^2 showed removal of components present in the milk solution responsible for the turbidity of the permeate, with a permeate flux stabilized at 30 L/h m^2 [62].

Effluents from the seafood industry may be classified as high-volume and low-strength wastewater primarily discharged from aquaculture farms and washings or low-volume and high-strength wastewater generated from various processing operations. The later type of effluents contain organic matter, fats, oils, grease (FOG) and proteins including excess nutrients (nitrogen and phosphorus) which, if not treated properly, may lead to the eutrophication process in water bodies causing harm to aquatic life [63]. Cooney et al. in an overview on various types of seafood reported the scope of purifying recovered protein/peptide from seafood using pressure diven membrane technology [64]. For processing of wastewaters rich in organic materials, biological treatment techniques have been found to be highly effective, particularly when it is generated from seafood processing operations [29].

Pharmaceutical wastewater is characterised by various types of organic and inorganic components, pharmaceutical pollutants (PPs) classified as emerging pollutants which are generated in course of the formulation and manufacture of drugs. There is an urgent need to use an advanced treatment system like membrane technology for effective removal of the pharmaceutical pollutants before discharging the effluent to the environment in order to avoid the irreversible consequences to human body [65]. Gurgenidze et al. reported detection of more than 200 pharmaceutical contaminants including antibiotics, hypoglycemic and hormonal drugs, etc. in surface waters [66]. Application of ceramic membranes in pharmaceutical sector have the benefits of being frequently bioinert, persistent against bacteria and their capability to withstand repeated chemical and steam sterilization at high temperatures and ease of cleaning with harsh chemicals [67, 68]. Ceramic UF has to be installed to prevent the growth of microorganisms that would lead to contamination of piping system and the purified water after electrode ionisation stage in pharmaceutical industries [69]. Bhattacharya et al. reported the scope of nanoparticle coated ceramic UF membrane over low cost clay-alumina support for ciprofloxacin removal and remediation of pharmaceutical components [70, 71].

Electroplating industry wastewater involves highly toxic cyanide (CN), heavy metal ions, oils and greases, organic solvents along with the complicated composition of effluents. A recent review mainly emphasizes the removal of heavy metals and cyanide from electroplating discharge using electrocoagulation and integrated membrane [72, 73].

High levels of suspended solids, BOD and COD along with chlorinated compounds, organic halides are generated in the wastewater from paper and pulp industries during different stages of processing like preparation of pulp, bleaching, deinking, etc. [74]. An integrated treatment process consisting of biodegradation followed by membrane filtration using a cost-effective ceramic membrane (1.01 μm pore size and 44% porosity) showed 92.7, 87.6 and 88.2% COD removal using wastewater from paper and pulp, biomass gasification and dairy industries respectively [75]. When treating papermaking wastewater, the presence of calcium carbonate was found to significantly reduce membrane flux and at elevated levels of calcium concentration (830 mg/L), the excessive presence of Ca^{2+} in the sludge resulted in severe inorganic fouling. Zhu et al. analysed the potential of intermittent operation in AnMBR system for treatment of high-calcium wastewater [76].

Despite the dominance of polymeric membranes, several reviews highlighted apparent benefits of ceramic membrane for treatment of industrial wastewaters to demonstrate their effectiveness in diverse industries, including textile, pharmaceutical, and petrochemical as well as for emerging contaminants removal [77–79]. Anorthite-mullite based low cost MF membrane made of fly ash and sugar scum (with comparatively much higher permeability of about 2357 L/h m^2 bar compared to other ceramic membranes with low cost precursors) exhibited 99.9% of dye (methylene blue) removal during the initial 40 min of at 1 bar [80].

4 MBR Processes

Membrane bioreactors (MBRs) have been developed through the integration of the traditional activated sludge process (ASP) with suspended biomass growth and microfiltration (MF) or ultrafiltration (UF) membrane systems. The biological part of the reactor is employed in the waste biodegradation and the membrane compartment is responsible for the physical filtration of the biologically treated water from the mixture [81, 82]. In MBR process, membranes act as a solid–liquid separation device, keeping the biomass within the bioreactor before discharging the treated effluent to the nature. MBRs have a smaller footprint, provide better biotreatment and can achieve a high-quality effluent. Three types of membrane geometries are used for MBRs, viz., Hollow fibre (HF); Flat sheet (FS); Tubular (or multi-tubular, MT) [83]. MBRs can be subdivided into side-stream MBRs (sMBRs) or first-generation system wherein membrane module is located externally aiming to reduce cake formation on the membrane surface and second-generation MBRs or immersed MBRs (iMBRs) with membranes submerged inside the reaction media, developed to reduce the energy demand significantly for wastewater treatment [84].

The hydraulic retention time (HRT) and sludge retention time (SRT) in an MBR are independent; thus, activated sludge can accumulate in the reactor [85]. While the sludge concentration is lower (typically 2–3 g/L suspended solids for 1–2 days SRT) in conventional activated sludge processes, it is comparatively higher (8–12 g/L suspended solids with longer SRT for 20–100 days) along with removal of micropollutants in case of traditional MBR systems. Owing to these advantages, MBR systems with a capacity of >10,000 m^3/d have been applied globally with more than 20 million m^3/d cumulative treatment capacity until 2016 [86].

The conventional gravitational sedimentation clarifier in the ASP is replaced by the membrane filtration system with a pore diameter in the range of 0.01–0.1 μm, which allows bacteria and particles to be filtered out of a permeate [87]. During early seventies of the past century, the external membrane modules were added in series with conventional activated sludge (CAS) treatments. Around 1989, the submerged modules, in which the membranes were placed inside the water treatment tank, was introduced to reduce the energy consumption for pumping feed water to the external module [88]. However, better removal of protozoa, bacteria and viruses is the key factor for the implementation of MBRs in the treatment of wastewater in rural areas. Although both MF and UF membranes can be used in MBR applications, UF membranes are preferred because of their superior separation characteristics to remove some colloids and viruses with lower fouling tendency and reduced risk of pore clogging. In the effluent from conventional wastewater treatment, the existence of *Giardia* cysts and/or *Cryptosporidium* oocysts necessitated tertiary treatment to fulfil the quality requirements as conventional activated sludge treatments is not suitable to provide a large reduction of this microbial contamination [89]. The characteristics of the bioreactor treatment have been addressed along with a comprehensive review of the membrane separation process with recent fouling mitigation methods [90]. With a view to recycle the wastewater from tanneries and leather industry, Dhiman and Mukherjee elaborated the scope of biotechnological approaches [91]. Goswami et al. discussed about three primary aspects of the new generation effluent such as colour, triclosan and surfactant content and reviewed the use of MBR for micropollutant removal from wastewater [92]. Liu et al. elaborated the performance of aerobic and anaerobic membrane bioreactors for trace organic contaminant removal in wastewater treatment [93]. High calcium papermaking wastewater challenges application of anaerobic biological technology, leading to sludge calcification and unstable reactor operation. This study employed an anaerobic membrane bioreactor (AnMBR) to treat high-calcium papermaking wastewater (1000 mg/L) and achieved an average COD removal efficiency of 90.1% [94].

Capital costs of a MBR process are substantially influenced by the membrane cost itself, which contributes around 50% of the total process capital cost. The feasibility of a submerged MBR fitted with low-cost NWPP (NonWoven Poly-Propylene) membranes was found to achieve only a 2 to 4-log reduction indicating that such polymeric membranes are possibly less suited to domestic wastewater reuse where disinfection is a prerequisite [95]. MBR using membranes with an appropriate pore size can provide a highly efficient removal of microbial pollutants, and thus could be the most suitable wastewater treatment to satisfy the conditions for more demanding requirements. Real domestic wastewater was treated directly using a gravity-driven membrane bioreactor (GDMBR) at 45 mbar without backwashing or chemical cleaning, and the effects of different membrane pore sizes (0.22, 0.45 μm, and 150 kDa) on flux development and contaminants removal were examined. The results showed that the flux initially decreased and then stabilized throughout long-term filtration and that the stabilized flux level of the GDMBR equipped the membranes with the pore size of 150 kDa and 0.22 μm was higher than that of 0.45 μm membrane and was in the range of 3.25–4.25 L m^{-2} h^{-1}. GDMBR system achieved efficient removals of chemical oxygen demand (COD), and ammonia, with average removal efficiencies of 60–80 and 70% respectively [96].

Integration of the membrane bioreactor with various wastewater treatment systems is a green technology with a suspended growth bioreactor for biochemical reactions (e.g., bio-oxidation, nitrification, and denitrification) and the membrane of appropriate pore size depends on the type and nature of contaminants required to be separated. The widespread use of MBRs has been due to its significant advantages such as high quality of produced water, higher biodegradation ability of pollutants for a lower cumulative footprint. Globally, MBR is being used in mitigation of both industrial and municipal wastewater. It has been reported that the annual growth rate of MBRs in the global market is around 15% [97]. Recently, anaerobic membrane bioreactors (AnMBR) have been promoted as a prominent alternative to conventional anaerobic digesters (CAD). Cengiz et al. presented an overview on the current situation about co-digestion application by AnMBRs for municipal WWTPs including the environmental and economic aspects along with the challenges and future perspectives related to the co-digestion applications by AnMBR [98]. Zhen et al. reported development of electrochemical membrane reactor (EC-AnMBR) for enhanced co-digestion of sewage sludge and food waste [99]. With a view to improve the nutrient removal and fouling control, Borea et al. suggested a combination of electrochemical processes with MBR [100].

4.1 Ceramic Membrane Based MBR

Most of the MBRs are based on polymeric membranes, while ceramic membranes represent a smaller part of the market owing to higher cost. Since the beginning of this century, ceramic membrane bioreactors (CMBR) have gained attention due to the advantages brought by the ceramic membranes and the full-scale market has grown with installations in USA, Japan, Netherlands, Singapore, etc. Nevertheless, the full commercial application of this technology requires more research on both reducing the cost of ceramic membranes, and fouling mitigation to achieve a stable long term operation [38].

In the MBR, the sludge concentration gradually increases, the dominant bacterial community is effectively enriched, and the biochemical reaction rate also increases gradually. Due to significant reduction in the amount of excess sludge and land occupation, application of MBR is expanding gradually from domestic sewage to high-concentration organic wastewater and refractory industrial wastewater. Pilot study on the ceramic flat membrane showed that MBR can retain high concentrations of biomass under the efficient separation of activated sludge and wastewater, which is a unique advantage in the treatment of toxic and refractory constituents in industrial wastewater [85, 102].

In the textile industry, more than 90% of the water consumption is transformed in to wastewater which contain a high concentration of salinity with 5–6% NaCl, color, solids, chemical oxygen demand (COD) and heavy metals. Recently, comparative evaluation between MBR and moving bed membrane bioreactor technology (MB-MBR) using a ceramic flat sheet membrane module (nominal pore size 0.1 μm) revealed the scope of MB-MBR as an alternative to reduce membrane fouling propensity for treatment of textile wastewater [102]. Investigation on AnMBR based on upflow anaerobic sludge blanket (UASB) using a flat sheet ceramic membrane revealed that increased salinity led to reduce the COD removal efficiency from 92 to 73%, but had a marginal effect on dye removal efficacy (90–96%). However, low salinity (5 g/L) boosted the biogas production whilst high salinity (>10 g/L) had a negative impact [103].

4.2 *Anaerobic Membrane Bioreactors (AnMBRs)*

For treatment of organics in wastewater, anaerobic digestion is more efficient compared to energy-intensive conventional activated sludge process which consume more than 50% of the total energy of a typical wastewater treatment plant to convert the organics in to CO_2 [24]. Therefore, anaerobic process has been integrated with membrane for development of the AnMBR which is characterized by several advantages like effective degradation of organic compounds to produce methane-rich biogas, removal of organic micropollutants for water reuse, longer solid retention time (SRT), shorter start-up time, complete biomass retention along with concentration and recovery nutrients compared to conventional anaerobic reactors that retain biomass in the form of granules or biofilm [104]. AnMBR exists in three different configurations: external cross-flow, internal submerged, and external submerged membrane and the membrane is separated from the reactor in the external cross-flow configuration. The main advantages of the AnMBR are its high treatment capacity, up to 30 g_{COD}/L/day, and complete removal of suspended solids allowing the permeate to be reused without any additional treatment.

Recently, AnMBRs have received increased attention for treating domestic wastewater due to several advantages as detailed above particularly for complete rejection of suspended solids by MF or UF membrane modules and removal of organic micropollutants. However, despite these merits, wastewater treatment by AnMBR is often limited due to the problem of severe fouling, which leads to high operation and maintenance costs. Investigation on the feasibility of an anaerobic ceramic membrane bioreactor (AnCMBR) in comparison with a conventional anaerobic membrane bioreactor (AnMBR) revealed that AnCMBR using alumina-based ceramic membrane achieved higher organic removal rates than AnMBR with PVDF polymeric membranes due to the ability of ceramic membranes to treat and retain a high concentration of biomass in the reactor along with higher organic removal ($91.0 \pm 13.8\%$) than that of AnMBR ($77.8 \pm 20.5\%$). Ceramic membrane could achieve higher COD removal maintaining a higher specific water flux than polymeric membrane [19].

Due to the beneficial effects of AnMBR for producing better effluent quality and scope of reusability, its performance has been evaluated for treatment of wastewater from food industries viz. livestock, brewery and meat industries including sewage, textile, etc. Jeong et al. carried out long term investigation to study the feasibility of enhancing biogas production using alumina-based ceramic MF membrane (0.1 μm nominal pore size and 0.2 m^2 total surface area) with increased organic loading by mixing domestic wastewater and food waste recycling wastewater [105].

The results on pilot scale treatment of distillery wastewater using AnMBR showed high COD removal efficiency of 96.9% along with a treatment capacity up to 3.97 kg_{COD}/m^3/day and at least 44 days of membrane cleaning cycle interval [106]. However, application of AnMBR in municipal wastewater treatment is a challenging task due to low substrate concentration and high particulate fraction. Although this process showed high COD removal for municipal wastewater, it requires energy for biogas sparging or high crossflow velocities for controlling membrane fouling. To control fouling of ceramic MF (0.4 μm) and UF membranes (0.05 μm, 100 kDa MWCO), Düppenbecker et al. reported improved performance of AnMBR with an external crossflow membrane using turbulence promoters to mitigate fouling at low crossflow velocities [107].

For sustainable wastewater treatment, attempts has been made to integrate bioelectrochemical systems (BES) like microbial fuel and electrolysis cells with AnMBR technology to address the pivotal challenge of fouling mitigation as highlighted in a recent review [3]. For effective nitrogen removal, the anoxic and oxic conditions can be operated alternatively in a cyclic mode for enhancing the nitrification and denitrification process. A combination of Anoxic–Oxic (A/O) process integrated with AnMBR was reported to have dual benefits like nitrogen and phosphorus removal and also reduces the operational costs of seafood processing effluents. A pilot-scale AnMBR-AO system of 0.5 m^3/day capacity using seafood processing wastewater showed 80 ± 5% and 98.3 ± 0.3% (27–41 mg/L) COD removal in case of AnMBR and combined AnMBR-AO respectively [108]. A recent review discussed development of various types of membrane technologies, viz. modified membrane bioreactor (MBR), hybrid MBR, anaerobic MBR, moving bed biofilm bioreactor (MBBR) and hybrid MBBR, osmotic membrane bioreactor (OMBBR) and reported that the MBBR technologies performed better than the existing MBR techniques among the currently employed MBR options. The review suggests a set of future research domains for sustainable treatment of tannery wastewater utilizing MBR [109].

5 Application of Low-Cost Ceramic Membranes in Wastewater Treatment

Low-cost ceramic membranes have the potential to provide high-volume filtration capabilities, which will help to provide safe water and clean environment [110]. Earlier studies on application of low cost ceramic membranes in MBR, have shown a promising performance using cylindrical-shaped filter made with a combination of 80% clay soil and 20% rice bran and fired at 900 °C [111] and macroporous ceramic filters (2–6 μm pore size) prepared from bagasse fly ash [112]. Majouli et al. prepared microfiltration membrane composed of two layers of perlite (mean pore size of 0.27 μm) over a macroporous support (with about 6.6 μm pore diameter and 42% porosity) for treatment of three types of wastewater (effluents from beam house section of tannery, textile effluent from jeans washing process and dicing wafer effluent generated by electronic industries). Tangential microfiltration perlite membrane (815 L/h m^2 bar water permeability) proved to be effective in removing pollutants from the three effluents with almost same efficiencies obtained with a commercial alumina membrane (pore diameter of 0.2 μm and a water permeability of 1022 L/h m^2 bar) [113]. In another work, S. Saja et al. reported the optimized preparation of low-cost ultrafiltration bentonite membrane deposited on ceramic perlite support with 0.75 wt% of bentonite by sintering at 500 °C. Performance evaluation of bentonite membrane under crossflow filtration mode at 4 bar showed 80.1% and 97.0% rejection of Rhodamine B and Direct Red 80 respectively [114]. Cross-flow microfiltration treatments investigated using domestic wastewater and Congo Red dye with the membranes sintered at an optimum temperature of 900 °C/3 h (with average pore sizes of 1.21, and 1.31 and 1.76 μm) showed 99% turbidity removal efficiency with 91–99% discoloration. The cost of the membranes was estimated to be around 21–23 $/$m^2$ due to the lower raw material cost of natural clay [115].

Bhattacharya et al. investigated the potential of ceramic MF and UF membranes in tubular configuration for grey water treatment and studied the effect of using treated and untreated greywater for uptake of essential nutrients in plant growth compared to fresh water. While MF membranes was found to retain turbid materials and suspended solids (almost 99% removal) and partial COD removal (75%), UF membranes exhibited good retention of turbidity and suspended solids (over 99%), oils, and microorganisms, in addition to 86% COD reduction [116]. Membrane bioreactors deliver a cleaner water quality than the conventional membrane separation process, however, they are more expensive, and the effluent that these processes deliver is always non-potable secondary use water [117]. Therefore, although the submerged membrane bioreactor (sMBR) is the most studied and widely used bio-reactor, it has the disadvantage of being expensive and difficult to operate, which makes it only profitable in small communities. Ding et al. [117] discussed gravity membrane reactors with microfiltration (MF) flat plate membrane, as an alternative to MBRs, which are less expensive because they do not require energy to propel the effluent [118]. However, several bottlenecks still limit their commercial scale-up in filtration and separation applications, including the high costs of conventional ceramic materials and fouling problems. While discussing the use of naturally occurring ceramic materials to mitigate the high costs of conventional ceramic materials, the strategies to control the fouling on the ceramic membrane surface were addressed by increasing membrane hydrophilicity and surface modification [43]. Jedidi et al. prepared a mineral porous tubular ceramic membrane based on coal fly ash and applied it in the treatment of the dying effluents generated in the washing baths of the textile industry [118]. A recent review focused on the factors that influence the properties and performance of coal fly ash based ceramic membranes in wastewater treatment and explored the cost analysis to evaluate their potential for large-scale applications [119].

To advance cost-effective strategies for utilizing waste mineral-based ceramic MF membranes, Jeong et al., carried out comparative evaluation with commercialized ceramic membranes which revealed the feasibility of using pyrophyllite-based membrane in the anaerobic ceramic membrane bioreactor (AnCMBR) for domestic wastewater treatment with 92.9 ± 5.5% COD removal and stable methane production by maintaining a slightly longer hydraulic retention time (HRT) [120]. Jeong et al. developed waste mineral-based ceramic MF membrane over pyrophyllite and alumina support coated with alumina powder to evaluate the feasibility of using it in MBR for low-strength domestic wastewater treatment and achieved consistently high organic removal efficiency under varying hydraulic retention times (HRTs). The performance of pyrophyllite-alumina composite ceramic membranes was also evaluated in a pilot-scale CMBR plant of 1 m^3/d capacity to demonstrate the efficiency of water resource recovery facility (WRRF) with improved nitrogen removal producing high quality effluent [121]. Ahmad et al. developed a novel type of submerged fluidized bed membrane reactor (SFMR) using low-cost pyrophillite ceramic membrane for dye wastewater treatment which showed more than 90% organic removal efficiency under GAC fluidization [122]. A comparative evaluation between low cost pyrophyllite based and pure alumina ceramic membrane showed almost 98% of organic rejection efficiency that could be achieved within initial 30 min of filtration during treatment of organic dye solutions using the submerged membrane reactor with the fluidization of GAC particles as fluidized agent. Under 10% packing ratio with GAC particles, membrane fouling was controlled successfully during 12 h filtration with almost complete UV removal efficiency within initial 4 h of filtration [123]. A pilot scale MBR using side stream low-cost clay-alumina based ceramic MF membrane in both single-channel and multichannel tubular configuration was evaluated for treatment of kitchen-sink wastewater (KSW) in cross-flow mode with considerable reduction in COD, BOD and suspended solids by adopting suitable process design and controlling the process parameters [124]. The colour removal efficiency for two types of low cost alumina coated UF membranes prepared by single step coating directly over clay-alumina support tubes with varying wall thickness (pore diameter distribution centred at 7.0 and 5.5 nm), was studied in the side stream mode with about 98% decolorization of synthetic dye solution and 95.6% colour removal of real textile effluent due to the combined effect of the UF membrane and bioreactor [125].

Two kinds of clay-based porous supports were prepared one of which was composed of 40 wt% clay with 20% each of chamotte, calcium carbonate and potato starch, while the other was made using residues from agricultural and industrial processes like 20 wt% each of almond shell, marble dust waste and fired tile scrap. It was projected that application of ceramic tile production technology for manufacturing ceramic membranes in the industrial scale may lead to achieve a competitive cost of around 25 €/m^2 for ceramic membranes against their polymeric counterpart. Accordingly, ceramic flat sheet supports manufactured by extrusion and sintering at 1150 °C were used for depositing different types of selective layer and sintered at 1120 °C. Low cost ceramic membranes developed using a layer of commercial TiO_2 powder was found to be promising for controlling TMP during

25 days operation producing the highest quality of effluent with respect to both physical, chemical and microbiological aspects. MBRs with the low-cost ceramic membranes were also reported to provide a removal efficiency of 99.99% for Giardia cysts and Cryptosporidium oocysts along with absence of *E. coli* in the permeate [88].

The potential of hybrid cross-flow microfiltration (CMF) was explored for biotreated kitchen-sink wastewater using a combination of biosorbent prepared from the dried roots of Eichhornia crassipes and tubular ceramic microfiltration membrane in single channel and multichannel configuration. The study showed better performance of the 19-channel ceramic membrane compared to the single channel membrane in terms of the permeate quality, viz. BOD, COD, turbidity, TSS etc., as well as, permeate flux [126]. Several studies on hybrid cross flow microfiltration process using low cost ceramic membranes appeared to be most promising in achieving better quality of the treated water [127–129]. During hybrid treatment combining adsorption on powder activated carbon (PAC) and ultrafiltration (150 KDa, Novasep, France), the permeate flux for aqueous solution of anionic AO7 dye was reported to remain stabilized in the range of 450–480 L/h m^2 at varying PAC doses (50–300 mg/L) along with a total retention of AO7 dye for PAC dose at 150 mg/L and colour removal at pH 6–7 [130].

6 Membrane Fouling

Fouling is an inevitable phenomena in the membrane separation process which occurs when the membrane surface and pores are covered and/or clogged by the contaminants present in the feed stream like microbial substances [131]. The fouling phenomena contribute to increase in transmembrane pressure (TMP) along with decrease in permeation flux leading to higher energy consumption and consequently higher operational costs. Several factors, such as operating conditions, membrane materials and their properties, flux, types of wastewater influent and biomass need to be optimized to reduce membrane fouling [132]. In a comprehensive review, Meng et al. discussed about different types of fouling control strategies like mechanically assisted aeration scouring, in-situ chemical cleaning, enzymatic and bacterial degradation of foulants, electrically assisted fouling mitigation, and nanomaterial-based membranes [133].

6.1 Types of Foulants, Classification and Treatment

Membrane fouling comprises of a significant obstacle to ensure cost-effective operation of membrane separation processes. Interactions between the membrane surface and foulants, play an important role in the performance of membrane. Foulants can be attached to the membrane by adsorption, crystallization and precipitation, by blocking the membrane pores or by making a more concentrated layer of alienated compounds on the surface of the membrane called gel layer or concentration polarization on the membrane surface. Fouling occurs when the convective transport of foulants from the feed solution to the membrane surface outweighs the back-diffusion of foulants from the membrane surface back to the bulk solution [134]. Fouling not only increases hydraulic resistance, but it also causes the formation of a cake layer with significant cake-enhanced concentration polarization (CECP) effects leading to blockage of the membrane pores.

Several potential membrane foulants have been identified and reported in the literature. Silica colloids, latex colloids, sodium alginate, colloidal aluminum oxide, humic acid, bovine serum albumin (BSA), fatty acids, surfactants, natural organic matters (NOM), and cellulose are among them [135]. Normally, foulants can be organic (humic and fulvic acids), hydrophobic and hydrophilic materials (polysaccharides and proteins), which would attach to the membrane by adsorption while inorganics (iron, manganese, and silica) tend to precipitate onto the membrane surface due to scaling (pH change) or through formation of colloidal iron or manganese oxides. Membrane fouling in mine wastewaters is due to inorganic foulants like calcium sulfate. Another category of foulants like algae and microorganism cause biofouling or biofilm formation on the membranes surface.

Based on the foulant nature, membrane fouling can be categorized into organic, inorganic, and biological fouling. Organic fouling refers to the deposition of biopolymers either as soluble microbial products (SMP) or extracellular polymeric substances (EPS) in the form of proteins and polysaccharides [136]. Biofouling occurs due to the attachment of microbial components and growth of biofilm on membrane surface and/or within membrane pores. Because of its complex nature different from particulate, organic and inorganic fouling, conventional fouling mitigation strategies like backwashing is not sufficient to tackle the problem of biofouling. Aslam et al. critically reviewed the causes for this inevitable phenomenon of biofouling in aerobic or anaerobic MBR systems and also addressed biofouling control/mitigation technologies like hybrid methods of treatment, use of functionalized media in particle-sparged MBR [137].

Municipal secondary wastewater often contains various types of foulants, resulting in combined fouling (e.g., organic–inorganic, organic–biological, inorganic–biological, and inorganic–organic–biological fouling) on high-pressure membranes. Compared to the membrane fouling induced by individual foulants, combined fouling sometimes leads to more pronounced decline in membrane flux and dynamic alterations in fouling layers [20]. Restoration of the permeate flux during membrane filtration of produced water is a major issue in petroleum industry and modification of the membrane surface has been studied in most cases with a view to either restricting oil fouling or forming an active layer suitable for cleaning easily. Tummons et al. reviewed the investigations on application of the shear force in the crossflow process to address membrane fouling by emulsified oil using MF and UF [138].

6.2 Fouling Mechanisms and Mitigation

In the membrane filtration system, whether it is pressure-driven or temperature-driven, membrane fouling can be controlled by optimizing the operation conditions or modifying the membrane surface properties, but it is not possible to avoid this completely. The factors which promote to foulant deposition and attachment on the membrane surface thereby contribute significantly to membrane fouling are described below [139]:

(i) Feed chemistry and composition, i.e., pH, ionic strength, and foulant concentration;
(ii) Concentration polarization (CP) which is broadly defined as the deposition of rejected solutes on the membrane's surface, resulting in the polarized layer, a region near the membrane with spatially varying concentrations. This increased resistance raises the osmotic pressure across the membrane, lowering the driving force of the process (TMP), permeate flux, and observed solute rejection, all of which increase the possibility of membrane failure;
(iii) Membrane properties like material type, hydrophobicity, surface charges, membrane morphology, porosity, poresize and membrane molecular weight cut-off (MWCO);
(iv) Process operating conditions such as temperature, pressure, aeration, permeate flux, and several other hydrodynamic conditions.

Four types of fouling mechanisms have been identified by Hermia, that can occur in membrane processes are: complete pore blocking, partial pore blocking, internal pore blocking, and cake formation [140]. Because more than one type of fouling can occur simultaneously in a process, the chemical nature of the foulants (e.g., organic, inorganic, or biological) must be known in order to inhibit or at least control such mechanisms [141]. Beyond the classification, the pollutant's nature can provide an overview of the type of fouling and its impact on membrane properties. Adsorption, accumulation, or precipitation can occur on the surface or within the pores of the membrane affecting separation performance, such as permeate flux and membrane selectivity, as well as membrane lifetime [142].

From a chemical engineering standpoint, the main challenge of membrane modification is to combine the desired property with materials that are economically feasible using more straightforward and effective techniques. Backpulsing has been used in a variety of low-pressure membrane configurations, including flat sheet, tubular, hollow fiber, and spiral-wound membranes while membrane surface modification by coating and grafting decreases flux decline as well as organic fouling on NF and UF membranes [143].

An understanding on the classification, mechanism, influencing factors can help to control of membrane fouling, online monitoring, prediction of membrane performance and life cycle. Membrane fouling can be classified into internal fouling, external fouling and concentration polarization fouling. The fouling caused by the deposition as well the adsorption of solutes and colloidal particles in the interior of the membrane pores is called internal fouling and sometimes referred to as pore blocking. The deposition of particles, colloids and macromolecules on the membrane surface is called external fouling which forms a fouling layer on the membrane surface. The fouling layer can be classified as the gel layer or cake layer. The gel layer is formed by the deposition of macromolecules, colloids and inorganic solutes on the surface of the membrane due to the pressure difference between the feed and permeate sides of the membrane. The cake layer is formed by the accumulation of solids on the membrane surface. Concentration polarization refers to the accumulation of solutes and ions in the thin liquid layer adjacent to the membrane surface, which increases the flow resistance and decreases the membrane flux. The concentration polarization layer is determined by the convective shear force. Increasing the convective velocity can alleviate membrane resistance caused by concentration polarization. Du et al. summarized the research progress of membrane fouling mechanism, control, prediction and detection in the MBR systems [144].

6.2.1 Pretreatment Processes for Membrane Fouling Mitigation

Pretreatment is carried out prior to the application of membrane separation processes which leads to change in the physical, chemical or biological properties of wastewater so as to make the membrane performance more efficient. Feed pretreatment plays an integral part in the success of the membrane process which not only reduces membrane fouling, but also contributes to energy utilization. Different physicochemical methods such as coagulation, adsorption and softening have been applied in several instances for pretreatment of wastewater before membrane separation.

(i) Permeate flux decline is a critical issue encountered during long term membrane fouling. To explore the potential advantages of applying intermittent ultrasound (US) to the UF processes in controlling long term membrane fouling, Hasani et al. showed that intermittent US-generated cavitation bubbles which can control membrane fouling achieving optimal flux at 28 kHz, 50 W, and 60 s interval due to creation of local turbulence and shear effect near the membrane surface to disrupt the concentration polarization and fouling layer [145].
(ii) The reduction and recovery of permeate flux after backwashing is an important indicator of membrane life. Presence of natural organic matter (NOM) is the key factor causing fouling and flux decline of ceramic membranes. Oxidative pretreatment using ozone can degrade the organic matter thereby improving membrane flux and reducing membrane fouling. With a view to propagate the application of ozonation and adsorption coupled with ceramic membrane technology, Qiao et al. investigated the mitigation mechanisms of ozone pre-oxidation, powdered activated carbon (PAC) pre-adsorption, and combined ozone-PAC pretreatment on membrane fouling caused by NOM [146].

Investigation on ceramic membrane bioreactors (CMBRs) revealed an optimal dosage of in-situ ozonation for long-term MBR operation (5 mg/L or 0.66 mg-ozone/g MLSS-mixed liquor suspended solid) with the optimal filterability of mixed liquor in wastewater treatment. Ceramic MBR with in-situ ozonation has not only significantly alleviated membrane fouling but also remarkably improved phosphorus removal performance [147].

To solve the issues in membrane filtration, heterogeneous catalytic ozonation (HCO) technology, a strategy of immobilizing catalysts on the membranes surface and pores has been discussed for dual functional properties. Consequently, membrane filtration and HCO processes can be well integrated in the hybrid ozonation membrane filtration (HOMF) process. Since ceramic membranes are commonly used in the HOMF, catalytic ceramic membranes (CCM) have been developed by loading the active catalyst onto the membrane to enhance the catalytic capacity. The removal efficiency of HOMF was also found to depend on several governing factors, including operational parameters (such as pH value, ozone dosage, and HRT), solution chemistry (dissolved organic matter and co-existing ions), and membrane properties (catalyst loading). A recent review discussed the HOMF process for water and wastewater treatment, including engineering applications, conclusions, and perspectives [148].

(iii) Advanced oxidation processes (AOPs) have been considered as a solution to address membrane fouling due to their ability to generate more reactive oxygen species (ROS) by activating oxidants. The most commonly used oxidants are hydrogen peroxide (H_2O_2), persulfate and ozone. In comparison, ozone outperformed as an environmentally friendly and powerful oxidant (2.07 V) which can be made insitu by an ozone generator. However, its utilization rate was found to be generally lower, resulting in limited oxidation capability and formation of toxic by products (aldehydes, ketones, and bromate). Thus, heterogeneous catalytic ozonation (HCO) has recently garnered increased attention, as it facilitates enhanced ozone decomposition and improves contaminants elimination. Nevertheless, the limited mass transfer and short-lived of hydroxyl radicals rendered HCO an inefficient utilization of radicals [148].

(iv) The strategy of using Micellar Enhanced Ultrafiltration (MEUF) was suggested as a clean technology for treatment and reclamation of textile effluent and dye house wastewater. The dye was removed from the aqueous stream using sodium dodecyl sulphate (SDS) and polysulfone membrane (MWCO 50 kDa) in a stirred ultrafiltration cell with an improvement in dye rejection by 15–20% compared to ultrafiltration which is within the limitations set by the Central Pollution Control Board (CPCB) and the Bureau of Indian Standards (BIS) [45].

(v) UF membrane fouling is the biggest obstacle affecting the wide application of UF technology although it exhibits higher removal efficiencies for suspended solids and organic macromolecules. Prior to UF processes, various pretreatment measures have been proposed like coagulation, adsorption, and advanced oxidation, mainly focusing on the contribution of natural macromolecular pollutants to UF membrane fouling. A recent review discussed on the removal efficiency and separation mechanism for emerging pollutants (EPs) using pretreatments combined with UF membrane separation technology along with the structural characteristics of the cake layer while addressing the problems for efficient separation of EPs [149]. For mitigation of membrane fouling and the long-term operation of membrane systems, its necessary to select effective pretreatment processes which represent UF (32.9%), MF (16.6%), activated carbon (AC) adsorption (16.0%), coagulation (14.3%), and ozonation (11.4%) as per statistical analysis of the publications available in literature about prevaling pretreatment techniques prior to high-pressure membrane systems [150].

6.2.2 Fouling and Methods of Membrane Cleaning

Fouling is an unavoidable hurdle in the performance of membrane processes for limiting the flux. Membrane system can be operated in two ways: constant permeate flux with variable TMP or variable permeate flux at constant TMP. Membrane fouling arises during an increase in TMP to maintain a flux. Deposition of foulants on the membrane surface or inside the pores affects membrane properties like hydrophilicity and surface charge which needs periodical cleaning for restoring steady performances. Broad classifications in membrane cleaning operations are generally based on physical, chemical, biological/biochemical or physico-chemical treatments. The insitu process implies cleaning of the membrane module when it remains fitted in place while exsitu cleaning is made separately after removing the membrane module [150–152]. Physical cleaning involves mechanical treatment of the membrane to dislodge the foulants from the membrane as described below:

(a) For back flushing, the permeate is forced back through the membrane from the permeate side to the feed side by applying higher pressure which causes displacement of the materials deposited on the membrane surface [153, 154]. Backwashing which involves removal of gel or cake layer formed on the membrane surfaces by deposition of the foulants is mostly used in industry to regenerate the permeate flux effectively. However, backwashing is ineffective for irreversible fouling due to clogging of pores by colloidal suspensions and dissolved materials [155].
(b) Air sparging, air lifting and air scouring is a pneumatic cleaning operation for removal of the foulants by applying a shear force on the membrane surface. Air bubble may be induced by pumping air in the feed to pass through the membrane to enhance permeate flow [151, 156].
(c) In the ultrasonic cleaning process, a turbulence is created over the membrane surface which is found to be very effective in cleaning the foulants. The variables in this physical cleaning technique are ultrasonic power, cleaning temperature, cross flow velocity and pulse duration [157–159].

(d) For scrubbing the membrane surface, sponge balls are used inside the tubular membrane for scrapping the foulants [153, 160].
(e) In case of irreversible fouling, cleaning is effected by using chemicals selected primarily based on the interaction with the foulant and/or membrane material where in the foulant is dissolved and the membrane material remains unaffected [161]. For cleaning in place (CIP) process, the retentate channel is filled the cleaning solution (detergent) to weaken the attachment between the membrane and the foulant which is washed out during crossflow operating mode [162].

Besides the acidic and alkaline chemicals, chelating agents or sequestrants, surfactants, or disinfectants are used as cleaning agents. Acid cleaning by HCl, H_2SO_4, HNO_3, H_3PO_4 removes inorganic foulants such as scaling or precipitates and metal oxides [151, 163]. Organic foulants are treated with alkalis like NaOH or alkali carbonates/phosphates [153, 160].

(f) Bioactive agents like enzymes, enzyme mixtures, etc. are employed for foulant removal in biological/biochemical cleaning which are more sustainable than other physical and chemical cleaning processes, particularly in membrane bioreactors [152, 163–165].
(g) Physico-chemical cleaning methods combines the addition of chemical agents during physical cleaning such as chemically enhanced backwashing (CEB) and use of ultrasound in chemical cleaning to enhance flux recovery of up to 95% [166, 167].

Regeneration of low cost apatite MF membrane coated over clay–alumina support was studied after color removal by varying H_2O_2 concentration and contact time which demonstrated the optimum efficiency at 1% H_2O_2 (v/v) with more than 60% regeneration of initial capacity after several separation cycles [168]. A low cost multichannel ceramic microfiltration (MF) membrane fabricated using clay-alumina was used for evaluation of the cleaning efficiency adopting different washing protocols after separation of dye-loaded activated carbon under varying operating conditions. Membrane fouling was found to be due to the cake layer formed by the activated carbon particles used for adsorption of common reactive dyes (reactive yellow, red, black and brown) used in the textile industry [128].

6.3 *Fouling in MBR*

Fouling is a self-accelerating process that deteriorates the membrane filtration process, and it is not only disastrous operationally but also economically. Membrane fouling mitigation has been the focus for researchers during the last two decades. Traditionally, based on the ease and extent of foulants removal as well as their interaction with the membrane surface, membrane fouling is broadly divided into two types viz. reversible fouling and irreversible fouling. It is generally believed that the fouling caused by a loose/soft deposition of the impurities on the membrane

Table 2 Onset of various types fouling in membrane bioreactor (MBR) [144]

Fouling type	Rate of fouling (Pa min^{-1})	Onset of fouling
Reversible fouling	10–100	10 min
Irreversible fouling (removed by maintenance chemical cleaning)	1–10	1–2 weeks
Irreversible fouling (removed by mandatory chemical cleaning)	0.1–1	6–12 months
Non-restorable fouling	0.01- 0.1	A few years

surface is attributed to reversible fouling, while irreversible fouling originates from the strong adhesion of contaminants which leads to blockage of membrane pores. It is transpired from several studies that the main form of membrane fouling is attributed to the formation and compaction of the cake layer compared to the pore blocking. Du et al. discussed about classification, mechanism, control of membrane fouling and influencing factors, online monitoring of membrane fouling, predicting membrane service life, etc. Table 2 shows the onset of various types of membrane fouling in MBR while Table 3 gives a comparison of membrane fouling control methods [144]. While it is feasible to restore the flux in case of reversible fouling by adopting backwashing or hydrodynamic scouring (surface washing), chemical cleaning is necessary for non-backwashable reversible fouling. Irreversible membrane fouling arises by chemisorption or by blocking of the membrane pores. The steady permeate flux cannot be recovered using hydrodynamical or chemical treatment in case of irreversible fouling by chemisorption or pore blocking for which extensive chemical cleaning or membrane replacement is required. Lin et al. reviewed physical and chemical cleaning and dealt with the fouling mitigation strategies and cleaning efficiencies [151].

6.3.1 Fouling Control Strategies in MBR Technology

To reduce operational and maintenance costs, it is necessary to improve the performance by adopting effective fouling control strategies. For development of cost-effective MBR technology, attempts have been made to use various cleaning operations in controlling membrane fouling which include backwashing, scouring aeration, chemical cleaning, etc. including membrane surface modification and functionalization [169].

For treatment of ammonium-rich wastewater, anaerobic ammonium oxidation (anammox) has been developed and to prevent membrane fouling while chemically enhanced backwash (CEB) has been adopted for promoting wider use of anammox MBR. CEB consists of backwashing in combination with a disinfectant chemical (NaClO) in varying concentrations in the range of 149–596 mg/L. Liao et al. [170] reviewed the key parameters affecting membrane fouling and related issues for low-pressure membranes and MBRs [171].

Table 3 Membrane fouling control methods [144]

Control methods	Controlling factors	Expected results	Precautions
Modification of membrane material	Improve membrane surface hydrophilicity	Reduce the adsorption of impurities on the membrane surface and membrane pores	The membrane material should be modified according to treatment objectives
Optimization of membrane components	Improve membrane surface water conditions	Improve the effect of membrane surface gas flow flushing and decontamination	High mechanical properties for membrane materials
Aeration, ultrasound	Remove membrane deposits and improve liquid properties	Gas–liquid flow flushes out membrane deposits to increase activated sludge activity	Excessive aeration or microwave vibration will break up the sludge flocs and increase the fouling of the membrane
Add flocculant or absorbent (PAC), ozone	Improve liquid properties	Improve sludge settling and reduce EPS and SMP in feed liquid	Inorganic flocculants change the pH of the feed, the absorbent itself may also become a contaminant, and ozone inhibits microbial activity
Intermittent suction	Improve film surface detachment properties	Conducive to the membrane surface gas flow flushing with pollutants	Too long stoppage will affect the amount of water produced, too short to achieve the desired result

Several studies have been carried out using ceramic membranes in MBRs [172, 173]. Investigations on the effect of backwashing as a mechanical method to mitigate biofouling have been conducted with up to 100% recovery of the initial flux in the submerged ceramic membrane bioreactor using alumina membrane with a pore size of 3 μm [174]. Beside membrane fouling in MBR process, another major issue to ensure cost-effectiveness to address higher cost of ceramic membrane by using low-cost membranes. Low-cost MF membranes made of clay and chamotte along with different of pore formers have been synthesized with two types of selective layers consisting of clay and/or TiO_2 with pore size in the range of 0.5–0.7 μm (d_{50}) which exhibited a removal efficiency of 99.99% for Giardia cysts and Cryptosporidium oocysts with no *E. coli* detected in the permeate. The study reveals the application of low-cost ceramic MF support with selective TiO_2 layer for removal of these bacteria and protozoa in a ceramic membrane bioreactor [88]. An innovative technique is based on the use of ceramic inkjet technology in low cost membrane manufacture and validation of this new membrane in both laboratory and pilot plant scale [175].

7 Emerging Issues

Emerging contaminants (ECs) are classified as pharmaceuticals and personal care products (PPCPs) which negatively impact ecology, human health and aquatic organisms [176–178]. The presence of emerging organic micropollutants including pharmaceuticals, endocrine disrupting compounds, and personal care products, require innovative technologies to address growing environmental issues in recent decades [179] Mitigating inorganic and organic contaminants, which is coined as contaminants of environmental and emerging concern (CEECs), is a big challenge as traditional physicochemical processes are not economically viable for managing mixed contaminants at low concentrations. As a result, low-cost materials must be designed to provide high CEEC removal efficiency [180, 181]. Increasing contamination of heavy metals, pharmaceuticals, is a emerging issue in the municipal and industrial wastewater treatment and biosorption has been considered as a promising approach using bio-materials like biofibers, biofilms, biopolymers, and biocomposites as biosorbents [182]. The scope of removal/remediation of emerging contaminants (ECs) in water has been reviewed using adsorption and membrane filtration as a single hybrid system [183].

8 Reusability Aspects

Application of ultrafiltration/microfiltration (UF/MF) membranes in industrial wastewater recycling has attracted attention due to their efficiency in removing suspended solids and macromolecular organic pollutants particularly in treating textile wastewater [46]. However, although there are several set guidelines worldwide for water reuse, it would be useful to follow an universal guideline for reuse of reclaimed water. Ahmed Abou-Shady and Heba El-Araby discussed the treatment technologies for water reuse and presented some recent guidelines for water reuse [184]. Recently, Alayande et al. reviewed the guidelines governing wastewater reuse in agriculture to identify the gaps and challenges associated with providing nutrients from decentralized AnMBR effluents to facilitate reuse in suitable, low land-footprint hydroponic controlled environment agriculture (CEA) systems [185].

Performance evaluation of low-cost ceramic membrane was investigated for treatment and reuse of high organic loaded grey water using the MF and UF process individually and as a two-stage treatment involving MF followed by UF. While about 73–90% COD reduction was achieved in the single-stage UF at an operating pressure of 2 bar, a minor improvement (about 84–94%) was observed for two-stage treatment. Reusability studies with the treated wastewater on a popular palm species showed significant potential for utilization of the MF and UF treated water which facilitated an enhanced uptake of most of the essential nutrients in plants compared with that of fresh water [116]. Saja et al. reported the application of a flat plate microfiltration membrane made of Moroccan ceramics (optimized by sintering at

950 °C with an average pore size of 1.70 μm and 52% porosity), for evaluation of the performance using two industrial wastewater effluents which showed turbidity removal of 96–97% for agro-food and tannery effluents, allowing the water to be used for secondary purposes such as agriculture while meeting water quality criteria [186]. Performance evaluation of the low cost ceramic MF membrane was investigated alone and in combination with a biological pretreatment for the treatment of cosmetic industrial effluent and municipal wastewater which showed potential of the combined process in terms of reduction in COD and BOD, turbidity, oil and grease, total suspended solids, etc. The treated water quality satisfied the discharge norms according to the physicochemical pollution parameters as prescribed by the Central Pollution Control Board, India for wastewater. Low cost ceramic membrane was found to be extremely stable under exhaustive cleaning conditions and initial permeability could be effectively restored which highlighted the potential of hybrid process for treatment of industrial effluent, as well as, municipal wastewater for reuse purpose [187]. Treatment of highly concentrated textile dyebath effluent (2220 mg/L COD) using clay-alumina based low cost ceramic MF membrane in combination with a biosorbent prepared from sugarcane bagasse showed 91% COD reduction with about 99% dye removal, >99% turbidity removal and >90% TSS reduction. The reusability study utilizing the treated wastewater in the cotton fabric dyeing process showed high potential with respect to water reclamation, as well as reduction of the associated chemicals consumption [188].

Zirconia UF membrane (40 nm mean pore diameter) coated over α-alumina-clay based support in 19-channel circular configuration showed about 82 and 92% COD removal for tannery wastewater and kitchen sink wastewater, respectively with turbidity level below 1 NTU for both the effluents. Complete removal of pathogenic organisms suggested reuse of treated wastewater for agricultural purposes [189]. Mzahma et al. studied the scope of reusing membrane treated textile effluents in agricultural sector and suggested that additional treatments by NF and RO improved the physico-chemical quality of textile effluents for reuse in agriculture [190]. Application of TiO_2-coated UF membrane over low-cost clay alumina based multichannel support showed reduction in the turbidity level from 51 to 1 (NTU) and microorganism content from 18 to 2 MPN per 100 ml during treatment of grey water from an university canteen with almost complete removal of O & G (0.002 from 8 g/l). Reduction in BOD and COD concentration to 15 and 50 (mg/l) in the permeate after UF treatment, from 892 and 4500 (mg/l) respectively indicated the quality of treated water within WHO-permissible limits [191].

Treatment of oilfield produced water presented multiple application avenues, from agricultural irrigation to industrial operations including reinjection using treated oilfield produced water. However, for potential resource recovery, further development is needed to address oil removal along with extraction of valuable metallic components [55].

9 Scope of Further Studies

Although membrane technology offers a promising solution due to its efficiency and minimal need for chemical additives or thermal inputs, the research on resource recovery still needs to be expanded, while addressing the challenges such as fouling, resistance to oil and organics including efficient cleaning protocols and economic viability. Dosing flocculant is the most promising method for controlling membrane fouling among current techniques, but developing an auto-flocculant dosing strategy is vital [192]. Research progress in the removal of emerging contaminants (ECs) showed an improved performance using an integrated hybrid system combining membrane filtration and adsorption as a single system compared to adsorption as the pre-treatment or post-treatment operation. However, further studies, particularly the scale-up issue, has been recommended for a wide range of ECs removals [183]. AnMBR technology is an attractive option for treatment of wastewater with recovery of resources like bioenergy in the form of renewable methane, fertilizing elements and water reuse. Maaz et al. reviewed the challenges and opportunities for further development of AnMBRs for wider spread and scale-up of the technology [193]. In order to improve the selectivity for recovery of valuables like vitamins, enzymes and antibiotics in the pharmaceutical industry sector, future research may focus on introducing new polymer-ceramic composite membranes with lower MWCO (below 450 Da) along with greater tolerance to aggressive conditions [68].

Analysis of the literature on fouling mitigation techniques revealed that the number of publications in the field of membrane surface modification is more compared to chemical cleaning, backflushing, ultrasound, etc. Surface modification is a relatively more effective option as compared with the fouling mitigation techniques since it deals with the foulant-membrane interaction [194]. Future research on the development of more cost effective fouling resistant membranes needs to be focused for treatment of different types of effluents [119]. Thus it is imperative to emphasize research efforts on the importance of developing antifouling membranes for sustainable waste management techniques while considering economic implications and market dynamics for resource recovery [55]. In this context, it is worthwhile to look into the prospect of developing antifouling alumina membrane by chemical grafting with organosilanes followed by oxidation which showed an significant enhancement of stable permeate flux, approximately eight times higher than that of Al_2O_3 membranes, exhibiting excellent antifouling performance in protein-containing systems [195].

Preparing low-cost membranes with high flux performance is another challenging area for future development of the inorganic membranes particularly for treatment of high volume of liquid effluent and removal of emerging contaminants (ECs). To intensify the concept of circular economy, it is imperative to undertake concerted research efforts globally to create potential revenue streams for local industrial waste in manufacturing cost efficient ceramic membranes especially for utilizing waste from local ceramic industry and agricultural sector [196].

10 Conclusions

Grey water being the most generated wastewater with lower pollutant load, its recycling of could be one of the foremost options that can be beneficially reused after suitable treatment. Anaerobic membrane bioreactors (AnMBRs) have received increased attention for treating domestic wastewater in view of several advantages including complete rejection of suspended solids, low sludge production, high organic removal, and biogas production.

Another attractive proposition is reuse of municipal wastewater and oilfield produced water after adequate treatment, which presents multiple application avenues, from industrial operations to agricultural and landscape irrigation. Reusing the reclaimed water after treating municipal wastewater is emerging as a strategy to tackle water scarcity by curtailing the discharge of untreated wastewater and conventionally-treated wastewater into natural ecosystems. However, the emerging contaminants such as pharmaceuticals and personal care products (PPCPs), antibiotics, micro-plastics, etc. pose a challenge for treatment and reuse of municipal wastewater.

Membrane fouling control necessitates selection of effective pre-treatment processes followed by adoption of appropriate cleaning protocol based on process optimization considering varying feed water characteristics, AOP and foulant-membrane surface interaction. Full commercial application of ceramic membrane bioreactor technology requires more attention on both reducing the cost of ceramic membranes and fouling mitigation to achieve a stable long term operation.

References

1. United Nations Sustainable Development Goal 6: ensure access to water and sanitation for all. https://www.un.org/sustainabledevelopment/water-and-sanitation/
2. Abbas MA, Iqbal M, Tauqeer HM et al (2022) Microcontaminants in wastewater. Environmental micropollutants. Elsevier, Amsterdam, pp 315–329
3. El Kik O et al (2024) Synergistic approach for enhanced wastewater treatment: harnessing the potential of bioelectrochemical systems in integration with anaerobic membrane bioreactor. J Environ Chem Eng 12(4):113162. https://doi.org/10.1016/j.jece.2024.113162
4. European Water Framework Directive (EU WFD), set more stringent nutrient and micropollutant removal requirements, reinforcing the need for advanced treatment facilities. https://environment.ec.europa.eu/topics/waste-and-recycling/waste-framework-directive_en
5. Artun G, Aşkin A (2022) Studies on production of low-cost ceramic membranes and their uses in wastewater treatment processes. Eur J Res Dev 2(2):126–140
6. Commission of European Communities. Communication no. 398 (2014) Towards a circular economy: a zero-waste programme for Europe. (COM 2014, 398). Commission of European Communities. Communication no. 614, 2015. Closing the loop—an EU action plan for the circular economy (COM 2015, 614)
7. Bandyopadhyay, Sibdas (eds) (2025) Low cost ceramic membranes: materials and fabrication process, eBook ISBN 978-981-96-1602-2, Springer
8. Neczaj E, Grosser A (2018) Circular economy in wastewater treatment plant–challenges and barriers. Proceedings 2(11):614. https://doi.org/10.3390/proceedings2110614

9. Mustafa HM et al (2022) Circular economy framework for energy recovery in phytoremediation of domestic wastewater. Energies 15(9):3075. https://doi.org/10.3390/en15093075
10. He Z, Lyu Z, Gu Q, Zhang L, Wang J (2019) Ceramic-based membranes for water and wastewater treatment. Colloids Surf A Physicochem Eng Asp 578:123513
11. https://www.who.int/news-room/fact-sheets/detail/drinking-water
12. Nielsen PH (2017) Microbial biotechnology and circular economy in wastewater treatment. Microb Biotechnol 10(1102–1105):13
13. Ranade VV, Bhandari VM (2014) Industrial wastewater treatment, recycling and reuse. Elsevier. https://doi.org/10.1016/C2012-0-06975-X
14. Li F, Wichmann K, Otterpohl R (2009) Review of the technological approaches for greywater treatment and reuses. Sci Total Environ 407(11):3439–3449
15. Ortiz J (2021) Advances in microfiltration and ultrafiltration technology for greywater treatment: a review. Nat Environ Pollut Technol 20(4):1601–1609
16. Chuanhong X, Eric T, Yi Q (1997) Inorganic membrane-bioreactor treatment of domestic sewage test. J Environ Sci 18(3):1–4
17. Mara D (2003) Domestic wastewater treatment in developing countries. Routledge, Oxfordshire, UK, 2013. ISBN 978-1-136-56792-6 (First Published 2003). https://doi.org/10.4324/9781849771023
18. Liu J, Zhou Y, Tang L et al (2007) Inorganic ceramic membrane treatment of domestic sewage in the membrane pollution and cleaning. J Anhui Inst Arch Nat Sci Ed 15(3):57–60. In: E3S web conference 2018, vol 53, p 04032; https://doi.org/10.1051/e3sconf/20185304032
19. Jeong Y et al (2018) Comparison of filtration and treatment performance between polymeric and ceramic membranes in anaerobic membrane bioreactor treatment of domestic wastewater. Sep Purif Technol 199(30):182–188. https://doi.org/10.1016/j.seppur.2018.01.057
20. Liu X-Y et al (2024) A review of combined fouling on high-pressure membranes in municipal wastewater reuse: behaviors, mechanisms, and pretreatment mitigation strategies. 485(1):150135. https://doi.org/10.1016/j.cej.2024.150135
21. Hamedi FHA et al (2023) Wastewater irrigation: a promising way for future sustainable agriculture and food security in the United Arab Emirates. Water 15(12):2284. https://doi.org/10.3390/w15122284
22. Tomczak W, Gryta M (2022) Energy-efficient AnMBRs technology for treatment of wastewaters: a review. Energies 15(4981). https://doi.org/10.3390/en15144981
23. Aslam A, Khan SJ, Shahzad HMA (2022) Anaerobic membrane bioreactors (AnMBRs) for municipal wastewater treatment potential benefits, constraints, and future perspectives: an updated review. Sci Total Environ 802:149612
24. Vinardell S et al (2020) Advances in anaerobic membrane bioreactor technology for municipal wastewater treatment: a 2020 updated review. Renew Sustain Energy Rev 130. https://doi.org/10.1016/j.rser.2020.109936
25. Rocco MJ, Hafuka A, Tsuchiya T, Kimura K (2023) Efficient recovery of organic matter from municipal wastewater by a high-rate membrane bioreactor equipped with flat-sheet ceramic membranes. Membranes 13(3):300. https://doi.org/10.3390/membranes13030300
26. Kramer FC, Shang R, Rietveld LC, Heijman SJG (2020) Fouling control in ceramic nanofiltration membranes during municipal sewage treatment. Sep Purif Technol 237(15):116373. https://doi.org/10.1016/j.seppur.2019.116373]
27. Lin F (2018) Application of ceramic membrane in water and wastewater treatment. In: E3S web conferences, vol 53, p 04032 ICAEER. https://doi.org/10.1051/e3sconf/20185304032
28. Akash FA, Shovon SM, Rahman W, Rahman MdA, Chakraborty P, Prasetya TAE, Monir MU (2024) Advancements in ceramic membrane technology for water and wastewater treatment: a comprehensive exploration of current utilizations and prospective horizons. Desalin Water Treat 319. https://doi.org/10.1016/j.dwt.2024.100569
29. Singh BJ et al (2023) A systematic review of industrial wastewater management: evaluating challenges and enablers. J Environ Manag. https://doi.org/10.1016/j.jenvman.2023.119230

30. 6.3 UN sustainable development goals. https://circularecology.com/sustainable-development-goals/clean-water.html#:~:text=Themainpurposeofgoal,tocommunitiesaroundtheworld
31. Bera SP, Godhaniya M, Kothari C (2022) Emerging and advanced membrane technology for wastewater treatment: a review. J Basic Microbiol 62(3–4, Special Issue: Emerging microbial technologies for wastewater treatment, March-April 2022, pp 245–259. https://doi.org/10.1002/jobm.202100259
32. Ezugbe EO, Rathilal S (2020) Membrane technologies in wastewater treatment: a review. Membranes (Basel) 10:89
33. Arumugham T, Kaleekkal NJ, Gopal S, Nambikkattu J, Rambabu K, Aboulella AM, Ranil Wickramasinghe S, Banat F (2021) Recent developments in porous ceramic membranes for wastewater treatment and desalination: a review. J Environ Manag 293, Article 112925. https://doi.org/10.1016/j.jenvman.2021.112925
34. He Z, Lyu Z, Gu Q, Zhang L, Wang J (2019) Ceramic-based membranes for water and wastewater treatment. Colloids Surf A Physicochem Eng Asp 578. https://doi.org/10.1016/j.colsurfa.2019.05.074
35. Zrelli A et al (2023) Eng Technol J 41(09):1130–1138
36. Kayvani Fard A, McKay G, Buekenhoudt A, Al Sulaiti H, Motmans F, Khraisheh M, Atieh M (2018) Inorganic membranes: preparation and application for water treatment and desalination. Materials 11:74. https://doi.org/10.3390/ma11010074
37. Issaoui M, Limousy L (2019) Low-cost ceramic membranes: synthesis, classifications, and applications. Comptes Rendus Chim 22:175–187. https://doi.org/10.1016/j.crci.2018.09.014
38. Asif MB, Zhang Z (2021) Ceramic membrane technology for water and wastewater treatment: a critical review of performance, full-scale applications, membrane fouling and prospects. Chem Eng J 418:129481. https://doi.org/10.1016/j.cej.2021.129481
39. Sandhya Rani SL, Kumar RV (2021) Insights on applications of low-cost ceramic membranes in wastewater treatment: a mini-review. Case Stud Chem Environ Eng 4:100149, https://doi.org/10.1016/j.cscee.2021.100149
40. Dong Y, Wu H, Yang F, Gray S (2022) Cost and efficiency perspectives of ceramic membranes for water treatment. Water Res 220:118629. https://doi.org/10.1016/j.watres.2022.118629
41. Vasanth D, Prasad AD (2019) Ceramic membrane: synthesis and application for wastewater treatment—a review. In: Rathinasamy M, Chandramouli S, Phanindra K, Mahesh U (eds) Water resources and environmental engineering II. Springer, Singapore. https://doi.org/10.1007/978-981-13-2038-5_10
42. Samaei SM et al (2018) The application of pressure-driven ceramic membrane technology for the treatment of industrial wastewaters—a review. Sep Purif Technol
43. Omar et al (2023) Recent progress and technical improvement strategies for mitigating ceramic membrane bottlenecks in water purification processes: a review. Int J Appl Ceram Technol 20(6):3327–3356. https://doi.org/10.1111/ijac.14503
44. De Gisi S, Notarnicola M (2017) Industrial wastewater treatment. In: Encyclopedia of sustainable technologies, pp 23–42. https://doi.org/10.1016/B978-0-12-409548-9.10167-8
45. Antad SG, Anekar SV (2023) Membrane treatment of industrial effluent. IJIRT 9(Issue 9)
46. Tang H, Shan M, Cheng Q, Wen R, Liu S, Zeng H, Yu J, Luo J (2024) Rapid pretreatment strategy to control nanofiltration membrane fouling in recycling of real textile wastewater: comparison and mechanisms. Chem Eng J 494:152913. https://doi.org/10.1016/j.cej.2024.152913
47. https://www.europarl.europa.eu/topics/en/article/20201208STO93327/the-impact-of-textile-production-and-waste-on-the-environment-infographics#:~:text]
48. Vo T, Bui X, Dang B, Nguyen T, Nguyen V, Tran DPH, Nguyen P, Boller M, Lin KA, Varjani S, Show PL (2021) Influence of organic loading rates on treatment performance of membrane bioreactor treating tannery wastewater. Environ Technol Innov 24:101810. https://doi.org/10.1016/j.eti.2021.101810
49. Yang F, Wang XB, Shan Y et al (2023) Research recap of membrane technology for tannery wastewater treatment: a review. Collag Leather 5:24. https://doi.org/10.1186/s42825-023-00132-8

50. Xiaodong X, Ji L, Yuming Y et al (2005) Membrane technology applied in the production of water treatment in the field. J Filter Sep 15(4):34–36
51. Naseri M, Omidkhah M, Mousavi SF (2024) A study on the efficient separation of oily water using mullite whiskers membrane through combined filtration and electrofiltration. https://doi.org/10.2139/ssrn.4628945
52. Rashad M, Logesh G, Sabu U, Balasubramanian M (2021) A novel monolithic mullite microfiltration membrane for oil-in-water emulsion separation. J Membr Sci 620
53. Omar NMA, Othman MHD, Tai ZS, Puteh MH, Jaafar J, Rahman MA, Ismail AF, Wong KY (2023) Effect of sintering temperature on functional properties of mullite-kaolinite and stainless steel composed hollow fibre membrane for oil-in-water emulsion separation. J Taiwan Inst Chem Eng 146:104859. https://doi.org/10.1016/j.jtice.2023.104859
54. Esmaili Z et al (2023) Anti-fouling and self-cleaning ability of BiVO4/rGO and BiVO4/g-C3N4 visible light-driven photocatalysts modified ceramic membrane in high performance ultrafiltration of oily wastewater. J Membr Sci. https://doi.org/10.1016/j.memsci.2023.122147
55. Siagian UUWR et al (2024) From waste to resource: membrane technology for effective treatment and recovery of valuable elements from oilfield produced water. Environ Pollut 340(Part 2):122717. https://doi.org/10.1016/j.envpol.2023.122717
56. Abuhantash F, Abuhasheesh YH, Hegab HM et al (2023) Hydrophilic, oleophilic and switchable Janus mixed matrix membranes for oily wastewater treatment: a review. J Water Process 56, id.104310. https://doi.org/10.1016/j.jwpe.2023.104310
57. Murić A et al (2014) Comparison of ceramic and polymeric ultrafiltration membranes for treating wastewater from metalworking industry. Chem Eng J 255(1):403–410. https://doi.org/10.1016/j.cej.2014.06.009
58. Barbera and Gurnari (2018) Wastewater treatment and reuse in the food industry. ISBN: 978-3-319-68441-3; https://doi.org/10.1007/978-3-319-68442-0
59. Tekerlekopoulou AG, Economou ChN, Tatoulis TI, Akratos CS, Vayenas DV (2020) Chapter 8—Wastewater treatment and water reuse in the food industry. In: Galanakis C (ed) The interaction of food industry and environment. Academic Press, pp 245–280. ISBN 9780128164495. https://doi.org/10.1016/B978-0-12-816449-5.00008-4
60. Costa CFFA et al (2022) Valorization of wastewater from food industry: moving to a circular bioeconomy. Rev Environ Sci Bio/Technol 21(13). https://doi.org/10.1007/s11157-021-09600-1
61. Gbadeyan OJ et al (2024) Waste-to-energy: the recycling and reuse of sugar industry waste for different value-added products such as bioenegy in selected countries—a critical review. Biofuels Bioprod Biorefin. https://doi.org/10.1002/bbb.2579
62. Lima LKS, Santana LNL, Lira HL, Rodríguez MA, Souza MYM, Júnior MGS, Lira BS (2024) Development of asymmetric ceramic membranes for dairy wastewater treatment—a comparison between co-sintering and conventional firing process. J Water Process Eng 57:104611. https://doi.org/10.1016/j.jwpe.2023.104611
63. Venugopal V et al (2021) Seafood industry effluents: environmental hazards, treatment and resource recovery. J Environ Chem Eng 9(Issue 2):104758. https://doi.org/10.1016/j.jece.2020.104758
64. Cooney R et al (2023) J Clean Prod 392:136283. https://doi.org/10.1016/j.jclepro.2023.136283
65. Murdoch K (2015) Pharmaceutical pollution in the environment: issues for Australia, New Zealand and Pacific Island countries. https://www.xn--hbo-tibble-15a.se/dokument/pdf/NTN-Pharmaceutical-Pollution-in-the-Environment-2015-05.pdf
66. Gurgenidze D, Romanovski V (2023) The pharmaceutical pollution of water resources using the example of the Kura River (Tbilisi, Georgia). Water 15(14):2574. https://doi.org/10.3390/w15142574
67. Baker RW (2012) Membrane technology and applications, 3 edn. Wiley. ISBN: 9780470743720
68. Pabby AK (2015) Handbook of membrane separations chemical, pharmaceutical, food and biotechnological applications, 2nd edn. Hoboken

69. Duscher S (2013) Ceramic membranes in chemical and pharmaceutical applications. Ceram Appl 1:19–23
70. Bhattacharya P, Mukherjee D, Deb N, Swarnakar S, Banerjee S (2020) Application of green synthesized ZnO nanoparticle coated ceramic ultrafiltration membrane for remediation of pharmaceutical components from synthetic water: reusability assay of treated water on seed germination. J Environ Chem Eng 8(3):103803
71. Bhattacharya P, Mukherjee D, Dey S, Ghosh S, Banerjee S (2019) Development and performance evaluation of a novel CuO/TiO_2 ceramic ultrafiltration membrane for ciprofloxacin removal. Mater Chem Phys 229(February):106–116
72. Zhang L, Qin L, Ma L, Shen Z, Jin Y, Chen S (2024) Treatment of electroplating wastewater using electrocoagulation and integrated membrane. Water Sci Technol 89(9):2538–2557. https://doi.org/10.2166/wst.2024.136
73. Hamdy A, Mostafa MK, Nasr M (2019) Techno-economic estimation of electroplating wastewater treatment using zero-valent iron nanoparticles: batch optimization, continuous feed, and scaling up studies. Environ Sci Pollut Res Int 26(24):25372–25385. https://doi.org/10.1007/s11356-019-05850-3]
74. Ashrafi O, Yerushalmi L, Haghighat F (2015) Wastewater treatment in the pulp-and-paper industry: a review of treatment processes and the associated greenhouse gas emission. J Environ Manage 158:146–157
75. Goswami L, Kumar RV, Pakshirajan K, Pugazhenthi G (2019) A novel integrated biodegradation—microfiltration system for sustainable wastewater treatment and energy recovery. J Hazard Mater 365:707–715. https://doi.org/10.1016/j.jhazmat.2018.11.029
76. Zhu M-C, Wang J-W, Lu Y-Z, Hu Z-X, Chen S-W, Li N (2024) Enhancement of membrane filtration performance and mitigation of sludge calcification in an intermittent anaerobic membrane bioreactor treating high-calcium wastewater. Sep Purif Technol 129028. https://doi.org/10.1016/j.seppur.2024.129028
77. Jarrar R, Abbas MKG, Al-Ejji M (2024) Environmental remediation and the efficacy of ceramic membranes in wastewater treatment—a review. Emerg Mater 7:1295–1327. https://doi.org/10.1007/s42247-024-00687-0
78. Chauhan N, Sharma R, Singh SP (2023) Synthesis of ceramic membranes and their application in wastewater treatment and emerging contaminants removal. In: Sinha A, Singh SP, Gupta AB (eds) Persistent pollutants in water and advanced treatment technology. Energy, environment, and sustainability. Springer, Singapore. https://doi.org/10.1007/978-981-99-2062-4_17
79. Chaukura N, Riet A, Mithi D, Manjoro M (2022) Ceramic nanocomposite membranes for dye removal. In: Muthu SS, Khadir A (eds) Membrane based methods for dye containing wastewater. sustainable textiles: production, processing, manufacturing & chemistry. Springer, Singapore. https://doi.org/10.1007/978-981-16-4823-6_11
80. Y. El maguana et al, Highly efficient ceramic membrane synthesized from sugar scum and fly ash as sustainable precursors for dyes removal, Heliyon 10 (2024) e27915. https://doi.org/10.1016/j.heliyon.2024.e27915
81. Judd S (2006) The MBR book: principles and applications of membrane bioreactors for water and wastewater treatment. Elsevier, Amsterdam
82. El Mekawy A, Mohanakrishna G, Srikanth S, Pant D (2015) The role of bioreactors in industrial wastewater treatment. In: Chandra R (ed) Environmental waste management. CRC Press, pp 157–183. https://doi.org/10.13140/RG.2.1.1139.2722
83. Membrane Bioreactors (MBR) for Wastewater Treatment, PCI Membranes (2022) https://www.pcimembranes.com/articles/membrane-bioreactors-mbr-for-wastewater-treatment/
84. Helmi A, Gallucci F (2020) Latest developments in membrane (bio)reactors. Processes 8(10):1239. https://doi.org/10.3390/pr8101239
85. Shi J, Wang Y, Lu S, Wang J, Liu J (2024) Pilot study on ceramic flat membrane bioreactor in treatment of coal chemical wastewater. Chemosphere 347. https://doi.org/10.1016/j.chemosphere.2023.140701
86. Li R, Kadrispahic H, Jørgensen MK, Berg SB, Thornberg D, Mielczarek AT, Bester K (2022) Removal of micropollutants in a ceramic membrane bioreactor for the post-treatment of municipal wastewater. Chem Eng J 427:131458. https://doi.org/10.1016/j.cej.2021.131458

87. Hoinkis J, Deowan SA, Panten V, Figoli A, Huang RR, Drioli E (2012) Membrane bioreactor (MBR) technology—a promising approach for industrial water reuse. Procedia Eng 33:234–241
88. Ugarte P et al (2022) Low-cost ceramic membrane bioreactor: Effect of backwashing, relaxation and aeration on fouling. Protozoa and bacteria removal. Chemosphere 306:135587
89. Ramo A, Cacho ED, Sánchez-Acedo C, Quílez J (2017) Occurrence and genetic diversity of Cryptosporidium and Giardia in urban wastewater treatment plants in north-eastern Spain. https://zaguan.unizar.es/record/70878/files/texto_completo.pdf
90. Al-Asheh S, Bagheri M, Aidan A (2021) Membrane bioreactor for wastewater treatment: a review. Case Stud Chem Environ Eng 4:100109. https://doi.org/10.1016/j.cscee.2021.100109
91. Dhiman S, Mukherjee G (2021) Chapter 12—Biotechnological approaches towards treatment and recycling of wastewater from tanneries and leather industry. In: Shah M, Rodriguez-Couto S (eds) Microbial ecology of wastewater treatment plants. Elsevier, pp 249–268, ISBN 9780128225035. https://doi.org/10.1016/B978-0-12-822503-5.00022-9
92. Goswami L, Kumar RV, Narayan S, Manikandan NA (2018) Membrane bioreactor and integrated membrane bioreactor systems for micropollutant removal from wastewater: a review. J Water Process Eng 26:314–328. https://doi.org/10.1016/j.jwpe.2018.10.024
93. Liu W, Song X, Huda N et al (2020) Comparison between aerobic and anaerobic membrane bioreactors for trace organic contaminant removal in wastewater treatment. Environ Technol Innov 17:100564. https://doi.org/10.1016/j.eti.2019.100564
94. Zhu M, Wang J, Pei D, Sun J, Lu Y, Hu Z, Chen S, Li N (2023) The performance and microbial community of anaerobic membrane bioreactor for high calcium papermaking wastewater treatment. J Water Process Eng 56:104311. https://doi.org/10.1016/j.jwpe.2023.104311
95. Chang et al (2001) Low-cost membranes for use in a submerged MBR. Trans IChemE 79(Part B)
96. Gong W et al (2023) A gravity-driven membrane bioreactor in treating the real decentralized domestic wastewater: flux stability and membrane fouling. Chemosphere 138948. https://www.sciencedirect.com/science/article/abs/pii/S0045653523012158
97. Neoh CH, Noor ZZ, Sabrina N et al (2015) Green technology in wastewater treatment technologies: integration of membrane bioreactor with various wastewater treatment systems. Chem Eng J. https://doi.org/10.1016/j.cej.2015.07.060
98. Cengiz AI, Guven H, Ozgun H, Ersahin ME (2024) Enhanced energy recovery in municipal wastewater treatment plants through co-digestion by anaerobic membrane bioreactors: current status and future perspectives. Rev Environ Sci Biotechnol. https://doi.org/10.1007/s11157-024-09691-6
99. Zhen G, Pan Y, Han Y, Gao Y, Gadow SI, Zhu X, Yang L, Lu X (2023) Enhanced co-digestion of sewage sludge and food waste using novel electrochemical anaerobic membrane bioreactor (EC-AnMBR). Bioresour Technol 377. https://doi.org/10.1016/j.biortech.2023.128939
100. Borea L, Naddeo V, Belgiorno V (2016) Application of electrochemical processes to membrane bioreactors for improving nutrient removal and fouling control. Environ Sci Pollut Res. https://doi.org/10.1007/s11356-016-7786-7]
101. Silva JA (2023) Wastewater treatment and reuse for sustainable water resources management: a systematic literature review. Sustainability 15(14):10940. https://doi.org/10.3390/su151410940
102. Erkan HS, Çağlak A, Soysaloglu A, Takatas B, Engin GO (2020) Performance evaluation of conventional membrane bioreactor and moving bed membrane bioreactor for synthetic textile wastewater treatment. J Water Process Eng 38:101631. https://doi.org/10.1016/j.jwpe.2020.101631
103. Wang K, Zhang H, Shen Y, Li J, Zhou W, Song H, Liu M, Wang H (2023) Impact of salinity on anaerobic ceramic membrane bioreactor for textile wastewater treatment: process performance, membrane fouling and machine learning models. J Environ Manag 345(1):118717. https://doi.org/10.1016/j.jenvman.2023.118717
104. Zielińska M, Ojo A (2023) Anaerobic membrane bioreactors (AnMBRs) for wastewater treatment: recovery of nutrients and energy, and management of fouling. Energies 16(6):2829. https://doi.org/10.3390/en16062829

105. Jeong Y, Hermanowicz SW, Park C (2017) Treatment of food waste recycling wastewater using anaerobic ceramic membrane bioreactor for biogas production in mainstream treatment process of domestic wastewater. Water Res 123:86–95. https://doi.org/10.1016/j.watres.2017.06.049
106. Deschamps L, Merlet D, Lemaire J, Imatoukene N, Filali R, Clément T, Lopez M, Theoleyre M-A (2021) Excellent performance of anaerobic membrane bioreactor in treatment of distillery wastewater at pilot scale. J Water Process Eng 41:102061
107. Düppenbecker B, Engelhart M, Cornel P (2017) Fouling mitigation in anaerobic membrane bioreactor using fluidized glass beads: evaluation fitness for purpose of ceramic membranes. J Membr Sci 537:69–82. https://doi.org/10.1016/j.memsci.2017.05.018
108. Hang TTT, Phat VV, Hanh HH, Luu TL, Thuan TH, Tuyen NV, Quang CX (2024) Improving organic and nutrient removal efficiencies in seafood processing wastewater using anaerobic membrane bioreactor (AnMBR) integrates with anoxic/oxic (AO) processes. Sci Tot Environ 954:176192. https://doi.org/10.1016/j.scitotenv.2024.176192
109. Abdul Moktadir Md, Maliha M, Tujjohra F, Munmun SA, Samaul Alam Md, Ariful Islam Md, Rahman MM (2024) Treatment of tannery wastewater by different membrane bioreactors: a critical review. Environ Adv 15:100478. https://doi.org/10.1016/j.envadv.2023.100478
110. Zhang Y, Tan Y, Sun R, Zhang W (2023) Preparation of ceramic membranes and their application in wastewater and water treatment. Water 15(19):3344. https://doi.org/10.3390/w15193344
111. Mahmudul Hasan Md, Shafiquzzaman Md, Shafiul Azam Md, Nakajima J (2011) Application of a simple ceramic filter to membrane bioreactor. Desalination 276(Issues 1–3):272–277. https://doi.org/10.1016/j.desal.2011.03.062
112. Tewari PK, Singh RK, Batra VS, Balakrishnan M (2010) Membrane bioreactor (MBR) for wastewater treatment: filtration performance evaluation of low cost polymeric and ceramic membranes. Sep Purif Technol 71(2):200–204. https://doi.org/10.1016/j.seppur.2009.11.022
113. Majouli A (2012) Elaboration of new tubular ceramic membrane from local moroccan perlite for microfiltration process. Application to treatment of industrial wastewaters. Ceram Int 38(5):4295–4303. https://doi.org/10.1016/j.ceramint.2012.02.010
114. Saja S et al (2020) Fabrication of low-cost ceramic ultrafiltration membrane made from bentonite clay and its application for soluble dyes removal. J Eur Ceram Soc 40(6):2453–2462. https://doi.org/10.1016/j.jeurceramsoc.2020.01.057
115. Iaich S et al (2021) Low-cost ceramic microfiltration membranes made from Moroccan clay for domestic wastewater and Congo Red dye treatment. Desalin Water Treat 235:251–271. https://doi.org/10.5004/dwt.2021.27618
116. Bhattacharya P, Sarkar S, Ghosh S, Majumdar S, Mukhopadhyay A, Bandyopadhyay S (2013) Potential of ceramic microfiltration and ultrafiltration membranes for the treatment of gray water for an effective reuse. Desalin Water Treat. https://doi.org/10.1080/19443994.2013.770198
117. Ding A, Liang H, Li G, Szivak I, Traber J, Pronk W (2017) A low energy gravity-driven membrane bioreactor system for grey water treatment: permeability and removal performance of organics. J Membr Sci 542:408–417. https://doi.org/10.1016/j.memsci.2017.08.037
118. Jedidi I, Khemakhem S, Saïdi S, Larbot A, Elloumi-Ammar N, Fourati A, Charfi A, Salah AB, Amar RB (2011) Preparation of a new ceramic microfiltration membrane from mineral coal fly ash: application to the treatment of the textile dying effluents. Powder Technol 208:427–432. https://doi.org/10.1016/j.powtec.2010.08.039
119. Sawunyama et al (2024) Application of coal fly ash based ceramic membranes in wastewater treatment: a sustainable alternative to commercial materials. Heliyon 10(2). https://doi.org/10.1016/j.heliyon.2024.e24344
120. Jeong Y, Cho K, Tsang YF, Rinklebe J, Park C (2017) Evaluating the feasibility of pyrophyllite-based ceramic membranes for treating domestic wastewater in anaerobic ceramic membrane bioreactors. Chem Eng J 328:567–573. https://doi.org/10.1016/j.cej.2017.07.080
121. Jeong Y, Lee S, Hong S, Park C (2017) Preparation, characterization and application of low-cost pyrophyllite-alumina composite ceramic membranes for treating low-strength domestic wastewater. J Membr Sci 536:108–115. https://doi.org/10.1016/j.memsci.2017.04.068

122. Ahmad R, Aslam M, Park E, Chang S, Kwon D, Kim J (2018) Submerged low-cost pyrophyllite ceramic membrane filtration combined with GAC as fluidized particles for industrial wastewater treatment. Chemosphere 206:784–792. https://doi.org/10.1016/j.chemosphere.2018.05.045
123. Ahmad R, Aslam M, Jing G, Kwon D, Kim J (2022) Comparison of pyrophyllite- and alumina-coated membrane treating industrial wastewater in aspect of membrane fouling and organic removal. Process Saf Environ Prot 168:1058–1066. https://doi.org/10.1016/j.psep.2022.10.057
124. Majumdar S, Sarkar S, Ghosh S, Bhattacharya P, Bandyopadhyay S, Saha A et al (2018) New trends for wastewater treatment and their reuse using ceramic membrane technology: a case study. In: Singh VP, Yadav S, Yadava R (eds) Water Quality management, water science and technology library, vol 79, pp 339–348. https://doi.org/10.1007/978-981-10-5795-3_29
125. Sarkar S, Ghosh S, Banerjee P, Larbot A, Cerneaux S, Bandyopadhyay S, Bhattacharjee C (2014) Ultrafiltration alumina membrane directly over porous clay-alumina tubular and capillary support for textile effluent treatment. Trans Ind Ceram Soc 73(3):1–8. https://doi.org/10.1080/0371750X.2014.882246
126. Ghosh S et al (2010) Comparative study on treatment of kitchen-sink wastewater using single and multichannel ceramic membrane. Int J Environ Technol Manag 13(3/4):336–347. https://doi.org/10.1504/IJETM.2010.038012
127. Sahoo G et al (2016) Preparation and characterization of microfiltration apatite membrane over low cost clay-alumina support for decolorization of dye solution. Desalin Water Treat 57(Issue 57):27700–27709. https://doi.org/10.1080/19443994.2016.1186565]
128. Karmakar S, Mondal M, Ghosh S, Bandyopadhyay S, Majumdar S, De S (2016) Removal of reactive dyes using a high throughput-hybrid separation process. J Desalin Water Treat 57(Issue 22):10295–10311
129. Santra B, Ramrakhiani L, Kar S et al (2020) Ceramic membrane-based ultrafiltration combined with adsorption by waste derived biochar for textile effluent treatment and management of spent biochar. J Environ Health Sci Eng. https://doi.org/10.1007/s40201-020-00520-w
130. Hammami A, Charcosset C, Amar RRB (2017) Performances of continuous adsorption-ultrafiltration hybrid process for AO7 dye removal from aqueous solution and real textile wastewater treatment. J Membr Sci Technol 7:171. https://doi.org/10.4172/2155-9589.1000171
131. Teng J, Shen L, He Y, Liao BQ, Wu G, Lin H (2018) Novel insights into membrane fouling in a membrane bioreactor: elucidating interfacial interactions with real membrane surface. Chemosphere 210:769–778. https://doi.org/10.1016/j.chemosphere.2018.07.086
132. Bagheri M, Mirbagheri SA (2018) Critical review of fouling mitigation strategies in membrane bioreactors treating water and wastewater. Bioresour Technol 258:318–334. https://doi.org/10.1016/j.biortech.2018.03.026
133. Meng F, Zhang S, Oh Y, Zhou Z, Shin HS, Chae SR (2017) Fouling in membrane bioreactors: an updated review. Water Res 1(114):151–180. https://doi.org/10.1016/j.watres.2017.02.006.]
134. Cheng X, Zhou W, Wu D, Luo C, Jia R, Li P, Zheng L, Zhu X, Liang H (2020) Pre-deposition layers for alleviating ultrafiltration membrane fouling by organic matter: role of hexagonally and cubically ordered mesoporous carbons. Sep Purif Technol 240:116599
135. Nthunya LN, Bopape MF, Mahlangu OT, Mamba BB, Van der Bruggen B, Quist-Jensen CA, Richards H (2022) Fouling, performance and cost analysis of membrane-based water desalination technologies: a critical review. J Environ Manag 301:113922
136. Zhang M et al (2020) Introduction to aerobic membrane bioreactors: current status and recent developments. Curr Dev Biotechnol Bioeng
137. Aslam M, Ahmad R (2018) Recent developments in biofouling control in membrane bioreactors for domestic wastewater treatment. Sep Purif Technol 206:297–315. https://doi.org/10.1016/j.seppur.2018.06.004
138. Tummons E, Han Q, Tanudjaja HJ, Hejase CA, Chew JW, Tarabara VV (2020) Membrane fouling by emulsified oil: a review. Sep Purif Technol 248:116919. https://doi.org/10.1016/j.seppur.2020.116919

139. Guo W, Ngo H-HH, Li J (2012) A mini-review on membrane fouling. Bioresour Technol 122:27–34
140. Kirschner AY, Cheng YH, Paul DR, Field RW, Freeman BD (2019) Fouling mechanisms in constant flux crossflow ultrafiltration. J Membr Sci 574:65–75
141. Goh PS, Lau WJ, Othman MHD, Ismail AF (2018) Membrane fouling in desalination and its mitigation strategies. Desalination 425:130–155
142. Pichardo-Romero D, Garcia-Arce ZP, Zavala-Ramírez A, Castro-Muñoz R (2020) Current advances in biofouling mitigation in membranes for water treatment: an overview. Processes 8:182
143. El Batouti M et al (2022) Review of new approaches for fouling mitigation in membrane separation processes in water treatment applications. Separations 9(1):1. https://doi.org/10.3390/separations9010001
144. Du X, Shi Y, Jegatheesan V, Ul Haq I (2020) A review on the mechanism, impacts and control methods of membrane fouling in MBR system. Membranes (Basel) 10(2):24. https://doi.org/10.3390/membranes10020024
145. Al-Hasani M, Doan H, Zhu N, Abdelrasoul A (2022) Optimal intermittent ultrasound-assisted ultrafiltration for membrane fouling remediation. Sep Purif Technol 303(15):122249. https://doi.org/10.1016/j.seppur.2022.122249
146. Qiao Y, Han X, Chen F, Sun C, Xu L, Liu R, Shen X (2024) Control behavior of pretreatment on ceramic membrane fouling caused by different organic substances. J Environ Chem Eng 12(Issue 1):111884. https://doi.org/10.1016/j.jece.2024.111884
147. Tang S, Li J, Zhang Z, Ren B, Zhang X, Sci X (2019) Comparison of long-term ceramic membrane bioreactors without and with in-situ ozonation in wastewater treatment: membrane fouling, effluent quality and microbial community. Tot Environ 20(652):788–799. https://doi.org/10.1016/j.scitotenv.2018.10.284. Epub 2018 Oct 22; PMID: 30380486
148. Xie X, Wang L, Wei J, He H, Guo Z, Wang C, Wen X, Song Y (2024) Progress in ceramic membrane coupling ozonation process for water and wastewater treatment: a critical review. Chem Eng J 493:152475. https://doi.org/10.1016/j.cej.2024.152475
149. Zhang J, Li G, Yuan X, Li P, Yu Y, Yang W, Zhao S (2023) Reduction of ultrafiltration membrane fouling by the pretreatment removal of emerging pollutants: a review. Membranes 13:77. https://doi.org/10.3390/membranes13010077
150. Ezugbe EO, Rathilal S (2020) Membrane technologies in wastewater treatment: a review. Membranes 10(5):89. https://doi.org/10.3390/membranes10050089
151. Lin JC-T, Lee D-J, Huang C (2010) Membrane fouling mitigation: membrane cleaning. Sep Sci Technol 45:858–872
152. Wang Z, Ma J, Tang CY, Kimura K, Wang Q, Han X (2014) Membrane cleaning in membrane bioreactors: a review. J Membr Sci 468:276–307
153. Zhao Y-J, Wu K-F, Wang Z-J, Zhao L, Li S-S (2000) Fouling and cleaning of membrane—a literature review. Environ Sci 12:241–251
154. Williams C, Wakeman R (2000) Membrane fouling and alternative techniques for its alleviation. Membr Technol 2000:4–10
155. Yigit NO, Civelekoglu G, Harman I, Koseoglu H, Kitis M (2009) Effects of various backwash scenarios on membrane fouling in a membrane bioreactor. Desalination 237:346–356
156. An Y (2010) Post-treatment of upflow anaerobic sludge blanket effluent by combining the membrane filtration process: fouling control by intermittent permeation and air sparging. Water Environ J WEJ 24:32–38
157. Li J, Sanderson RD, Jacobs EP (2002) Ultrasonic cleaning of nylon microfiltration membranes fouled by Kraft paper mill effluent. J Membr Sci 205:247–257
158. Wan M-W, Reguyal F, Futalan C, Yang H-L, Kan C-C (2013) Ultrasound irradiation combined with hydraulic cleaning on fouled polyethersulfone and polyvinylidene fluoride membranes. Environ Technol 34:2929–2937
159. Kyllönen HM, Pirkonen P, Nyström M (2005) Membrane filtration enhanced by ultrasound: a review. Desalination 181:319–335

158. José Miguel A, Beatriz G-F, María S (2011) Membrane cleaning. In: Ning RY (ed) Expanding issues on desalination, vol 3. IntechOpen, Valencia, Spain
161. Liu C, Caothien S, Hayes J, Caothuy T, Otoyo T, Ogawa T (2001) Membrane chemical cleaning: from art to science. Pall Corporation, Port Washington, NY, USA, p 11050
162. Shorrock CJ, Bird MR (1998) Membrane cleaning: chemically enhanced removal of deposits formed during yeast cell harvesting. Food Bioprod Process 76:30–38
163. Porcelli N, Judd S (2010) Chemical cleaning of potable water membranes: a review. Sep Purif Technol 71:137–143
164. Maartens A, Swart P, Jacobs EP (1996) An enzymatic approach to the cleaning of ultrafiltration membranes fouled in abattoir effluent. J Membr Sci 119:9–16
165. Allie Z, Jacobs EP, Maartens A, Swart P (2003) Enzymatic cleaning of ultrafiltration membranes fouled by abattoir effluent. J Membr Sci 218:107–116
166. Popović S, Djurić M, Milanović S, Tekić MN, Lukić N (2010) Application of an ultrasound field in chemical cleaning of ceramic tubular membrane fouled with whey proteins. J Food Eng 101:296–302
167. Maskooki A, Mortazavi SA, Maskooki A (2010) Cleaning of spiralwound ultrafiltration membranes using ultrasound and alkaline solution of EDTA. Desalination 264:63–69
168. Sahoo GC, Halder R, Jedidi I, Oun A, Nasri H, Roychoudhurry P, Majumdar S, Bandyopadhyay S, Amar RB (2016) Preparation and characterization of microfiltration apatite membrane over low cost clay-alumina support for decolorization of dye solution. Desalin Water Treat. https://doi.org/10.1080/19443994.2016.1186565
169. Sun J, Liang P, Yan X, Zuo K, Xiao K, Xia J, Qiu Y, Wu Q, Wu S, Huang X, Qi M, Wen X (2016) Reducing aeration energy consumption in a large-scale membrane bioreactor: process simulation and engineering application. Water Res 93:205–213. https://doi.org/10.1016/j.watres.2016.02.026
170. Jiang C et al (2019) Effect of scrubbing by NaOCl backwashing on membrane fouling in anammox MBR. Sci Tot Environ 670:149–157. https://doi.org/10.1016/j.scitotenv.2019.03.170
171. Liao B, Bokhary A, Cui L, Lin H (2017) A review of membrane technology for integrated forest biorefinery. J Membr Sci Res 3(Issue 3):120–141. https://doi.org/10.22079/JMSR.2016.22839
172. Ninomiya Y, Kimura K, Sato T, Kakuda T, Kaneda M, Hafuka A, Tsuchiya T (2020) High-flux operation of MBRs with ceramic flat-sheet membranes made possible by intensive membrane cleaning: tests with real domestic wastewater under low-temperature conditions. Water Res 181. https://doi.org/10.1016/j.watres.2020.115881
173. Wang C, Ng TCA, Ng HY (2021) Comparison between novel vibrating ceramic MBR and conventional air-sparging MBR for domestic wastewater treatment: performance, fouling control and energy consumption. Water Res 203. https://doi.org/10.1016/j.watres.2021.117521
174. Jankowska K, Domingo-Félez C, Prado-Rubio OA, Skiadas IV, Woodley JM, Pinelo M (2024) Effect of backwash as a strategy for biofouling control in the submerged ceramic membrane bioreactor for high-density cultivations: process optimization and fouling mechanism at pilot scale. Sep Purif Technol 338:126428. https://doi.org/10.1016/j.seppur.2024.126428
175. Luján-Facundo MJ, Mendoza-Roca JA, Bes-Piá A, Zuriaga-Agustí E, Mestre S, Palacios M-D (2023) Low-cost ceramic membranes manufacture using INKJET technology for active layer deposition and validation on membrane bioreactors. Process Saf Environ Prot 176:618–626. https://doi.org/10.1016/j.psep.2023.06.045
176. Rashed MN (2013) Adsorption technique for the removal of organic pollutants from water and wastewater. In: Nageeb Rashed M (ed) Organic pollutants—monitoring, risk and treatment. https://doi.org/10.5772/54048; https://www.intechopen.com/chapters/42059
177. Rasheed T, Bilal M, Nabeel F, Adeel M, Iqbal HMN (2019) Environmentally-related contaminants of high concern: potential sources and analytical modalities for detection, quantification, and treatment. Environ Int 122:52–66. https://doi.org/10.1016/j.envint.2018.11.038

178. Bilal M, Adeel M, Rasheed T, Zhao Y, Iqbal HMN (2019) Emerging contaminants of high concern and their enzyme-assisted biodegradation—a review. Environ Int 124:336–353. https://doi.org/10.1016/j.envint.2019.01.011
179. Arias A, Alvarino T, Allegue T, Suárez S, Garrido JM, Omil F (2018) An innovative wastewater treatment technology based on UASB and IFAS for cost-efficient macro and micropollutant removal. J Hazard Mater 359:113–120. https://doi.org/10.1016/j.jhazmat.2018.07.042
180. Krishnani KK, Boddu VM, Singh RD et al (2023) Plants, animals, and fisheries waste-mediated bioremediation of contaminants of environmental and emerging concern (CEECs)—a circular bioresource utilization approach. Environ Sci Pollut Res. https://doi.org/10.1007/s11356-023-28261-x
181. Ramrakhiani L, Halder A, Majumder A, Mandal AK, Majumdar S, Ghosh S (2017) Industrial waste derived biosorbent for toxic metal remediation: mechanism studies and spent biosorbent management. Chem Eng J 308:1048–1064. https://doi.org/10.1016/j.cej.2016.09.145
182. Dele-Afolabi TT, Azmah Hanim MA, Ojo-Kupoluyi OJ, Atoyebi EO (2023) Biobased materials for wastewater treatment. In: Advanced applications of biobased materials. Elsevier, pp 593–624
183. Dewi R, Shamsuddin N, Abu Bakar MS, Santos JH, Bilad MR, Lim LH (2021) Progress in emerging contaminants removal by adsorption/membrane filtration-based technologies: a review. Indonesian J Sci Technol 6(3):577–618
184. Abou-Shady A, El-Araby H (2023) Treatment technologies and guidelines set for water reuse. In: Taşeli BK (ed) Sewage management. https://doi.org/10.5772/intechopen.109928; https://www.intechopen.com/chapters/85878
185. Alayande AB et al (2024) Use of reclaimed municipal wastewater in agriculture: comparison of present practice versus an emerging paradigm of anaerobic membrane bioreactor treatment coupled with hydroponic controlled environment agriculture. Water Res. https://doi.org/10.1016/j.watres.2024.122197
186. Souad S et al (2018) Elaboration and characterization of low-cost ceramic membrane made from natural Moroccan perlite for treatment of industrial wastewater. J Environ Chem Eng 6(1):451–458. https://doi.org/10.1016/j.jece.2017.12.004
187. Bhattacharya P et al (2012) Operational performance of ceramic microfiltration process for reclamation of cosmetic and municipal wastewater. Int J Biotechnol Chem Environ Eng (IJBCEE) 1(1)
188. Bhattacharya P, Ghosh S, Swarnakar S, Mukhopadhyay A (2014) Reuse of textile effluent for dyeing using combined technology of ceramic microfiltration and surface treated sugarcane bagasse: toxicity evaluation using Channa punctatus as model. Desalin Water Treat 54:715–735. https://doi.org/10.1080/19443994.2014.887035
189. Dey S et al (2018) Single step preparation of zirconia ultrafiltration membrane over clay-alumina based multichannel ceramic support for wastewater treatment. J Membr Sci Res 4:28–33
190. Mzahma S, Duplay J, Souguir D, Amar RB, Ghazi M, Hachicha M (2023) Membrane processes treatment and possibility of agriculture reuse of textile effluents: study case in Tunisia. Water 15(7):1430.https://doi.org/10.3390/w15071430
191. Pandit S et al (2024) Greywater treatment and reuse trends using membrane bioreactor technology. Int J Adv Res Trends Sci 3(1):4–13
192. Yang J (2024) Pathways and challenges in effectively controlling membrane fouling in anaerobic membrane bioreactors. Crit Rev Environ Sci Technol 1–21.https://doi.org/10.1080/10643389.2024.2346446
193. Maaz M, Yasin M, Aslam M, Kumar G, Atabani AE, Idres M, Anjum F, Jamil F, Ahmad R, Khan AL, Lesage G, Heran M, Kim J (2019) Anaerobic membrane bioreactors for wastewater treatment: novel configurations, fouling control and energy considerations. Bioresour Technol 283:358–372. https://doi.org/10.1016/j.biortech.2019.03.061
194. Ullah A, Tanudjaja HJ, Ouda M, Hasan SW, Chew JW (2021) Membrane fouling mitigation techniques for oily wastewater: a short review. J Water Process Eng 43:102293. https://doi.org/10.1016/j.jwpe.2021.102293

195. Qi T, Yang D, Chen X, Ke W, Qiu M, Fan Y (2023) Sulfonated ceramic membranes with antifouling performance for the filtration of BSA-containing systems. Sep Purif Technol 313:123513. https://doi.org/10.1016/j.seppur.2023.123513
196. https://cordis.europa.eu/article/id/241021-ceramic-solution-opens-recycled-wastewater-possibilities

Part I
Wastewater Management and Industrial Wastewater Treatment

Chapter 2
Membrane Technology for Wastewater Management: Global Scenario and Sustainable Solutions

Angshuman Roy Choudhury, Piyali Roy Choudhury, and Sibdas Bandyopadhyay

1 Introduction

Urbanisation and rural development have increased drastically in the last few decades. Development is foreseen to occur at an exponential rate in the coming future. This will result in increased demand for power, infrastructure, oil and gas, steel, pharmaceuticals, and chemical industries. Worldwide, the faster growth of industries has also contributed to the generation of enormous waste in different forms. Among all wastes, wastewater is a major concern. Wastewater quality from industries like steel, power, petroleum, and refineries is predictable. But wastewater from the chemical industry, pharmaceuticals, textiles, and leachate from solid waste management yards is unpredictable. These wastewaters contain very complex and toxic contaminants. Only about 0.3% of the freshwater available on Earth is easily accessible to humans. These wastewaters are creating a health hazard by contaminating surface water, groundwater, the sea, and the ocean as well. Unhygienic conditions and the discharge of untreated sewage into natural waterbodies cause health hazards. Diseases due to poor quality water, lack of sanitation, and ignorance about hygiene contribute about 4% of total deaths and about 5.7% of total diseases globally [1]. So

A. R. Choudhury (✉)
BGR Energy Systems Limited, Guna Complex, 443, Anna Salai, Teynampet, Chennai, Tamil Nadu 600018, India
e-mail: bgr.arc2021@gmail.com

P. R. Choudhury
Department of Rubber and Plastics Technology, Anna University MIT Campus, Chromepet, Chennai, Tamil Nadu 600044, India

S. Bandyopadhyay
Department of Environmental Studies, Siksha-Bhavana, Visva-Bharati, Santiniketan, West Bengal 731235, India

S. Bandyopadhyay (ed.), *Application of Low Cost Ceramic Membranes in Wastewater Treatment*, Materials Horizons: From Nature to Nanomaterials,
https://doi.org/10.1007/978-981-97-5842-5_2

the need for sewage treatment plants (STP) has increased. Therefore, to allow further modernization, treating and reusing wastewater has become a necessity.

Today, demand and the lack of responsibility of humans towards environment have converted the most abundant substance on earth into a commodity. Fortunately, wastewater treatment technologies are also continuously upgraded. Membrane technologies like membrane bioreactors (MBR), microfiltration (MF),ultrafiltration (UF), nanofiltration (NF), reverse osmosis (RO), forward osmosis (FO), demineralization (DM), ceramic membranes (CM), electrodeionization (EDI), and zero liquid discharge systems (ZLD) are available today [2–10]. These technologies are being applied to treat wastewater, depending on various operational constraints and design limitations. Permutations of combinations of these technologies are applied in various wastewater treatment plants to achieve better-quality water. But the operation and maintenance of a technology and its design limitations are not the only influential factors. There are many important factors influencing the acceptance and implementation of a technology. Factors like government policies, discharge water quality norms, demand from a demographic, level of awareness, and commercial aspects also play an important role in it. Industries like iron and steel, oil and gas, refineries, and power plants are major consumers of fresh water. About 70% of their consumption is used for cooling water recirculation. This recirculation process mostly contributes to the total dissolved solids (TDS) in blowdown water. Blowdown water can be treated and reused using membrane technology. In order to meet demand and curtail competition in the global market, industries needed to become more cost-competitive to survive. Membrane technologies for recycling wastewater focus on reducing the operational costs of industries. This also helps to meet the stringent discharge norms.

In municipal sectors, wastewater has the highest possibility of infiltrating and contaminating surface water and groundwater. Such wastewater requires membrane technologies for reuse or groundwater recharge. However, commercial aspects are the primary criteria for adapting membrane technologies. Another important factor in the selection of membrane technology for wastewater treatment is the availability of a safe place far from residential areas due to hygiene, foam, and odour issues. In such situations, MBR, which is about 1.5–1.8 times more expensive than conventional treatment, has been preferred. MBR has higher operational expenses, but it requires 30–50% less footprint than conventional STP. In recent developments in technologies like bioculture consortiums, clarification units, pressure sand filtration systems, and UF, the footprint requirement has drastically reduced compared to the conventional system. Issues of odour and foam common in STP have also been reduced due to these technologies.

A low-capital-cost membrane with minimum operational expenses and a maximum lifespan is highly expected. As per the recent trend, CM is expected to meet future expectations. At the same time, advancements in primary and tertiary treatment technologies are also required as a pre-treatment to membrane technology. Such a technological solution is foreseen to be the most sustainable, acceptable, and affordable in wastewater management in the industrial and municipal segments globally.

In this chapter, various water consumers in the industrial and municipal sectors are described broadly. The process of wastewater generation is also explained. Today's best available membrane-based wastewater management is highlighted. The basic design specifications of original equipment manufacturers (OEM) for membrane are explained in detail. The global scenario, various influential factors, and future prospects of membrane technology for wastewater management with low-cost membranes are also briefly discussed.

2 Membrane Based Technology and Wastewater Treatment

Membrane technology is unavoidable if wastewater needs to be recycled. There are various membrane technologies available for wastewater treatment. Conventional membrane technology has always faced many obstacles. This constraint has influenced its efficiency, effectiveness, and quality performance. The fouling of membranes is the biggest constraint for membrane technology. Many combinations and hybrid or integrated technologies have been incorporated in industries to treat wastewater and avoid fouling problems [11]. This integrated or hybrid membrane process has shown many benefits. It is noted that it reduces contaminants, improves product quality, and reduces power consumption. This also helped industries optimise their capital and operating costs. An example of such hybrid or integrated technology is MBR, where membranes are combined in a bioreactor. The selection of the membrane depends on the desired water quality. Frequently used membrane technologies are described below.

2.1 Microfiltration

Pore sizes of membranes vary from 0.1 to 10 m in MF [2, 3]. MF can work up to any customised capacity. Normally, it is designed to handle up to 200 m^3/h with a feed pressure ranging from 0.5 to 1.5 bar [2, 3]. MF helps to remove solids like colloidal based on pore size. MFs are used to allow proteins to pass through the membrane but not colloidal solids. However, MF allows monovalent and multivalent ions to pass through its membranes. These membranes are capable of removing all microorganisms, like bacteria. Microorganisms like viruses are much smaller than the pore size of MF. A part of the virus is restricted in MF due to its attachment to the biofilm formed by bacteria. MF normally finds its application in pharmaceuticals, food and beverage industries, STP, flue gas wastewater treatment, industrial effluent treatment, etc.

Generally, two types of fouling are noticed in MF. External fouling of MF is the result of cell fragments and the accumulation of restricted particles on the membrane surface. Internal fouling is caused by the deposition of macromolecules and small particles inside the membrane. External fouling is reversible, but internal fouling is noted to be irreversible. To reduce fouling in MF, many methodologies are proposed such as dynamic filtration, vortex systems, backflushing and backpulsing, and constant filtrate flux operation [12].

2.2 Ultrafiltration

Pore sizes of UF membrane range from 0.001 to 0.1 m [2, 4, 5]. UF is capable of removing organic compounds, depending on molecular cut-off, and microorganisms like viruses. UF works with a little more feed pressure than MF. The feed pressure required is an average of two bars. Feed pressure is designed in the range of 0.3–6.5 bar. In industries, tertiary treatment is recommended as a pre-treatment for UF. Tertiary treatment helps reduce contaminants like O&G, organic compounds, and TSS. Like MF, UF finds its use in dairy farms, food and beverage, metal, and textile industries.

In industries, filtration flux is generally considered to be up to 60 LMH. The backwash flux is designed in the range of 230–300 LMH. An average backwashing time of 160 s is required in UF. The main difference between MF and UF is pore size. Both of these MF and UF are used as pre-treatments for the NF and RO membranes. These help reduce the silt density index of RO.

2.3 Nano Filtration

NF is a pressure-driven process. The pore size of the NF membrane ranges between 0.2 and 2 nm. The molecular weight cut-off of NF is 200–1000 Da [6]. The NF process is a treatment in between the UF and RO membranes. NF membranes are capable of removing colouring agents, microorganisms, organic compounds, including multivalent ions, and solids based on their molecular weight cut-off [13]. In recent years, NF has found applications in drinking water treatment, the removal of pesticides, heavy metals, and nitrates from wastewater, recycling, etc.

2.4 Reverse Osmosis

The opposite operation of the natural osmosis process is known as reverse osmosis. In osmosis, water with a lower concentration moves through a semipermeable membrane towards high-concentration water under osmotic pressure. When pressure exceeds osmotic pressure in RO, water with a higher concentration starts penetrating the semipermeable membrane towards water with a lower concentration. RO applications are famous in drinking water treatment, seawater desalination, and effluent

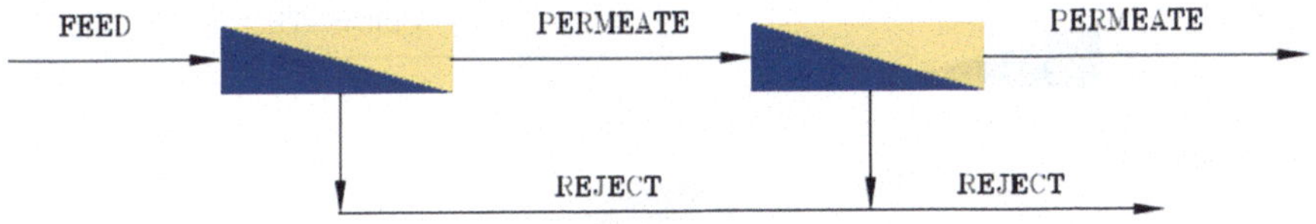

Fig. 1 Schematic of RO pass configuration

recycling. Today, RO has found application in almost every field of the industry. In industries, ROs are generally designed in two configurations [7].

A configuration where permeate of one RO enters another RO; this system is known as the pass. Such a configuration is applicable when a very high TDS has to be reduced to a very low TDS. This configuration also helps with the high recovery of water. This is applied to seawater desalination and wastewater recycling. In another configuration, the concentrate from a RO is fed to another RO. This configuration is known as a stage. Stages are applied to increase the recovery rate of water while maintaining all design criteria of the semi-permeable membrane. In stages, membranes are configured in a Christmas tree pattern

- Water Users
  - Industrial Sector
    - Power Plant
    - Iron & Steel Industry
    - Textile Industries
    - Paper & pulp
    - Oil & Gas and Refineries
    - Pharmaceutical & Chemical Industries
  - Municipal Sector
    - Private sector

A typical schematic of various configurations is shown in Figs. 1 and 2. Different dimensions of membranes, like 18″, 8″, 4″, etc., are available with OEM. However, 8-inch spiral-wound polymer membranes are in demand in industries. To increase the life cycle of membranes from colloidal and organic fouling, pre-treatment with UF is preferred prior to RO.

3 Users and Wastewater

The presence of toxic metal ions, heavy metals, high TDS, total hardness (TH), pH, and organic contaminants make wastewater different from water. Before discussing the various membrane-based wastewater management processes used in different sectors, we need to understand a few things first like the characteristics of wastewater

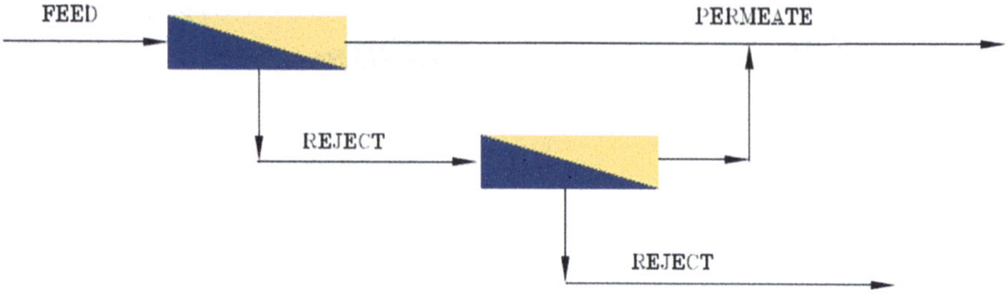

Fig. 2 Schematic of RO stage configuration

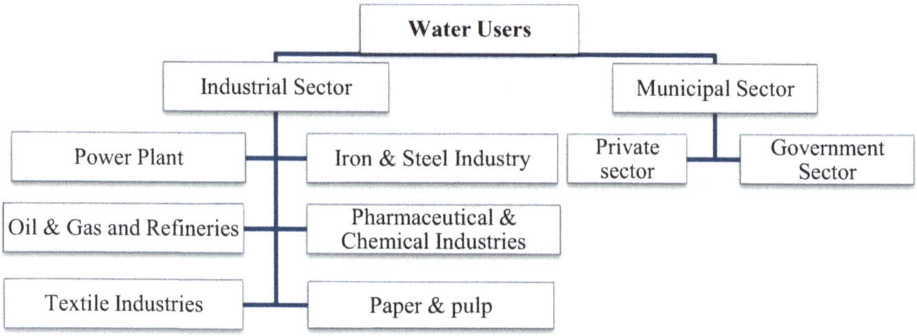

Fig. 3 Types of water users

generated, the need for treated water quality, and the process of wastewater generation. The limitation of other technology installed upstream of membrane technology. The same is briefed here in the following paragraphs. Out of the numerous users, some of the major wastewater contributors are listed below (Fig. 3).

3.1 *Industrial Sector*

Major water consumers and the biggest wastewater generators are identified as power plants, the iron and steel industry, oil and gas refineries, pharmaceutical and chemical industries, textile industries, and paper and pulp industries. Many technologies other than membranes are used to treat and recycle wastewater. The following discussion will briefly explain how membrane technology is of great significance for wastewater management and recycling. The capability, importance, and limitations of other technologies installed prior to membrane technology are also discussed.

3.1.1 Power Plant

Water is one of the most important process requirements for a thermal power plant. The consumption is about 3.5–4 m^3/h per MW [14]. The majority of water is required for cooling the equipment and boiler units. A negligible quantity of TDS is most desirable in water to minimise the scaling or deposition of salts. To generate the desired water, either an ion exchange process or RO membrane is implemented. This, in turn, contributes to a significant amount of wastewater rich in TDS. Water is also required in huge quantities for ash removal and flue gas desulfurization. In the downstream process, a vast quantity of wastewater is produced. About 80% of this total freshwater consumption is makeup water. The typical water consumption distribution is tabulated in Table 1 [15]. Wastewater is typically sludge, waste backwash water, blowdown from cooling tower and boiler units, and wastewater from the regeneration of demineralization (DM), condensate polishing units (CPU), or concentrate from an RO.

A typical water mass flow diagram of a 4 × 350 MW steam power plant is shown in Fig. 4. The typical wastewater characteristics of power plants are tabulated below in Table 2 [16]. A typical water balance scheme in around 4 × 350 MW power plants is illustrated in Fig. 4. The conversion of water to wastewater starts with the intake system. Freshwater intake is generally done from rivers or the sea. Water is stored in any reservoir or settling pond with a retention time of around 3–15 days. The use of water depends on the water quality required by technology. Like boiler blowdown, quenching water can be directly pumped from a settling pond. Water is pumped from this settling pond to the water treatment plant (WTP). WTP in general comprises a flash mixer, a flocculator, a clarifier, sand filters, and a storage tank or fire water reservoir. Users like firewater, FGD makeup water, and service water need primarily treated water with a suspended load of less than 10–20 ppm. Filtered water is also

Table 1 Water consumption distribution in industries

Sl. no	Users	Power plant (%)	Integrated steel plant (%)	Refinery sector (%)
1	Cooling water	85	66.7	84.6
2	Ash handling plant	2	–	–
3	Coal handling plant	2	–	–
4	DM plant	2	1.1	8.6
5	Service water	2	–	–
6	Potable water	1	–	–
7	Effluent/ recycling	5	32.2	6.8
8	FGD (flue gas desulfurization)	1	–	–

used as makeup water in cooling towers for auxiliary units. Underflow from clarifiers and other wastewater generated in WTP is routed to the thickener unit. All cooling tower blowdown water is routed to an effluent pond.

In case water intake is from the sea, the wastewater generated is huge. To illustrate, let's consider an example. The TDS level in the sea is around 35,000 mg/l. And in general, the TDS requirement for drinking water is recommended as less than 500 mg/l. At present, seawater membranes (SWRO membranes) available with OEM can provide a comfortable recovery of 35–40%. If the freshwater requirement is about 814 m^3/h, considering 40% recovery, the seawater intake capacity shall be around 2035 m^3/h. So the concentrate from SWRO is 1,221 m^3/h with a TDS level of around 58,000 mg/l. These concentrates are routed back to the sea as outfall.

The recovery depends on the optimisation of power requirements. In general, the pressure required by SWRO starts at 35 bars. High-pressure booster pumps for SWRO are one of the biggest power-consuming units in any industry. About 60% rejection in the SWRO plant counts as a huge loss in power consumption.

The process water in the boiler shall contain negligible TH and TDS. So, filtered water from WTP shall be treated again in a RO plant to reduce TDS below 500 mg/l. To reduce TDS, RO is the best available technology today. To increase the life span of RO membranes, pre-treatment of the water is required. UF is used as a pre-treatment prior to RO membranes. UF membranes reduce the organic load and colloidal particles. This helps minimize the chances of membrane fouling. To have more recovery and fewer losses, RO can be designed in 2–3 stages to recover up to 80%. In general, brackish water membranes (BWRO membranes) are used in this process. The feed pressure requirement for these membranes starts at 15 bar. Concentrates from BWRO membranes contain around 2,500 mg/l TDS. BWRO concentrate, UF, and media filter backwash water are routed to the wastewater pond. Permeate from BRWO is collected in an RO permeate tank. The permeate is further treated in a mixed bed (EDI) and demineralization plant to achieve the desired TH and TDS for boiler use. A portion of UF permeates also passes through an activated carbon filter (ACF) and is used for drinking and potable purposes. These users also generate sewage. Generally, sewage generated is treated using MBR. Treated water from MBR is collected in an effluent pond.

Oily wastewater generated in the process is treated in an oil separator and corrugated plate interceptor (CPI). Treated water from CPI and chemical wastewater is routed to a wastewater pond. Wastewater from the wastewater pond is treated in a clarifier. Overflow from the wastewater clarifier is collected in an effluent pond. Underflow from the wastewater clarifier is treated in the thickener unit. Overflow from the thickener is routed back to the wastewater pond. Underflow sludge from thickeners is dewatered, and solids generated are disposed off.

Wastewater generated in FGD is treated in a pre-clarifier. After clarification, overflow is stored in the FGD pre-clarifier pond, followed by the FGD wastewater pond. Then wastewater is clarified again in a clarifier. Overflow from this clarifier is routed to the effluent pond. Underflow from clarifiers is treated in thickener units and dewatered. All overflows are routed back to the FGD wastewater pond for recycling. Solid waste generated is disposed off.

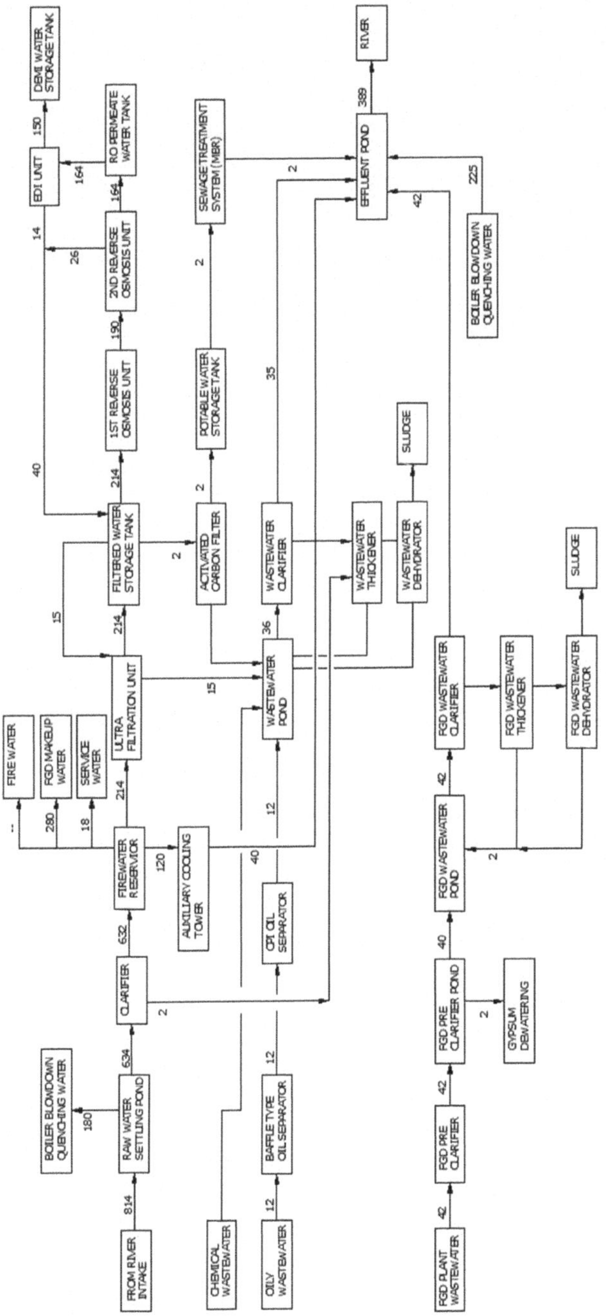

Fig. 4 Typical water mass flow diagram of 4 × 350 MW steam power plant

Table 2 Wastewater characteristics in power plant

Sl. no	Description	Unit of measurement (UoM)	Waste water quality
1	pH (20 °C)	pH	7–9
2	Electrical conductivity	ds/m	<0.9
3	Total solids (TS)	mg/l	<1200
4	TDS	mg/l	<650
5	Total suspended solids (TSS)	mg/l	<1000
6	Dissolve oxygen (DO)	mg/l	<3.5
7	Bio chemical oxygen demand (BOD)	mg/l	<130
8	Chemical oxygen demand (COD)	mg/l	<70
9	TH	mg/l	<215

Now, the effluent generated is about 389 m^3/h in the effluent pond. This is 47.8% of the freshwater produced. Effluent is discharged to the river according to discharge norms. As illustrated in Table 1, effluent generation can be reduced to 5% by recycling using membrane technology. Referring to Table 2, the level of TDS can be reduced by using RO membranes. Wastewater can be recovered up to 75% by RO and reused. The concentrate from the RO can be used for coal dust suppression. BOD and COD elimination and improvement in DO can be done by incorporating biological reactors followed by UF. Some CM is also capable of taking BOD and COD in feed water. To recycle the whole amount of effluent, ZLD can be implemented.

3.1.2 Iron and Steel Industry

Numerous processes are involved, from ironmaking to steelmaking. Processes involve sintering and palletisation, coke ovens, calcinations, blast furnaces, basic oxygen furnaces, steel melting, casting machines, rolling mills, colour coating lines, galvanising lines, etc.

The major water consumption is for cooling water recirculation in most of the processes. Along with the product, a significant quantity of wastewater is produced. One of them is that cooling towers blowdown, containing high TDS and temperatures [17]. Wastewater generated in gas scrubbing systems, blast furnaces, and coke ovens is contaminated with cyanides, sulphur, phenol, dust, metal ions, ashes, slags, and ore particles. Wastewater produced in the smelting of pig iron and open-hearth furnaces contains mostly oil and grease (O&G), dust, acid, iron, and metals. Approximately 2–7 m^3/ton of wastewater is generated in the caster machine. Water comes into direct contact with the surface of hot steel during the water spray and cooling of rollers and motors. Around 12–72 m^3/ton of process water turns into wastewater in this process. This wastewater contains high TDS, temperature, TSS, and O&G. This needed recycling is to make the operation cost-effective and commercially feasible. A typical water consumption distribution is presented in Table 1 [18, 19].

In Fig. 5, one of the most important water recirculation processes in the steel industry is illustrated. There are three types of water recirculation processes explained here: direct contact water (DCW) circuits, laminar cooling water (LCW) circuits, and indirect contact water (ICW) circuits. In the DCW circuit, water comes into direct contact with hot steel in the caster machine and hot strip mill. As a result, water gets contaminated with O&G and scale particles. As water comes into direct contact with hot steel at temperatures between (1000° to 1400°) C, a considerable amount of water is evaporated. So the TDS level in the wastewater increases. In general, this water is taken into a scale pit to eliminate the larger iron scale particles. From the scale pit, hot wastewater is pumped to the settling tank. In a settling tank, sedimentation of fine iron-scale particles takes place. After sedimentation, hot clarified water is pumped to cooling towers via pressure sand filters (PSF). This helps to restrict the scale particles up to 10 mg/l and O&G up to 5 mg/l. In a cooling tower, the temperature of filtered water comes down to around 35 °C from 70 °C. Cold water is stored in a cold basin. From cold basin water is pumped to caster machine and hot strip mill for recirculation in the process. But due to evaporation loss in cooling towers and continuous recirculation of water, TDS increases gradually in the water. So to maintain the desired level of TDS in recirculation water, a portion of the water is taken out of the cooling tower as blowdown water and collected. To maintain a balance in the process, the cycle of concentration (COC) is maintained around 3–4 in the DCW circuit. The same amount of fresh makeup water is added to the cold basin.

In an LCW circuit, water comes into direct contact with steel. The wastewater is not as contaminated as the DCW circuit. LCW is incorporated when the temperature of steel is reduced and it is about to get rolled as a final product. Wastewater from LCW is collected in a hot water tank.

About 40% of the water is pumped to cooling towers via pressure sand filters. While remaining 60% is pumped directly to cooling towers. As the contamination load is less in this water. Dilution of contamination is done by part filtration. After cooling the tower, water is stored in a cold basin. Then pumped to a strip mill for recirculation. But due to water evaporation in cooling towers and steel cooling, TDS increases in water. A portion is taken out as blowdown and collected in a blowdown water tank. A similar portion of fresh water is added as makeup.

In an ICW circuit, water never comes into contact with steel. These waters are used for jacket cooling. Hot water comes out of the process and is cooled in a cooling tower. Due to evaporation, TDS increases in cold water stored in a cold basin. Blowdown is required to maintain a COC of 2–3 in cooling towers. Same portion of makeup water is also added in the system.

All the blowdown water is collected in a blowdown water tank. If TSS is high in blowdown water, clarification may be required. In general, DCW circuits of caster machines and rolling mills require TDS 200 mg/l. Considering a COC of 4 in DCW cooling towers, TDS in blowdown shall be around 800 mg/l. In the blowdown tank, when all circuits get mixed, TDS remains at a level of around 500 mg/l, as indicated in Table 3 [20]. In general, blowdown water is filtered into PSF. The filtrate is pumped

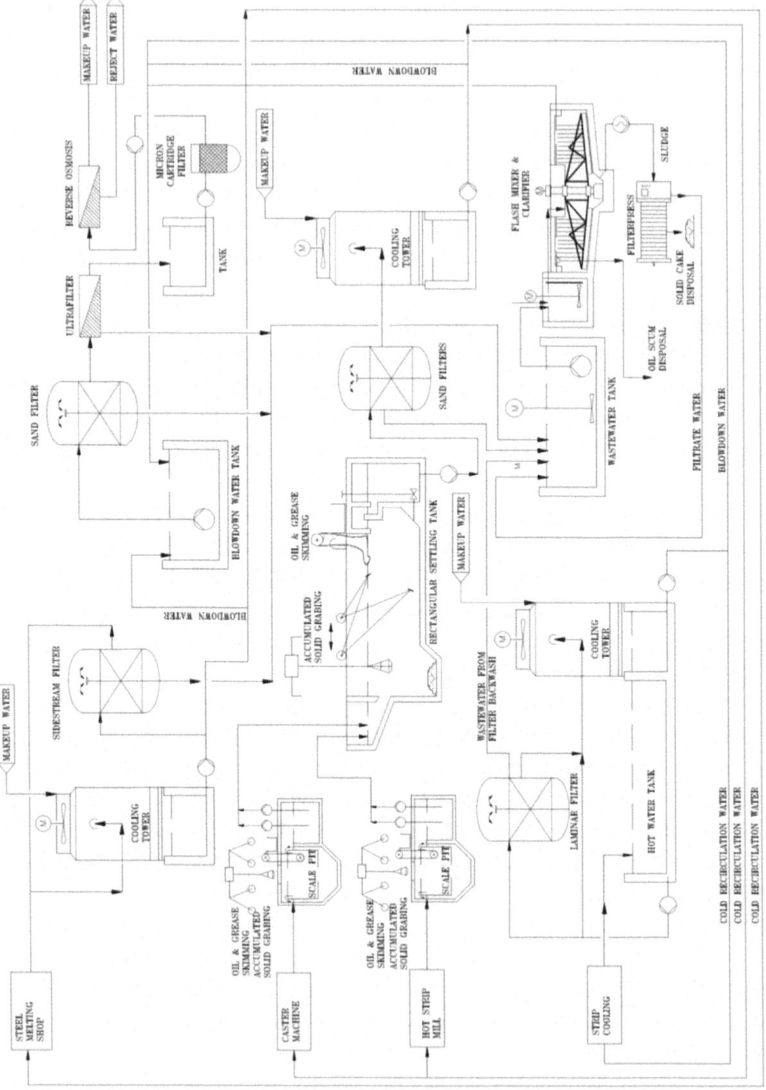

Fig. 5 Typical wastewater treatment and recycling system in steel plant

Table 3 Wastewater characteristics in steel plant

Sl. no	Description	UoM	From blast furnace	From slag crushing	From rolling mills	From pig iron
1	pH (20°C)	pH	7–9	–	–	7–8
2	TSS	mg/l	320–675	<600	<1500	<3500
3	Volatile suspended solids (VSS)	mg/l	<200	<50	<100	<350
4	TDS	mg/l	<4000	<600	<500	<2000
5	Cyanide (CN)	mg/l	<1.5	Nil	–	–
6	Phenol	mg/l	<1.2	<2.5	–	–
7	Total iron content	mg/l Fe	<1200	–	–	–
8	Alkalinity	mg/l $CaCO_3$	–	<4	<5	–
9	Sulphates	mg/l SO_4	–	<150	<150	<650
	O&G	mg/l	–	–	<50	–

through UF as a pre-treatment for RO membranes. It helps reduce suspended loads and restrict O&G, or organic contaminants. Up to 75% recovery is possible in RO in two stages for blowdown water treatment. Permeate from RO is collected and recycled as makeup water for cooling towers. This helps reduce the drastic amount of fresh water consumption in steel plants. With 75% recovery, concentrate from RO contains a TDS of 2,000 mg/l. The cooling water requirement is 66.7% of total consumption, as indicated in Table 1. UF and RO play an important role in reducing effluent generation to 32.2% in steel plants. If RO concentrate can be used in molten slag quenching, this also helps steel plant in zero liquid discharge and helps to avoid many statutory norms and pollution norms.

Subsequently, rejects from the steel production plant are reduced to approximately 0.055 m^3/ton. Water recovered by RO membranes can be reused in the steelmaking process. In colour coating and galvanising lines, the major contaminants in wastewater are phenol, chromium, ammonia, nitrogen, etc. Chemical oxidation or biological treatment, followed by clarification, media filtration, UF, and RO, are suitable to eliminate these contaminants. Recently, ZLD has become mandatory due to stringent pollution norms. However, ZLD requires huge chemical consumption to protect the RO membranes and evaporators from scaling. ZLD helps save the environment from the adverse effects of toxic contaminants.

3.1.3 Oil and Gas and Refineries

A significant quantity of fresh water is used to extract oil and gas (O&G) from reservoirs under the earth's surface. In the extraction process, water is injected through the injection well into the production well. This water floods the production well, later known as produced water. Produced water is a by-product of oil and gas extraction [21]. It is saline in nature and contains a high TDS concentration. Sometimes it contains heavy metals like zinc, lead, manganese, iron, and barium other than O&G and TSS based on the type of rock [22]. In earlier days, the produced water was stored in a pond for evaporation. However, this disposal method is no longer acceptable due to its toxic nature. Ideally, produced water is supposed to be reused as injection water or in agriculture. For disposal, treated water should match the governing discharge norms. Industries treat produced water to minimise various constraints, like scarcity of water, commercial optimisation, and social acceptance. Therefore, the treatment of produced water depends on the objectives of industry. In general, to reduce operating costs and environmental hazards, industries prefer to treat produced water and reuse it. The treatment process of the produced water varies depending on the different chemicals and contaminants found after the extraction process.

The de-oiling process of removing dispersed O&G by using the American Petroleum Institute (API), corrugated plate interceptor (CPI), or dissolved air flotation (DAF) is the primary treatment of produced water. Removal of soluble organics and dissolved gases (DO) is done by chemical or biological oxidation processes. Then a filtration system helps to remove TSS and sand, followed by disinfection and softening. High- and medium-pressure de-oiling hydro cyclones followed by degassing vessels are provided in stages with flocculent and reverse de-emulsifier dosing before disposal.

Membrane technology is implemented to reduce the TDS in treated water [23]. Some of these technologies are discussed in this section. The typical wastewater quality of the produced water is tabulated in Table 4 [23]. For recovery of around 85–90% in higher volumes of produced water, CPI, induced gas flotation (IGF), a nutshell filter with disinfectants, and an oxygen scavenger dosing system are provided. Desired water quality contains O&G below 10 mg/l and TSS below 30 mg/l. A typical produced water treatment plant is shown in Fig. 6.

Table 4 Typical produced water quality

Sl. no	Description	UoM	Produced water quality
1	pH		4–10
2	O&G	mg/l	<600
3	TSS	mg/l	Up to 1000 (Variable)
4	Conductivity	μs/cm	Around60000
5	Calcium as Ca	mg/l	<26,000
6	Magnesium as Mg	mg/l	<6000
7	Sodium as Na and Potassium as K	mg/l	<97,000
8	Bicarbonates as HCO_3	mg/l	<4000
9	Chloride as Cl	mg/l	<200,000
10	Sulphates	mg/l	<1700

Fig. 6 Typical produced water treatment plant

To enhance water quality further, dissolved gas flotation (DGF) is implemented to separate oil and various contaminants from the produced water then followed by treatment in the Walnut Shell Filter (WSF). These filters can deliver superior water quality, longer filtration runs, and greater throughput efficiency. Walnut shell media eliminates free oil and TSS. After media filtration, ceramic membranes are incorporated. Permeate from the CM will be directed to the de-aeration system in a sequential manner. In de-aeration, Henry's Law predicts the concentration of water-insoluble gases like oxygen and nitrogen. Typical produced water treatment process is illustrated in Fig. 6.This treatment process may provide treated water quality with oil content less than 0.5 mg/l and turbidity less than 5 NTU.

Oil refineries have numerous critical processes like distillation of crude oil, catalytic cracking, catalytic reforming, alkylation, hydro treating, and other utility systems, etc. [24]. All these processes need water in various ways. Cooling water recirculation uses around 80–85% of the total water consumption. However, once-through recirculation is rare today. In coastal areas, seawater is still used as once-through cooling water. Cooling tower blowdown water is treated in coagulation, flocculation, clarifiers, media filters, UF, and RO systems for recycling. Typical water consumption distribution in refineries is shown in Table 1 [25, 26].

Around 20–22% of total fresh water intake is converted to wastewater other than recirculation systems. Wastewater is also generated in boilers, desalting, sanitary or storm water networks, etc. Based on crude oil quality, its treatment process varies. So the wastewater produced in treating crude oil has fluctuating characteristics. This is a challenge when treating such wastewater. Oily wastewater is treated with API, DAF, or both and then transferred to a collection tank. All other wastewater streams [27] are also collected in a collection tank. Biological treatment is being applied as a primary treatment to eliminate complex contaminants. In coastal areas, wastewater after primary treatment is disposed of in the sea. In water-scarce or land-locked areas, wastewater is treated with API or CPI, DAF, clarifiers, media filters, and UF. UF permeate is then passed through RO to produce reusable water. Typical wastewater characteristics of refineries are listed in Table 5 [28].

3.1.4 Pharmaceutical and Chemical Industries

Wastewater characteristics are always unpredictable in the pharmaceutical and chemical industries. End products and production capacity in these industries vary. Accordingly, wastewater characteristics and volume are variable. Wastewater contains commonly toxic compounds, traces of drugs, antibiotics, BOD, COD, hormones, phenol, and other organic compounds, etc. [29]. After production, a considerable

Table 5 Wastewater characteristics in oil refineries

Sl. no	Description	UoM	Desalter effluent	Sour water	Crude tank bottom sediments	Cooling tower blowdown
1	COD	mg/l	<1000	<1200	<1000	<150
2	Free hydrocarbon	mg/l	<1000	<10	<1000	<5
3	TSS	mg/l	<500	<10	<500	<200
4	Sulphide	mg/l	<100	<10	<100	
5	Phenol	mg/l	<100	<200	–	
6	Benzene	mg/l	<15	–	–	
7	Ammonia	mg/l	–	<100	–	
8	TDS	mg/l	–	–	–	<700

quantity of wastewater is generated by washing equipment. These wastewaters contain O&G, toxic chemicals and complexes, soaps, surfactants, detergents, etc. Laboratories also contribute wastewater with very complex contaminants and microbial toxicity. Even water treatment plants (WTP) produce wastewater containing high TDS, TSS, a wide pH range, etc. Designing a treatment plant for such fluctuating wastewater characteristics is a challenge. However, treatment processes are designed depending on the type of contaminants and their elimination processes. Typical wastewater treatment processes in the pharmaceutical and chemical industries are illustrated in Fig. 7. Wastewater is generated in production units, laboratories, and WTPs.

After every batch of production, wastewater is also generated from the washing of equipment. Generally, wastewater treatment plants (WWTP) are divided into three parts: primary, biological, and tertiary treatment [30]. Wastewater started with screening. Fine, medium, or coarse screens are used to remove floating particles, sand, and other objects, depending on the particle sizes. Wastewater neutralisation follows screening. A homogeneous pH in the range of 6–7 is maintained in the neutralisation tank by adding lime or aluminium bisulphate. Sometimes neutralisation is done in an equalisation tank as well. A primary clarifier with flocculation and coagulation is provided to reduce TSS. After primary treatment, the oxidation process is incorporated.

Typical physiochemical characteristics of wastewater in pharmaceutical industries are listed in Table 6 [31]. High BOD and COD concentrations are noted in wastewater. TN and TP are also noted to be on the higher side. Sometimes the contaminants are difficult to biodegrade in biological treatment. Then chemical oxidation like ozonation, Fenton's reagent, H_2O_2, photo-catalysis, and UV is incorporated [32]. However, these treatments are not cost-effective. Anaerobic digesters are capable of handling high concentrations of BOD and COD. Aerobic systems can be of different types, like activated sludge processes (ASP), moving bed bioreactors (MBBR), sequential batch reactors (SBR), or anaerobic digestion systems [33]. The retention time of the biological reactor can go up to 24 h, depending on the concentration of BOD,

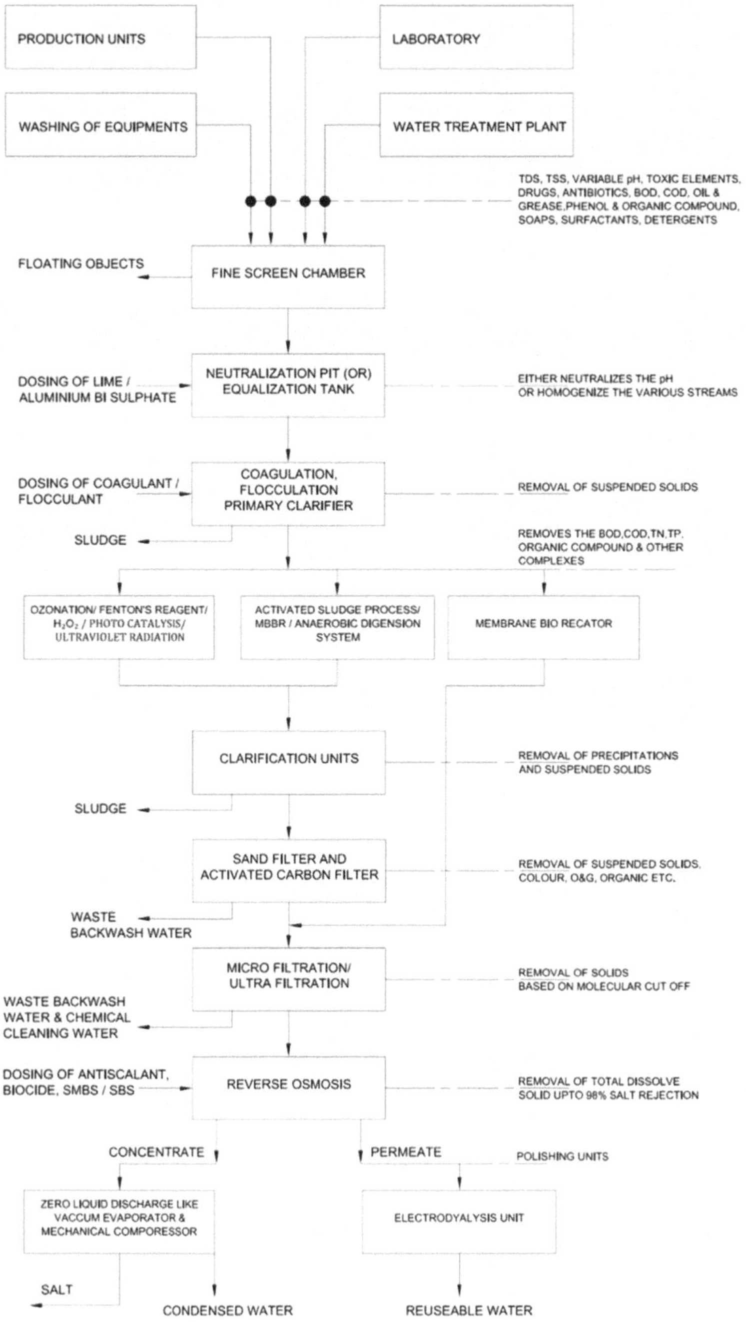

Fig. 7 Typical wastewater treatment schemes in pharmaceutical and chemical industry

Table 6 Wastewater characteristics of pharmaceutical industries

Sl. no	Description	UoM	Wastewater quality
1	pH		1–8
2	BOD	mg/l	Up to 2500
3	COD	mg/l	Up to 10,000
4	Total nitrogen (TN)	mg/l	Up to 1500
5	TSS	mg/l	Up to 500 (variable)
6	Total phosphorus (TP)	mg/l	Up to 250

COD, TKN, phenol, and organic compounds. Sludge production is higher in the biodegradation process. In some cases, COD reduction efficiency is not significantly affected by biological processes. In membrane technology, MBR is the best available technology to treat such contaminants. MBR is incorporated to achieve significant reductions in COD, BOD, TKN, and TP [34]. MBR works with a high MLSS like 13,000 in comparison to biological treatments like ASP, MBBR, and SBR. MLSS maintained in ASP, MBBR, and SBR ranges from (3000 to 5000). Biological treatment is followed by a secondary clarifier and pressure sand filter to remove TSS. After media filtration, the traces of organics and O&G are arrested in the activated carbon filter (ACF). For the removal of colloidal particles, microorganisms, and organics, MF or UF are incorporated. If MBR is installed in lieu of a biological system, units from the biological reactor until UF can be omitted. MBR helps reduce footprint, use fewer process units, and achieve consistent water quality.

To achieve reusable water quality, RO followed by mixed bed (MB) or EDI are used. RO helps to ensure the elimination of contaminants and microorganisms. Permeate can be used in pharmaceuticals manufacturing. Sometimes the contaminants are very complex and not biodegradable. Such wastewater and RO concentrate are evaporated in evaporators and mechanically compressed by ZLD. The water vapour is condensed for further polishing before recycling.

3.1.5 Textile Industries

The textile industry is one of the fastest-growing industries. Water is used as a processing medium to colour and finish the textile product. Water that remains unused becomes wastewater. Textile production generates large quantities of wastewater.

The wastewater characteristics depend on the facilities used in the textile industry for production processes. Different stages of the textile process contribute different contaminants to wastewater. Different impurities contributed to wastewater during various textile processes are listed in Table 7 [35–37].

Textile wastewater contains biodegradable, hazardous, and toxic chemicals. Azo dye is the most common colouring agent used for dyeing [38]. Typical wastewater characteristics in the textile industry are tabulated in Table 8 [39]. Dye mixed in wastewater can cause a serious health issue.

Table 7 Production process and wastewater in textile industry

Sl. no	Process	Contaminants
1	Fibre preparation	The wastewater generation is very low or negligible
2	Yam, spinning	The wastewater generation is very low or negligible
3	Sizing	Metals, BOD, COD, size and cleaning waste
4	Weaving	The wastewater generation is very low or negligible
5	Knitting	The wastewater generation is very low or negligible
6	Tufting	The wastewater generation is very low or negligible
7	Desizing	BOD, soluble size, synthetic size, foam, lubricants, Starch, biocides, enzymes, wax, NH_3 and antistatic compound
8	Scouring	Disinfectant and insecticides, detergents, soaps, NaOH, surfactants, oil, wax, fats, pectin, lubricants, spin finishes, spent solvents and enzymes
9	Bleaching	High pH, Hydrogen Peroxide, organic stabilizers / sodium silicate and AOX
10	Singeing	The wastewater generation is very low or negligible
11	Mercerizing	NaOH and high pH
12	Heat settling	The wastewater generation is very low or negligible
13	Dyeing	Various metals, salts, colour, toxic elements, sulphide, BOD, COD, alkali and acids, surfactants, spent solvents, formaldehyde and organic compound
14	Printing	TSS, Metals, solvents, BOD, heat, foam, colour and urea
15	Finishing	BOD, COD, wax, TSS, toxic chemical, acetate, resins, spent solvents, chlorinated compound, softeners etc
16	Product fabrication	Wastewater generation is very low or negligible

Table 8 Wastewater characteristics in textile industry

Sl. no	Description	UoM	Wastewater quality
1	pH		6.0–12.0
2	BOD	mg/l	Up to 6000
3	COD	mg/l	Around 30,000
4	Total kjeldahl nitrogen (TKN)	mg/l	50–80
5	TSS	mg/l	<8000
6	TDS	mg/l	Up to 12,000
7	Chloride	mg/l	Up to 6000
8	Colour	Pt–Co	Up to 2500

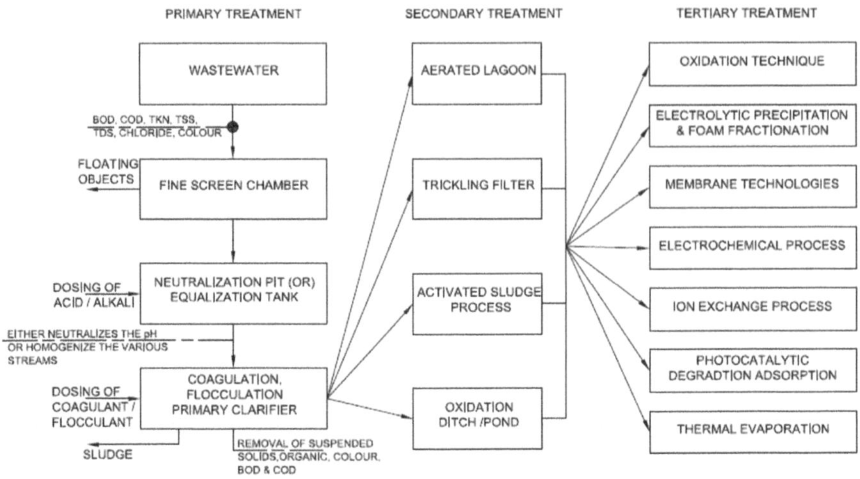

Fig.8 Wastewater treatment scheme in textile industry

Various physiochemical treatments are used to remove impurities from these wastewaters. The capital investment for such treatment is significantly high. A typical wastewater treatment scheme in the textile industry is shown in Fig. 8. In textiles, wastewater treatment includes primary, secondary, and tertiary stages. In primary treatment, screening helps arrest floating TSS. Then the wastewater is homogenized in the equalization tank. For pH correction, a neutralization pit is provided then primary sedimentation unit via the coagulation and flocculation units is followed. Some of the TSS, organic colour, BOD, and COD are removed at this stage.

Biological treatment is incorporated as a secondary treatment. Biological treatment is implemented due to cost effectiveness, lower operational costs, less sludge production, and COD removal efficiency. Various biological reactors are also useful for such treatment like an aerated lagoon, ASP, trickling filter, or oxidation pond. In biological treatment, BOD removal is the primary objective. Other than organic matter, COD and toxicities were also removed. In some cases, inefficiency in the complete removal of dyes is noted. Dye, pesticides, phenol, cyanide, organics, and toxic chemicals are removed by adsorption with activated carbon. After biological treatment, secondary clarification is required. Recently activated adsorption system and ozonation or ultraviolet radiation units are installed in lieu of biological treatment.

Water plays an important role in the production processes of the textile industry. So for operational cost optimization, wastewater recycling is recommended. Treated water can either be recycled back into the process or used for irrigation [40]. To reuse water, membrane technologies, ion exchange processes, electrolytic precipitation and foam fractionation, photocatalytic degradation adsorption, and thermal evaporation are incorporated [41]. Tertiary treatment aims at the elimination of TDS, residual colour, and odour. Membrane filtration plays an important role in the removal of microorganisms, TDS, sugar, organics, dyes, and other contaminants. According to recent trends, demand for RO is growing. RO permeate can be recycled into the production process.

As indicated in Table 8, TDS in wastewater can go high as 12,000 mg/l and chloride is around 6000 mg/l. At present the best available technology to eliminate TDS is NF or RO membranes. As the pore size of RO is much smaller than NF, RO is preferred in TDS removal. Best part of the membrane is the elimination of TDS from water as desired. But in the DM process the TDS are replaced by desired compounds. So the desired TDS level can only be achieved by membrane technology only. Wastewater generation in membrane technology is noted less compared to other technologies like DM. Moreover to eliminate the toxic chemical and undesired substances from water, NF and RO membrane has shown successful results. For selective passages of elements based on particle size or molecular cut-off NF and RO plays an important role. Footprint required by membrane technology is also noticed to be lesser compared to DM plant.

3.1.6 Pulp and Paper

There are three segments in this industry [42]. One part is the raw material preparation segment called pulp. The second segment produces paper and is called the paper industry. The last segment is paper recycling. All segments produce a significant quantity of wastewater. Contaminants in wastewater are fully dependent on the raw material and the production process [43]. In general, the raw materials required are wood, straw, hemp, bagasse, etc. The colour of the paper depends on the type of wood or raw material used. The process may include bleaching. Chemical procedures like Kaft processes and sulfite processes are used in pulp production [44]. The former is an alkaline process, while the latter is an acidic process. The Kaft process consumes the most water. Due to the presence of thiols and sulphides, odour is a disadvantage in this process. Wastewater from such processes contains high BOD, COD, SO2, and an alkaline pH. Other processes, such as neutral sulfite semi-chemical, thermo-mechanical, and chemi-thermo-mechanical, are also in use. In pulp production, wastewater is generated from washing, cleaning, condensates, and bleaching solutions. Wastewater generation in pulp and paper production is typically about 20–25 m^3/ton and 150–200 m^3/ton in mechanical and chemical processes, respectively [45]. Wastewater generated accounts for about 85% of the water consumed in this industry. Paper production depends on the quality of the pulp and the colour of the paper.

Depending on these two factors, chemicals like talc, TiO_2, $CaCO_3$, kaolin, latex, starch, aluminium sulphate, and colours can be added to the pulp [46]. Chlorine, or chlorine dioxide, is noted to be more effective than any other oxidising agent in the bleaching process. However, the bleaching process increases AOX in wastewater. The typical wastewater characteristics in pulp and paper production are listed below in Table 9 [47].

Wastewater is also generated from boiler operations, cooling towers, and WTP. To reduce environmental impact and optimise costs, many measures are taken to

Table 9 Typical wastewater characteristic in pulp and paper production

Sl. no	Description	UoM	Thermo-mechanical process	Bleaching	Pulping
1	pH		4	9.5	–
2	BOD	mg/l	3200	5000	9000
3	COD	mg/l	7200	46,000	22,000
4	TSS	mg/l	560	23	2400
5	Carbohydrate	mg/l	1650	Nil	3350
6	Acetic acid	mg/l	580	Nil	1400
7	Methanol	mg/l	25	2250	60
8	Nitrogen (N)	mg/l	12	230	55
9	Phosphorus (P)	mg/l	2.5	1	20
10	Sulphur (S)	mg/l	75	120	450

save water. WWTP starts with screening, grid removal, DAF, and sedimentation units. It reduces the coarse, bulky, fibrous, and settable particles [48]. Generally, lamella clarifiers are used in WWTP. Underflow sludge from clarifiers collected is dewatered to dry solid cake or moves to incineration. Biological treatment is preferred due to its eco-friendliness after primary treatment. This is a reliable and sustainable technology. Microorganisms decompose pollutants by using them as nutrients. Among biological reactors, anaerobic digesters have been very popular for treating wastewater in such industries. It can treat high BOD and produce low sludge. Anaerobic reactors produce biogas. It helps reduce sludge disposal costs. However, aerobic reactors are also in use. In secondary treatment, the objective is to eliminate the BOD, COD, TN, organics, TPs, and TSS. Filtration is used to remove traces of suspended solids, colloidal, organic, etc. Recently, MBR has also gained popularity due to its minimal footprint and efficiency. MBR technology is noted to perform well by eliminating the biological reactor, secondary clarifier, sand filter, and UF. This helps reduce wastewater generation by backwashing sand filters and UF. MBR is also optimised for power savings. Due to membranes in MBR, output water quality is less fluctuating. With reference to Table 9, membrane filtration should be recommended with MF, UF, or NF, or RO with ozonation to recycle and reuse the treated water. All primary and secondary treatments have limitations in eliminating the contaminants found in wastewater. To polish and eliminate the maximum possible contaminants, a membrane is needed. NF treatment for the effluent has been in use in newsprint paper mills for decades. NF is widely used in the paper mill for the recovery of water. This recovered water is used in the recirculation of the paper mill. The commercial aspects and processes of NF and RO are mostly similar. As RO produces better-quality water than NF, RO gets more preference. Membrane technology helps industries reduce fresh water consumption in the production process. Many patented technologies were introduced by TNO, like filtration-assisted crystallisation technology and multifo-softening technology by Veolia for WWTP [49]. The evaporation technique is noted to be very efficient, flexible, and reliable in treating different types of impurities in high concentrations.

3.2 Municipal Sector

Wastewater generated from domestic usage is called sewage. Sewage quantity and characteristics may vary from place to place based on the infrastructure of the sewerage collection system. According to the norms of the Indian government, the sewage disposal quality after treatment shall be as below in Table 10.

Following Table 11 explains the requirement for potable water in premises in the Indian region.

Table 10 Treated sewage quality

Sl. no	Description	Concentration [50]
1	pH	6.5 8.5
2	BOD	<10
3	COD	<50
4	TSS	<10
5	Ammonia nitrogen	<5
6	TN	<5
7	Nitrate nitrogen	5
8	Fecal coliform	<100 MPN/100 ml

Table 11 Potable water requirement in various premises

Sl. no	Description	Parameter [51]	UoM
1	Hospital (>100 Beds)	450	Litre per bed
2	Hospital (≤100 Beds)	340	Litre per bed
3	Lodge/Hotels	180	Litre per bed
4	Hostels/Nurses Home/Medical Quarter/Boarding School/College	135	lpcd
5	Restaurants/Airport/Sea Port, duty Staff/Train/Bus Station, duty staff	70	Litre per seat
6	Day School/College/Offices/Factories, duty staff	45	lpcd
7	Airport/Sea Port, boarding person/Train/Bus Station, boarding person/Cinema, concert halls and Theatres	15	lpcd

About 80% of potable water converts to sewage [52]. There are various treatment technologies for STP. This discussion will elaborate on the need for membrane technology in sewage treatment and recycling. The basic design is based on the concepts of anaerobic and aerobic bioreactors. But certainly this technology has its own limitations. Membranes play an important role in recycling and sustainable treatment. Municipal sectors can be divided into two major sub-divisions based on sewage characteristics and treatment approaches.

3.2.1 Private Sector

Private sector STP covers residential, commercial, community, and industrial sectors, etc. Sewage is treated in an aerobic reactor in skid-mounted or modular form. The selection of treatment process units depends on the utilisation of treated water. Recently, discharge norms were amended to be more stringent. Moreover, in areas of water scarcity, reuse of water for utility services other than drinking is encouraged. Treated sewage water is stored in tanks for use in gardening, flushing, and cleaning. Even artificial groundwater recharge is recommended, depending on the contaminants concentration and treatment involved [53]. MBBR, Extended Aeration System (EA), and ASP are most common in housing complexes and small communities. This is due to the cost-effectiveness of the process.

In many places, secondary treatment is followed by UF modules. This helps to achieve better water quality. However, very few places like Nambia, Australia, Singapore, California, New Maxico, and Virginia are recycling treated sewage for drinking purposes [54]. In power plants and refineries, modular MBR technologies are normally in demand as STP. Treated sewage from the MBR is mixed with treated effluent in a CMB. From CMB, water is treated using membrane technology and used in industrial processes.

3.2.2 Government Sector

Generally, these STPs are designed for clusters of areas, districts, towns, or cities. In a few cities, treated water is routed for industrial use. These help reduce freshwater and groundwater supplies for industries. According to the Manual on Sewage and Sewage Treatment Systems by CPHEEO in collaboration with JICA, STPs are designed with the sewage characteristics listed below in Table 12 [52].

STP can be divided into three parts. The first part includes screening, grit removal, O&G removal, and collection in an equalisation tank. Air circulation in equalisation tanks helps homogenise different streams of sewage. A primary clarifier is provided to reduce TSS. Around 30% of BOD and COD are also reduced in this primary clarification. In the second part, biological treatment based on anaerobic, anoxic, and aerobic reactors is provided. After the biological treatment, clarification is performed. Then chlorine is dosed as a disinfectant. Clarified water is filtered through a sand filter, followed by ACF. UF is used to retain all TSS and solids based on molecular

Table 12 Typical sewage characteristics

Sl. no	Description	UoM	Wastewater quality
1	BOD	mg/l	250.0
2	COD	mg/l	425.0
3	TSS	mg/l	375.0
4	VSS	mg/l	262.5
5	TN	mg/l	50.0
6	Organic nitrogen	mg/l	12.5
7	Ammonia nitrogen	mg/l	32.5
8	Nitrate nitrogen	mg/l	5.0
9	TP	mg/l	7.1
10	Ortho phosphorus	mg/l	5.0

cut-off. UF is considered a pre-treatment for RO. UF permeate is pumped with high pressure to RO modules to make usable water.

Depending on discharge quality norms, backwashes of UF and rejects of RO are further treated. STP's typical scheme is indicated in Fig. 9. Intervention of industrial effluent in sewage is a major concern in the government sector. This changes the sewage's characteristics drastically. In general, the ratio of BOD to COD in raw sewage is found in the range of 0.5 to 0.9.

However, due to industrial effluent intervention, the ratio becomes unpredictable. Figure 10 shows a 45 MLD Tertiary Treated Reverse Osmosis Plant (TTRO) plant installed by M/s BGR Energy Systems Limited in Kodungaiyur, Chennai, for M/s Chennai Metropolitan Water Supply and Sewerage Board (CMWSSB), India. 45 MLD of secondary treated sewage water is treated in the TTRO plant for recycling in industries in Chennai. The water received is filtered through rapid gravity sand filters and stored in a filtered water tank. Filtered water is treated with UF to ensure the elimination of organic compounds.

UF is a hollow fibre dead-end membrane with an in–out configuration. But considering the wastewater generation in UF due to its cleaning and backwash, the recovery in UF is noted at around 90% as optimum. UF permeate are stored in UF permeated tank. Then water is pumped to the RO skids via micron cartridge filters. RO recovery for this service is noted 78–80%. The RO permeate with around 70 mg/l

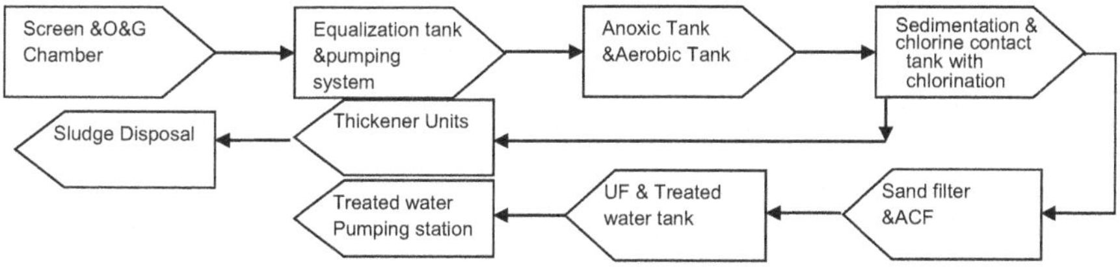

Fig. 9 Process flow diagram in sewage treatment

Fig. 10 45 MLD TTRO, Kodingaiyur plant, Chennai, India. Pic courtesy: BGR energy systems limited

TDS [55] is stored and pumped to Industrial areas as processed water. This helped Chennai in reducing ground water consumption. This also helped Chennai to reduce supply of fresh to industries. Solid waste is dumped in city solid waste management areas. These solid wastes degrade, and leachate is produced. Typical characteristics of the leachate can be tabulated below in Table 13 [56]. Depending on leachate characteristics, biological treatments like ASP, nitrification, or denitrification, chemical oxidation processes, activated adsorption, air stripping, or membrane technology can be implemented.

Table 13 Typical characteristics of leachate

Sl. no	Landfill leachate quality	UoM	Range
1	pH		About 7.7
2	COD	mg/l	<2000
3	BOD	mg/l	<900
4	TN	mg/l	<1200
5	Ammonical nitrogen	mg/l	<700
6	Chlorides	mg/	<3200
7	Sulphates	mg/l	<80
8	Total chromium	mg/l	<0.8
9	Iron	mg/l	<40
10	Nickel	mg/l	<0.6
11	Arsenic	mg/l	<0.75
12	Cadmium	mg/l	<0.6
13	Lead	mg/l	<12.0

In many countries, leachate treatment is done successfully with STP [57]. BOD, COD, and nitrogen in leachate are evaluated before co-treatment with STP. The treated water after the secondary treatment is filtered through a sand filter and an ACF. To reuse filtered water, plate-and-frame RO or CM is implemented. The concentrate from membranes is collected and treated in ZLD systems like evaporators.

4 Global Market and Technologies

About 99% of liquid freshwater available in earth is contributed by groundwater. Seven out of ten countries in Asia Pacific extract maximum groundwater. These countries are Bangladesh, China, India, Indonesia, Iran, Pakistan and Turkey. It contributes around 60% of total World's groundwater extraction. Around 3% of the total cultivated land in Sub-Saharan Africa is under irrigation. Surprisingly only 5% of that is irrigated with groundwater. Less than 30% of the freshwater is contributed by groundwater abstraction in Latin America and the Caribbean. It is due to the availability of abundant surface water. About 50% of the water used in industries is groundwater in Latin America. Over extraction of ground water in Arabian countries has made it most water-scarce in the world [57]. This is an increasing concern for countries. According to statistics, out of 100% freshwater withdrawn, 70% is used in the agriculture, 20% is used by industry and 10% is used for domestic purpose [58]. Due to drastic population growth worldwide, food demand is going to grow drastically in coming years. If freshwater scarcity arises food scarcity is also unavoidable. These figures are dependent on the demography and economic condition of the country. For instance, in a statistics, it has been noted that 55% of total fresh water is used in industries in Europe, 29% in agriculture, and 16% in domestic activities [59]. The statistics of water consumption in some Asian countries are noted below in Table 14 [60–63].

Globally, demand for fresh water varies a lot. Freshwater withdrawals are foreseen to increase by a few folds with the increase in global population. Now, the major source of water contamination is domestic and industrial wastewater. Globally, agriculture is the major consumer of freshwater. But to reduce water scarcity, domestic and industrial wastewater needs treatment and recycling. Recently, government policies enforced wastewater treatment and recycling. Insufficient quantities

Table 14 Water consuming sector in Asian countries

Water consumption area	Bangladesh (%)	India (%)	Pakistan (%)	Singapore (%)	South Korea (%)	Sri Lanka (%)	Thailand (%)
Agriculture	87.8	80	94	4.0	62	87.4	90.4
Domestic	10	8	5.3	51.0	30	624	4.8
Industry	2.1	12	0.8	45.0	8	6.4	4.8

of freshwater are not the only concern for global water scarcity. Good water quality is also an important factor. Freshwater sources like rivers, lakes, and groundwater are also a matter of concern. Moreover, maintaining proper drinking water quality is always a problem in both developed and developing countries.

About 75% of drinking water in Africa comes from groundwater [64]. This issue forces us to rethink both point-source and non-point-source pollution. Around 82% and 55% of urban areas in Africa have access to safe water and sanitation facilities, respectively. 85% of drinking water in South Asia comes into contact with human faeces, causing many more health hazards and environmental problems. Due to many reasons, proper quality water distribution throughout Southeast Asia is difficult. Microbial contamination is not only an issue here; high concentrations of arsenic in groundwater are also a major environmental concern in Southeast Asia. In addition to this, industrial pollution also plays a major role in contaminating water. As most Southeast Asian countries are in the developing stage water consumption occurs in the industrial sector rather than agriculture. Therefore, to understand future demand for water as well as wastewater treatment and recycling, the demand trend for necessities requires analysis.

As water consumption and wastewater generation are directly proportional to energy production like electricity, power, crude oil, and natural gas, the demand trend for energy is tabulated in Figs. 11 and 12 [65–67]. It clearly indicates the steady growth in water consumption and wastewater generation every year. Almost all countries have initiated seawater desalination projects to quench the need for industry and drinking water. The trend of seawater desalination plants globally is shown in Fig. 13 [68]. It has been noted that two-thirds of the global desalination plants are used in the municipal sector, while one-third are used for industrial purposes.

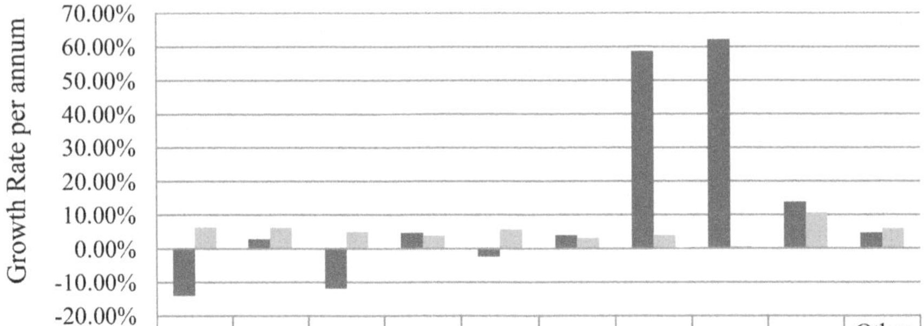

	Bangladesh	India	Indonesia	Malaysia	Philippines	South Korea	Sri Lanka	Thailand	Vietnam	Other Asia Pacific
Consumption	-14.00	2.90%	-11.90	4.70%	-2.30%	4.00%	58.60%	62.10%	13.90%	4.70%
Generation	6.30%	6.20%	5.00%	3.80%	5.70%	3.10%	4.00%	0.50%	10.60%	6.00%

Fig. 11 Hydroelectricity consumption vs. electricity generation in Asian countries (till 2019 until the covid-19 strike)

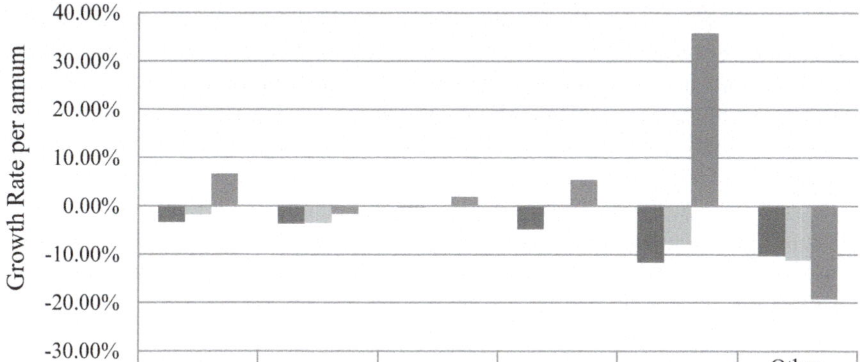

	India	Indonesia	Malaysia	Thailand	Vietnam	Other Asia Pacific
Crude Oil	-3.30%	-3.60%	-0.20%	-4.80%	-11.60%	-10.20%
Oil	-1.70%	-3.50%	-0.10%	0.30%	-7.90%	-11.20%
Natural Gas	6.70%	-1.60%	1.90%	5.40%	35.80%	-19.30%

Fig. 12 Demand trend of crude oil, oil and natural gas liquid in Asian countries (till 2019 until the covid-19 strike)

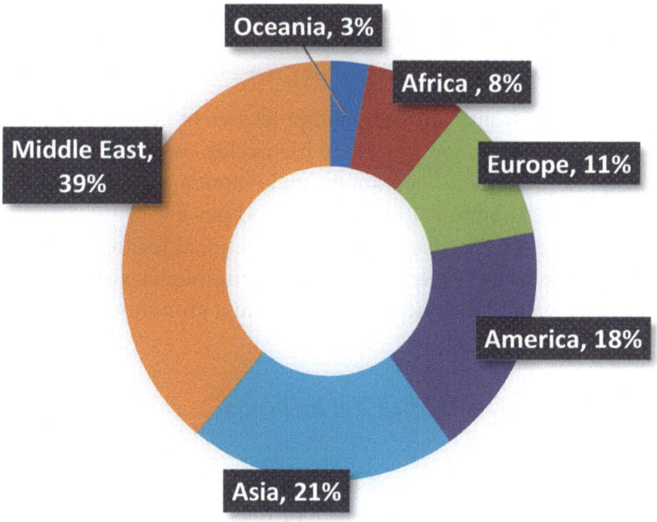

Fig. 13 Seawater desalination plant across the globe

The remarkable part is to note that domestic and industrial demand is growing at the same rate. Rapid growth is visible in countries like the Middle East, Africa, USA, Latin America, China, Taiwan, and Hong Kong. Industries like oil and gas, power, and mining are growing in these parts of the world. Therefore, to understand the growing trend of water consumption globally, we need to understand the demand for power and energy that is directly proportional to urbanisation.

In countries like India, treating sewage in communities and government sectors is becoming compulsory due to stringent pollution norms [69]. Treated sewage after UF is now commonly recycled for toilet flushing. This has largely reduced freshwater consumption.

5 Sustainable Technology and Its Acceptability

The implementation of newly invented technologies depends on various factors. Remarkable among them are contaminants, water quality standards, and cost [70]. Considering the last century of water treatment technologies, treatment depended mostly on chemical precipitation and clarifiçation, followed by media filtration and chlorination. In the last few decades, the water treatment industry has evolved with alternative, membrane technologies. With the advancement of membrane technology, users have raised a big question about the acceptability of the emerging membrane technology. Sustainable membrane technology must have the following advantages:

1. Advantages over conventional treatment processes in quality and quantity.
2. Reliability and sustainability with more life cycles.
3. Lower in capital investment, operations, and maintenance costs.
4. Should demand the minimum consumables and power.
5. Higher efficiency and effectiveness than conventional technologies,
6. Operation must be easier, safer, and free of negligible hazards.
7. Minimum wastewater and solid generation with a minimum footprint.

5.1 Demography, Health and Technology

Technology can improve or change a lifestyle. Similarly, demography, socioeconomic life can also influence the acceptance or rejection of a technology. Recently, awareness for odour issues and safe water is increasing among people. Water scarcity and recycling of wastewater are becoming a part of life. No liquid discharge norms for mining, metal, or chemical industries are also imposed by pollution control board. This initiative is to improve environmental and health condition of living beings. Considering hygiene and the health of living beings, discharge norms for treated wastewater becoming stringent every day. Conventional treatment processes have reached the limit of treating water to maximum improved quality. The acceptance of cost-effective Membrane technology is now a necessity.

5.2 Industry, Economy and Technology

Globalisation has affected all types of economies. In also provided vast opportunities in the open market. Due to this, price competition has drastically increased globally. These never-ending completions among industries globally are increasing day by day. As a result, more advanced technologies with minimal commercial impacts are in demand. Like sewage, it can be treated either by conventional ASP, EA, FAB, or MBR. From a commercial point of view, the cost of MBR is almost 1.8 times that of conventional, followed by UF. Cost is an important factor in the acceptance of technology by users. This is how socioeconomics can influence the acceptance of technology.

Acceptance of technology depends on similar performance and a very low cost difference. Similarities in performance [71–73] between MF vs. UF and NF vs. RO are indicated in Table 15.

Another major constraint for acceptance of membrane technology is the process of standardisation in EPC or technology companies. Due to heavy competition among technology providers, all suppliers are encouraging standardisation of engineering processes, items, and equipment. So acceptance of membrane technology is not necessary unless the situation calls for investment and re-engineering. This decision plays an important role in the implementation of new technology.

6 Conclusion

The need is driving the development of membrane technologies in wastewater management. As the need grows, advanced membrane technologies are also evolving with new facilities and advantages. Existing technologies are continuously upgraded along with evolving membrane technologies. However, demand and feasibility are looking for more cost-effective and efficient membrane technologies. Membrane technologies shall be capable of addressing the maximum possible constraints for users. Water consumption by users will not be reduced or optimised drastically. Rapid urbanisation will certainly increase the consumption of water. So, wastewater generation is also foreseen to increase with time. Present technologies can manage wastewater treatment. However, the growing competition in businesses and industries and the need for cost-effectiveness require sustainable membrane technology. Membrane technologies should be low-cost solutions with longer life spans. Ceramic membranes have vast potential for the membrane treatment of wastewater. More research and development are expected to make membranes with higher production capacity, lower operation and maintenance costs, and a longer life with fewer consumables.

Table 15 Similarities of MF versus UF AND NF versus RO

Sl. no	Similarities		Similarities	
	MF	UF	NF	RO
1	Semi permeable membrane—pore size 0.1–10 μm and MWCO—1,00,000 to 5,000,000 Da	Semi permeable membrane—pore size 0.1–0.01 μm MWCO—1,000 to 1,000,000 Da	Semi permeable membrane—pore size 0.2–2 nm and MWCO—200 to 1000 Da	Semi permeable membrane—pore size0.1 nm and MWCO—100 to 200 Da
2	Remove finer particles (in general < 2 μm particle size)	Remove finer particles (in general < 2 μm particle size)	Eliminate particles of 0.0008–0.0006 micron diameter	Eliminate particles of 0.0001 micron diameter
3	Remove various particulates, microorganisms like bacteria, viruses, dyes, organic compounds, colour and odour etc	Remove various particulates, microorganisms like bacteria, viruses, dyes, organic compounds, colour and odour etc	Remove dissolve organic compounds and total dissolve salts	Remove dissolve organic compounds and total dissolve salts
4	Used as a pre-treatment prior to ROtreatment to reduce fouling	Used as a pre-treatment prior to ROtreatment to reduce fouling	Pre-treatment with MF or UF is required to prevent fouling	Pre-treatment with MF or UF is required to prevent fouling
5	May eliminate clarification and sand filtration units (in certain cases)	May eliminate clarification and sand filtration units (in certain cases)	High Pressure driven process	High Pressure driven process
6	Foot print almost same as UF system	Foot print almost same as MF system	Foot print almost same as RO system	Foot print almost same as NF system
7	Configuration • Media filter • Membrane module • Pump • Clean in place system (CIP) • Water tanks	Configuration • Media filter • Membrane module • Pump • Clean in place system (CIP) • Water tanks	Configuration • UF • Pump • Cartridge filter • High pressure pump • Membrane module • Clean in place system (CIP) • Chemical dosing system • Water tanks	Configuration • UF • Pump • Cartridge filter • High pressure pump • Membrane module • Clean in place system (CIP) • Chemical dosing system • Water tanks

(continued)

Table 15 (continued)

Sl. no	Similarities		Similarities	
	MF	UF	NF	RO
8	Membrane Construction features can be • Hollow fibre • Plate and frame • Spiral-wound • Tubular	Membrane Construction features can be • Hollow fibre • Plate and frame • Spiral-wound • Tubular	Membrane Construction features can be • Plate and frame • Spiral-wound	Membrane Construction features can be • Plate and frame • Spiral-wound
9	Membrane material can be polypropylene, cellulose acetate, polysulfone, ceramic or stainless steel	Membrane material can be polypropylene, cellulose acetate, polysulfone, ceramic or stainless steel	Membrane material can be polypropylene, cellulose acetate, polysulfone, ceramic or stainless steel	Membrane material can be polypropylene, cellulose acetate, polysulfone, ceramic or stainless steel

Acknowledgements Author ARC would like to thanks M/s BGR Energy Ltd., India. And author PRC would like acknowledge Department of Rubber and Plastics Technology, Anna University MIT campus, Chennai, India.

References

1. Khambete AK (2019) When water kills, India waterportal. https://www.indiawaterportal.org/articles/when-water-kills#:~:text=Waterborne%20diseases%20in%20India,of%20%24600%20million%20a%20year
2. Lenntech, Micro filtration and ultra-filtration. https://www.lenntech.com/microfiltration-and-ultrafiltration.htm#ixzz6W1XwaNBU
3. Huisman IH (2000) Membrane separations|microfiltration. In: Wilson ID (eds) Encyclopedia of separation science. Academic Press, pp 1764–1777
4. Shkinev V, Spivakov B (2005) Membrane techniques. Ultrafiltration. In book: Encyclopedia of analytical science, pp 524
5. Singh R (2005) Hybrid membrane systems for water purification. In: Singh R (eds) Introduction to membrane technology, chapter 1. Elsevier Science, pp 1–56
6. Van derBruggen B, Vandecasteele C (2003) Removal of pollutants from surface water and groundwater by nanofiltration: overview of possible applications in the drinking water industry. Environ Pollut 122:435–445. https://doi.org/10.1016/S0269-7491(02)00308-1
7. Kucera J (2015) Reverse osmosis industrial application and processes, 2nd edn. ISBN: 978-1-118-63974-0
8. Choudhury P, Mondal P, Majumdar S, Saha S, Sahoo GC (2018) Preparation of ceramic ultrafiltration membrane using green synthesized CuO nanoparticles for chromium(VI) removal and optimization by response surface methodology. J Clean Prod 203:511–520. https://doi.org/10.1016/j.jclepro.2018.08.289
9. Roy Choudhury P, Majumdar S, Sahoo GC, Saha S, Mondal P (2018) High pressure ultrafiltration CuO/hydroxyethyl cellulose composite ceramic membrane for separation of Cr(VI) and Pb(II) from contaminated water. Chem Eng J 336:570–578. https://doi.org/10.1016/j.cej.2017.12.062

10. Roy Choudhury P, Majumdar S, Sarkar S, Kundu B, Sahoo GC (2019) Performance investigation of Pb(II) removal by synthesized hydroxyapatite based ceramic ultrafiltration membrane: bench scale study. Chem Eng J 35:510–519. https://doi.org/10.1016/j.cej.2018.07.155
11. Khan AA, Boddu S (2021) Hybrid membrane process: an emerging and promising technique toward industrial wastewater treatment. In: Shah MP, Rodriguez-Couto S (eds) Membrane-based hybrid processes for wastewater treatment, chapter 13. Elsevier, pp 257–277
12. Charcosset C (2012) Membrane processes in biotechnology and pharmaceutics, microfiltration, chapter 3. Elsevier, pp 101–141
13. Mohammad AW, Teow YH, Ang WL, Chung YT, Oatley-Radcliffe DL, Hilal N (2015) Nanofiltration membranes review: recent advances and future prospects. Desalination 356:226–254. https://doi.org/10.1016/j.desal.2014.10.043
14. Central electricity authority (2012) Report on minimisation of water requirement in coal based thermal power stations, New Delhi—110066
15. Water optimization & wastewater usage in thermal power plants by confederation of Indian industry (2021) Standard water balance, pp 30
16. Afroza A, Naseer HS, Arif I (2019) Characterization of thermal power plant discharged wastewater and city sewage wastewater: their potential use as irrigants for urban agriculture. J Indus Pollut Control 35:2313–2322
17. Guru I (2015) Managing usage of water in an integrated steel plant. https://www.ispatguru.com/managing-usage-of-water-in-an-integrated-steel-plant/
18. Walling FB, Otts Jr LE (1967) Water requirements of the iron and steel industry by Table 3. Water use in the steel industry. United States government printing office, Washington
19. Erika A, Ismael M, Giacomo FP, Valentina C (2014) Wastewater treatment in iron and steel industry: process integration for water re-use. In: Conference: 1st international process integration forum for the steel industry at: Lulea (Sweden). https://www.researchgate.net/publication/262377191
20. Jørgensen SE (1979) Waste water from the iron and steel industry and mining. Stud Environ Sci 5:217–227. https://doi.org/10.1016/S0166-1116(08)71605-4
21. Mohammad A, Al-Ghouti MA, Al-Kaabi M, Dana YA (2019) Produced water characteristics, treatment and reuse: a review. J Water Process Eng 28:222–239. https://doi.org/10.1016/j.jwpe.2019.02.001
22. Wikipedia, the free encyclopaedia, Produced water. https://en.wikipedia.org/wiki/Produced_water#:~:text=Much%20produced%20water%20is%20brine,contains%20oil%20and%20suspended%20solids.&text=Metals%20found%20in%20produced%20water,manganese%2C%20iron%2C%20and%20barium
23. Igunnu ET, Chen GZ (2012) Produced water treatment technologies. Int J Low-Carbon Technol 9:157–177. https://doi.org/10.1093/ijlct/cts049
24. Pingping S, Amgad E, Michael W, Jeongwoo H, Robert H (2018) Estimation of U.S. refinery water consumption and allocation to refinery products. Fuel 221. https://doi.org/10.1016/j.fuel.2017.07.089
25. Otts Jr LE (1963) Water requirements of the petroleum refining industry, Geological survey water-supply paper 1330-G, United States Government Printing Office, Washington
26. Jacob Consultancy (2016) Water consumption in US petroleum refineries final report prepared for Argonne national labs
27. Rahi M, Jwied A, Al-Maamori A (2021) Treatment of petroleum refinery effluents and wastewater in Iraq: a mini review. IOP Conf Ser Mater Sci Eng 1058:1–11
28. Petroleum refining water/waste water use and management (2010) IPIECA. www.ipieca.org
29. Pal P (2017) Treatment and disposal of pharmaceutical wastewater: towards sustainable strategy. Sep Purif Rev. https://doi.org/10.1080/15422119.2017.1354888
30. Choudhary A (2013) Wastewater treatment process (effluent treatment), Pharmaceutical guidelines. https://www.pharmaguideline.com/2013/07/wastewater-effluent-treatment-procedure.html
31. Guo Y, Qi PS, Liu YZ (2017) A review on advanced treatment of pharmaceutical wastewater. IOP Conf Ser Earth Environ Sci 63

32. Rekhate CV, Srivastava JK (2020) Recent advances in ozone-based advanced oxidation processes for treatment of wastewater—a review. Chem Eng J Adv 100031. https://doi.org/10.1016/j.ceja.2020.100031
33. Condorchemenvitech, Wastewater treatment in pharmaceutical industry. https://blog-en.condorchem.com/wastewater-treatment-in-pharmaceutical-industry/#.X1XfKtIzbIU
34. GE Power & water, water & process technologies (2011) Pharmaceutical wastewater treatment, case study. https://www.ge.com/in/sites/www.ge.com.in/files/GE-ZeeWeed-pharma-treat.pdf
35. Berhe G (2015) Chapter three, Textile effluent treatment. https://www.slideshare.net/GranchBerheTseghai/textile-effluent-treatment
36. Vandana T (2013) Waste water and its treatment in textile industry. https://www.slideshare.net/textilevandana/waste-water-and-its-treatment-in-textile-industry
37. EPA (1995) Best management practices for pollution prevention in the textile Industry, office of research and development, ATMI, Comments on draft document, 1997b
38. Engr. Uddin M (2022) Azo dye: a popular but hazardous dye should be eliminated from industrial process. https://www.textiletoday.com.bd/azo-dye-popular-hazardous-dye-eliminated-industrial-process/
39. Yaseen DA, Scholz M (2019) Textile dye wastewater characteristics and constituents of synthetic effluents: a critical review. Int J Environ Sci Technol 16:1193–1226. https://doi.org/10.1007/s13762-018-2130-z
40. Singh RP, Singh PK, Gupta R, Singh RL (2018) Treatment and recycling of wastewater from textile industry. Advances in biological treatment of industrial waste water and their recycling for a sustainable future, pp 225–266. https://doi.org/10.1007/978-981-13-1468-1_8
41. Buckley T, Xu X, Rudolph V, Firouzi M, Shukla P (2021) Review of foam fractionation as a water treatment technology. Sep Sci Technol 929–958. https://doi.org/10.1080/01496395.2021.1946698
42. Lenntech BV (2015) Pulp and paper industry water treatment. https://www.lenntech.com/pulp-and-paper-industry-water-treatment.htm
43. Haddaway A (2014) Pulp & paper: a look at wastewater treatment trends and technologies. Water technology. https://www.watertechonline.com/wastewater/article/16211172/pulp-paper-a-look-at-wastewater-treatment-trends-and-technologies
44. Ferraz A, Guerra A, Souza-Cruz PB, Mendonca R (2002) Attempts to correlate biopulping benefits with changes in the chemical structure of wood components and enzymes produced during the wood biotreatment with ceriporiopsissubvermispora—[33 refs.], Prog. Biotechnol. 21, pp 73–80; Dep.Biotec., Fac. Engenharia Quim. Lorena, 12600 Lorena, Brazil; Eng.)—Lindner
45. Gove GW (1980) Pulp and paper industry wastewater management. Water Pollut Control Fed 52:1386–415
46. Hubbe MA, Gill RA (2016) Fillers for papermaking: a review of their properties, usage practices, and their mechanistic role. BioRes 11:2886–2963
47. Ji-WhanAhn ML (2009) Characteristics of wastewater from the pulp paper industry and its biological treatment technologies. J Korean Inst Resour Recycl 18
48. Hubbe MA, Metts JR, Hermosilla D, Angeles Blanco M, Yerushalmi L, Haghighat F, Lindholm P, Khodaparast Z, Kamali M, Elliott A (2016) Wastewater treatment and reclamation: a review of pulp and paper industry practices and opportunities. Bioresources 11:7953–8091. https://doi.org/10.15376/biores.11.3.Hubbe
49. Pulp and paper technology, Waste water treatment solutions in pulp and paper industry. https://www.pulpandpaper-technology.com/articles/waste-water-treatment-solutions
50. Nasir (2021) CPCB guidelines for sewage treatment plants (STPs). https://www.logicladder.com/cpcb-guidelines-sewage-treatment-plant/
51. Central Public Health and Environmental Engineering Organization in collaboration with Japan International cooperation agency (2013) Manual on sewerage and sewage treatment systems (edn) Table 3.4 of CPHEEO manual 2013
52. Central Public Health and Environmental Engineering Organization in collaboration with Japan International cooperation agency, Manual on sewerage and sewage treatment systems, 2013 Edition. Table 5.4 of CPHEEO manual 2013

53. I am Renew (2019) SC Allows Bengaluru to use sewage water for irrigation, Recharge groundwater. https://www.iamrenew.com/policy/sc-allows-bengaluru-to-use-sewage-water-for-irrigation-recharge-groundwater/
54. Cho R (2011) WATER from wastewater to drinking water. https://blogs.ei.columbia.edu/2011/04/04/from-wastewater-to-drinking-water/#:~:text=But%20a%20few%20countries%20like,and%20help%20ease%20water%20shortages.
55. https://timesofindia.indiatimes.com/city/chennai/sewage-recycling-will-save-50mld-of-chennais-water/articleshow/71367136.cms
56. Zhang Q-Q, Tian B-H, Zhang X, Ghulam A, Fang C-R, He R (2013) Investigation on characteristics of leachate and concentrated leachate in three landfill leachate treatment plants. Waste Manag 33:2277–2286. https://doi.org/10.1016/j.wasman.2013.07.021
57. Central public health and environmental engineering organisation, Municipal solid waste management manual (2015)
58. UNESCO—The United nations world water development report 2022. GROUNDWATER making the invisible visible
59. Flörke M, Teichert E, Bärlund I, Eisner S, Wimmer F, Alcamo Joseph (2013) Domestic and industrial water uses of the past 60 years as a mirror of socio-economic development: a global simulation study. Glob Environ Chang 23:144–156. https://doi.org/10.1016/j.gloenvcha.2012.10.018.
60. Odey G, Adelodun B, Lee S, Adeyemi KA, Cho G, Choi KS (2022) Environmental and socioeconomic determinants of virtual water trade of grain products: an empirical analysis of South Korea using decomposition and decoupling model. Agronomy 12(12):3105. https://doi.org/10.3390/agronomy12123105
61. https://www.drishtiias.com/daily-updates/daily-news-editorials/growing-water-stress-in-india#:~:text=However%2C,India%20is%20used%20for%20drinking.&text=Remaining%2012%25%20goes%20to%20industrial,people%20facing%20acute%20water%20shortages
62. Nasima C (2010) The relative efficiency of water use in Bangladesh agriculture. The South Asian network for development and environmental economics. Working papers. 49
63. https://tableau.apps.fao.org/views/AQUASTATDashboard/result_country?%3Aembed=y&%3AisGuestRedirectFromVizportal=y
64. Somaya A (2011) Water quality in developing countries, South Asia, South Africa, Water quality management and activities that cause water pollution. In: international conference on environmental and agriculture engineering, IPCBEE vol. 15. IACSIT Press, Singapore
65. BP Energy Outlook 2019 edn. https://www.bp.com/content/dam/bp/business-sites/en/global/corporate/pdfs/energy-economics/energy-outlook/bp-energy-outlook-2019.pdf
66. Energy Outlook 2020 edn. https://www.bp.com/content/dam/bp/business-sites/en/global/corporate/pdfs/energy-economics/energy-outlook/bp-energy-outlook-2020.pdf
67. Mordor Intelligence, oil refining market - growth, trends, covid-19 impact, and forecasts (2021–2026). https://www.mordorintelligence.com/industry-reports/oil-refining-market
68. Eke J, Yusuf A, GiwaA, Sodiq A (2020) The global status of desalination: an assessment of current desalination technologies, plants and capacity. Desalination 495:114633. https://doi.org/10.1016/j.desal.2020.114633
69. The Economic Times Politics, New TTRO plant to quench the thirst of industries in Chennai (2019). https://economictimes.indiatimes.com/news/politics-and-nation/new-ttro-plant-to-quench-the-thirst-of-industries-in-chennai/articleshow/71410991.cms?from=mdr
70. Najm I, Trussell RR (1999) National research council. Identifying future drinking water contaminants. Washington, DC, chapter: 11 New and emerging drinking water treatment technologies. The National Academies Press, pp 220–243. https://doi.org/10.17226/9595
71. Membrane filtration. https://www.mrwa.com/WaterWorksMnl/Chapter%2019%20Membrane%20Filtration.pdf

72. Samco (2017) Microfiltration vs ultrafiltration processes: what is the difference? https://www.samcotech.com/microfiltration-vs-ultrafiltration-processes-what-is-the-difference/
73. Samco (2017) Reverse osmosis vs nanofiltration membrane process: what is the difference? https://www.samcotech.com/reverse-osmosis-vs-nanofiltration-membrane-process-what-is-the-difference/#:~:text=NF%20delivers%20slightly%20coarser%20filtration,to%200.005%20%CE%BCm%20in%20diameter.&text=NF%20removes%20harmful%20contaminates%2C%20such,that%20RO%20would%20otherwise%20remove.

Chapter 3
Low-Cost Ceramic Clay-Based Ultrafiltration Membranes for Dyes Removal

Brahim Achiou, Abdelmjid Bouazizi, Youness Kouzi, Majda Brieda, José Ignacio Calvo, Mohamed Ouammou, and Saad Alami Younssi

1 Introduction

Ultrafiltration (UF), a membrane-based technology with pore diameters ranging from 2 to 50 nm, facilitates in removing viruses, bacteria and soluble organic compounds [1]. Polymers, metal oxides and ceramics are the main materials used to fabricate UF membranes. Recently, polymeric membranes are generally utilized in wastewater treatment. However, these membranes face some limitations such as fast fouling, less chemical and thermal stability, and short lifetime [2]. Ceramic membranes

B. Achiou (✉) · Y. Kouzi · M. Brieda · M. Ouammou · S. Alami Younssi
Laboratory of Materials, Membranes and Environment, Faculty of Sciences and Technologies of Mohammedia, Hassan II University of Casablanca, Casablanca, Morocco
e-mail: brahim.achiou@fstm.ac.ma

M. Ouammou
e-mail: mohamed.ouammou@fstm.ac.ma

S. Alami Younssi
e-mail: saad.alamiyounssi@fstm.ac.ma

A. Bouazizi
Laboratory of Materials, Membranes and Nanotechnology, Department of Chemistry, Faculty of Sciences, Moulay Ismail University, Zitoune, PB 11201, Meknes, Morocco

Department of Sciences, Ecole Normale Superieure, Moulay Ismail University of Meknes, Meknes, Morocco

J. I. Calvo
SMAP (Surfaces and Porous Materials Group), Institute of Sustainable Processes, University of Valladolid, C/Dr. Mergelina, s/n, 47011 Valladolid, Spain
e-mail: joseignacio.calvo@uva.es

Applied Physics Department, ETSIIAA, University of Valladolid, Campus la Yutera, 34004 Palencia, Spain

S. Bandyopadhyay (ed.), *Application of Low Cost Ceramic Membranes in Wastewater Treatment*, Materials Horizons: From Nature to Nanomaterials,
https://doi.org/10.1007/978-981-97-5842-5_3

possess some special features in comparison with their organic counterparts notably higher thermal, mechanical and chemical stability, excellent porosity, longer lifetimes, higher hydrophilicity and lower fouling [3]. On a commercial level, ceramic membranes occupy a large space owing to unique structure of ceramic materials in the academic as well as in research field. Ceramic UF membranes in flat and tubular configuration consist of an active layer on a porous support [4]. The active layer is mainly responsible for the separation of dyes from solution while the support provides the needed mechanical strength to resist the transmembrane pressure during filtration.

Cost-efficient approach for preparation of ceramic membranes can be achieved using cheaper raw materials or precursors, fast and reproducible processing stages, particularly lower sintering temperature, during fabrication of support, intermediate and active layers. A new generation of low-cost ceramic membranes were successfully fabricated using geomaterials such as clays, phosphate, pozzolan, magnetite and perlite [5]. Among these natural materials, clays exhibit promising physicochemical and structural features (suitable plasticity, high porosity and low sintering temperature, etc.) compared with commonly industrial oxides such as alumina and silica. This could justify the great interest made by both industrial and academic researchers to develop clay-based UF membranes [6].

In this chapter, geomaterials used for the preparation of low-cost membrane supports are reviewed and categorized into clays (e.g., bentonite, cordierite, illite and kaolinite) and other geomaterials with diverse sorts of structural units (e.g., phosphate, perlite and pozzolan) [7]. This chapter presents an overview of the preparation of low-cost ceramic UF membranes using Moroccan clay for effective removal of dyes from wastewater. The preparation and characterization of clay-based membranes have been thoroughly described in this chapter. The mechanisms of dyes removal by clay-based membranes have been explained in details, considering the factors that influence the membrane performance such as membrane structure and operational conditions. Furthermore, the chapter suggests future directions towards the development of clay-based UF membranes for large-scale water treatment and decontamination of wastewater.

2 Characterization of Natural Clay

Prior to utilizing a low-cost material in the preparation of UF membranes, it is crucial to carry out materials characterization and testing to determine their feasibility in the fabrication process. In the light of this approach, Achiou et al. undertook the characterization of a natural clay from the Fez region in Morocco for its application in the ceramic membrane technology [8]. This step was vital in ensuring the success of the fabrication and the cost-effectiveness of the material.

Table 1 Chemical composition of natural clay [8]

Oxides	SiO_2	Al_2O_3	CaO	Fe_2O_3	MgO	K_2O	Na_2O	TiO_2	LOI[a]
Wt%	41.28	12.29	14.23	4.13	2.80	1.73	1.02	0.62	20.20

[a] Loss of ignition at 950 °C

2.1 *Chemical Composition*

The chemical composition presented in Table 1 indicates that the raw natural clay is mainly composed of silica (SiO_2) and alumina (Al_2O_3), which are likely associated with clay minerals such as illite, kaolinite and smectite. The SiO_2/Al_2O_3 ratio is equal to 3.35, which could be explained by the high concentration of quartz and amorphous SiO_2. The high amount of calcium oxide (CaO) in the natural clay might indicate the presence of a significant amount of carbonate, which was confirmed by Bernard calcimeter analysis, indicating that the raw clay contains about 14.23 wt% of carbonate. In addition, raw clay also contains some minor constituents like magnesia (MgO), potassium oxide (K_2O), sodium oxide (Na_2O), and titania (TiO_2). They are generally related to exchangeable cations in the interlayer space of aluminosilicate structure (kaolinite/smectite). The LOI at a temperature of 950 °C is found to be 20.20 wt%, which is associated with the decomposition of organic matter and inorganic carbonates in addition to the dehydroxylation of clayey minerals. It should be noted that it is possible to quantify the amount of organic matter by measuring the LOI at 550 °C [8].

2.2 *Mineralogical Composition*

Clay minerals were identified by XRD of the raw material which shows that all the peaks correspond only to quartz and calcite Fig. 1a. However, identifying peaks associated with clay minerals can be challenging because they have lower intensity compared to the peaks of quartz and calcite. Additionally, clay minerals often exhibit low crystallinity, which further reduces the intensity of their peaks. Consequently, even if the clay minerals are present in the sample, their peaks could not be detected. Therefore, to identify and quantify clay minerals, crystalline impurities should be removed from natural clay Fig. 1b. Illite and chlorite are easily recognized by their characteristic peaks at 8.87° and 26.74° for illite and at 25.19° for chlorite. A peak at 6.31° can be attributed to smectite or chlorite. As is well-known, smectite is a swelling mineral. Therefore, to prove the presence of swelling clay minerals, a sample of the clay is treated with ethylene glycol. This treatment causes a shift in the reflection angle only for swelling clay minerals. In the XRD pattern of ethylene glycol treated sample, it was noticed that the peak has shifted from 2θ of 6.31° to 5.19° which proves that the reflected peak corresponds to smectite. Another peak was observed at 12.45° that could correspond to kaolinite and/or chlorite. At temperatures around

550 °C, kaolinite is transformed into metakaolin (an amorphous phase) that can't be detected. For this reason, a heated sample was prepared by thermally treating the clay at 550 °C. In the XRD pattern of heated sample, it was noticed that the peak has collapsed which proves that the peak was reflected by kaolinite mineral. Quantification of the clay minerals was done by calculation of profile parameters such as intensity and integrated area. The amounts of each clay mineral are presented in Fig. 1c.

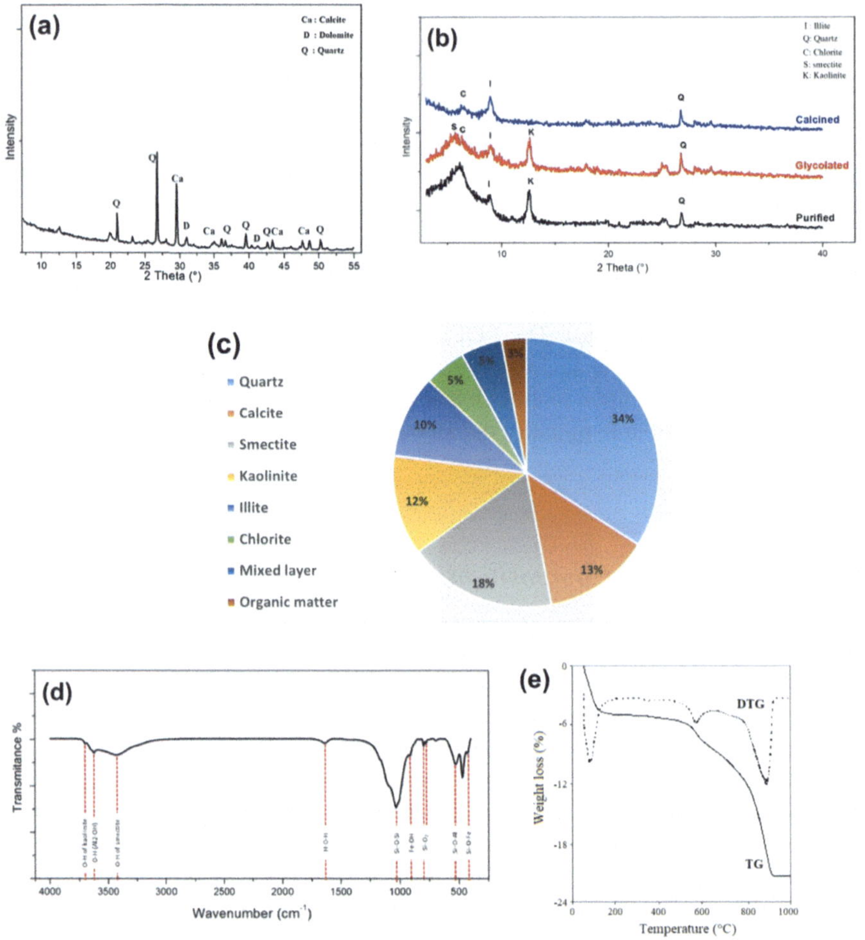

Fig. 1 Characterization of raw natural clay: **a** XRD patterns, **b** XRD patterns of treated clay, **c** mineralogical composition, **d** FTIR spectra, and **e** thermoanalytic curves [8, 9]

2.3 FTIR

Figure 1d of the raw natural clay FTIR shows the presence of 12 characteristic bands. Three bands, one at 420 cm^{-1} corresponding to Si–O–Fe, another at 470 cm^{-1} corresponding to Si–O–Mg, and the third at 525 cm^{-1} corresponding to Si–O–Al, appeared close to each other. An intense and sharp band at 1020 cm^{-1} corresponds to Si–O–Si, while another band observed at 3622 cm^{-1} might be contributed to Al_2-OH. Additionally, a band at 920 cm^{-1} appears, corresponding to the bending bonds characteristic of phyllosilicates. A band located at 3422 cm^{-1} is contributed to –OH stretching of smectite, while another one observed at 3700 cm^{-1} corresponds to –OH stretching of kaolinite. The peak at 1630 cm^{-1} is attributed to water hydration, the band at 878 cm^{-1} is characteristic of iron hydroxide, and the peak at 1429 cm^{-1} is attributed to carbonate. Furthermore, the two peaks at 797 and 779 cm^{-1} correspond to quartz. Overall, these results suggest that all the observed characteristic bands match the chemical and mineralogical composition of the clay sample.

2.4 Thermal Analysis

Thermal analysis of the sample was conducted using the TGA technique. The sample was heated from room temperature to 1000 °C in an oxidizing atmosphere with a heating rate of 10 °C min^{-1}. TG and its derivative (DTG) show four peaks, as displayed in Fig. 1e. The first loss at 100 °C corresponds to the removal of physisorbed and interlayer water, resulting in a 5 wt% loss due to dehydration. The second peak, noticed only in DTG at 400 °C, could be caused by the decomposition of organic matter present in small quantities. The third peak, with a loss of 2.5 wt%, corresponds to the dihydroxylation of structural OH at 570 °C. The last peak, displayed from 800 to 900 °C, corresponds to the decarbonation of calcite, resulting in an 8 wt% loss. The results obtained can be employed to optimize the sintering temperature and residence time at each segment of the thermal treatment program during support fabrication to avoid structural defects.

2.5 Classification of the Natural Clay

In exhibiting their inherent physical properties like plasticity, sinterability, porosity, strength, adsorption ability, etc. each clay mineral has diverse characteristics with respect to air-drying and shrinkage behaviour. The virtuous chemical and physical responses on the surface of clay minerals occur because of their high surface area and smaller crystallite size of these minerals. The classification of natural clay was performed according to NF P 94-056 and NF P 94-057 standards [8]. NF P 94-056 standards consist in wet sieving of the material with fraction above 80 μm in

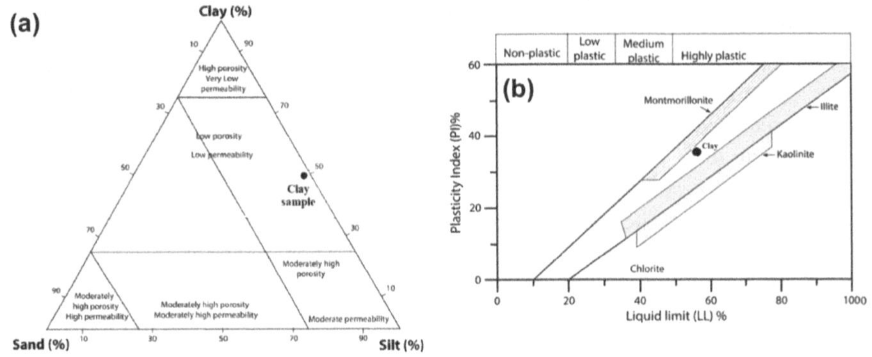

Fig. 2 **a** Ternary diagram of studied clay sediments, **b** position of clay samples on the Holtz and Kovacs diagram [8]

diameter while NF P 94-057 standard uses the sedimentation method to quantify fraction with diameter less than 80 μm. Size distribution of the material reveals that 50% of particles have diameter under 2 μm (clay minerals), 49% have diameter between 2 and 80 μm (silt) and only 1% of particles are bigger that 80 μm (sand). After plotting the obtained results in McManus ternary diagram Fig. 2a, it shows that the extracted material belongs to low porosity and low permeability materials category.

2.6 *Plasticity*

The limit of liquidity (LL) was determined using the Casagrande method (NF P 94-052-1 standard), while the limit of plasticity (LP) was obtained by applying the rolled sample technique (NF P 94-051 standard) [8]. The plasticity index was calculated as the difference between LL and LP. The plastic limit of the studied material was found to be 20%, with a limit of liquidity of 56%, resulting in a plasticity index of 36%. Plotting the plasticity index and limit of liquidity on the Holtz and Kovacs diagram Fig. 2b showed that the clay material belonged to the highly plastic region, which was consistent with the mineralogical analysis. Based on the results of the plasticity tests, this clay could be a suitable material for preparing plastic paste with excellent rheological properties, which could be used to make membrane supports using the wet route.

3 Purification of Natural Clay for Membrane Manufacturing

It is worth noting that the purification of clay is a critical step particularly in preparing coating formulations for use in the development of membrane layers over macroporous supports. The impurities present in raw clay can significantly impact the properties and performance of the resulting membrane, including its selectivity, permeability, and stability. Typically, the purification of natural clay involves several steps like pretreatment, decomposition of carbonates, dissolution of hydr(oxides), oxidation of organic matter, separation of silica, etc. However, the specific steps can vary depending on the targeted properties required for a particular application. Therefore, the selection of purification steps should be based on the desired properties of the clay. The scope of application of purified clay in developing the UF top layer has been demonstrated later by coating over a pozzolan support [9].

3.1 Interest of Purification

The purification of natural clay offers numerous advantages, particularly in the identification of clay minerals. Purification helps to eliminate non-clay materials such as carbonate, (hydr)oxides, organic matter, and amorphous silica, which can interfere with the identification process [10]. Additionally, purification can enhance the physical properties of the clay, such as colloidal properties, as observed in purified bentonite clay that was used as a dispensing agent in water-based drilling fluids [11]. Furthermore, scientific studies have demonstrated that the purification process significantly impacts the rheological properties of the clay paste, making it more stable and controllable. This is due to the removal of impurities that can destabilize the paste characteristics. Moreover, the elimination of carbonate impurities can result in a higher density material, further enhancing its properties with respect to porosity.

3.2 Pretreatment

The natural clay undergoes an initial pretreatment step by dispersing it in NaCl solution (2 mol L^{-1}) with continuous stirring for 4 h. This process helps to eliminate exchangeable cations from the interlayer space and soluble impurities that may exist in the natural clay. The resulting mixture is then subjected to centrifugation to separate the solid particles from the aqueous phase. The obtained material is washed repeatedly with water and centrifuged again to remove excess salts. Finally, a suspension of the prepared clay is placed in a graduated glass tube, allowing the fine fraction (<2 μm) to remain in the suspended state and settling of coarser particles such as quartz, silts, etc., in accordance with Stokes' law (Eq. 1).

$$t = 190 \times \frac{x}{d^2} \quad (1)$$

where t refers to sedimentation time (min), x to the depth (cm) and d to particle size (μm).

3.3 Decomposition of Carbonate

During the sintering process of clay, the carbonate decomposes at temperatures above 600 °C, leaving a highly porous structure in the final clay product. To avoid such an outcome, carbonate compounds are removed by acid treatment. A suspension of the clay with the fine fraction is prepared and the pH is adjusted to 4.5 using HCl.

3.4 Dissolution of (Hydr)Oxide

In a clay suspension, (hydr)oxides release trivalent cations that cause coagulation and prevent the complete dispersion of the clay minerals. The (hydr)oxides found in natural clays are mainly Fe_2O_3, $Fe(OH)_3$, Al_2O_3, $Al(OH)_3$, and Mn_2O_3, and they are removed through a complexation process. Sodium citrate ($Na_3C_6H_5O_7$) and sodium bicarbonate ($NaHCO_3$) were used to complex the multivalent cations (Fe^{3+}, Al^{3+} and Mn^{3+}), while sodium dithionite ($Na_2S_2O_4$) was added to reduce Fe^{3+} to Fe^{2+}, which forms a more stable citrate complex. Sodium dithionite was continuously added until the suspension changed color from brown to blue. Then, the suspension was stirred for 30 min at 75 °C. The clay was separated from the solution and washed at least twice with NaCl.

3.5 Oxidation of Organic Matter

Organic matter present in soil contains a high proportion of humus, which has a negative influence on the mechanical properties and creates structural defects. Organic matter is removed through an oxidation reaction. Hydrogen peroxide (H_2O_2) is commonly used in this process, but it has been observed that complete oxidation of organic matter cannot be achieved with hydrogen peroxide alone. In some cases, other compounds such as bromine (in aqueous solution) or a mixture of peroxodisulphate ($Na_2S_2O_8$) with sodium bicarbonate ($NaHCO_3$) have been used, and they show higher oxidation efficiency. After adding the oxidizing agent, the solution was heated to temperatures between 75 and 80 °C. Then, the solid portion was separated, washed with NaCl, and dried.

3.6 Dissolution of Silica

Amorphous silica has a high melting point, and when it melts, it forms a highly dense structure. Silica removal can be achieved through dissolution in a boiling solution of 5 wt% sodium carbonate. However, this procedure can also harm clayey minerals, as they are also attacked. An alternative method to treat clay while avoiding damage to clayey minerals involves the use of acid oxalate in a dark environment with a pH of 3.5 for dissolution of amorphous silica while leaving clay minerals intact.

4 Preparation of Low-Cost Ceramic UF Membranes

In order to reduce the cost for development of UF membranes with pore size ranging from 2 to 50 nm, two types of low-cost clay based supports were used in place of using costly commercial supports made of α-alumina. A bentonite based support [13] was selected for preparation of the UF membrane in the higher pore size range with an active separating layer of TiO_2. Another support was fabricated from Moroccan clay in a tubular configuration using extrusion technique and sintered at 1225 °C [15]. Following that, ZrO_2 was coated and sintered at 300 °C to produce a microfiltration (MF) layer that acts as an intermediary layer, preventing small particles from infiltrating the large support pores during the deposition UF layer leading to the formation of smaller pore size range.

4.1 Nano-TiO_2/Bentonite Clay

Utilizing oxide-based materials on a bentonite support for UF membrane deposition has emerged as a promising method for efficient water treatment. It is worth noting that, based on literature, there is a widely accepted consensus that the chemical stability of materials decreases in the order of $TiO_2 > ZrO_2 > Al_2O_3 > SiO_2$ [12]. Therefore, TiO_2 was chosen to prepare the UF active layer. A nano-TiO_2 thin film was deposited on a bentonite-based support [13]. The nanoparticle powder of TiO_2 used is in the anatase phase with an average diameter of 21 nm. 3 wt% of nano-TiO_2 powder were dispersed in water containing polyvinyl alcohol to prepare the coating solution. The TiO_2 film was deposited using the spin-coating method of a colloidal solution followed by calcination at 800 °C for 2 h. The resulting membrane was characterized by SEM as shown in Fig. 3. The top- and cross-sectional views reveal that the membrane layer is homogeneous with good adhesion to the support, and a thickness of 18 μm. Additionally, the mean pore diameter was estimated to be 72 nm. The water permeability of the prepared TiO_2 membrane was calculated using Eqs. (2) and (3). The UF membrane exhibits a water permeability of 33 $L\ h^{-1}\ m^{-2}\ bar^{-1}$.

$$J_w = \frac{V}{A \times t} \tag{2}$$

$$J_w = L_p \times \Delta P \tag{3}$$

where J_w is the permeate flux (L h^{-1} m^{-2}) and ΔP is the transmembrane pressure (bar).

In another study, Bouazizi et al. used nano-TiO_2 anatase to prepare a UF layer. They deposited the nano-TiO_2 layer on a bentonite support using the colloidal solution destabilization (CSD) technique via dip-coating [14]. Subsequently, they added specific amounts of cetytrimethylammonium bromide (CTAB) and PVA as surfactant and peptizing agents, respectively (Fig. 4). The membrane calcinated at 750 °C had a surface area of 64 m^2 g^{-1}. The SEM analysis revealed that the obtained membrane was homogeneous and had no surface cracks. The layer exhibited good adhesion between the bentonite support and the active TiO_2 layer. The pore size distribution analysis using SEM images and N_2 adsorption–desorption isotherm from the desorption branch of the isotherms demonstrated the efficient formation of a UF

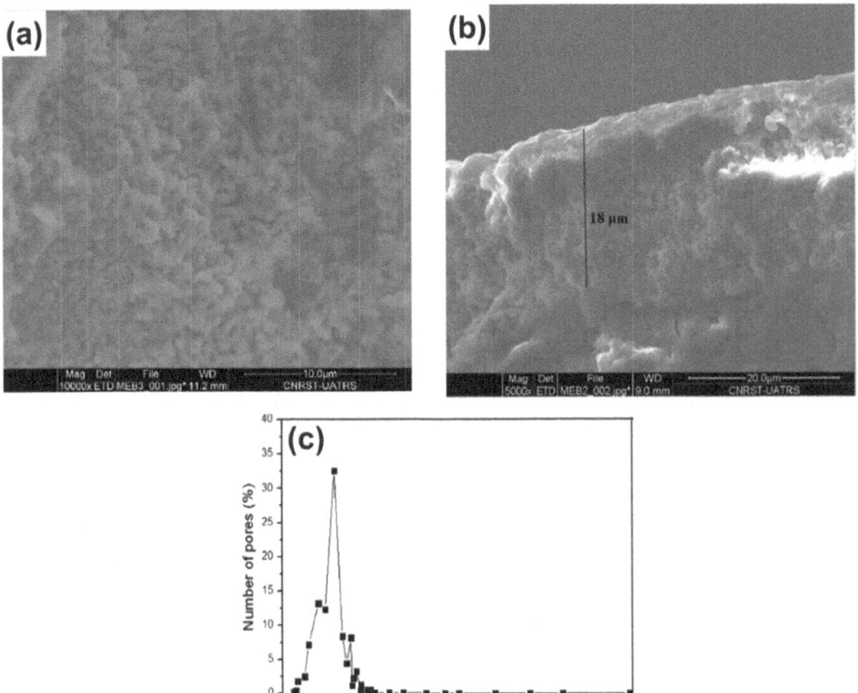

Fig. 3 SEM micrographs of **a** top-surface and **b** cross-section, **c** pore size distribution of the prepared Bentonite/nano-TiO_2 membrane [13]

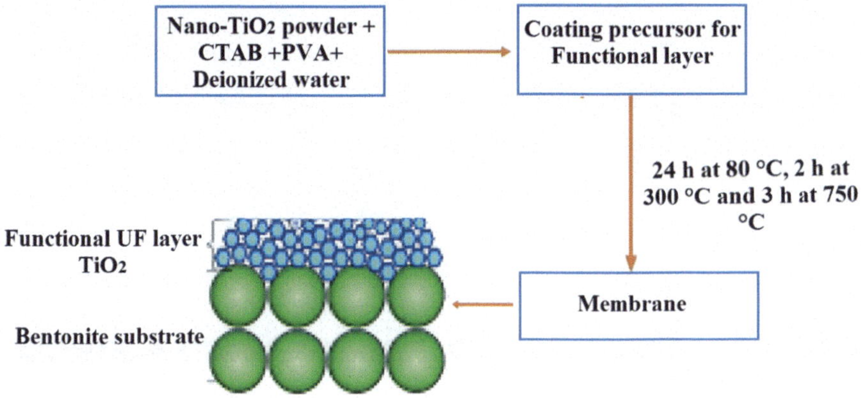

Fig. 4 Procedure for the coating process and schematic representation of the UF membrane showing active layer on the bentonite porous support

selective layer with an average pore diameter of 10 nm. Surface roughness is a key characteristic of ceramic membranes because it influences the fouling phenomenon. The surface roughness of the nano-TiO_2 UF deposited on bentonite support via dip-coating was found to be in the range of 14.30–20 nm, which is lower than that of TiO_2 deposited by sip-coating on bentonite-based support. Furthermore, the water permeability was determined to be 16 L h^{-1} m^{-2} bar^{-1}. Figure 5 shows the SEM micrograph, pore size distribution, and AFM micrographs of the prepared nano-TiO_2/bentonite.

4.2 *TiO_2–$ZnAl_2O_4$/Kaolinite*

TiO_2–$ZnAl_2O_4$ UF membrane was prepared on a low-cost clay support [15]. To achieve the TiO_2–$ZnAl_2O_4$ UF membrane (Fig. 6), $ZnAl_2O_4$ and TiO_2 sols were prepared. Two solutions were required to form the $ZnAl_2O_4$ sol. The first solution was a boehmite suspension prepared by hydrolyzing $Al_2O_3{\cdot}H_2O$ using deionized water and adding nitric acid (HNO_3) solution as a peptizing agent. The second solution was made of zinc nitrate ($Zn(NO_3)_2$) prepared by mixing $Zn(NO_3)_2{\cdot}6H_2O$ with HNO_3 and deionized water. The former solutions were then mixed to form the $ZnAl_2O_4$ sol. On the other hand, the TiO_2 sol was prepared by adding titanium isopropoxide ($Ti(iOPr)_4$) to deionized water, and the solution was stirred for 1 h. Subsequently, nitric acid (1 mol L^{-1}) was added, and the solution was stirred for 24 h at a temperature of 60 °C to obtain the TiO_2 sol. The TiO_2–$ZnAl_2O_4$ sol was then prepared by mixing the obtained sols in a proportion of 50 wt% of each and adding 10 g of hydroxyethyl cellulose as a binder (2 wt%). The obtained TiO_2–$ZnAl_2O_4$ sol was coated on the MF membrane using the slip-casting method with a casting time of 2 h. The membrane was finally dried at room temperature for 24 h, then heated at 250 °C for 2 h to remove

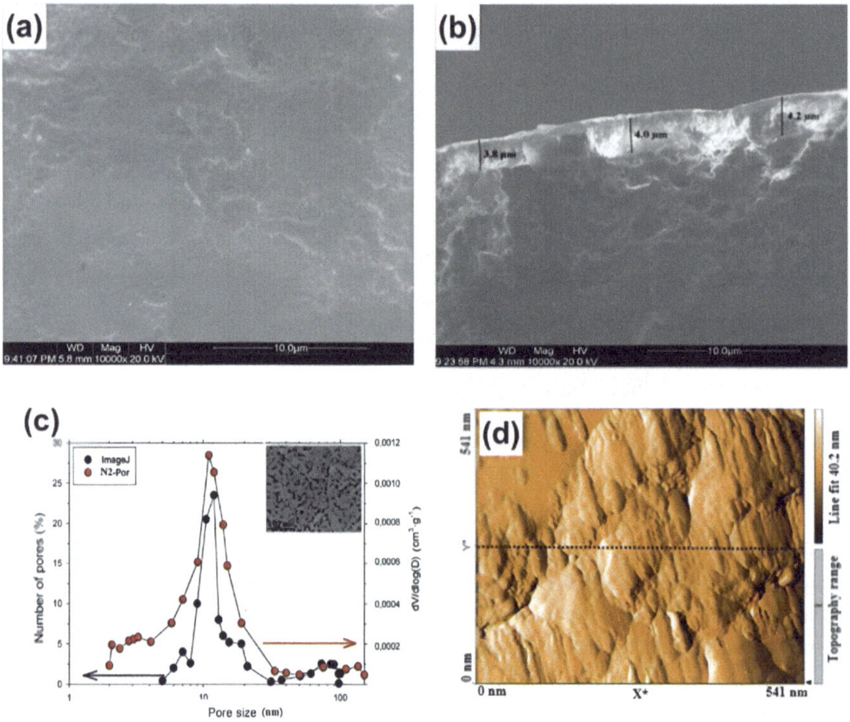

Fig. 5 SEM micrographs of **a** top-surface and **b** cross-section, **c** pore size distribution **d** AFM micrographs of the prepared nano-TiO_2/bentonite [14]

organic compounds and sintered at 400 °C for 2 h. The resulting TiO_2–$ZnAl_2O_4$ UF membrane was analyzed by SEM technique Fig. 7a, b, showing that the deposited layer presented a thickness of 1 μm with a homogeneous and defect-free morphology. The pore size of the layer was measured using the nitrogen adsorption–desorption method, and the result confirms the UF character of the prepared layer with an average pore diameter of 5 nm. The permeability of the membrane was determined to be 9.41 L h^{-1} m^{-2} bar^{-1}. The molecular weight cutoff of the membrane was determined by filtration of polyethylene glycol with a molecular weight ranging from 600 to 5000 Da, showing that the rejection achieved 90% for MW of 4500 Da.

4.3 *Purified Clay/Pozzolan*

Clay minerals are composed of layered silicates that contain a mixture of silicon and oxygen elements. Ouaddari et al. developed a UF ceramic membrane using purified clay and pozzolan [9]. The selective UF layer was made by coating purified clay onto a pozzolan support. The flat support was prepared from natural pozzolan

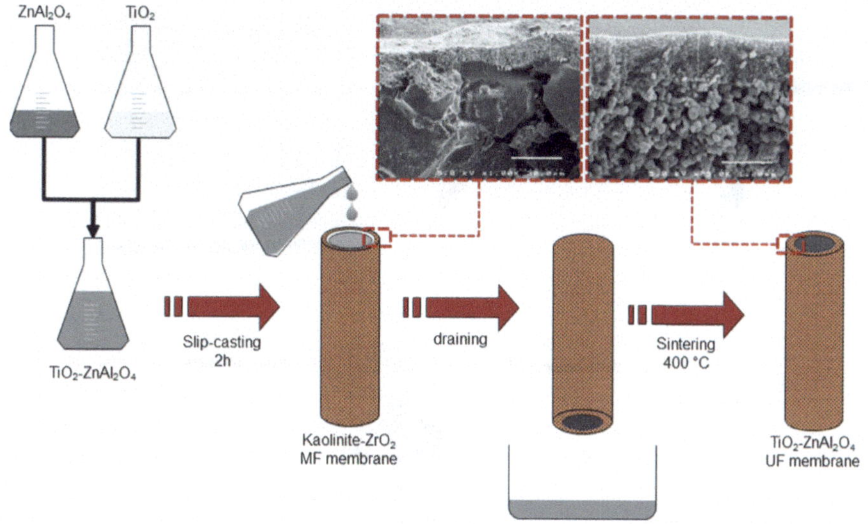

Fig. 6 Schematization of TiO_2–$ZnAl_2O_4$ UF membrane preparation [15]

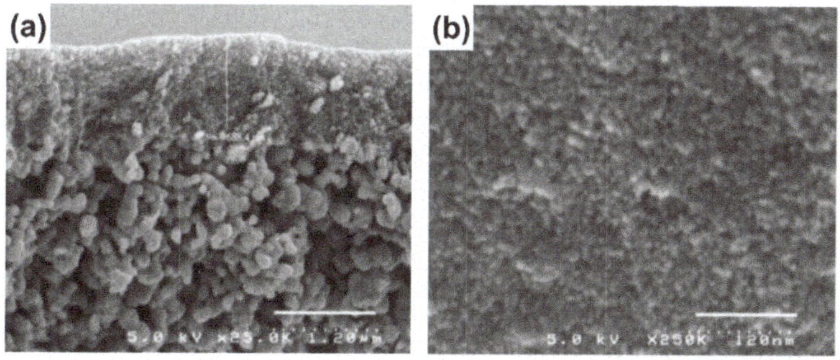

Fig. 7 **a** SEM micrographs of the TiO_2–$ZnAl_2O_4$ UF layer, **b** Pore size distribution of TiO_2/$ZnAl_2O_4$ layer [15]

using a dry route and sintered at 950 °C, resulting in a support with a pore size of approximately 2.8 μm, 32% porosity, mechanical strength of 14.8 MPa, and permeability of 1424 L h^{-1} m^{-2} bar^{-1} [16].

The purified clay was used to prepare the UF top layer (Fig. 8). A suspension of 3 wt% purified clay, 67 wt% water, and 30 wt% PVA (12 wt% in aqueous solution) was prepared as a binder. The suspension was deposited on the MF membrane surface using a spin-coating technique and then dried overnight at room temperature. The membrane was heated at 300 °C for 2 h to remove the PVA and sintered at 800 °C for 2 h to consolidate the clay layer. SEM images of the prepared UF membrane showed a homogeneous surface free from defects and a good distribution of pores (Fig. 9a). The resulting layer had a much denser structure, with clay particles highly

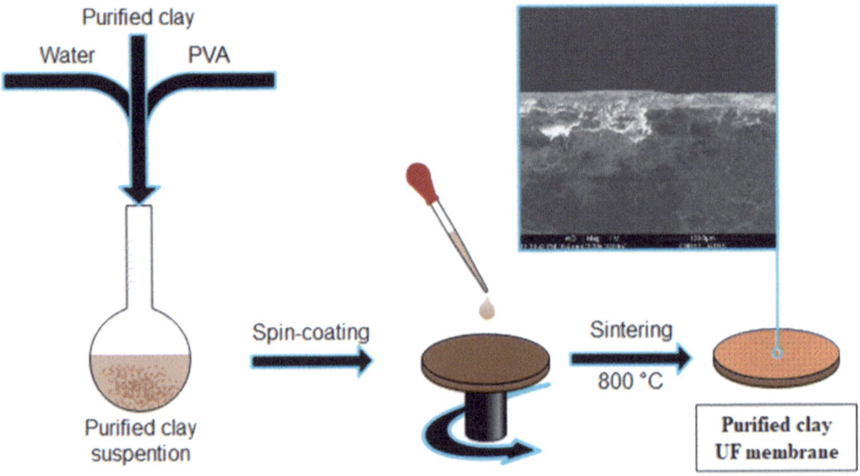

Fig. 8 Preparation of purified clay UF membrane [9]

consolidated with each other and a thickness of 13.2 μm (Fig. 9b). The developed structure resulted from the sintering of the clay layer, rich in clayey minerals such as illite and smectite with a very small amount of carbonate. The average pore size was around 75.0 nm (Fig. 9c), which may be attributed to the prior elimination of pore-forming compounds, mainly carbonate and organic matter, from the clay through the purification process described earlier. The water permeability of the membrane was found to be 14.72 L h^{-1} m^{-2} bar^{-1}, which is reasonable considering the membrane pore size and the top layer thickness.

4.4 *Bentonite/Perlite*

Bentonite clay is a naturally occurring material with the molecular formula $Al_2H_2Na_2O_{13}Si_4$, and it comprises several clay minerals such as montmorillonite, beidellite, and nontronite [17]. Saja et al. have developed a cost-effective UF bentonite film deposited on a ceramic perlite support [18]. Mineralogical analysis of the bentonite revealed that it is mainly composed of montmorillonite as the main clay mineral, along with small quantities of quartz and calcite as impurities. The film of bentonite was prepared using the spin-coating method of a colloidal solution with varying bentonite concentrations ranging from 0.25 to 1.50 wt%, followed by calcination at 500 °C. Figure 10 shows the coating process of the bentonite layer on the perlite substrate. To ensure proper performance (good correlations between permeability and selectivity) of selective UF layers-based on bentonite, microstructural studies were conducted using SEM to highlight the absence of defects such

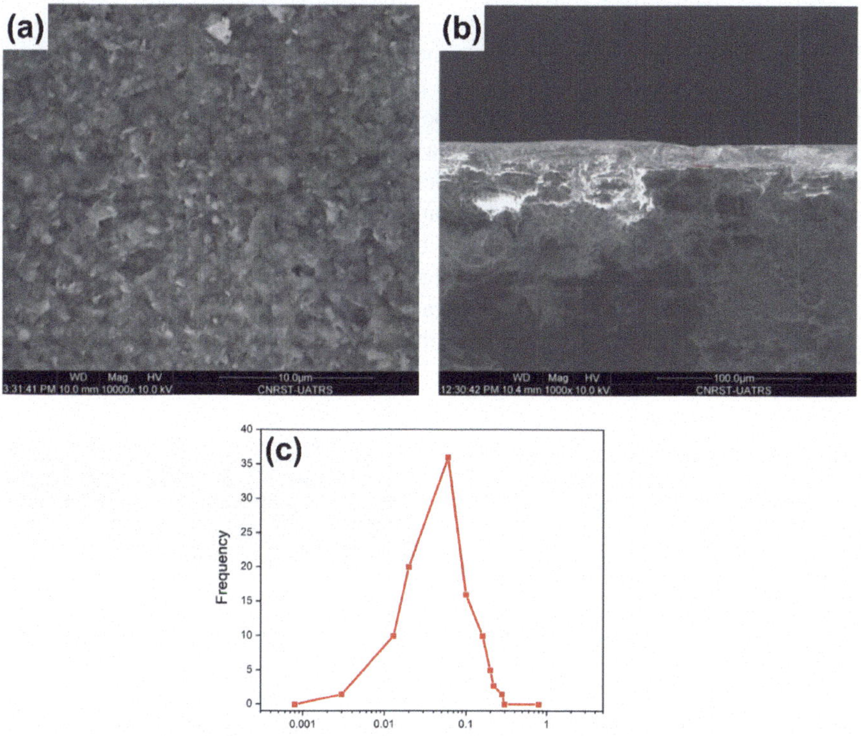

Fig. 9 **a** SEM micrographs of top-surface, **b** cross-section of clay membrane and **c** pore size distribution of prepared membrane [9]

as pinholes, coarse grains, and/or cracks in the sintered membrane layers (coated surface and support).

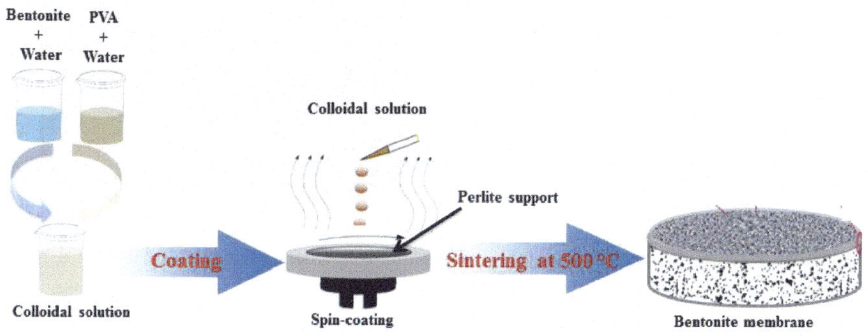

Fig. 10 Procedure for the dip-coating process with schematic representation for preparation of bentonite based UF membrane

The layer prepared by mixing 0.75 wt% of bentonite exhibits an efficient bond on the perlite support, with a uniform layer and no surface cracks (Fig. 11a). Moreover, the thin film has a width of 6 μm (Fig. 11b). The average linear roughness of the bentonite UF layer was determined to be 20 nm using atomic force microscopy (AFM), as shown in Fig. 11d, indicating a smoother surface compared to TiO_2 and other similar membranes. The hydrophilicity of the membrane was determined by measuring the contact angle of the bentonite UF membrane. The membrane has a contact angle of 63°, indicating that it is hydrophilic. The water permeability and hydraulic pore size of the prepared membrane are 30 L h^{-1} m^{-2} bar^{-1} and 13 nm, respectively. Based on the membrane characterization results (e.g., SEM and AFM), it was found that the bentonite-based membrane layer effectively achieves UF range on low-cost perlite support.

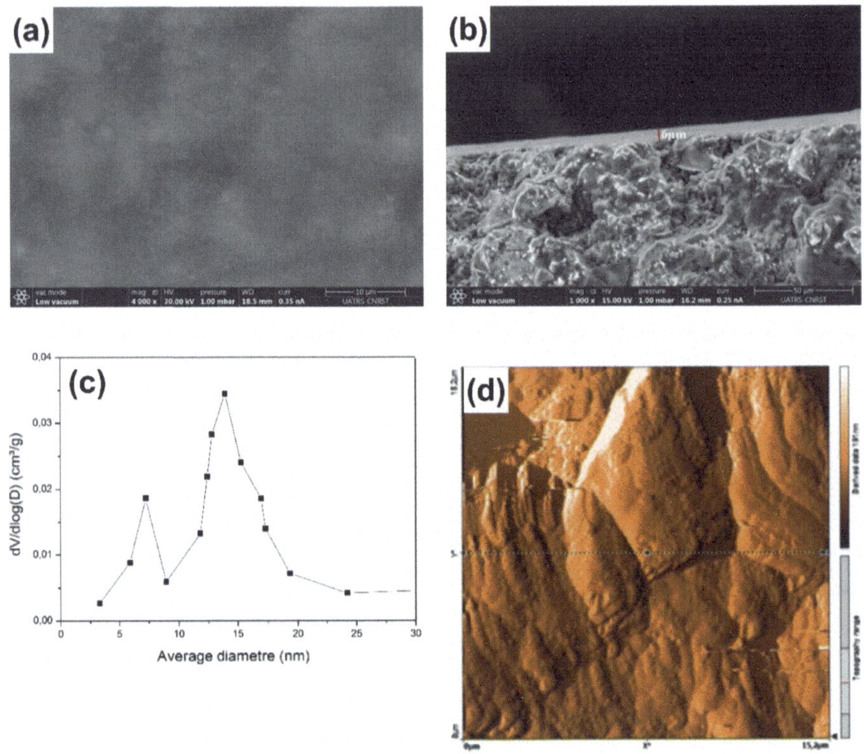

Fig. 11 SEM micrographs of **a** top-surface and **b** cross-section, **c** pore size distribution **d** AFM micrographs of the prepared bentonite/perlite [17]

4.5 Zirconia/Clay

Zirconia is a ceramic material that exhibits high mechanical strength, thermal stability, and chemical inertness, making it suitable for a variety of applications such as membranes. Zirconia membranes offer exceptional features such as high permeability, selectivity, fouling resistance, and stability under harsh operating conditions, which make them suitable for water treatment [19]. In this regard, Elomari et al. developed a UF-based zirconia layer deposited on a low-cost clay support [20]. In the present study, Moroccan clay was used to make a flat membrane support using uniaxial pressure, and its performance was enhanced by adding 10% cornstarch and sintering it at 950 °C [21], resulting in a mechanical strength of 14 MPa and a pore size of 1.95 μm. The support was coated with a zirconia UF layer using the dip-coating method (Fig. 12). The sol–gel method was used to synthesize zirconium oxalate sol ($ZrOC_2O_4$), whereby solutions of oxalic acid ($H_2C_2O_4 \cdot 2H_2O$) and zirconium oxychloride ($ZrOCl_2 \cdot 8H_2O$) were mixed at 60 °C. The aqueous sol was agitated for 3 h before being dried at 100 °C to prepare a stable $ZrOC_2O_4$ gel, which was then ground to obtain $ZrOC_2O_4$ powder. The coating formulation was made by combining 3 wt% $ZrOC_2O_4$ with 67 wt% water and 30 wt% PVA and the support was dipped in the coating suspension for 30 min. The membrane was allowed to dry at room temperature before being heated at 250 °C to remove the PVA and at 600 °C to convert $ZrOC_2O_4$ to ZrO_2 based membrane layer. The resulting zirconia/clay membrane is free of defects and has good morphology. The UF layer is uniformly deposited on the support, and there is no obvious zirconia particle penetration into the support structure (Fig. 13). The membrane has a water permeability of 36 L h^{-1} m^{-2} bar^{-1}, and its pore size was calculated using the Hagen Poiseuille formula, yielding a value of 80 nm.

Fig. 12 Preparation process of zirconia UF membrane [20]

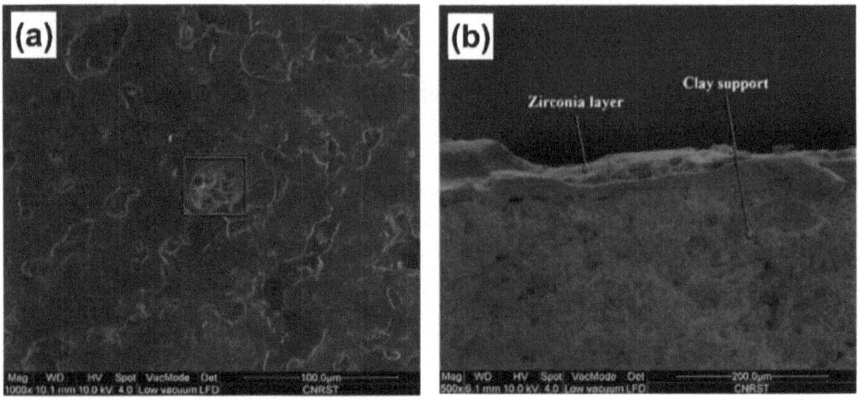

Fig. 13 SEM images of zirconia UF membrane: **a** top-view and **b** cross-section [20]

5 Removal of Dyes

Currently, various industries such as textiles, paper, and plastics use synthetic dyes, which have caused significant environmental pollution. It is estimated that about 7×10^5 tons of dyes are produced annually, and more than 10,000 varieties of dyes are commercially available worldwide [22]. Several important colors found in wastewater fall under the category of durable dyes, the removal which is essential due to their toxicity, mutagenicity, carcinogenicity, and high solubility. Conventional methods of dye removal cannot fully eliminate the majority of these colors, and they persist until extinction due to the aromatic rings present in their chemical structures, resulting in high physicochemical, thermal, and optical stability. Therefore, there is a need to find an effective way to reduce dye concentration in effluents. In this perspective, membrane separation techniques are currently viewed as economically and technologically lucrative options for industrial wastewater treatment, and the textile industry is one of the principal beneficiaries of membrane processes [23].

5.1 Anionic Dyes Removal

The characteristics of several low cost UF membranes developed are tabulated in Table 2 which have been investigated for separation and removal of various synthetic dyes, including Direct Red 80 (DR80), Acid Orange 74 (AO74), Methylene Blue (MB), and Rhodamine B (RB) while Table 3 summarizes the main properties of these dyes.

Table 2 Characteristics of low cost UF membranes developed using clays

Membrane	Materials used		Shape	UF-layer deposition techniques	Parameters				Refs.
	Clay mineral	Other materials			Mechanical strength (MPa)	Pore size (nm)	Thickness (μm)	Permeability (L h^{-1} m^{-2} bar^{-1})	
Bentonite/TiO_2	Bentonite (MF)	Micronized phosphate (MF) – Nano-TiO_2 (UF)	Flat	Spin-coating	14.6	72	18	33	[13]
Bentonite/ nano-TiO_2	Bentonite (MF)	Nano-TiO_2 (UF)	Flat	Dip-coating	22	10	4	16	[14]
TiO_2–$ZnAl_2O_4$/ kaolinite	Kaolinite (support)	ZrO_2 (MF) – TiO_2 (UF) – $ZnAl_2O_4$ (UF)	Tubular	Slip-casting	10	5	1	9.42	[15]
Pozzolan/purified clay	Natural clay (UF)	Pozzolan (MF)	Flat	Spin-coating	14.8	75	13.2	14.72	[9]
Perlite/Bentonite	Bentonite (UF)	Perlite (MF)	Flat	Spin-coating	21.68	30	13	6	[17]

Table 3 Dye characteristics

Dyes	Molecular formula	Chemical structure	Molecular weight (g mol^{-1})	Charge
Direct Red 80	$C_{45}H_{26}N_{10}$ $Na_6O_{21}S_6$	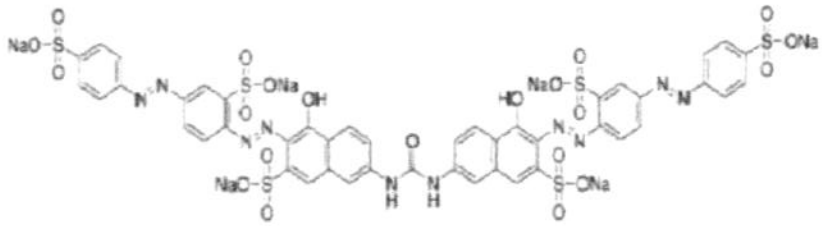	1373.09	Anionic
Rhodamine B	$C_{28}H_{31}ClN_2O_3$	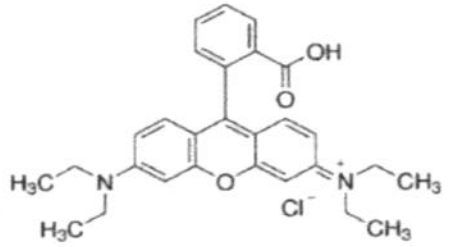	479.01	Cationic
Acid Orange 74	$C_{16}H_{11}Cr$ N_5NaO_8S	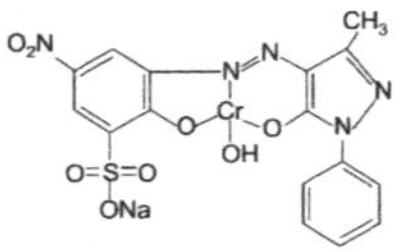	508.34	Anionic
Methylene Blue	$C_{16}H_{18}ClN_3S$	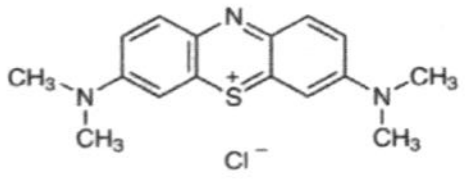	319.85	Cationic

5.1.1 Direct Red 80 (DR80)

DR80 is a polyazo dye commonly used in textile coloring. It is discharged into the environment and is a substantial source of pollution, as it limits sunlight penetration, resists photochemical and biological action, and produces hazardous or even mutagenic and carcinogenic degraded products [24]. Bouazizi et al. conducted an experiment to determine the effect of pH solution values ranging from 3.0 to 9.5 on DR80 retention and flux [14]. The results showed a continuous increase in retention with a decrease in flux as the pH is increased (Fig. 14a). The TiO_2 UF membranes used in these experiments were characterized by a zero point of charge (PZC) of 5.7. Therefore, the membrane surface is positively charged when the pH is less than 5.7 and becomes negatively charged when the pH is greater than 5.7. At low pH, a strong attraction force between the positive charge of the membrane surface and the negative charge of the dye molecules facilitates the dye adsorption as well as the passage of molecules, leading to a low dye rejection. In contrast, the electrostatic repulsion of anionic dyes caused by the negative charge of the membrane at pH greater than 5.7 leads to higher rejection and a flux reduction.

In another study, Ouaddari et al. separated DR80 by means of purified clay-based UF membranes [9]. They developed a purified clay based UF layer deposited on a

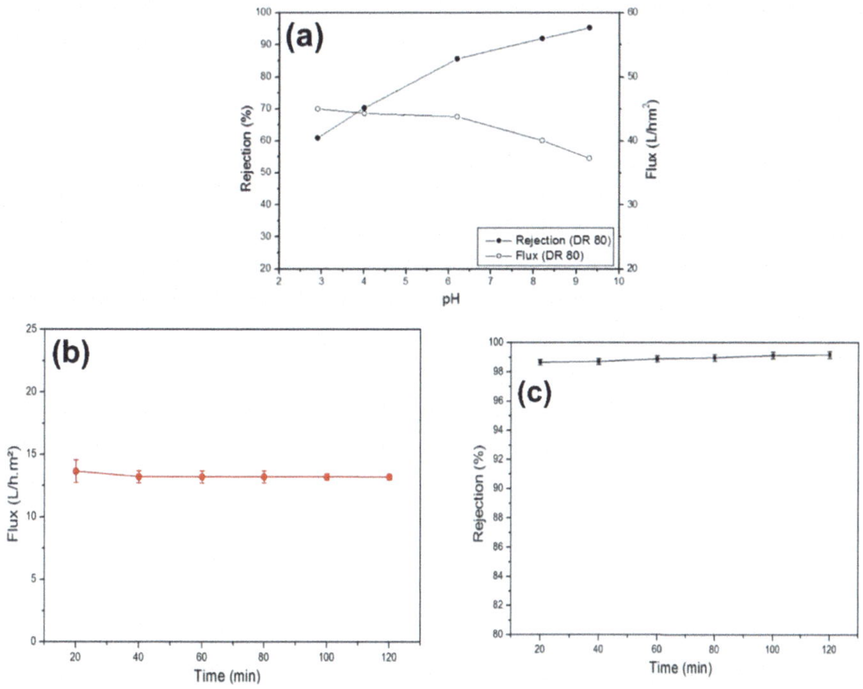

Fig. 14 **a** Permeate flux and rejection of DR80 as function of pH [14], **b** permeate flux of DR80 as function of time, **c** rejection of DR80 as function of time [9]

low-cost pozzolan support and assessed its efficiency using a 50 ppm solution of DR80 in a cross-flow mode at 3 bar. The effect of time on flux and rejection was studied. As shown in Fig. 14b, c, the flux decreased slightly from approximately 14.0 to 13.5 L h^{-1} m^{-2} after 40 min of filtration and stabilized afterwards. This decrease could be explained by either the adsorption of dye molecules within the membrane pores or accumulation of dye molecules on the membrane surface. On the other hand, the rejection of dye was quite stable during 2 h of filtration with a rejection rate of about 99%. The high rejection rate is in agreement with the steric transport of dye molecules through the membrane, as DR80 molecules are much larger than the diameter of the membrane pores.

5.1.2 Acid Orange 74 (AO74)

AO74 is a widely used azo dye in the color development of wool, wood, and textiles [25]. It is known to be harmful and can damage both human health and the environment. Elomari et al. prepared UF zirconia/clay membrane and tested its performance in the filtration of AO74 [20]. At a given pressure of 3 bar, the effect of concentration ranging from 50 to 150 ppm on flux and rejection was investigated. According to the results shown in Fig. 15a, increasing the concentration in AO74 causes a flux reduction which is primarily explained by the stacking of rejected dye molecules on the zirconia membrane, which is accelerated as concentration rises resulting in rapid blockage of the pores. Increase in dye concentration also leads to increase in rejection from around 90 to 98%, which may be attributed to the formation of the polarization layer due to the accumulation of dye molecules near the membrane surface which acts as a secondary filtration layer, causing in better removal. They also explained the high membrane rejection owing to the repulsive interactions as the pH_{PZC} value was found to be about 2 (Fig. 15b), indicating that the membrane has a negative charge at pH values more than 2 while pH of the AO74 feed solution is 7.3. As a result, repulsive interactions between the anionic dye and the negatively charged membrane surface are expected. Likewise, Saffaj et al. studied the effect of applied pressure on AO74 removal at concentration of 50 ppm using $ZnAl_2O_4$–TiO_2/kaolinite [15]. Results in Fig. 15c show that increase in pressure has a positive effect on rejection, increasing the value from 80% to above 93%.

5.2 Cationic Dyes Removal

Rhodamine B (RhB) is a fluorescent dye that produces a pink to red color when excited by light. It is commonly used in the textile, biological, and food industries, but has been identified as a carcinogenic, teratogenic, and mutagenic substance, leading to restrict its usage in several developed countries [26]. Due to its widespread use in the textile industry, RhB is often found in dye effluent. Methylene blue (MB) is a cationic dye commonly used in the textile industry as a synthetic dye [27]. It is

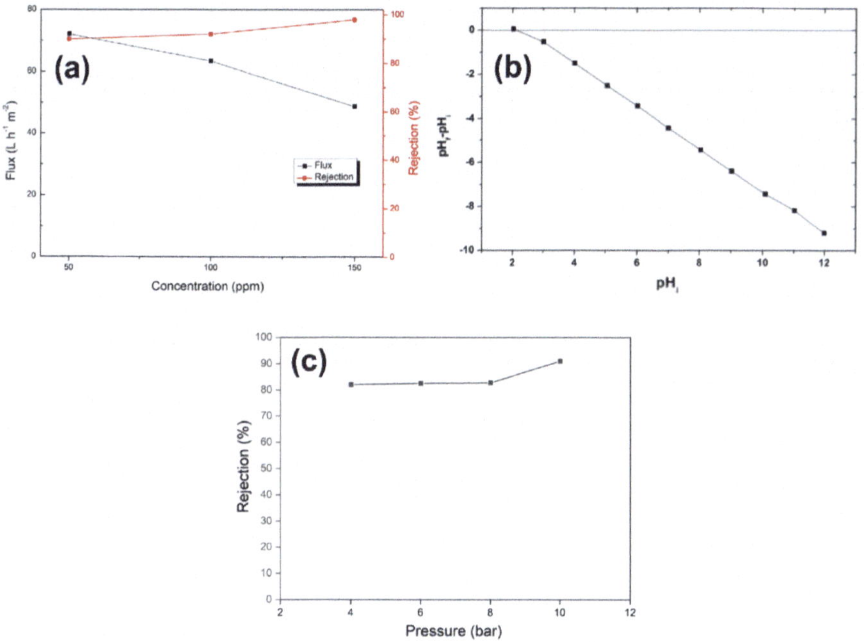

Fig. 15 **a** Variation of flux and rejection of AO74 as a function of concentration, **b** determination of pH_{PZC} of zirconia/clay membrane, **c** variation of AO74 rejection as a function of pressure [15, 20]

a heterocyclic aromatic chemical containing nitrogen and sulfur, which gives it a positive charge when ionized. However, due to its ecotoxicity, various efforts have been made to eliminate it. Removal of two types of cationic dyes commonly used in textile industry was demonstrated using UF membranes deposited over low cost ceramic support. The performance of the membrane separation was influenced by variations in process parameters and dye concentration as elaborated below.

5.2.1 Rhodamine B (RhB)

In a study conducted by Saja et al., an UF membrane composed of a bentonite film deposited on a perlite support was used to separate RhB solutions [17]. The effect of both time and dye concentration on flux and rejection of colorant was studied as shown in Fig. 16. Under a pressure of 4 bar, the permeate demonstrated a rejection of 80.1% (Fig. 16a). These results were attributed to dye accumulation on the membrane surface, followed by absorption on the bentonite layer. Compared to initial permeability, a drop in flux to 12.35% was observed with time as shown in Fig. 16b and Saja et al. concluded that selectivity for RhB was enhanced with time.

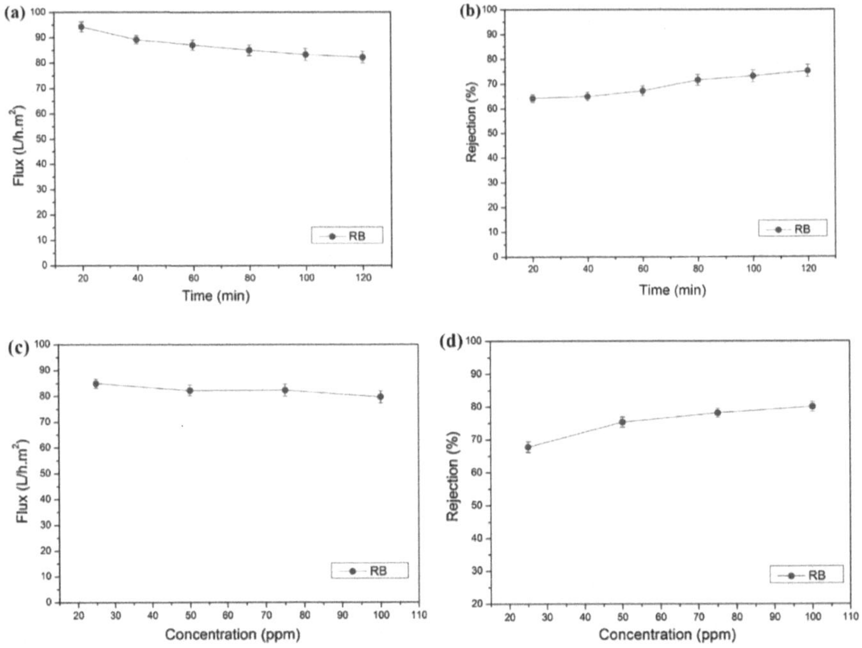

Fig. 16 Separation performances of RhB as a function of **a** time on flux, **b** time on rejection, **c** concentration on flux and **d** concentration on rejection [17]

The effect of the initial concentration of dyes (ranging from 25 to 100 ppm) on the flux and color rejection was also studied. The authors reported that with an increase in feed dye concentration resulted in an increase in rejection and a decrease in flux Fig. 16c, d. The high rejection values were attributed to the increase in osmotic pressure, while the reduction in flux could be explained by the dye absorption on the membrane pores. A decrease in flux was observed over filtration time due to the formation of a polarization layer.

5.2.2 Methylene Blue (MB)

Saffaj et al. prepared a UF $ZnAl_2O_4$–TiO_2/kaolinite membrane deposited on a kaolin-based support with a tubular configuration and tested its performance in the separation of MB [15]. The pressure effect was tested by filtering a solution of 50 ppm of MB at pH 5.8 (Fig. 17a). The results showed that as the pressure increased from 4 to 10 bar, the flux increased from approximately 20.0 to 42.5 L h^{-1} m^{-2}. The rejection rate was also found to increase from 35.0 to 75.0% after 2 h of operation. The increase in pressure has a positive impact on separation performance, which can be explained by the adsorption of dye molecules on the membrane surface at higher pressure. The effect of pH value was studied by filtering a solution of 50 ppm of MB under

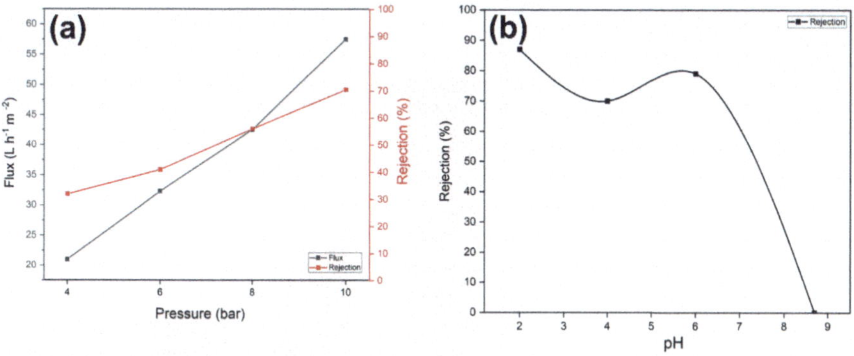

Fig. 17 Variation of flux and rejection of methylene blue **a** as a function of pressure, C = 50 ppm, pH 5.8; **b** as a function of pH at AP = 10 bar, C = 50 ppm [15]

a pressure of 10 bar (Fig. 17b). The rejection dropped from 86.0 to 0.0% as pH value increased from 2.0 to 9.0. When the pH value is lower than the isoelectric point, surface gets a similar charge as dye molecules which creates a repulsive force between the dye molecules and the membrane surface, thereby enhancing rejection. Increase in pH above the isoelectric point, the surface gets an opposite charge and rejection decreases.

5.3 Antifouling Study

UF is an effective process for removing organic compounds with high flux and rejection rates. However, organic fouling can be a persistent issue during membrane separation process, resulting in a decline in membrane performance, particularly in terms of flux. This fouling phenomenon occurs when organic molecules interact with the membrane, either through covalent or non-covalent bonds, causing the membrane pores to become blocked [28]. Ouaddari et al. used the following approach to determine fouling parameters in order to examine the membrane's fouling resistance [9]. Firstly, the water flux (J_{w0}) and the flux of the solution (J_{wp}) (DR 80, 50 ppm in this case) were measured using a virgin membrane. Afterward, the membrane is cleaned with distilled water for 20 min and the water flux is measured again (J_{w1}). Using the flux values obtained, fouling parameters are calculated. Flux recovery ratio (FRR) using Eq. (4) measures irreversible fouling, while total flux decline ratio (TFR) following Eq. (5) gives an indication of the reversible fouling. The reversible flux decline ratio (FRF) based on Eq. (6) helps to determine the fouling caused by the formation of the polarization layer, while irreversible flux decline ratio (IFR) as calculated from Eq. (7) provides an assessment of fouling caused by blocking of membrane pores.

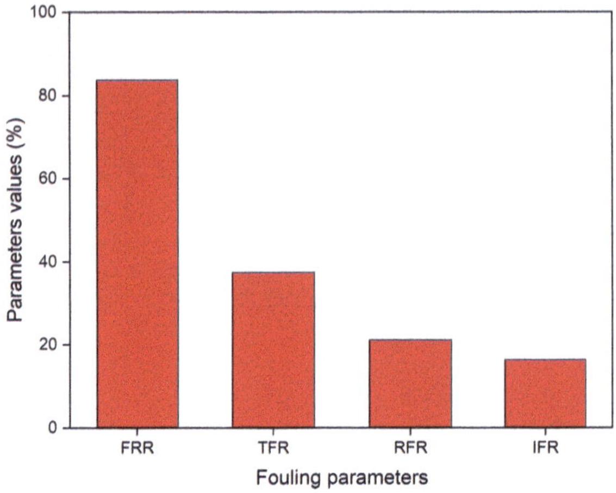

Fig. 18 Fouling parameters of purified clay-based UF membrane [9]

$$FRR = \frac{J_{w1}}{J_{w0}} \times 100 \tag{4}$$

$$TFR = \left(1 - \frac{J_p}{J_{w0}}\right) \times 100 \tag{5}$$

$$RFR = \left(\frac{J_{w1} - J_p}{J_{w0}}\right) \times 100 \tag{6}$$

$$IFR = \left(\frac{J_{w0} - J_{w1}}{J_{w0}}\right) \times 100 \tag{7}$$

The flux results for the membrane used are 14.7, 12.3 and 9.2 L m^{-2} h^{-1} bar^{-1} for J_{w0}, J_{w1} and J_p respectively resulting in fouling parameters of 83.67% for FRR, 37.41% for TFR, 21.08% for RFR and 16.32% for IFR Fig. 18. The fouling parameters illustrate that the UF membrane has strong resistance to fouling, which might be explained by the hydrophilic nature of the clay, since Essate et al. found that as the membrane get more hydrophobic, it become less resistant to fouling [29] and that IFR is mostly produced by dye molecule adsorption on the surface or clinging of these molecules inside the membrane.

6 Future Pathway Towards the Development of Clay-Based Membranes

Ceramic membranes have several advantages over polymer membranes, which make them more suitable for wastewater treatment [30]. However, a commercial ceramic membrane may cost from \$500 to \$3000 m^{-2}, while a polymer membrane costs only

\$20 to \$200 m^{-2}. Due to the significant difference in their prices, use polymeric membranes is preferred for wastewater treatment despite their inferior properties.

Several factors affect the production cost of ceramic membranes, such as the choice of raw materials, sintering temperature, and preparation process. To reduce costs, researchers have replaced expensive raw materials with natural and abundant low-cost materials, such as clays and volcanic rocks, resulting in a lower final price [31]. Typically, low-cost ceramic membranes cost less than \$200 m^{-2}, which is almost comparable with that of polymer membranes.

Low-cost membranes have shown high separation performance in many applications, such as textile industries, heavy metal removal, and dairy wastewater treatment. Economically, significant progress has been made in developing membranes with almost the same properties as commercial ceramic membranes but at a lower cost. However, the microstructure of these membranes is difficult to control due to the use of natural raw materials containing variety of impurities with different structures and behaviors. Thus, optimizing membrane parameters require suitable processing and conditioning of raw materials. Although low-cost ceramic membranes have achieved high filtration performances in many applications, their use is still limited to the laboratory scale. They are mostly used as membrane supports or for MF filtrations, and in rare cases, they have been used to prepare the top layer of UF membranes.

Researchers may explore the use of different types of clay, such as layered double hydroxides, to develop membranes with unique properties. Another important area of research is the development of functionalization techniques to modify the surface of the clay-based membrane to improve its performance for specific applications. The future pathway for the development of clay-based membranes includes a combination of material optimization and functionalization approaches to improve their performance for a wide range of applications. Recent studies have focused on the synthesis of clay-based UF membranes with various types of clay, such as montmorillonite, kaolinite, and sepiolite. These studies have shown that the properties of the resulting membranes, such as pore size, pore distribution, and surface charge, can be tuned by adjusting the type and amount of clay used, as well as the intercalation conditions.

One important area of research is the modification of clay-based UF membranes with functional groups, such as carboxyl, amine, or sulfonic acid, to enhance their adsorption and separation performance for specific applications. Another area of research is the use of clay-based UF membranes in combination with other separation techniques, such as adsorption or oxidation, to achieve more efficient dye removal. Overall, the development of clay-based UF membranes has the potential to provide a sustainable and cost-effective solution for the treatment of dye-containing wastewater. The field continues to advance with the development of newer methodologies and materials, as well as the integration of advanced characterization techniques to understand the properties and performances of these membranes.

7 Conclusion

Clays are promising raw materials for preparing ceramic membranes due to their abundance and unique mineralogical and chemical properties that can enhance membrane performances. Furthermore, clays require lower sintering temperatures compared to industrial oxides, resulting in energy savings during the manufacturing process. The chapter provided an overview of the preparation of low-cost ceramic UF membranes made from Moroccan clay, including the formulation of selective layers, coating techniques and operation conditions. Investigation on the separation of soluble dyes was demonstrated using clay-based UF membranes while considering the role of operating parameters and separation mechanisms. Finally, a future pathway for developing clay-based UF membranes in large-scale was suggested for wastewater treatment application to meet the environmental regulations and reduce the ecological impact.

References

1. Shah V, Wang B, Li K (2021) High-performance PVDF membranes prepared by the combined crystallisation and diffusion (CCD) method using a dual-casting technique: a breakthrough for water treatment applications. Energy Environ Sci 14:5491–5500. https://doi.org/10.1039/D1EE02009A
2. Asif MB, Zhang Z (2021) Ceramic membrane technology for water and wastewater treatment: a critical review of performance, full-scale applications, membrane fouling and prospects. Chem Eng J 418:129481. https://doi.org/10.1016/j.cej.2021.129481
3. Liang D, Huang J, Zhang H, Fu H, Zhang Y, Chen H (2021) Influencing factors on the performance of tubular ceramic membrane supports prepared by extrusion. Ceram Int 47:10464–10477. https://doi.org/10.1016/j.ceramint.2020.12.235
4. Malekshahi M, Sabbaghi S, Rasouli K (2022) Preparation of α-alumina/γ-alumina/γ-alumina-titania ceramic composite membrane for chloride ion removal. Mater Chem Phys 287:126218. https://doi.org/10.1016/j.matchemphys.2022.126218
5. Machtani Idrissi DE, Elidrissi ZC, Achiou B, Ouammou M, Younssi SA (2023) Fabrication of low-cost kaolinite/perlite membrane for microfiltration of dairy and textile wastewaters. J Environ Chem Eng 109281. https://doi.org/10.1016/j.jece.2023.109281
6. Addich M, El Baraka N, Laknifli A, Saffaj N, Fatni A, El Hammadi A, Alrashdi AA, Lgaz H (2022) New low-cost tubular ceramic microfiltration membrane based on natural sand for tangential urban wastewater treatment. J Saudi Chem Soc 26:101512. https://doi.org/10.1016/j.jscs.2022.101512
7. Grba N, Baldermann A, Dietzel M (2023) Novel green technology for wastewater treatment: geo-material/geopolymer applications for heavy metal removal from aquatic media. Int J Sediment Res 38:33–48. https://doi.org/10.1016/j.ijsrc.2022.08.002
8. Achiou B, Omari HE, Bennazha J, Albizane A, Daoudi L, Saadi L, Ouammou M, Younssi SA, Maadi AE, Chehbouni M (2016) Physicochemical and mineralogical characterizations of clays from Fez region (basin of Saiss, Morocco) in the perspective of industrial use. J Mater Environ Sci 7:1474–1484
9. Ouaddari H, Karim A, Achiou B, Saja S, Aaddane A, Bennazha J, El Amrani El I, Hassani M, Ouammou AA (2019) New low-cost ultrafiltration membrane made from purified natural clays for direct Red 80 dye removal. J Environ Chem Eng 7:103268. https://doi.org/10.1016/j.jece.2019.103268

10. Bergaya F, Lagaly G (2013) Purification of natural clays. In: Developments in clay science. Elsevier, pp 213–221. https://doi.org/10.1016/B978-0-08-098258-8.00008-0
11. Pereira IDS, Silva VC, Duarte Neto JF, Neves GA, Ferreira HC, Menezes RR (2018) Influence of the purification of bentonite clay from new deposits in the state of Paraíba-Brazil for use in water-based drilling fluids. Cerâmica 64:538–546. https://doi.org/10.1590/0366-69132018643722402
12. Lee M, Wu Z, Li K (2015) Advances in ceramic membranes for water treatment. In: Advances in membrane technology for water treatment. Elsevier, pp 43–82. https://doi.org/10.1016/B978-1-78242-121-4.00002-2
13. Bouazizi A, Breida M, Karim A, Achiou B, Ouammou M, Calvo JI, Aaddane A, Khiat K, Younssi SA (2017) Development of a new TiO_2 ultrafiltration membrane on flat ceramic support made from natural bentonite and micronized phosphate and applied for dye removal. Ceram Int 43:1479–1487. https://doi.org/10.1016/j.ceramint.2016.10.118
14. Bouazizi A, Breida M, Achiou B, Ouammou M, Calvo JI, Aaddane A, Younssi SA (2017) Removal of dyes by a new nano-TiO_2 ultrafiltration membrane deposited on low-cost support prepared from natural Moroccan bentonite. Appl Clay Sci 149:127–135. https://doi.org/10.1016/j.clay.2017.08.019
15. Saffaj N, Persin M, Younsi SA, Albizane A, Cretin M, Larbot A (2006) Elaboration and characterization of microfiltration and ultrafiltration membranes deposited on raw support prepared from natural Moroccan clay: application to filtration of solution containing dyes and salts. Appl Clay Sci 31:110–119. https://doi.org/10.1016/j.clay.2005.07.002
16. Achiou B, Elomari H, Ouammou M, Albizane A, Bennazha J, Aaddane A, Younssi SA (2018) Study of added starch on characteristics of flat ceramic microfiltration membrane made from natural Moroccan pozzolan. J Mater Environ Sci 9:1013–1021
17. Saja S, Bouazizi A, Achiou B, Ouaddari H, Karim A, Ouammou M, Aaddane A, Bennazha J, Younssi SA (2020) Fabrication of low-cost ceramic ultrafiltration membrane made from bentonite clay and its application for soluble dyes removal. J Eur Ceram Soc 40:2453–2462. https://doi.org/10.1016/j.jeurceramsoc.2020.01.057
18. Saja S, Bouazizi A, Achiou B, Ouammou M, Albizane A, Bennazha J, Younssi SA (2018) Elaboration and characterization of low-cost ceramic membrane made from natural Moroccan perlite for treatment of industrial wastewater. J Environ Chem Eng 6:451–458. https://doi.org/10.1016/j.jece.2017.12.004
19. He Z, Lyu Z, Gu Q, Zhang L, Wang J (2019) Ceramic-based membranes for water and wastewater treatment. Colloids Surf Physicochem Eng Asp 578:123513. https://doi.org/10.1016/j.colsurfa.2019.05.074
20. Elomari H, Achiou B, Beqqour D, Khaless K, Beniazza R, Ouammou M, Aaddane A, Younssi SA, Benhida R (2022) Preparation and characterization of low-cost zirconia/clay membrane for removal of acid orange 74 dye. Mater Today Proc 51:1948–1956. https://doi.org/10.1016/j.matpr.2021.03.674
21. Elomari H, Achiou B, Karim A, Ouammou M, Albizane A, Bennazha J, Younssi SA, Elamrani I (2017) Influence of starch content on the properties of low cost microfiltration membranes. J Asian Ceram Soc 5:313–319. https://doi.org/10.1016/j.jascer.2017.06.004
22. Ibrahim GPS, Isloor AM, Lakshmi B (2020) Chapter 3—Synthetic polymer-based membranes for dye and pigment removal. In: Ismail AF, Salleh WNW, Yusof N (eds) Synthetic polymeric membranes for advanced water treatment, gas separation, and energy sustainability. Elsevier, pp 39–52. https://doi.org/10.1016/B978-0-12-818485-1.00003-4
23. Behera M, Nayak J, Banerjee S, Chakrabortty S, Tripathy SK (2021) A review on the treatment of textile industry waste effluents towards the development of efficient mitigation strategy: an integrated system design approach. J Environ Chem Eng 9:105277. https://doi.org/10.1016/j.jece.2021.105277
24. Ben Jeddou K, Bouaziz F, Ben Taheur F, Nouri-Ellouz O, Ellouz-Ghorbel R, Ellouz-Chaabouni S (2021) Adsorptive removal of direct red 80 and methylene blue from aqueous solution by potato peels: a comparison of anionic and cationic dyes. Water Sci Technol 83:1384–1398. https://doi.org/10.2166/wst.2021.075

25. Li A, Weng J, Yan X, Li H, Shi H, Wu X (2021) Electrochemical oxidation of acid orange 74 using Ru, IrO2, PbO2, and boron doped diamond anodes: direct and indirect oxidation. J Electroanal Chem 898:115622. https://doi.org/10.1016/j.jelechem.2021.115622
26. Liang L, Cheng L, Zhang Y, Wang Q, Wu Q, Xue Y, Meng X (2020) Efficiency and mechanisms of rhodamine B degradation in Fenton-like systems based on zero-valent iron. RSC Adv 10:28509–28515. https://doi.org/10.1039/D0RA03125A
27. Iwuozor KO, Ighalo JO, Ogunfowora LA, Adeniyi AG, Igwegbe CA (2021) An empirical literature analysis of adsorbent performance for methylene blue uptake from aqueous media. J Environ Chem Eng 9:105658. https://doi.org/10.1016/j.jece.2021.105658
28. Xu H, Xiao K, Wang X, Liang S, Wei C, Wen X, Huang X (2020) Outlining the roles of membrane-foulant and foulant-foulant interactions in organic fouling during microfiltration and ultrafiltration: a mini-review. Front Chem 8. https://www.frontiersin.org/articles/10.3389/fchem.2020.00417. Accessed 6 Mar 2023
29. Essate A, Achiou B, Benkhaya S, Chakraborty S, Ouammou M, Younssi SA (2022) Low-cost polysulfone/polystyrene ultrafiltration membrane with efficient azoic dyes removal and excellent antifouling performance for colored wastewater. Polym Adv Technol pat.5969. https://doi.org/10.1002/pat.5969
30. Belgada A, Charik FZ, Achiou B, Kambuyi TN, Younssi SA, Beniazza R, Dani A, Benhida R, Ouammou M (2021) Optimization of phosphate/kaolinite microfiltration membrane using Box–Behnken design for treatment of industrial wastewater. J Environ Chem Eng 9:104972. https://doi.org/10.1016/j.jece.2020.104972
31. Charik FZ, Achiou B, Belgada A, Elidrissi ZC, Ouammou M, Rabiller-Baudry M, Younssi SA (2022) Optimal preparation of low-cost and high-permeation NaA zeolite membrane for effective ethanol dehydration. Micropor Mesopor Mater 112229. https://doi.org/10.1016/j.micromeso.2022.112229

Chapter 4
Reuse and Renovation of Tannery Effluent Using Clay-Alumina Membrane

Mithu Naskar, Priyankari Bhattacharya, Sourja Ghosh, and Sibdas Bandyopadhyay

1 Introduction

Leather industry comprises one of the most important manufacturing sector across the world associated with the production of various types of the raw, processed and finished articles from animal hides. In the processing of pre- and post-leather tanning, large volume of fresh water is consumed at various steps starting from raw hide to finished leather products, along with the association of several chemical agents such as acids, alkalis, salts, tanning agents, chromium salts, surfactants, metallorganic dyes, solvents, sulphonated oils and fats, etc. resulting in production of large volume of highly contaminated and toxic effluent. About 30 L of wastewater is produced per kg of hide or skin processing with generation of about 50,000 m^3/day of effluent in India [12]. The processing of raw hide/skin to finished leather undergoes through two major steps, beam house operations and tanning processes. In the beam house operations trimming and desalting of skins/hides, soaking followed by liming is done which in turn produce considerable volume of wastewater rich in salts, limes, initial dirt, blood and impurities associated with the skin, soluble proteins, detergents, biocides, sodium sulphide, ammonia, various degreasing agents and enzymes, etc. [16]. In the various steps of the tanning processes, namely pickling, chrome tanning/vegetable

M. Naskar · S. Ghosh (✉)
Membrane and Separation Technology Division, CSIR-Central Glass and Ceramic Research Institute, Kolkata 700032, India
e-mail: sourja@cgcri.res.in

P. Bhattacharya
Department of Metallurgical and Material Engineering, Jadavpur University, Kolkata 700032, India

S. Bandyopadhyay
Formerly Ceramic Memrae Division, CSIR-Central Glass and Ceramic Research Institute, Kolkata 700032, India

S. Bandyopadhyay (ed.), *Application of Low Cost Ceramic Membranes in Wastewater Treatment*, Materials Horizons: From Nature to Nanomaterials,
https://doi.org/10.1007/978-981-97-5842-5_4

tanning, dyeing, fat liquoring, and finishing, etc. large volume of complex wastewater is evolved associated with various types of chemicals such as, sulphuric acid, common salt, chromium sulphate, sodium formate, sodium bicarbonate, dyes, formic acid, synthetic tannins, etc. Hence treatment of such toxic and hazardous industrial effluent is a critical issue, not only from the pollution impact point of view, but also from the recycling opportunity of the large volume of wastewater to minimize the industrial water consumption and protect the precious fresh water reserves. Accordingly, there have been continuous research efforts to establish advanced and cleaner technologies for effective treatment of tannery wastewater for recycling purpose.

1.1 Integrated Approaches for Tannery Wastewater Treatment

While individual processes such as, biological treatments, coagulation-flocculation, advanced oxidation processes, electrochemical treatment and membrane driven separation processes have been employed with varying degree of efficiency depending on the nature of the effluent, often a combination of such processes have been found effective, as well as, sustainable.

Aerobic and anaerobic mode of biological treatment processes often show efficiency for high organic loading tannery wastewater subject to a prior physico-chemical pretreatment, however the presence of sulphides, chromium and other recalcitrant compounds are the major inhibiting factors in such processes [13]. Anaerobic-Sequence Batch Reactor combined with hydrolysis and constructed wetland was employed as an innovative treatment process to explore the potential of reusing the tannery effluent in irrigation purpose [1]. About 82–99% removal efficiency was obtained. Concentration of various toxic metal ions such as Cr, Pb, Cd and Co accumulated in the vegetable samples were measured which were observed lower the FAO/WHO guidelines. The study revealed no considerable health risk with respect to the chromium concentration observed in the obtained vegetable samples by estimating the target hazard quotients.

The efficiency of the bioelectrochemical cell was studied for tannery wastewater treatment as a clean and environmentally sustainable process with renewable energy production [13]. Compared to the toxic raw tannery wastewater, incorporation of an electrochemical pretreatment showed better microbial activity and thus, an effective treatment involving COD (90%), BOD_5 (84%) and sulfate (96%) removal along with complete removal of chromium. Additionally, the process showed potential for bioelectricity generation. Combination of electrocoagulation and fungal induced biological treatment was found effective for COD (96%) and Cr^{+6} (98%) removal in tannery wastewater [11]. Composite raw tannery effluent was treated by ozonation combined with phycoremediation [28]. The process enabled removal of COD (84%), inorganic carbon (90%), PO_4-P (100%), NH_4^+-N (82%) and chromium (97%) along with value-added microalgal biomass (*Nannochloropsis oculata*) production.

1.2 Membrane Based Separation Processes for Tannery Wastewater Treatment

Due to the heterogeneous nature of tannery wastewater with accumulation of effluent from various subprocesses having large portion of organic/biodegradable materials, inorganic pollutants (salts, acids, alkalis), as well as, toxic and recalcitrant compounds (dyes, heavy metals), etc. no single process can provide adequate treatment. Although combination of various physico-chemical separation processes has been used with biological treatment, however such processes are often associated with high chemical consumption, energy requirement, large volume of sludge and toxic byproducts production and further, inefficiency to reach the discharge norms required for recycling of the treated wastewater. Membrane technology can play a key role in this context. Membrane based separation processes carry some unique features of no phase change involvement, elimination of external chemicals and thermal energy input, low energy requirement, reduced environmental impact, ease in collection and reuse of sludges and high degree of separation making them ideal candidates for recovery and recycling of complex natured tannery wastewater [10]. Accordingly, pressure driven membrane processes with different pore sizes and molecular cut-offs are increasingly used in single/multi-stages with or without combination of other physico-chemical/biological treatment processes and also in association with the membrane bioreactor (MBR), electrodialysis and membrane distillation processes, etc.

1.3 Application of Polymeric Membranes

The feasibility of membrane technology was assessed as a substitute of secondary treatment for a pre-treated tannery wastewater [30]. Nanofiltration and reverse osmosis membrane modules of spiral wound configuration [Osmonics] were used in series at pilot scale in feed and bleed configuration. The optimum condition for the NF membrane module was observed as 6 bar operating pressure with permeate flux of 4.4 L h^{-1} m^{-2}, RO module was operated with 22 bar pressure with no significant fouling issues. Technoeconomic analysis revealed that membrane technology can potentially substitute other conventional secondary treatment processes with about 21% saving in the total cost. Significantly, the process allows discharge of 90% of the original wastewater stream in surface waters along with reuse of 5% wastewater as chromium-rich concentrate without any cost.

Ultrafiltration membrane with high hydrophilicity was prepared with cellulose acetate, nanochitosan and polyethyelene glycol having 83% porosity and clean water flux of >25 l/m^2h [32]. The membrane could effectively reduce COD, BOD, EC, TDS, TS and salinity. In addition, upto 95% removal of chromium was achieved using an operating pressure of 100 kPa and 0.2 mm of membrane thickness.

Reduction of the salinity in tannery wastewater is a major limitation of the conventional treatment processes which inhibit its reuse in industrial, as well as, agricultural and other applications. The RO membrane technology can provide the solution enabling reuse/recycle of the large volume of tannery wastewater. To reduce the biofouling of RO membrane and enhance the overall performance, MBR process may be employed as pretreatment of mixed tannery effluents [29]. The combined process was found highly efficient in organic loading and suspended solids reduction (91–100%) along with salt rejection of 97%, allowing the treated water suitable for agricultural reuses. For recycling purpose, a combination of powdered activated carbon and ultrafiltration was used prior to RO process for a conventionally pretreated tannery wastewater [18]. The integrated process was found effective in reduction of COD, colour, turbidity, silt density index, as well as, dissolved solids with ~75% of RO recovery and the treated water quality was suitable for reuse at various purposes.

To achieve the desired degree of separation particularly in terms of the dissolved solids, integration of RO is essential along with the MF/UF process. A combination of microfilter, polymeric ultrafiltration membranes of different cut-offs (Orelis) and ESPA-1 based RO membranes (Hydranautics) were used for the treatment of tannery effluent obtained from conventional physical–chemical treatment for reuse purpose [15]. RO process showed permeate flux of ~40 L m^{-2} h^{-1} at an operating pressure of 30 bar. Membrane fouling and flux recovery in RO process had been suggested as the critical aspects for industrial implementation of such kinds of tannery effluent. Tannery wastewater obtained from biological treatment process containing high TDS and residual organic loading was further treated using nanofiltration (NF) and RO membranes (polyamide) in a pilot unit of 1 m^3/h to achieve high water recovery and better quality suitable for reuse [31]. About 98% TDS removal was obtained in RO process with 78% recovery of the permeate stream. The RO treated water was recycled in the wet finishing processes in the tannery industry while the RO reject was subjected to solar evaporation plants. The study indicated potential of NF and RO membrane based processes for water recovery and reuse subject to selection of suitable pretreatment which can increase the membrane life and process output.

The composite effluent accumulated from various upstream units except chrome tanning process of a leather plant was collected and treated with NF followed by RO. Prior to membrane filtration the effluent was allowed to settle and thereafter pretreated with coagulation and cloth filtration [17]. Mathematical modeling was done to predict the process efficiency using osmotic pressure and solution diffusion model. The combined process enabled reduction of the organic loading (BOD and COD) adequately to attain the discharge norms. Treatment of the simulated and industrial tannery wastewater was carried out using membranes of various pore sizes using a vibratory shear-enhanced process [36]. Nine membranes involving MF (teflon), UF (regenerated cellulose) and RO (polyamide/polyester) were used. An average removal rate of 50% was observed in the MF process which increased to 85% in the UF process and finally attained to 99% in the RO process in terms of the various pollution parameters such as, COD, tannins, total nitrogen, color and turbidity using vibration amplitude of 31.75 mm.

Tannery effluent from fat liquoring process was treated by a combination process of gravity settling followed by chemical coagulation, NF and RO [25]. COD reduction in the effluent with high initial COD (13,688 mg/L) obtained in the chemical assisted (ferrous sulfate and calcium oxide) coagulation process was about 64% which was increased to 85% after the NF process. To achieve the COD value beyond the permission level of effluent discharge, RO process was employed as the final step wherein, the COD value could be reduced to 117–174 mg/L under operating pressure range of 1313–1724 kPa. Performance of the NF and RO process were modeled which indicated a close agreement of the experimental and the model predicted values of the permeate concentration and permeate flux values.

1.4 Application of Ceramic Membranes

Inorganic membranes, namely the ceramic membranes possess considerably high chemical, thermal and mechanical strength due to which such membranes are often considered as attractive and reliable options for complex tannery wastewater treatment. MF membrane prepared by zirconia coating on cordierite based macroporous support, as well as, commercial alumina membrane was used for treating beamhouse effluent [20, 21]. The process enabled TSS removal of >98% along with 60–65% COD removal. The research findings indicated that ceramic MF membranes can be combined with other separation processes to achieve the adequate treatment to reach the discharge/reuse norms. Ceramic membranes of varying pore sizes (0.2, 3 and 10 μm) were prepared by solid state ion sintering process and applied for tannery wastewater treatment [34]. Depending on pore size differences, about 63% reduction in hardness was obtained along with the removal of COD, Cr, sulfide and NH_3-N upto 83%. The process was found suitable to obtain the national discharge standard and also for industrial recycling.

At CSIR-Central Glass and Ceramic Research Institute, India, low-cost ceramic membranes in tubular configuration have been developed indigenously using clay and alumina based ingredients in optimum proportion. Such membranes show excellent stability towards harsh chemical environment, longer life and have well defined pore size distributions making them potential candidates for treatment and recycling of various industrial wastewater either alone or in combination with other physico-chemical/membrane separation processes [4]. A dual stage process was proposed involving the clay-alumina based MF membrane in combination with polymeric RO membrane for treatment of composite tannery wastewater. The combined process showed complete reduction of COD and BOD5 along with turbidity reduction to 0.025 NTU. The reuse possibility of the treated water was explored further in the leather processing with cow hide and compared with that of the fresh water system. Interestingly in comparison with the fresh water based system, the leather sample tanned with treated wastewater indicated improved mechanical properties such as, tensile strength, grain crackness property and stich tear strength with higher dye uptake. Thus, application of the ceramic MF membrane brings promises for recovery/ recycling of tannery wastewater in combination with the RO process.

Zirconia and alumina were coated on tubular clay-alumina support and used to explore the possibility for clarification and subsequently recycling of the sectional stream and composite tannery effluent from beamhouse operation [26]. In addition with the direct membrane filtration, two types of pre-treatment such as, gravity settling and chemical coagulants addition followed by gravity settling were used. The alumina coated membrane showed considerable removal of the suspended solids with turbidity of the permeate stream <1 NTU, with much higher flux of 350 L/m^2h for the chemically pre-treated effluent at 1.2 bar operating pressure. The study indicates possibility for recycling of the various sectional stream in tannery industry by ceramic membrane-based filtration after addition of the balanced amount of required chemicals which is crucial from minimization of the fresh water usage volume.

Further, clay-alumina MF membranes were employed to treat composite tannery effluent sample from CETP after a biological pre-treatment [6]. The pre-treatment was done using activated sludge obtained from the CETP. While MF process was highly effective for turbidity and suspended solids removal from various types of feed sample with high initial loading, incorporation of the biological pre-treatment with MF process was found quite effective for COD (67–92%) and sulfide (98–99%) reduction along with substantial enhancement in the permeate flux (125%). Performance of the clay-alumina based MF membrane process was investigated further for biologically treated tannery wastewater using a pilot prototype with 1000 L capacity [8]. While the study shows promising application of the low-cost membrane materials in recycling of the post-CETP discharge of tannery effluent, to establish the optimum process design for producing repeated cycles of recycling grade permeate stream and economic feasibility, further investigations are required. Ceramic UF membranes were prepared on clay-alumina substrates using coating of bentonite clay and used in MBR process for tannery effluent treatment with different COD values [2]. The process was found highly effective in terms of organic loading, chromium, TKN and oil and grease removal apart from turbidity and TSS reduction and the treated water characteristics were suitable for industrial recycling. Various influencing parameters in the MBR process, such as HRT, influent concentration (COD) and MLSS concentration in the bioreactor were optimized using statistical tools. For industrial scale operation, an appropriate surface modification of the UF membranes would be beneficial to minimize the bio-fouling effects and maintain the process output.

In view of the various studies performed on different types of tannery effluent samples, the clay-alumina based robust ceramic membranes appear as potential candidates for effective and economic treatment of tannery wastewater towards reuse/ recycling in industrial/other applications. This chapter elaborates on various modes of the low-cost membrane applications for tannery wastewater: as pre/post treatment, either alone or in combination with other physico-chemical separation processes, dual stage MF/RO process, membrane bioreactor process and clay-alumina as substrate for UF membrane preparation. The upscaling aspects, techno-economic evaluation and future prospects are also discussed that would be beneficial for designing efficient treatment strategy for large volume of the toxic and higher loading tannery effluent aiming for resourceful reuse perspectives and approaching towards zero emission initiatives.

2 Performance Efficiency of Clay-Alumina Membranes Alone and in Combination with Other Separation Processes

2.1 Application of Ceramic Membranes of Different Origins for Tannery Wastewater Treatment

Considering the enormous and heterogeneous loading nature of tannery wastewater, the multifunctional treatment approaches need to be focused with the aim of process intensification along with recycling of water, as well as, various salts and metal ions [35]. Due to their high material stability, extent of surface regeneration and high durability, ceramic membranes often have been the choices. The commercially available ceramic membranes (alumina, zirconia and titania based) being costly, various kinds of cost-effective raw materials (mostly clay based) have been employed in the preparation of ceramic membranes along with an optimization in the manufacturing process, mainly sintering temperature towards cost reduction of such membranes for industrial wastewater treatment in comparison to their polymeric counterparts [22]. Table 1 shows application of ceramic membranes prepared from various types of ingredients for remediation of tannery wastewater.

2.2 Application of Clay-Alumina Based MF Membranes Alone and in Combination with Other Separation Processes for Various Types of Tannery Wastewater Treatment

Ceramic MF membranes developed from the low cost raw material composition of clay and alumina can serve as an efficient tool in both pre- and post-treatment combined with other separation processes for remediation of tannery wastewater originated from various sources in CETP, such as composite effluent, primary and secondary clarifier effluent and mixed effluent from the equalization tank. The efficiency of such processes in terms of reduction of various parameters such as, COD, TSS, sulphide, TDS and Cr concentration, as well as, process output in terms of the permeate flux is represented in Table 2. Figure 1 shows the schematic of a treatment scheme involving clay-alumina based ceramic MF membrane process in the cross-flow mode in combination with a biosorbent derived from waste fruit peel

Table 1 Performance evaluation of different ceramic membranes for treatment of tannery industrial effluent

Sl. No.	Membrane type	Pore size (nm)	Clean water permeability ($L\ m^{-2}\ h^{-1}\ bar^{-1}$)	Performance efficiency for tannery wastewater treatment (removal %)	Refs.
1.	γ-alumina based tubular UF membrane	10	57	Color: >95%, COD: 78% from raw tannery wastewater	Kaplan-ekaroglu and Gode [19]
2.	Natural clay and Moroccan phosphate based flat MF membrane	2500	928	Turbidity: 99.8% from beamhouse effluent	Mouiya et al. [23]
3.	(i) Cordierite/zirconia tubular MF membrane (ii) Commercial Al_2O_3 membrane	(i) 230 (ii) 200	(i) 1136 (ii) 1033	Turbidity: 98.5% for (i) and 100% for (ii) COD: 60% for (i), 65% for (ii) TKN: 57.34% for (i), 58.82% for (ii) from beamhouse effluent	Majouli et al. [20, 21]
4.	Natural clay based tubular MF membrane	1500	13.86	COD: 73.28%, BOD_5:73.39%, Conductivity: 26.48% from raw tannery wastewater	Khalfaouyet al. [14]
5.	(i) Clay/banana peels based flat support (ii) Clay based UF membrane prepared on (i)	(i) 5500 (ii) 450	(i) 1150 (ii) 550	Turbidity: 95.95% for (i) and 99.99% for (ii) COD: 50.36% for (i), 90.38% for (ii) Absorbance: 55.10%for (i), 96.70% for (ii)	Mouiya et al. [24]
6.	Natural Moroccan perlite based flat MF membrane	1700	1433.46	Turbidity: 96%	Saja et al. [27]
7.	Micronized phosphate incorporated pozzolan MF membrane	1330	1732.50	Turbidity: 97%	Beqqour et al. [3]
8.	Clay-alumina based tubular MF membrane	1000	300–320	COD: 77–80%, Turbidity: 95–98% from secondary clarifier effluent	Bhattacharya et al. [8]

(continued)

Table 1 (continued)

Sl. No.	Membrane type	Pore size (nm)	Clean water permeability (L m^{-2} h^{-1} bar^{-1})	Performance efficiency for tannery wastewater treatment (removal %)	Refs.
				COD: 68%, Turbidity: >99%, Sulphide: 95% from equalization tank in CETP	Bhattacharya et al. [6]
9.	Bentonite clay based tubular UF membrane prepared on clay-alumina MF membrane	4.85	18.54	COD:92–97%, Turbidity: >99.5%, Chromium:91–92.4% used in MBR process for composite tannery effluent treatment	Banerjee et al. [2]

[5]. The combined mode of treatment enabled an effective reduction of chromium (~99.9%), sulphide (>97%), COD (91.5–96.6%) and TSS (86–99%). Depending on the initial loading of the effluent, permeate flux values obtained were in the range of 12 (composite effluent), 19 (primary clarifier effluent) and 25 (secondary clarifier effluent) L m^{-2} h^{-1}, respectively (Table 2).

Integration of biosorption with the membrane filtration process resulted in adequate reduction of various organic and inorganic contaminants present in tannery wastewater represented in terms of the COD and BOD for organic loading, turbidity and TSS for turbid and suspended contaminants, Cr, Ni, Pb, etc. and other inorganic anions such as sulphide.

Figure 2 shows removal (%) of the above parameters for different effluent streams, a (composite effluent), b (primary clarifier effluent) and c (secondary clarifier effluent). It may be observed that about 95.22% reduction in BOD_5 was achieved for a, for b and c the reduction were 95.19% and 86.8%, respectively. Figure 2 also indicates that apart from chromium, complete removal (~100%) of lead was obtained in all the three types of effluent samples, while nickel removal was 98.6% (a), 89% (b) and 60% (c). Overall, it may be opined that the combination of biosorption with ceramic MF membrane based microfiltration process is suitable for treating tannery effluent with varying degree of organic, as well as, inorganic pollutants loading and the treated effluent attained discharge norms for most of the parameters, however, for very high initial loading wastewater such as composite effluent used in the study with COD loading of ~12,000–13,000 mg/L, an additional post-treatment may be required. In Fig. 3 permeate flux values of wastewater samples at fixed transmembrane pressure (TMP) of 1 kg/cm^2 measured for various days has been shown. Among different kinds of effluent samples, sample c having less initial loading showed higher flux while the concentrated effluent samples (a) showed low permeation rate which is the expected trend.

Table 2 Efficiency of clay-alumina based ceramic MF membranes in combination with other processes for tannery wastewater treatment

Sl. No.	Effluent type	MF process alone/in combination with pre/post-treatment	Performance efficiency (removal %)	Permeate flux (L m^{-2} h^{-1}) at 1 kg/cm^2 TMP	Refs.
1.	(i) Composite effluent	MF process alone	COD: 36% Sulphide: 95% TSS: 99.6%	32	Bhattacharya et al. [6]
		Biological pretreatment + MF	COD: 81% Sulphide: 99.5% TSS: 99.7%	72	
	(ii) Effluent from equalization tank	MF process alone	COD: 62% Sulphide: 95.4% TSS: 96%	88	
		Biological pretreatment + MF	COD: 93% Sulphide: 99.6% TSS: 97%	125	
	(iii) Effluent from primary clarifier	MF process alone	COD: 74% Sulphide: 95.42% TSS: 96.2%	104	
		Biological pretreatment + MF	COD: 94% Sulphide: 99.56% TSS: 97%	139	
2.	(i) Composite effluent	Biosorption + MF	COD: 96.6% Sulphide: 97% TSS: 99.8% Cr: 99.96%	12	Bhattacharya et al. [5]
	(ii) Effluent from primary clarifier	Biosorption + MF	COD: 93.3% Sulphide: 97.8% TSS: 99.6% Cr: 99.9%	19.5	
	(iii) Effluent from secondary clarifier	Biosorption + MF	COD: 91.5% Sulphide: 98% TSS: 86% Cr: 99.9%	25.5	

(continued)

Table 2 (continued)

Sl. No.	Effluent type	MF process alone/in combination with pre/post-treatment	Performance efficiency (removal %)	Permeate flux ($L\ m^{-2}\ h^{-1}$) at 1 kg/cm^2 TMP	Refs.
3.	Composite effluent	MF	COD: 90.85% Sulphide: 99.9% TSS: 96.2% TDS: 26.8% Cr: 99.86%	35–50	Bhattacharya et al. [4]
		MF + RO	COD: 90.99% Sulphide: 99.99% TSS: 94.72% TDS: 94.72% Cr: 99.98%		

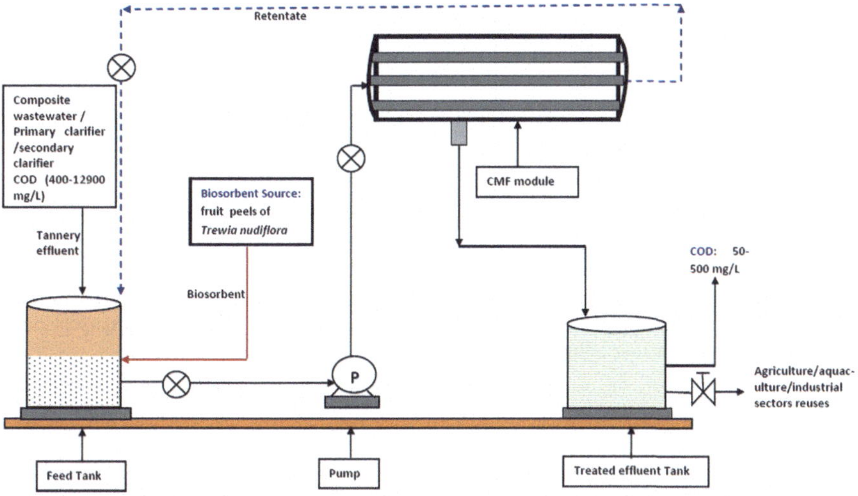

Fig. 1 Schematic representation of combined treatment process involving cross-flow microfiltration with clay-alumina based ceramic MF membrane and waste biomass derived biosorbent [5]

Apart from adsorptive treatment, biological pretreatment can be an effective option for tannery industrial effluent prior to the ceramic MF membrane driven filtration. To investigate the efficiency of such combined mode of treatment, three types of effluent samples, such as composite effluent (E1), effluent from equalization tank of CETP (E2) and primary clarifier effluent (E3) was selected [6]. Activated sludge collected from the CETP was used as a source of microorganisms in

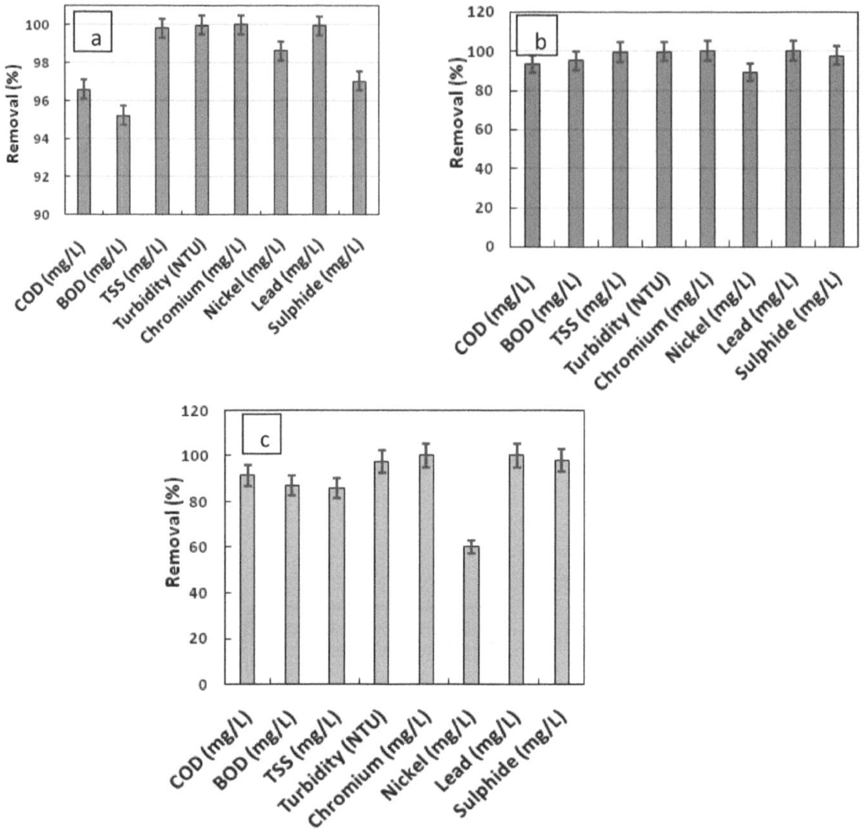

Fig. 2 Removal of various parameters in tannery wastewater samples collected from CETP with varying loading in combined adsorption and MF process: **a** composite effluent; **b** primary clarifier effluent; **c** secondary clarifier effluent

the biological pretreatment process. Figure 4 shows the schematic of the combined treatment strategy employed in the study. It may be noted from Fig. 5 that in the single stage MF process COD removal was quite low, ~35.8% for E1 with very high initial COD loading (~18,250 mg/L) which increased to 62% for E2 with initial COD loading of ~3375 mg/L and upto 74% removal was achieved for feed with lower COD loading (E3, 1950 mg/L). Inclusion of biological pretreatment facilitated in significant high removal of COD which ranged from 81% (E1) to 94% (E3). Similarly, BOD removal also increased effectively in the combined process: from 51.6 to 85.23% (E1), 56.84 to 96.84% (E2) and 71.77 to 97.21% (E3) compared to the single-stage MF process. In addition to the organic loading, considerable high reduction was achieved for sulphide, phosphorus and TSS content in the combined process for all the three samples. Phosphorus removal was 97.4% (E1), 96.9% (E2) and 96.44% (E3), respectively while sulphide removal was >99.5% in all cases.

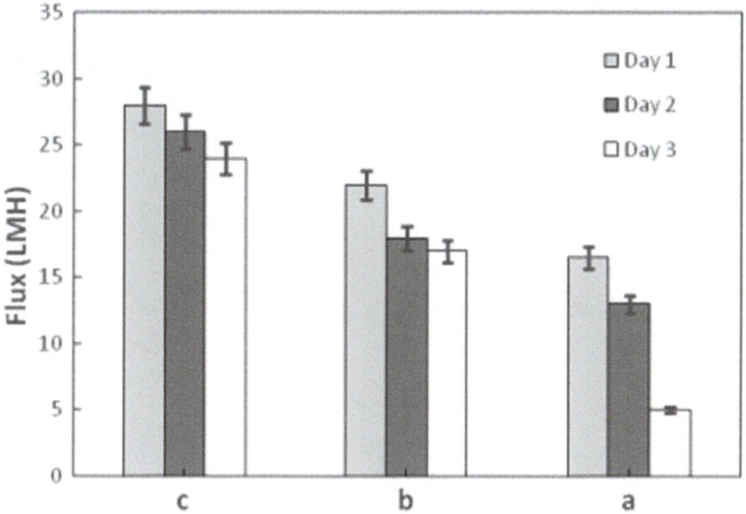

Fig. 3 Permeate flux at 120 min of tannery effluent samples with varying loading in combined adsorption and MF process at TMP 1 kg/cm^2: **a** composite effluent; **b** primary clarifier effluent; **c** secondary clarifier effluent

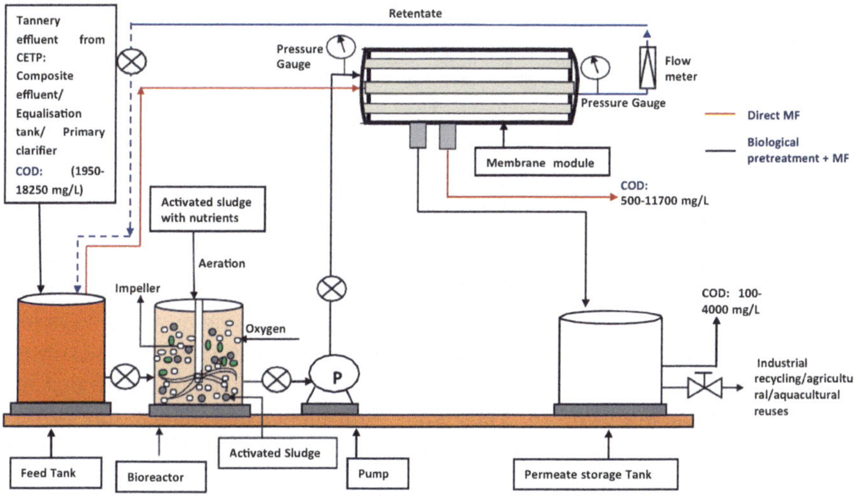

Fig. 4 Schematic representation of combined treatment process involving biological pretreatment along with cross-flow microfiltration with clay-alumina based ceramic MF membrane [6]

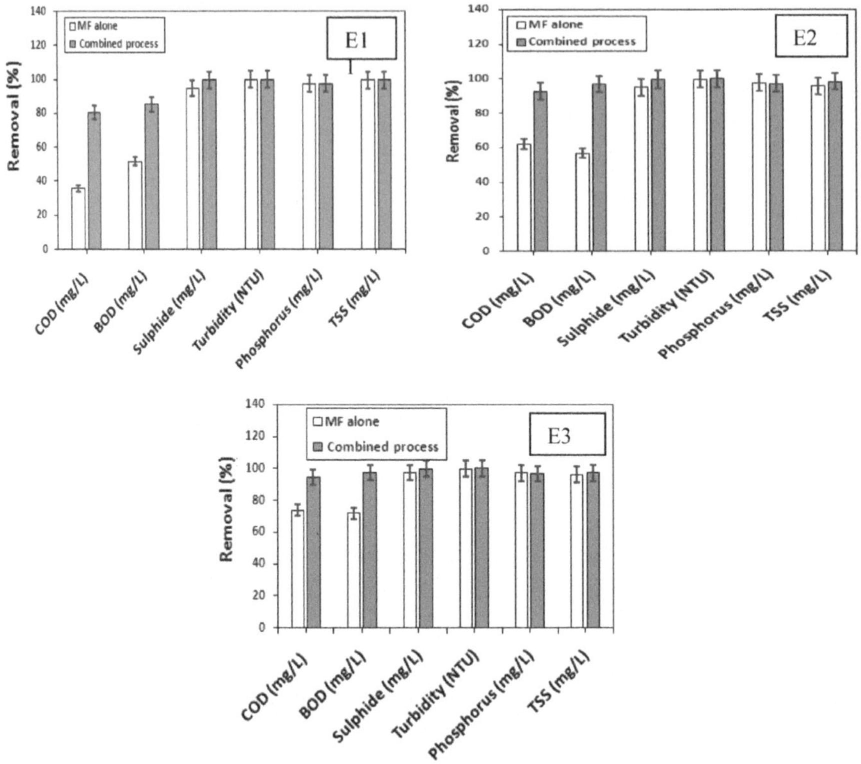

Fig. 5 Removal of various parameters in tannery wastewater samples collected from CETP with varying loading in the single stage MF and biological pretreatment combined with MF process: composite effluent (E1); effluent from equalization tank (E2); primary clarifier effluent (E3)

Apart from the treatment efficiency, permeate flux was analyzed for both the processes. As expected, in the single stage MF process flux value was always low compared to the combined process since the biological pretreatment provided considerable reduction of various dissolved organic and inorganic loading, suspended and turbid pollutant loading (Fig. 6). Thus along with higher flux values, rate of flux decline was less due to the formation of cake growth and associated membrane pore blocking to a reduced extent.

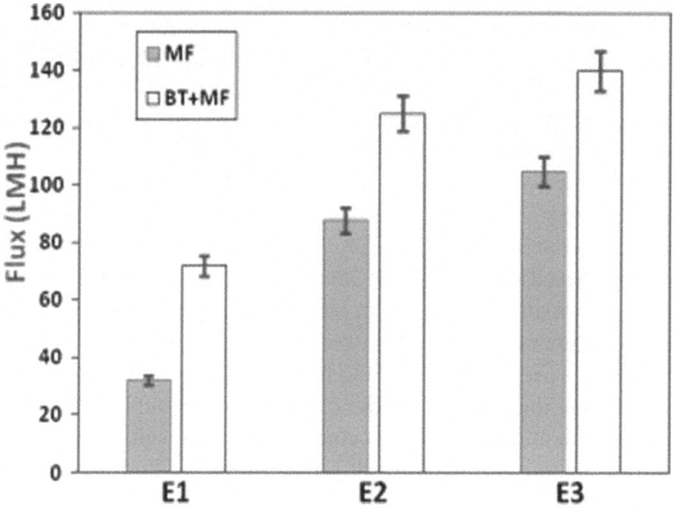

Fig. 6 Permeate flux at 120 min of tannery effluent samples with varying loading in the single stage MF and biological pretreatment combined with MF process at TMP 1 kg/cm^2: composite effluent (E1); effluent from equalization tank (E2); primary clarifier effluent (E3)

3 Reuse Prospect of the Clay-Alumina Membrane Treated Effluent in the Tannery Industry

3.1 Performance Evaluation of the Membrane Treated Effluent in Tanning Process

Based on the high degree of separation obtained in the dual stage treatment process involving the clay-alumina membrane based microfiltration followed by polymeric RO process for composite tannery effluent with respect to the various parameters such as, COD, TSS, TDS, chromium and sulfide content (Table 2), the reuse potential of the treated effluent in tanning process was explored using cow hide samples [4]. The fresh water based system was considered as control while the treated effluent obtained from the MF+RO process was used as the experimental one. The obtained leather samples were assessed in terms of various physical properties such as, tensile strength, grain crackness, dye uptake and stich tear strength, etc. Interestingly, the study indicated positive results for the experimental set in comparison with that of the control set. Most of the measured parameters were observed to improve in the experimental set, for example, tensile strength was 24% higher, grain crackness (just crack) was 37.6% higher and stitch tear strength was ~61% more compared to the control set. In the two-stage treatment process the clay-alumina based ceramic MF membranes could reduce the turbid and colloidal contaminants, suspended solids and organic loading of composite effluent significantly enabling production of permeate stream suitable to be used as feed for the RO process. The subsequent RO process facilitated adequate removal of various dissolved salts, cations and anions present in the effluent, rendering the treated water suitable to be recycled in the tanning operation.

The recycling potential of clay-alumina membranes were studied in another work wherein, wastewater sample received from the secondary clarifier of CETP was subjected to filtration by clay-alumina membranes (MF) and the permeate samples obtained from the process was used in the leather wet processing and the various physical and chemical properties were measured and compared with that of the control, i.e. the fresh water based system [8]. The study showed values of different parameters measured such as, tensile strength, grain bursting strength and tongue tear strength as comparable or higher in the treated effluent based system in comparison with the control set. The improved qualities of the leather samples obtained from the recycling studies indicate the potential of the clay-alumina membrane based processes as sustainable treatment option either in combination with suitable pretreatments or with RO as post operation for industrial recycling of the high loading composite tannery wastewater.

3.2 Techno-economic Assessment

Cost analysis of the combination process of clay-alumina based MF and RO was carried out with respect to feed volume of 3000 L [4, 5]. Capital cost of the pilot prototype was evaluated as 58,622 USD while the cost of treated water (L/day) was estimated as 0.01 USD both of which were observed lower compared to that of the conventional wastewater treatment process at the industrial scale (1167 USD, 592 USD and 0.07 USD, respectively). Inclusion of suitable pretreatment and appropriate fouling mitigation strategies, such as optimized backwashing protocol can enhance the process output and thereby, increase the cost-effectiveness of the process further. The low cost clay-alumina membranes with high shelf life can support the process to a significant extent in comparison with the commercially available counter parts (alumina/zirconia/titania based) considering the sustainability aspects.

4 Analyzing the Toxic Impacts of Tannery Effluent During Course of Treatment

4.1 Comparative Assessment of the Untreated Tannery Effluent and Membrane Treated Effluents

It had been observed that tannery effluent treated by the secondary clarifier in CETP which is usually disposed to the nearby water bodies, still possess some organic and inorganic loading (COD: 180–980 mg/L; TOC: 75–135 mg/L; TSS: 56–118 mg/L, etc.). Hence, the prospect of reuse of such effluent in irrigation/aquaculture activity was studied by employing clay-alumina membrane-based microfiltration treatment [7]. The process was successful in reduction of COD (70–86%) and TOC (85%). Further, possible toxic impact of the effluent in the environment was evaluated selecting two systems of aquatic plant and animal models, *Pistia* sp. and *Poecilia* sp., respectively. Various oxidative stress biomarkers such as, superoxide dismutase, guaiacol peroxidase, etc. as observed in both the living models were found to increase significantly for the untreated effluent while the values were alike for the treated wastewater and fresh water (control) based systems. Thus, ceramic MF membrane induced treatment could be beneficial in reducing the toxic load in tannery effluent obtained from the secondary clarifier, thus could be used as a tertiary treatment for potential reuse applications like aquaculture, agriculture and horticulture, etc.

Environmental impact of tannery wastewater subject to the clay-alumina based ceramic MF treatment in combination with waste biomass derived the biosorption process was assessed further considering *Euphorbia hirta* as a model plant which is found to grow profusely in the CETP complex [5]. The membrane-based treatment process was found to reduce the toxic load of the effluent effectively as measured in terms of the various stress enzyme activities in *E. hirta*, such as superoxide dismutase, peroxidase and catalase. The combined process provides a simple and cost-effective approach of tannery effluent treatment and thus, opens an avenue for reuse of the effluent in agricultural or related applications.

Environmental impact of the composite tannery effluent of CETP was further assessed during various modes of treatment with respect to an aquatic organism, *Pila globosa* [9]. In this study three different systems were selected, (i) the untreated effluent, (ii) treated effluent from single stage MF process (clay-alumina membrane) and (iii) treated effluent from two stage process involving MF (clay-alumina membrane) + RO (polymeric). The whole body tissues, mantle and gonad of snails were used to study the various antioxidant enzyme activities such as SOD, CAT, GST, GR, etc. along with the characterization of various biochemical parameters such as, protein, amino acid and carbohydrates. In addition, the mantle and gonad tissues were subjected to histological analysis and comet study was performed to understand the DNA damage in snails due to the exposure of various types of wastewater. Clearly distinct findings were observed in various parameters for the untreated effluent and membrane treated effluents. The activities of oxidative stress enzymes measured in various tissues of *P. globosa* were considerably reduced in the MF treated effluent and further low in the MF + RO treated effluent in comparison with the untreated effluent. The biochemical parameters were much low on exposure of untreated effluent while for the membrane treated effluents, the parameters were found close to that of the fresh water based system (control). Significantly, genotoxicity and DNA damage as observed from the comet assay was quite prominent for the untreated effluent based system which were not observed in both the membrane treated systems indicating clay-alumina based membranes can facilitate detoxification of hazardous tannery effluent towards safe discharge in the environment or reuse in various purposes.

5 Potential of the Clay-Alumina Membrane Based Technology in Pilot Scale Treatment of Tannery Effluent

Based on the encouraging results obtained in the lab mode studies, the clay-alumina based ceramic MF process was upscaled to the pilot level [8]. Figure 7 shows the membrane modules used in the pilot study. It involves two numbers of membrane modules (SS) accommodating seven tubular membranes (19 channels) in each. The tubular membranes were of 1 m length, OD 34 mm having 3.55 m^2 filtration area of. Both the feed tank (1000 L) and permeate tank (500 L) were made of PVC. The setup was operated with 150–200 L/h feed flow rate using a 2 HP centrifugal recirculation pump and 1–1.2 kg/cm^2 of the transmembrane pressure. The pilot unit was installed at CETP of Kolkata leather complex. The objective of the inclusion of the ceramic MF membrane process was to reduce the organic loading and other trace contaminants present in the discharge of secondary clarifier to an adequate extent, so as to enable the treated water to be recycled in various steps of leather processing.

Figures 8 and 9 shows variation in the permeate flux values after 2 h of operation for various samples of secondary clarifier effluent collected at different days with varying feed COD values operated under TMP of 1 kg/cm^2 in the labscale MF unit and pilot MF unit, respectively. As observed from the figures, the flux values were almost in similar range in both types of filtration units for the effluent samples with similar COD values. To minimize the fouling effect, the pilot unit was operated with high cross flow velocity (2.74 m/s) which prevented flux reduction to a considerable extent. After 4 h of operation, feed sample with relatively high COD value (820 mg/L) showed flux reduction of ~8% [6, 7]. In the MF process up to 80% COD removal was noted overall

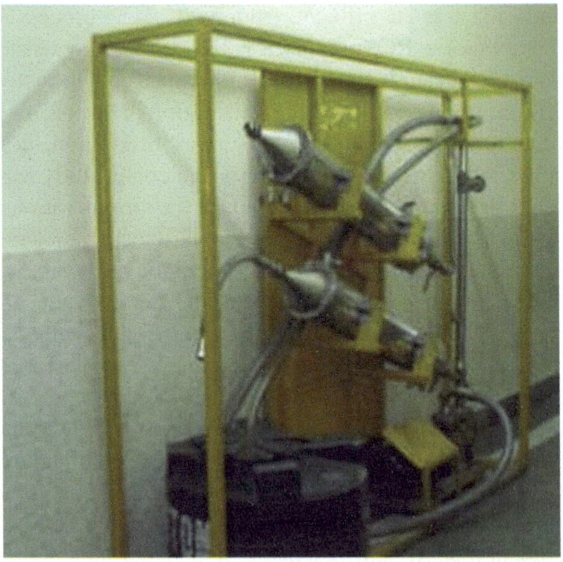

Fig. 7 Photograph of the clay-alumina membrane based pilot prototype unit

with COD value in the permeate samples in the range of 138–189 mg/L, BOD values 22–79 mg/L, total chromium <0.1 mg/L and sulphide <0.01 mg/L. Due to the MF range of pore sizes, only partial reduction of TDS could be achieved (~20–54%). The feasibility of recycling the treated water in the wet finishing process of leather tanning was investigated and compared with that of the fresh water based system (control). The various characteristics obtained for the leather samples processed with the treated water was found comparable with that of the fresh water based system, indicating potential of the proposed treatment process for industrial recycling purpose.

Disposal of the concentrated stream produced in the membrane process is a critical issue which needs to be addressed carefully for treating complex industrial wastewater. In the present pilot study, volume concentration ratio (VCR) was observed to be ~2.3. The increase of retentate COD values were in the range of 400–700 mg/L corresponding to the feed COD being 330–520 mg/L [8]. Thus, an additional treatment of retentate stream is required before its discharge into the environment. The membrane reject stream may be recycled into the feed tank (secondary clarifier) as sustainable option to manage issues associated with disposal of the concentrated waste stream from environmental point of view.

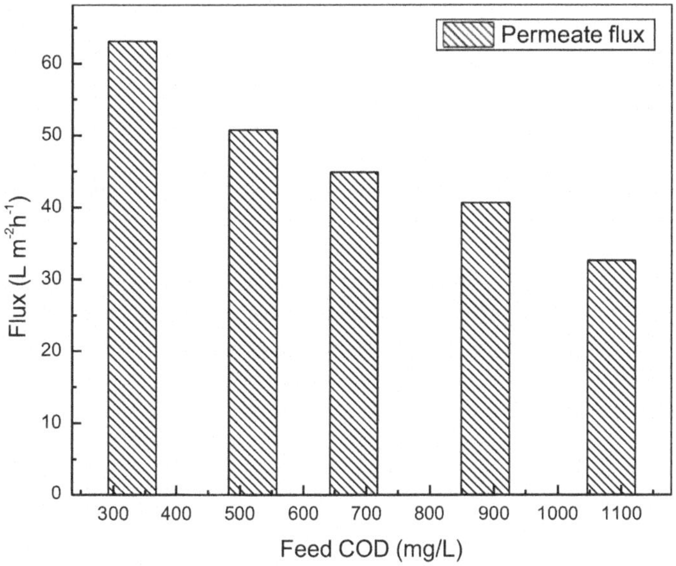

Fig. 8 Variation of permeate flux with feed COD values of secondary clarifier effluent in labscale MF unit after 2 h of operation (TMP 1 kg/cm^2)

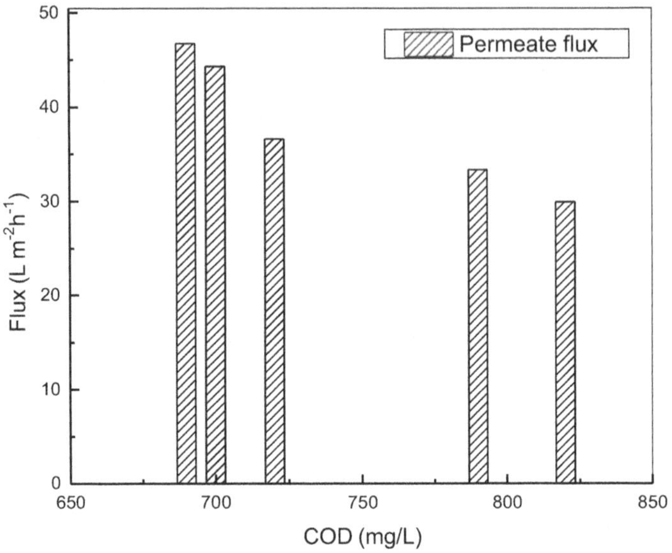

Fig. 9 Variation of permeate flux with feed COD values of secondary clarifier effluent in pilot scale MF unit after 2 h of operation (TMP 1 kg/cm^2)

6 Conclusions

Performance efficiency of the clay and alumina based membranes have been assessed in various modes for treatment of tannery effluent with varying characteristics and pollutant loading produced in different sections of the common effluent treatment plant of the leather complex. The membranes with microfiltration range of pore sizes can efficiently eliminate turbid and suspended contaminants, organic loading and partial reduction of the dissolved solids/cations and anions present in the tannery effluent. The high stability of such membranes promote them as suitable candidates to handle complex tannery effluent with high loading with appropriate pre/post treatment strategies to manage the membrane fouling issues which are critical in designing industrial effluent treatment processes. A two-stage treatment process involving biological pretreatment followed by the ceramic MF process for tannery effluent with high organic loading can be an effective option to address the membrane fouling and restore the permeate flux. On the other hand, integration of a suitable adsorbent system with the MF process can contribute to significant reduction of toxic inorganic ions such as chromium, nickel and lead along with adequate removal of suspended and organic contaminants rendering the treated effluent to be superior quality free from toxic ingredients, thus safe to be discharged in the environment and suitable for reuses in various purposes. For complete recycling of tannery effluent integration of RO would be necessary as post-treatment of the MF process for reduction of the total dissolved solids. Clay-alumina based MF membranes can be prepared in multichannel (19) configuration with 1 m length for pilot scale implementation wherein, appropriate design can accommodate a series of membrane modules with high surface area in low foot-print area. Compared to the widely available polymeric membranes, the ceramic based membranes show greater durability, however, the material cost need to be reduced appropriately for larger scale industrial implementation of such membranes. Overall, the excellent chemical, mechanical and thermal resistance of such membranes facilitates operation under high pressure and feed flow rate, rigorous chemical cleaning and membrane regeneration prospects which can make them potential candidates for sustainable treatment of tannery industrial effluent with recycling objectives.

7 Future Research Perspectives

In spite of high removal efficiency offered by clay-alumina based membranes, several challenges are there for pilot scale implementation in tannery industries. Few critical perspectives of the future direction of research are projected here.

- Enhancement of the surface porosity, pore connectivity and other permeation characteristics of the membranes is much essential to compete with the polymeric membrane based treatment processes.
- Incorporation of the defect-free and stable coating layer on clay-alumina MF membranes and process optimization for large scale coating process is crucial for improved treatment.
- Management of the concentrated retentate stream produced from the membrane process by applying advanced treatment processes, such as AOP.
- Development of ceramic-polymer composite membranes with more hydrophilic surface and reduced pore sizes can be attractive options to obtain high permeate flux with enhanced separation characteristics.
- Overall, design of the treatment plant with ceramic MF/UF membranes followed by polymeric RO membranes can be a viable alternative to address the effluent recycling and zero water discharge issues with low volume of sludge production, hence further studies in this direction with pilot process optimization and in-depth techno-economic analysis would be beneficial towards sustainable technology implementation.

References

1. Alemu T, Mekonnen A, Leta S (2019) Integrated tannery wastewater treatment for effluent reuse for irrigation: encouraging water efficiency and sustainable development in developing countries. J Water Process Eng 30:100514
2. Banerjee S, Santra B, Banerjee D, Ghosh S, Majumdar S (2022) Performance assessment of the indigenous ceramic UF membrane in bioreactor process for highly polluted tannery wastewater treatment. Environ Sci Pollut Res 29:48620–48637

3. Beqqour D, Achiou B, Bouazizi A, Ouaddari H, Elomari H, Ouammou M, Bennazha J, Alami Younssi S (2019) Enhancement of microfiltration performances of pozzolan membrane by incorporation of micronized phosphate and its application for industrial wastewater treatment. J Environ Chem Eng 7:102981
4. Bhattacharya P, Roy A, Sarkar S, Ghosh S, Majumdar S, Chakraborty S, Mandal S, Mukhopadhyay A, Bandyopadhyay S (2013) Combination technology of ceramic microfiltration and reverse osmosis for tannery wastewater recovery. Water Resour Ind 3:48–62
5. Bhattacharya P, Ghosh S, Mukhopadhyay A (2013) Combination technology of ceramic microfiltration and biosorbent for treatment and reuse of tannery effluent from different streams: response of defence system in Euphorbia sp. Int J Recycl Org Waste Agric 2:1–11
6. Bhattacharya P, Ghosh S, Majumdar S, Roy SN, Bandyopadhyay S (2015) Ceramic membrane based microfiltration for treatment of highly contaminated tannery wastewater. Environ Eng Manag J 14(9):2139–2148
7. Bhattacharya P, Ghosh S, Swarnakar S, Mukhopadhyay A (2015) Tannery effluent treatment by microfiltration through ceramic membrane for water reuse: assessment of environmental impacts. Clean Soil Air Water 43(5):633–644
8. Bhattacharya P, Majumdar S, Bandyopadhyay S, Ghosh S (2016) Recycling of tannery effluent from common effluent treatment plant using ceramic membrane based filtration process: a closed loop approach using pilot scale study. Environ Prog Sustain Energy 35:60–69
9. Bhattacharya P, Swarnakar S, Mukhopadhyay A, Ghosh S (2016) Exposure of composite tannery effluent on snail, Pila globosa: a comparative assessment of toxic impacts of the untreated and membrane treated effluents. Ecotoxicol Environ Saf 126:45–55
10. Cassano A, Molinari R, Romano M, Drioli E (2001) Treatment of aqueous effluents of the leather industry by membrane processes: a review. J Membr Sci 181(1):111–126
11. Deveci EÜ, Akarsu C, Gönen Ç, Özay Y (2019) Enhancing treatability of tannery wastewater by integrated process of electrocoagulation and fungal via using RSM in an economic perspective. Process Biochem 84:124–133
12. Doble M, Kruthiventi AK (eds) (2007) Chapter 9—Industrial examples. In: Green chemistry and engineering. Academic Press, pp 245–296. https://doi.org/10.1016/B978-012372532-5/50010-9. ISBN 9780123725325
13. Elabed A, Abed SE, Ibnsouda S, Erable B (2019) Sustainable approach for tannery wastewater treatment: bioelectricity generation in bioelectrochemical systems. Arab J Sci Eng ffhal-02284196:1–10. https://doi.org/10.1007/s13369-019-04050-y
14. El Khalfaouy R, Elabed A, Khallouk K, El Knidri H, Belaabed R, Addaou A, Laajeb A, Lahsini A (2017) Microfiltration process for tannery wastewater treatment from a leather industry in Fez-Morocco area. JMES 8(7):2276–2281
15. Fababuj-Roger M, Mendoza-Roca JA, Galiana-Aleixandre MV, Bes-Piá A, Cuartas-Uribe B, Iborra-Clar A (2007) Reuse of tannery wastewaters by combination of ultrafiltration and reverse osmosis after a conventional physical-chemical treatment. Desalination 204:219–226
16. Ganesh R, Sousbie P, Torrijos M, Bernet N, Ramanujam RA (2015) Nitrification and denitrification characteristics in a sequencing batch reactor treating tannery wastewater. Clean Technol Environ Policy 17:735–745
17. Jain SK, Purkait MK, De S, Bhattacharya PK (2006) Treatment of leather plant effluent by membrane separation processes. Sep Sci Technol 41(15):3329–3348. https://doi.org/10.1080/01496390600915080
18. Jang A, Jung J-T, Kang H, Kim H-S, Kim J-O (2017) Reuse of effluent discharged from tannery wastewater treatment plants by powdered activated carbon and ultrafiltration combined reverse osmosis system. J Water Reuse Desalin 7(1):97–102. https://doi.org/10.2166/wrd.2016.001
19. Kaplan-Bekaroglu SS, Gode S (2016) Investigation of ceramic membranes performance for tannery wastewater treatment. Desalin Water Treat 57:17300–17307. https://doi.org/10.1080/19443994.2015.1084595
20. Majouli A, Tahiri S, Alami Younssi S, Loukili H, Albizane A (2012) Elaboration of new tubular ceramic membrane from local Moroccan Perlite for microfiltration process. Application to treatment of industrial wastewaters. Ceram Int 38(5):4295–4303. https://doi.org/10.1016/j.ceramint.2012.02.010

21. Majouli A, Tahiri S, Alami Younssi S, Loukili H, Albizane A (2012) Treatment of effluents coming from beamhouse section of tannery by microfiltration through cordierite/zirconia and alumina tubular ceramic membranes. J Mater Environ Sci 3(5):808–815
22. Mestre S, Gozalbo A, Lorente-Ayza MM, Sánchez E (2019) Low-cost ceramic membranes: a research opportunity for industrial application. J Eur Ceram Soc 39:3392–3407. https://doi.org/10.1016/j.jeurceramsoc.2019.03.054
23. Mouiya M, Abourriche A, Bouazizi A, Benhammou A, Hafiane YE, Abouliatim Y, Nibou L, Oumam M, Ouammou M, Smith A, Hannache H (2018) Flat ceramic microfiltration membrane based on natural clay and Moroccan phosphate for desalination and industrial wastewater treatment. Desalination 427:42–50
24. Mouiya M, Bouazizi A, Abourriche A, Benhammou A, El Hafiane Y, Ouammou M, Abouliatim Y, Younssi SA, Smith A, Hannache H (2019) Fabrication and characterization of a ceramic membrane from clay and banana peel powder: application to industrial wastewater treatment. Mater Chem Phys 227:291–301
25. Prabhavathy C, De S (2010) Treatment of fatliquoring effluent from a tannery using membrane separation process: experimental and modelling. J Hazard Mater 176(1–3):434–443
26. Sahoo GC, Ghosh S, Majumdar S, Bandyopadhyay S (2014) Clarification of tannery wastewater from beam house operation using ceramic MF membrane. Appl Sci Adv Mater Int 1(1):12–15
27. Saja S, Bouazizi A, Achiou B, Ouammou M, Albizane A, Bennazha J, Younssi SA (2018) Elaboration and characterization of low-cost ceramic membrane made from natural Moroccan perlite for treatment of industrial wastewater. J Environ Chem Eng 6:451–458
28. Saranya D, Shanthakumar S (2020) An integrated approach for tannery effluent treatment with ozonation and phycoremediation: a feasibility study. Environ Res 183:109163
29. Scholz WG, Rougé P, Bódalo A, Leitz U (2005) Desalination of mixed tannery effluent with membrane bioreactor and reverse osmosis treatment. Environ Sci Technol 39(21):8505–8511. https://doi.org/10.1021/es050330p
30. Stoller M, Sacco O, Sannino D, Chianese A (2013) Successful integration of membrane technologies in a conventional purification process of tannery wastewater streams. Membranes 3. https://doi.org/10.3390/membranes3030126,126-135
31. Suthanthararajan R, Ravindranath E, Chits K, Umamaheswari B, Ramesh T, Rajamam S (2004) Membrane application for recovery and reuse of water from treated tannery wastewater. Desalination 164:151–156
32. Vinodhini PA, Sudha PN (2017) Removal of heavy metal chromium from tannery effluent using ultrafiltration membrane. Text Cloth Sustain 2:5. https://doi.org/10.1186/s40689-016-0016-3
34. Wang JG, Liu YH (2012) Study on the preparation of ceramic membrane and the treatment of tanning wastewater. Adv Mater Res 487. https://doi.org/10.4028/www.scientific.net/amr.487.903,903-907
35. Zhao J, Wu Q, Tang Y, Zhou J, Guo H (2022) Tannery wastewater treatment: conventional and promising processes, an updated 20-year review. J Leather Sci Eng 4:10. https://doi.org/10.1186/s42825-022-00082-7
36. Zouboulis AI, Peleka EN, Ntolia A (2019) Treatment of tannery wastewater with vibratory shear-enhanced processing membrane filtration. Separations 6:20. https://doi.org/10.3390/separations6020020

Chapter 5
Oily Wastewater Treatment Using Mullite and Mullite-Alumina Membranes

Fariba Sadat Kamelian, Toraj Mohammadi, and Mohammad Sadegh Nabavi

Abbreviations

AC	Activated carbon
AGR	Aerobic granular reactors
Al_2O_3	Aluminium oxide
CFV	Cross-flow velocity
COD	Chemical oxygen demand
EPS	Extracellular polymeric substances
MBR	Membrane bioreactor
MF	Microfiltration
O/W	Oil in water emulsion
PES	Polyethersulfone
PVDF	Polyvinylidene fluoride
TMP	Trans-membrane pressure
UF	Ultrafiltration
W/O	Water in oil emulsion

F. S. Kamelian · T. Mohammadi (✉)
Center of Excellence for Membrane Research and Technology, Iran University of Science and Technology (IUST), Narmak, Tehran, Iran
e-mail: torajmohammadi@iust.ac.ir

F. S. Kamelian
e-mail: f_kamelian@alumni.iust.ac.ir

M. S. Nabavi
Energy & Process Department, Norconsult Sverige AB, P.O. 11221, Stockholm, Sweden
e-mail: Sadegh.Nabavi@norconsult.com

S. Bandyopadhyay (ed.), *Application of Low Cost Ceramic Membranes in Wastewater Treatment*, Materials Horizons: From Nature to Nanomaterials,
https://doi.org/10.1007/978-981-97-5842-5_5

1 Introduction

In different chemical and petroleum industries, large volumes of oily wastewater are produced daily. Because of endangering human health and water resources, direct discharge of this wastewater is restricted. With the purpose of recycling and discharge to the environment, it should be treated to meet the specifications for relevant standards. However, such standards varies in different countries, which is becoming stricter every year [1, 2]. The treatment of wastewater also plays a decisive role in evaluating operational costs for addressing the associated environmental problems. Therefore, the performance of the treatment techniques with respect to the economical and eco-friendly features is essential for their application under field conditions [1–3].

Different methods such as heating, air flotation, hydrocyclone, centrifuge as well as biological processes are applied to treat oily wastewater [1, 4, 5]. There are some disadvantages to treating with conventional methods, like high operating and capital costs as well as low operational efficiency. They also are unable to separate oil droplets in the micron and submicron size range of less than 20 μm. Membrane technologies can be considered as alternative methods, which can overcome these shortcomings [1, 3, 6]. Production of the final product with desirable quality is enumerated as an important advantage of membrane-based technologies [1]. Furthermore, the advantages of microfiltration techniques are easy preparation method, less energy demanding, and minimum requirement of space and need for extra chemicals etc. [3]. Both ceramic and polymeric membranes can be used for microfiltration applications. Compared to polymeric ones, ceramic membranes have better mechanical strength and chemical resistance along with the ease of cleaning and regeneration; however, their use is restricted due to higher capital costs [1, 7]. The development of inexpensive ceramic membranes with better performance is a challenge for the separation of various types of oily wastewater [8]. Accordingly, the potential of mullite and mullite-alumina-based ceramic membranes as low-cost alternatives for the treatment of oily wastewater has been dealt with in this chapter.

2 Oily Wastewater

Because of high amounts of heavy metals, aromatic compounds, salts, grease, kerosene, oil, crude oil, and recalcitrant materials, the United States Environmental Protection Agency (USEPA) classified oily wastewater as the most dangerous sewage, which is difficult to purify. Huge amounts of oily wastewater are produced daily in different industries related to hydrocarbon processing like natural gas, oil, and petrochemical production and refining processes [1, 2, 5]. There are three different kinds of oily wastewater such as water in oil (W/O) and oil in water (O/W) emulsions with droplet sizes lower than 20 μm and immiscible O/W mixture. The immiscible mixture is less stable than the emulsion one, therefore it is easier to separate [2, 9].

The majority of industries produce the O/W emulsion with a concentration between 50–1000 ppm. The acceptable value for discharging the O/W emulsion in land and water is 5–10 ppm; however, the value of 20 ppm is acceptable for entering into public sewage [1–3, 8].

3 Treatment of Oily Wastewater

The selection of appropriate methods for treating oily wastewater depends on oil droplet size and concentration of oil and other chemical compounds. The utilization of appropriate methods decreases operational cost and time as well as leads to better compatibility with the environmental standards [2]. Separation procedures according to the density differences cannot separate emulsified oil. Chemical treatments are unable to de-stabilize O/W emulsion [10]. Among various methods, the separation of O/W emulsion with nano and micron-scale droplets especially faces severe challenges [4, 10, 11]. Membrane technology is an outstanding process for treating oil droplets <20 μm. Figure 1 illustrates publication numbers regarding different technologies used to treat oily wastewater. Accordingly, membrane technologies, due to their less footprint and energy needs, no requirement for extra chemicals and higher separation efficiency, have attracted lots of attention [10].

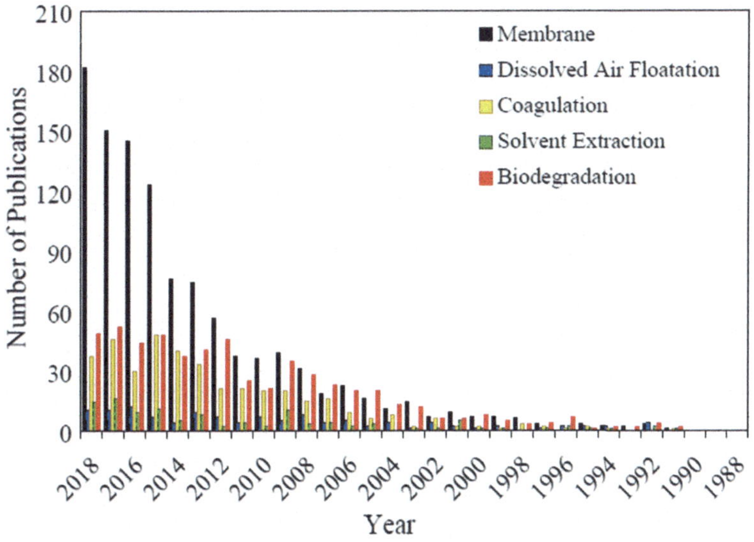

Fig. 1 Publication numbers related to different technologies to treat oily wastewater [10]

3.1 Membranes Processes

Both microfiltration (MF) and ultrafiltration (UF) can be applied to separate oil from water. However, deformation of bigger oil droplets through the membrane pores can occur and consequently, the oil rejection reduces [2, 12]. In stabilized emulsion with an oil droplet size greater than 1 μm, MF illustrates a high separation efficiency of more than 98.5%. However, the separation efficiency of UF and MF significantly decreases for oil droplet sizes smaller than 1 μm. Because of superior selectivity and ease of continuous and automatic operational conditions, MF is recommended as an alternative to traditional methods [5, 13]. Furthermore, the formation of fouling and concentration polarization layers is the main challenge for the membrane processes [2]. Considering all these factors, both ceramic and polymeric membrane types can be used [1, 7].

3.1.1 Polymeric Membranes

Different polymeric membranes like polyvinylidene fluoride (PVDF) and polyethersulfone (PES) have been widely applied. The majority of the polymeric membranes used have higher costs for their raw materials and preparation methods [14]. Due to the presence of oleophilic properties of the polymeric membranes, it is necessary to enhance their hydrophilicity and anti-oil adhesion properties. However, various surface modification methods involve high costs and complex procedures. Frequently, the modified membranes can separate O/W emulsion in the microscale, however, they have failed in the nanometer scale of O/W emulsion [4]. Polymeric membranes suffer from deterioration and swelling in the presence of organic compounds, especially under the operating temperature above 80 °C [15]. Furthermore, membrane fouling is a severe constraint regarding the use of polymeric membranes [8]. Therefore, the use of ceramic membranes is recommended in most cases [15].

3.1.2 Ceramic Membranes

Different types of ceramic membranes such as zeolite membranes like NaA, etc., having micropores within the lattice, as well as alumina and mullite membranes, with intra-particulate pores, have been used for oily wastewater treatment. The most important advantage of ceramic membranes is their ability to achieve the treatment objectives without the need for pretreatment [5–7]. They also benefit from other merits such as higher flux due to their higher porosity, better hydrophilicity, and higher resistance to operating conditions particularly regarding chemical, mechanical, and thermal aspects. The latter merit results in better performance in the case of ceramic membranes. However, ceramic membranes suffer from sealing problems because of their module housing. Furthermore, in some specific cases, when a layer, like a zeolite, is coated on a tubular ceramic membrane, it is difficult to get a fully coated layer without defects which makes it not an easy candidate [16].

Generally, oily wastewater contains different salt compounds, heavy chemical contents, and high acidic and alkaline compounds most of the time. Each of the such compounds can have severe influences on the membrane performance. Accordingly, ceramic membranes, based on their superior stability and longer lifetime in the harsh environment, are better candidates [1, 7, 17]. Compared to polymeric ones, ceramic membranes also have better mechanical strength; however, they have usually higher capital costs [4, 7]. Consequently, in this chapter, the properties, preparation, and performances of mullite and mullite-alumina membranes as low-cost alternatives are thoroughly surveyed.

4 Preparation of Mullite and Mullite-Alumina Membranes

Among different ceramic membranes, mullite and mullite-alumina membranes are inexpensive due to their low-cost raw materials, additives, and preparation methods [18].

4.1 Low-Cost Ceramic Membranes

The main disadvantage of ceramic membranes is their high cost in large-scale application [19]. Therefore, use of low-cost materials to prepare ceramic membranes has been expanded recently [14]. The preparation cost of mullite, α-alumina, and stainless steel membranes are approximately 7, 500, and 3000 $\$/m^2$, respectively [14, 19]. As clearly observed, the mullite membrane is more cost-effective than the conventional ceramic membranes [14]. Using inexpensive natural alumino-silicalite minerals like bauxite, kaolin clay, pozzolan, sepiolite, bentonite, attapulgite, and industrial waste such as fly ash, rice husk ash, BF slag, etc. can be recommended as low-cost materials for the preparation of the mullite membranes [19].

The author's group synthesized mullite and mullite–alumina based symmetric MF membranes from kaolin clay and α-alumina powder for comparison of performance for treatment using real and synthetic wastewaters under best operating conditions (3 bar pressure, 1.5 m/s cross-flow velocity and 35 °C temperature). The results showed that the mullite membrane has the highest rejection (93.8%) and the lowest fouling resistance (28.97%) while the mullite–alumina membrane with 75% alumina performs high permeation flux (244 L/m^2 h) and lower rejection (81.3%) [18].

Mohammadi et al. [20] prepared tubular mullite membranes with an average pore radius is 0.476 μm by extruding a mixture of about 62–69% kaolin clay (containing 64% kaolinite, 2.4% illite, 27% quartz, and 6.6% feldspar) and distilled water followed by drying at room temperature and calcination at 1250 °C for 3 h. However, mullite-alumina (50% alumina content) based MF membranes with 49%

porosity were synthesized at 1350 °C for 3 h from kaolin clay and commercial grade α-alumina with 99.6% purity. The performance of these lab-made membranes was evaluated in the hybrid PAC-MF system with a synthetic wastewater sample of oil-in-water emulsions (1000 ppm oil in water) having a mean droplet size of 1.09 μm, prepared by mixing condensate gas (C8-C12) and distilled water using Triton X-100 (0.01 wt-%) as an emulsifier. The results showed that PAC at low concentration improved the permeate flux of the MF system from 79.76 L m–2h–1 for the mullite membrane to 95.43 L m–2h–1 for the PAC-MF system with 200 ppm PAC when the fouling resistance decreased from 29.87 to 13.91%. By adding PAC at the optimum concentration (200 ppm for the mullite and 400 ppm for the mullite-alumina membranes) and operating conditions of pressure 3 bar, cross-flow velocity 1.5 m s^{-1}, and temperature 35 °C, permeate flux increases 19.65% for the mullite and 61% for the mullite-alumina membranes. Also, by addition of PAC up to 1200 ppm, TOC rejection increases from 93.8 to 97.4% for the mullite and from 89.6 to 92.4% for the mullite-alumina membranes [20].

Fang et al. [21] reported a new kind of spherical-fly-ash-based ceramic membrane which revealed a high permeate flux of 159 L m^{-2} h^{-1} in the microfiltration of oil-in-water emulsion at a low transmembrane pressure of 0.10 MPa after continuously running for 10 h without backwashing. The rejection of oil droplets of more than 95% exhibited lower membrane fouling resistances. The formation of the cake layer and internal pore blocking was attributed to the fouling mechanisms for the spherical-fly-ash-based membrane [21].

Chen et al., 2020 [22] elucidated a waste-to-resource strategy for the development of cost-effective and high-performance robust functional ceramic membranes for the treatment of oil-in-water emulsion wastewater using low-cost $NiAl_2O_4$ spinel-based hollow fiber ceramic membranes with excellent surface hydrophilicity from a mixture of nickel-laden sludge (NiO) and naturally abundant bauxite mineral [22].

Fu et al. [23] developed low-cost high-performance ceramic membranes with novel structures by fabricating a highly porous whisker-structured mullite derived from waste coal fly ash (containing 55.05 wt.% SiO_2 and 31.97 wt.% Al_2O_3, together with 1.58 wt.% CaO) and $Al(OH)_3$ as raw materials and MoO_3 as a single sintering additive. Without any pore-forming agent, the open porosity was found to increase significantly from 42 to 58% after sintering at 1300°C, without degradation of the mechanical properties, with increasing MoO_3 content from 0 to 20 wt.%, while shrinkage and pore size decreased [23].

*I*nexpensive whisker-constructed ceramic membranes made from coal fly ash and bauxite with WO_3 additive after sintering at 1200 °C, exhibited as high as 94% mullite phase which resulted in highly efficient treatment performance of real oily emulsions with exceptional permeance and less energy input under key operating parameters. Such low-cost mullite membranes with a novel whisker-constructed structure show higher porosity and water permeance, and better surface oleophobicity in water compared with common ceramic membranes [24].

In a recent literature, low-cost porous mullite whisker membranes were synthesized from kaolin and bauxite and the optimized sample showed needle-shaped interlocked network structures with 95.4 wt % of mullite phase. The oil rejection was found to be 95.4% at 3 bar along with an apparent porosity of 50.37%, biaxial flexural strength of 75.21 MPa, pure water permeance of 838.64 L/m2 h bar prepared under optimum conditions of sintering at 1300 °C using 4 wt % of AlF3.3H2O and 6 wt % of MoO3 as additives. The monolithic mullite membrane exhibited excellent performance and lower production costs in comparison to monolithic ceramic MF membranes reported in earlier studies [25].

4.2 Low-Cost Additives

Additives can not only take part in phase formation, but also help as pore formers. Different additives have been reported in the literature, as listed in Table 1. For instance, calcium carbonate and quartz help in pore formation and mechanical stability, respectively. At low temperatures, feldspar has been employed for forming a glassy phase and as a fluxing agent or sintering aid. To provide handling strength of the green body and, the inclusion of ball clay helps to enhance the plasticity of mix composition [26].

4.3 Preparation Methods

For membrane fabrication, kaolin clay is a good candidate due to its smaller particle size. Milling can also help to decrease the particle size readily. Most commonly, kaolin particles are sintered in temperatures ranging between 850 and 1550 °C for the membrane production. Different methods such as uniaxial pressing, slip casting, tape casting, extrusion, sol–gel coating, and solid-state sintering have been used to prepare ceramic membranes [37]. Although higher temperatures lead to improved mechanical strength by better bonding between ceramic particles, it reduces overall porosity due to the formation of the glassy phase. Mechanical strength is also controlled by the additives, as mentioned before. There are different approaches to test the stability of the membranes, i.e., three-point bending, compression, biaxial flexural strength, and diametrical compression tests. In general, the three-point bending test is mostly used to facilitate comparative analysis. Different kinds of silica and alumina sources can be used to prepare mullite [40]. The preparation of mullite from kaolin via high-temperature calcination is presented below. Kaolin as the source of silica and alumina is converted to mullite according to the following reaction (Eq. 1) through the transformation of kaolinite in to amorphous metakaolinite phase.

$$3(Al_2O_3 \cdot 2SiO_2) \rightarrow 3Al_2O_3 \cdot 2\,SiO_2 + 4SiO_2 \quad (1)$$

Before sintering, different phases like quartz, calcium carbonate, pyrophyllite, feldspar, and kaolin can be observed in the mix composition by XRD analysis. Various kinds of phase transformation occur during the sintering process, in which

Table 1 Utilization of different additives in mullite and mullite-alumina membranes in various applications [26]

Additive	Membrane type	Feed stream contents	References
Boric acid, calcium carbonate, sodium carbonate, and Quartz	MF	–	[27]
Sodium metasilicate, sodium carbonate, calcium carbonate, quartz, boric acid	MF	Mosambi juice	[28]
Quartz, calcium carbonate, sodium carbonate, boric acid, and sodium metasilicate	MF	O/W emulsion	[29]
Quartz, ball clay, pyrophyllite, and feldspar	MF	O/W emulsion	[29]
Quartz and calcium carbonate	MF	Oil and bacteria	[30]
Limestone	Support layer	–	[31]
Lime	Support layer	–	[32]
Boric acid, sodium metasilicate nanohydrate, and feldspar	MF	–	[33]
Dolomite	Support layer		[34]
Calcium carbonate	Support layer		[35]
Calcite	Support layer		[36]
Bentonite, talc, sodium borate, and carbon black	MF	O/W emulsion	[37]
Bauxite	MF	–	[38]
Calcium carbonate, alumina, quartz, and Ball clay	MF	–	[39]
Ball clay, feldspar, calcium carbonate, and pyrophyllite	Support layer	–	[39]

the reaction between the inherent silica and alumina contents is the main reaction for mullite formation with the presence of excess silica or alumina present in the raw materials. A porous mullite body is prepared by leaching out free silica. Leaching time is controlled according to silica content, initial porosity of mullite body, leaching solution concentration, and temperature [14, 24]. Mullite-alumina membranes are prepared by adding alumina or due to the presence of excess alumina in the raw materials used for support preparation in addition to the top layer coating using alumina-based formulations.

5 Performance Evaluation of Membranes for Oily Wastewater Treatment

Oily wastewater is generated in various domestic and industrial processes such as oil and gas, food and beverage, and metal processing. Besides the major composition emulsified oil, real oily wastewaters also contain other complicated compositions such as surfactants, organic matters, inorganic salts, and hydrocarbons. Low-cost ceramic membranes based on mineral or industrial waste such as mullite and spinel are increasingly being considered for economic treatment of O/W emulsion wastewater. Considering their moderate performance, it is required to further improve water treatment efficiency via enhancing their water permeance and antifouling ability. Membrane fouling is still a challenge that needs to be solved due to the intrinsic hydrophilic property of these ceramic membranes. Nanostructured coatings have been fabricated to modify the ceramic membrane with superhydrophobic/oleophilic surfaces [41]. Superhydrophilicity enables the membrane to alleviate oil-fouling issues while exhibiting excellent separation efficiency above 98% and fascinating permeation flux during oil/water separation. The introduction of a rough surface with hierarchical structures is crucial in constructing a superhydrophilic surface. Although some advanced membranes with nanostructures have been proven effective in separating oil/water mixtures, the cost of constructing nanostructures becomes a tremendous burden for large-scale applications [42].

As evident from above, utilization of the membrane processes on industrial scale is expected to face some challenges. Membrane fouling is always a major drawback for long-term filtration. Besides separation capability, rational tailoring of membrane surface wettability is the core of the lower fouling membrane separation process. Inorganic membranes are hydrophilic in the air but become oleophobic after immersing in water, the membranes with such an underwater superoleophobicity can prevent the oil from really touching the membrane surface. Wu et al. demonstrated a novel protocol for efficient treatment of both synthetic and real oily wastewaters with low energy consumption via employing whisker-structured ceramic membranes featuring lower processing cost, higher porosity, and permeance. Particularly, enhanced underwater oleophobicity was realized for this novel structure with higher surface roughness, improving the possibility of anti-oil-fouling ability [24].

In view of the above, some appropriate models should be developed on the laboratory scale, which can be used to forecast the membrane performances on the industrial scale. Some researchers have focused on the characterization of the impacts of different parameters on membrane performances. Different models have been developed to characterize the effects of different parameters on the membrane behavior. The results were used to evaluate the membrane quality for the treatment of a real emulsion on an industrial scale. The trend of flux variation in the treatment of W/O and O/W emulsions was surveyed by some researchers [2]. To validate these models, experimental data are needed to characterize the effects of different parameters. Accordingly, the effects of operational conditions as influential parameters on the performances of mullite and mullite-alumina membranes are surveyed in the following sub-section.

5.1 Effects of Operational Conditions

Different operational conditions like pressure, temperature, cross-flow velocity, oil concentration, and salt concentration are effective in the treatment of oily wastewater [43].

5.1.1 Pressure

Permeate flux is enhanced by increasing pressure based on Darcy's law. Enhancing pressure provides more compaction of oil droplets on the membrane surface, which can block the membrane pores. Critical pressure is defined as the lowest pressure, which is required to penetrate inside the membrane pores. Minimum membrane fouling occurs when operational pressure is lower than critical pressure. Permeate flux also has a straight relation with trans-membrane pressure (TMP). Therefore, critical pressure as the best operational pressure can prevent oil droplet breakup [44, 45]. Increasing pressure causes to increase in the thickness of concentration polarization on the membrane surface; consequently, permeate flux decreases. However, total rejection is not commonly affected by increasing TMP [15]. Abbasi et al. [44] surveyed the critical pressure of permeation for oily wastewater treatment via microscopic modeling. Their results showed that surface tension, shear rate, contact angle, as well as pore and droplet diameters are effective variables. Among these variables, the contact angle was introduced as the most influential parameter to increase the membrane performances and decrease membrane fouling.

5.1.2 Cross-Flow Velocity

The thickness of concentration polarization is reduced by increasing cross-flow velocity (CFV), which also increases the permeate flux. Higher CFV creates better mixing and more turbulence in the diffusion-formed boundary on the membrane surface. At higher CFV, the greater separation performance results from the concentration polarization restriction. The low solute concentration in the permeate side is created by decreasing their accumulation on the membrane surface. On the other hand, the enhanced adsorption ability of the membrane surface leads to form a severe fouling layer. Alumina shows high adsorption ability for highly viscous organic compounds such as bitumen and asphaltenes, which are available in the oil sands produced water. Use of mullite-alumina membranes may enhance the probability of the formation of a concentration polarization layer and decrease in permeate flux.

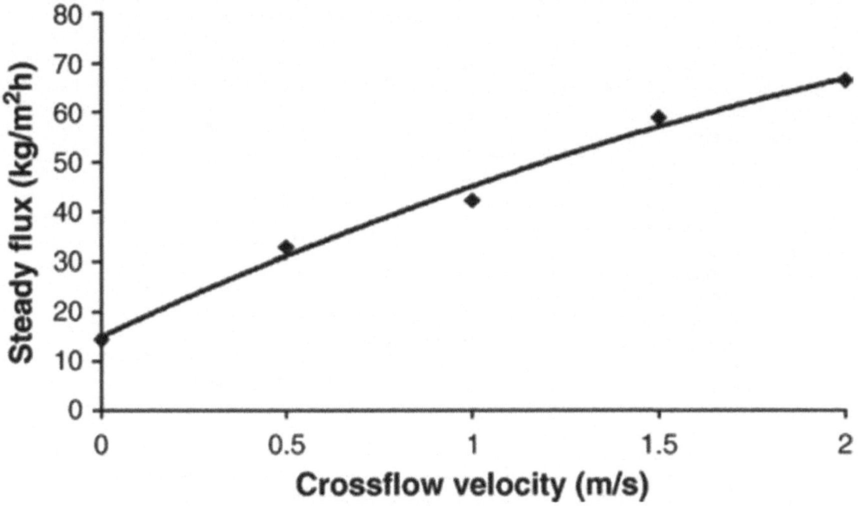

Fig. 2 Steady flux versus CFV (Oil concentration, temperature, and pressure with values of 1000 ppm, 25 °C, and 3 bar, respectively.) [45]

Therefore, higher CFV is needed in mullite-alumina membranes compared to mullite ones [15, 18].

Abbasi et al. [45] studied steady flux versus CFV, as shown in Fig. 2. Their results illustrated that steady flux and CFV have a direct correlation between these two process parameters. An increase in CFV enhances the mass transfer coefficient in the concentration boundary layer and causes better mixing on the membrane surface. By increasing CFV, the aggregation of compounds on the membrane surface is decreased and oil droplets diffuse back to the bulk solution. Furthermore, concentration polarization and fouling on the membrane surface are reduced. In general, high CFV causes increased energy consumption; consequently, high CFV is not cost-effective and cannot be recommended but should be optimized [18].

5.1.3 Temperature

Due to the presence of compounds with high viscosity, temperature shocks or partial heat recovery results in sudden alteration of viscosity of the formed layer. Decreasing the viscosity of the fouling layer simultaneously enhances its detachment as well as increases permeation flux and separation efficiency. According to Fig. 3, increasing temperature provides higher solubility of the oil phase [15, 46]. Abbasi et al. [45] recommended an operating temperature of 35°C because of increasing operational costs with increasing temperature. The steam injection into the membrane feed channels was also found effective in restricting fouling. Because of the thermal stability

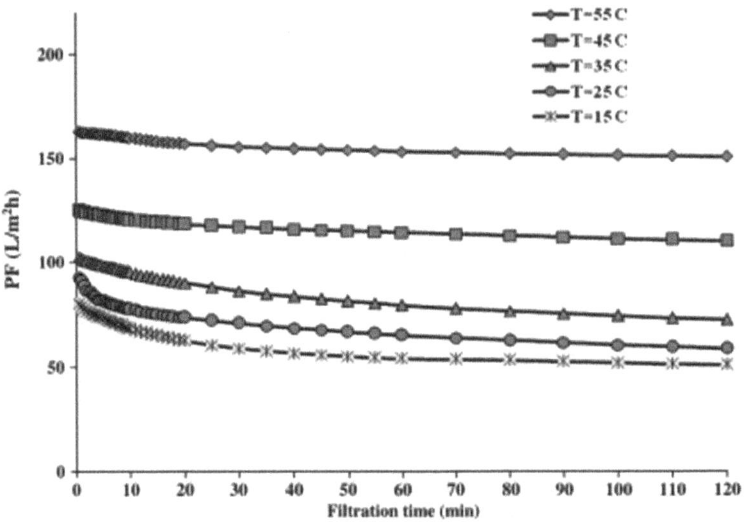

Fig. 3 Temperature versus time (Oil concentration, CFV, and pressure with values of 1000 ppm, 1.5 m/s, and 3 bar, respectively.) [18]

of ceramic membranes, thermal shock can be used without creating defects during the membrane process [15, 46].

5.1.4 Oil Concentration

Permeate flux can be affected by oil concentration. The formed oil layer can be removed with hydrodynamic flow on the membrane surface at low oil concentrations. However, complete removal of the layer is not feasible at high oil concentrations [42]. Abbasi et al. [45] reported that oil concentration variation from 250 to 3000 ppm causes sudden decreased permeate flux due to the formation of the oil layer (see Fig. 4). However, the bond between the oil layer and the mullite membrane is very weak and can be removed easily because of hydrophilicity of the membrane surface. Thus, the formation of the oil layer as the fouling layer cannot be the main problem during O/W separation using the mullite membrane.

5.1.5 Salt Concentration

In the presence of salt, properties and stabilities of O/W emulsions are changed as the concentration of ionic compounds in the solutions is varied. Hydration of the surfactant is reduced in high ionic solutions; thus, the emulsion becomes unstable [17]. Nanofiltration (NF) illustrates great performances for salt rejection, even in the case of highly salty water. Because of the shielding of the electrical double layer,

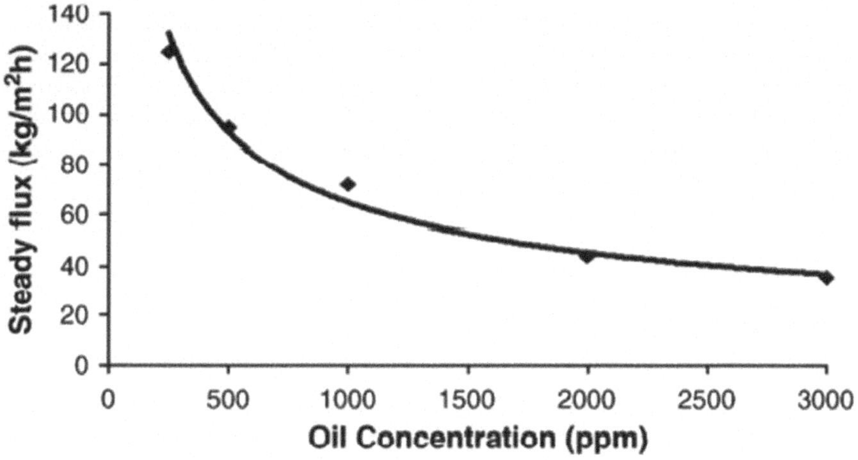

Fig. 4 Steady flux versus oil concertation (temperature, pressure, and CFV with values of 25 °C, 3 bar, and 1.5 m/s respectively.) [18]

which influences ionic repulsion, salt rejection decreases with the increase in their concentration. This is, however, less effective than that expected in the case of salt concentration higher than 40 g/l. Rejection of all salts increases with an increase in TMP due to the enhancement of convective water transport among the membrane [46]. Thickness of the double layer around the emulsion droplets is reduced by high ionic concentration in the presence of low salt concentration. Thus, the electrostatic barrier to coalescence leads to higher permeate flux and lower recovery flux. The emulsion viscosity is increased and the salt crystal accumulates on the membrane surface due to the salt concentration polarization. Furthermore, the average size of the oil droplet increases in the presence of salt. Therefore, two different effects on the permeate flux are created in the presence of salt. Formation of salt crystal and increasing oil droplet size have incremental and detrimental effects on permeate flux, respectively. Consequently, optimum salt concentration should be used. Usually, rejection is almost constant, and variation of salt concentration from 25 to 200 g/l leads to minor variation from 94.0 to 95.5% of rejection [17, 46].

5.2 Membrane Bioreactors (MBRs)

Membrane bioreactors (MBRs) performance almost relies on sludge properties and microbial species, which decompose pollutants. The oil removal directly depends on operating conditions and biological performances. Biological treatment of oily wastewater is carried out by a consortium of microorganisms [47]. Pendashteh et al. [48] reported the formation of a fouling layer in the case of hypersaline oily wastewater treatment in an MBR system. The fouling layer is formed by the deposition of organic and inorganic constituents present in the feed stream, which contain hydrocarbon compounds and extracellular polymeric substances (EPS). The overall results showed that the major part of the fouling layer is related to the presence of microorganisms. They also reported the presence of Ca, K, Mg, Fe, and Al in the fouling layer. Ghosh et al. [49] surveyed the biological treatment of oily wastewater in an aerobic granular reactor (AGR). They reported that the removal efficiency of oil, nitrate, and chemical oxygen demand (COD) is around 61.34, 99.9, and 99.9, respectively [49].

6 Future Outlook

The adhesion of oil and microorganisms on the membrane surfaces poses big challenges due to the reduced membrane service life and increased operational cost. The reusability and separation efficiency of membranes have a direct relation with their fouling and biofouling behaviors [17, 50]. The mechanism of membrane fouling is related to the oil droplet behavior like accumulation, coalescence and wetting, and deposition. The interaction between membrane and droplet determines the membrane fouling, which is affected by the charge, structure, and chemistry of the membrane surface. The concentration and composition of wastewater and operational conditions are also effective [51–53].

Another challenge is the presence of surfactant and its effects on promoting fouling [17]. Superhydrophilic/ underwater superoleophobic and superhydrophobic/ superoleophilic membranes can separate oily wastewater based on intrinsic immiscibility among oil and water phases and therefore selective separation of oil and water occurs [8]. For example, simultaneous use of fouling release and fouling resist mechanisms is recommended in the case of superhydrophilic membranes. The low surface energy of the membrane surface decreases adhesion energy among surface and oil in the fouling release case, resulting in a better release of the adsorbed oil. Formation of the hydration layer based on the hydrophilicity prevents contact between oil and membrane surface in the fouling resist case. Coupling these mechanisms provides a powerful fouling strategy [54].

Low-cost ceramic membranes can be used as porous substrates to fabricate superhydrophilic and superhydrophobic membranes. These membranes can adsorb and separate to treat oily wastewater in a one-step process. These membranes illustrate high separation efficiency up to 99%. However, using low-cost and simple preparation methods to form one layer with superhydrophilic and superhydrophobic properties on the ceramic membrane is very crucial [55]. Kayvani Fard et al. [17] reported that superhydrophilicity of hybrid Al_2O_3/AC membrane leads to high-quality treatment with separation efficiency higher than 99% of oil in an O/W emulsion in the presence of surfactant. In Fig. 5, the superior performances of superhydrophilic Al_2O_3/AC membrane compared to hydrophilic Al_2O_3 membrane are illustrated. Similarly, Ma et al. [56] demonstrated superior thermal and chemical stability and great resistance against high-salt concentration for superhydrophobic polyimide nanofibers. High permeate flux and separation efficiency and extraordinary anti-fouling behavior

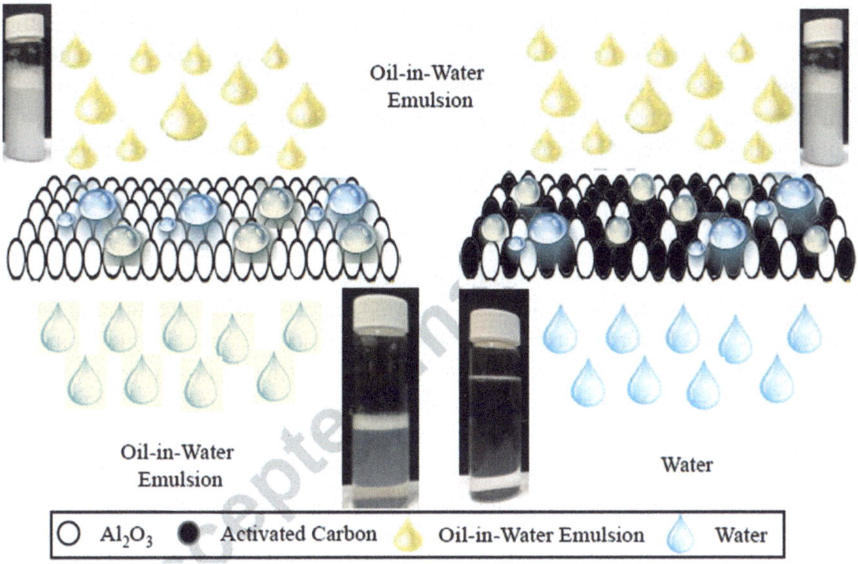

Fig. 5 Influence of hydrophilicity on treatment of oily wastewater [17]

were reported in the case of oil and water contact angles of 0° and 154°, respectively [56]. Underwater superolephobic membranes with rough micro-nanoscale by low surface free energy also were used to separate micro-nanoscale O/W emulsions in the presence of surfactant-stabilized with oil droplet size lower than 1 μm. These membranes could prevent contact between oil and membrane surface according to the presence of a stable hydration layer on the membrane surface, resulting in anti-oil fouling characteristics [12]. Accordingly, the preparation of low-cost superhydrophilic/ superhydrophobic ceramic membranes based on the low-cost materials, preparation, and modification should be considered as future research [17, 56].

Furthermore, the use of nanomaterials for membrane preparation is proposed in the application of O/W separation. Because of tunable pore size and very thin active separation layer, high permeate flux and separation efficiency can be concluded in the case of nanometer and micrometer-sized emulsions [4, 9]. Accordingly, the preparation of membranes by nanoscale raw material with high permeate flux and separation efficiency should also be considered worthy of further investigation as future research.

7 Comparison Literature

Different researches in the fields of oily wastewater treatment using mullite and mullite-alumina membranes are classified in Table 2. Ceramic membranes in various modules like disk, tubular, and hollow fiber have been used [7]. As seen, the majority of researches are focused on using mullite membranes. This can be resulted from the high cost of alumina as the starting material and its preparation process (sintering temperature up to 1700 °C) [5–7]. However, the average permeate flux of mullite membranes is less than that of mullite-alumina membranes even in the case of tubular or hollow fiber. These results specify the need for surface modification of mullite membranes to increase their permeate flux besides keeping their rejection percentage constant. Firstly, the presence of additives with appropriate concentration can provide all said needs. According to the literature, the preparation of superhydrophilic and superhydrophobic membranes can also be proposed. In the case of alumina membranes, Kayvani Fard et al. reported increased permeate flux and rejection from 3.0 to 7.5 L/m^2.h and 50 to 99% due to reduced contact angle from 13° to 0°, respectively [17]. Furthermore, the utilization of superhydrophilic and superhydrophobic membranes can restrict the formation of the fouling layer. Although, low-cost and facile modification methods should be used [57]. Kamelian et al. [57] surveyed the preparation of superhydrophobic mullite membrane for in-situ recovery of biofuels. According to the acceptable performances of mullite membranes, as shown in Table 2, the preparation of these membranes with special characteristics such as superhydrophilicity and superhydrophobicity should be considered for future research.

8 Conclusion

Mullite and mullite-alumina ceramic membranes as low-cost membranes illustrate perfect performances in the case of oily wastewater treatment with TOC rejection higher than 94% for oil concentrations lower than 1000 ppm. Compared to mullite membranes, alumina-mullite has higher permeation flux and lower rejection because of its higher porosity and average pore diameter. The permeate flux is enhanced with increasing CFV, temperature, pressure, and alumina concentration. However, increasing oil concentration decreases permeate flux. On the other hand, oil rejection slightly enhances with increasing pressure, oil concentration, salt concentration, and alumina concentration. To provide better performances, the utilization of superhydrophilic/ superhydrophobic mullite and mullite-alumina membranes should be surveyed in future research.

Table 2 Comparison of performances of different researches in the field of oily wastewater treatment regarding mullite and mullite-alumina membranes

Feed type	A	M	OC	O. P	T	CFV	M. S	P	M. P. S	R	P. F	References
			(mg/L)	(bar)	(°C)	(m/s)	(MPa)	(%)	(μm)	(%)	(kg/m^2.h)	
Mullite membranes												
Synthetic oily wastewater	–	F	200	0.7	–	–	10	34	0.46	96	–	[14]
	TiO_2	F	200	0.7	–	–	–	48	1.32	99.2	–	[58]
	–	F	50	1.6	25	–	–	42	0.55	97.3	0.12	[59]
	–	F	400	2	–	–	25	37.4	2.16	97.9	0.48	[60]
	TiO_2	H.F	200	0.25	–	0.15	–	51	1.02	97	3.75	[61]
	TiO_2	F	200	0.7	–	–	10	23	0.25	98.4	3.9	[62]
	CNT	H.F	1000	1	25	–	–	62.4	–	100	6.5	[62]
	–	T	1000	4	25	1	–	48	–	–	40	[45]
	–	T	800	2	35	1	–	33	0.54	99.3	50	[63]
	–	T	200	3	25	1.5	21.6	49.4	1.79	97.2	61.7	[6]
	–	T	1000	3	25	1	–	41	0.3	93.8	72.7	[18]
	Z	T	800	2	35	1	–	36	0.81	98	100	[18]
	–	T	1500	1	–	–	–	35	0.35	98.5	100	[43]
	PVA	T	200	2	25	1.5	–	40	1	97.3	123	[5]
	Z	T	1500	1	–	–	–	50	0.38	98.2	150	[43]
	Z–AC	T	1500	1	–	–	–	80	0.32	97.4	200	[43]
Real oily wastewater	–	T	1000	3	25	1	–	41	0.3	84	41.3	[18]
Mullite-alumina membranes												
Synthetic oily wastewater	–	T	1000	3	25	1	–	41	0.3	81.3	244	[18]
	–	T	800	2	35	1	–	34	0.74	99.2	250	[43]
	Z	T	800	2	35	1	–	46	1.01	98.5	400	[43]
Real oily wastewater	–	T	1000	3	25	1	–	41	0.3	70.8	91.5	[18]

A: additives; PVA: polyvinyl alcohol; Z: Zeolite; AC: Activated carbon; CNT: Carbon nanotube; M: Modules; T: Tubular; F: Flat; H.F: Hollow fiber; O. C: Oil concentration; O. P: operating pressure; M. S: Mechanical strength; P.F: Permeation flux; R: Rejection; P: Porosity; M. P. S.: Mean pore size; T: temperature; CFV: Cross flow velocity

References

1. Othman NH, Alias NH, Shahruddin MZ, Hussein SNCM, Dollah A (2017) Supported graphene oxide hollow fibre membrane for oily wastewater treatment. AIP Conf Proc 1901:1–7. https://doi.org/10.1063/1.5010445

2. Sutrisna PD, Candrawan J, Tangguh WW (2019) Microfiltration of oily waste water: a study of flux decline and feed types. IOP Conf Ser Mater Sci Eng 543(1). https://doi.org/10.1088/1757-899X/543/1/012079
3. Shoba B, Jeyanthi J, Vairam S (2020) Synthesis, characterization of cellulose acetate membrane and application for the treatment of oily wastewater 41(12). Taylor & Francis
4. Yan L et al (2019) Shrimp shell-inspired antifouling chitin nanofibrous membrane for efficient oil/water emulsion separation with in situ removal of heavy metal ions. ACS Sustain Chem Eng 7(2):2064–2072. https://doi.org/10.1021/acssuschemeng.8b04511
5. Sheikhi M, Arzani M, Mahdavi HR, Mohammadi T (2019) Kaolinitic clay-based ceramic microfiltration membrane for oily wastewater treatment: assessment of coagulant addition. Ceram Int 45(14):17826–17836. https://doi.org/10.1016/j.ceramint.2019.05.354
6. Arzani M, Mahdavi HR, Bakhtiari O, Mohammadi T (2016) Preparation of mullite ceramic microfilter membranes using Response surface methodology based on central composite design. Ceram Int 42(7):8155–8164. https://doi.org/10.1016/j.ceramint.2016.02.022
7. Arzani M, Mahdavi HR, Sheikhi M, Mohammadi T, Bakhtiari O (2018) Ceramic monolith as microfiltration membrane: preparation, characterization and performance evaluation. Appl Clay Sci 161(February):456–463. https://doi.org/10.1016/j.clay.2018.05.021
8. Wei G, Dong J, Bai J, Zhao Y, Qin C (2020) Turning tissue waste into high-performance microfiber filters for oily wastewater treatment gaoliang. Materials (Basel) 13(2). https://doi.org/10.3390/ma13020378
9. Zhang W, Liu N, Cao Y, Lin X, Liu Y, Feng L (2017) Superwetting porous materials for wastewater treatment: from immiscible oil/water mixture to emulsion separation. Adv Mater Interfaces 4(10). https://doi.org/10.1002/admi.201700029
10. Tanudjaja HJ, Hejase CA, Tarabara VV, Fane AG, Chew JW (2019) Membrane-based separation for oily wastewater: a practical perspective. Water Res 156:347–365. https://doi.org/10.1016/j.watres.2019.03.021
11. Yu L, Kanezashi M, Nagasawa H, Tsuru T (2020) Phase inversion/sintering-induced porous ceramic microsheet membranes for high-quality separation of oily wastewater. J Memb Sci 595:117477. https://doi.org/10.1016/j.memsci.2019.117477
12. Shi G et al (2020) Effective separation of surfactant-stabilized crude oil-in-water emulsions by using waste brick powder-coated membranes under corrosive conditions. Green Chem 22(4):1345–1352. https://doi.org/10.1039/c9gc04178h
13. Pei B, Chen J, Liu P, He T, Li X, Zhang L (2019) Hyperbranched poly(amidoamine)/TMC reverse osmosis membrane for oily saline water treatment. Environ Technol (United Kingdom) 40(21):2779–2788. https://doi.org/10.1080/09593330.2018.1452982
14. Vasanth D, Suresh K, Pugazhenthi G (2014) Fabrication of circular shaped ceramic membrane using mixed clays by uniaxial compaction method for the treatment of oily wastewater. Int J Nano Biomater 5(1):75–88. https://doi.org/10.1504/IJNBM.2014.061812
15. Atallah C, Mortazavi S, Tremblay AY, Doiron A (2019) In-process steam cleaning of ceramic membranes used in the treatment of oil sands produced water. Ind Eng Chem Res. https://doi.org/10.1021/acs.iecr.9b02257
16. Kamelian FS, Mohammadi T, Naeimpoor F (2019) Fast, facile and scalable fabrication of novel microporous silicalite-1/PDMS mixed matrix membranes for efficient ethanol separation by pervaporation. Sep Purif Technol 229. https://doi.org/10.1016/j.seppur.2019.115820
17. Kayvani Fard A, Bukenhoudt A, Jacobs M, McKay G, Atieh MA (2018) Novel hybrid ceramic/carbon membrane for oil removal. J Memb Sci 559:42–53. https://doi.org/10.1016/j.memsci.2018.05.003
18. Abbasi M, Mirfendereski M, Nikbakht M, Golshenas M, Mohammadi T (2010) Performance study of mullite and mullite-alumina ceramic MF membranes for oily wastewaters treatment. Desalination 259(1–3):169–178. https://doi.org/10.1016/j.desal.2010.04.013
19. Aziz MHA et al (2019) Pretreated aluminium dross waste as a source of inexpensive alumina-spinel composite ceramic hollow fibre membrane for pretreatment of oily saline produced water. Ceram Int 45(2):2069–2078. https://doi.org/10.1016/j.ceramint.2018.10.110

20. Abbasi M, Sebzari MR, Mohammadi T (2011) Enhancement of oily wastewater treatment by ceramic microfiltration membranes using powder activated carbon. Chem Eng Technol 34(8):1252–1258. https://doi.org/10.1002/ceat.201100108
21. Fang J, Qin G, Wei W, Zhao X, Jiang L (2013) Elaboration of new ceramic membrane from spherical fly ash for microfiltration of rigid particle suspension and oil-in-water emulsion. Desalination 311:113–126. https://doi.org/10.1016/j.desal.2012.11.008
22. Chen et al (2020) Spinel-based ceramic membranes coupling solid sludge recycling with oily wastewater treatment. Water Res 169:115180. https://doi.org/10.1016/j.watres.2019.115180
23. Fu M et al (2019) Waste recycling of coal fly ash for design of highly porous whisker-structured mullite ceramic membranes. J Eur Ceram Soc 39:5320–5331. https://doi.org/10.1016/j.jeurceramsoc.2019.08.042
24. Wu H, Sun C, Huang Y, Zheng X, Zhao M, Gray S, Dong Y (2012) Treatment of oily wastewaters by highly porous whisker-constructed ceramic membranes: Separation performance and fouling models. Water Res 211:118042. https://doi.org/10.1016/j.watres.2022.118042
25. Naseri M, Omidkhah MR (2023) Optimizing the fabrication conditions of monolithic mullite whisker membrane from kaolin and bauxite using the Taguchi method. Ceram Int 49:23612–23626. https://doi.org/10.1016/j.ceramint.2023.04.196
26. Abdullayev A, Bekheet MF, Hanaor DAH, Gurlo A (2016) Materials and applications for low-cost ceramic membranes. Membranes (Basel) 9(9). https://doi.org/10.3390/membranes9090105
27. Boudaira B, Harabi A, Bouzerara F, Zenikheri F, Foughali L, Guechi A (2016) Preparation and characterization of membrane supports for microfiltration and ultrafiltration using kaolin (DD2) and CaCO3. Desalin Water Treat 57(12):5258–5265. https://doi.org/10.1080/19443994.2015.1021998
28. Emani S, Uppaluri R, Purkait MK (2013) Preparation and characterization of low cost ceramic membranes for mosambi juice clarification. Desalination 317:32–40. https://doi.org/10.1016/j.desal.2013.02.024
29. Emani S, Uppaluri R, Purkait MK (2014) Cross flow microfiltration of oil-water emulsions using kaolin based low cost ceramic membranes. Desalination 341(1):61–71. https://doi.org/10.1016/j.desal.2014.02.030
30. Vasanth D, Pugazhenthi G, Uppaluri R (2011) Fabrication and properties of low cost ceramic microfiltration membranes for separation of oil and bacteria from its solution. J Memb Sci 379(1–2):154–163. https://doi.org/10.1016/j.memsci.2011.05.050
31. Harabi A, Guechi A, Condom S (2012) Production of supports and filtration membranes from Algerian kaolin and limestone. Procedia Eng 33(2011):220–224. https://doi.org/10.1016/j.proeng.2012.01.1197
32. Guechi A et al (2016) Elaboration and characterization of tubular supports for membranes filtration. Desalin Water Treat 57(12):5246–5252. https://doi.org/10.1080/19443994.2015.1021096
33. Das B, Chakrabarty B, Barkakati P (2016) Preparation and characterization of novel ceramic membranes for micro-filtration applications. Ceram Int 42(13):14326–14333. https://doi.org/10.1016/j.ceramint.2016.06.125
34. Kouras N et al (2017) Macro-porous ceramic supports for membranes prepared from quartz sand and calcite mixtures. J Eur Ceram Soc 37(9):3159–3165. https://doi.org/10.1016/j.jeurceramsoc.2017.03.059
35. Boudaira B, Harabia A, Bouzerara F, Condom S (2009) Preparation and characterization of microfi ltration membranes and their supports using kaolin (DD2) and CaCO3. Desalin Water Treat 9(1–3):142–148. https://doi.org/10.5004/dwt.2009.764
36. Harabi A, Zenikheri F, Boudaira B, Bouzerara F, Guechi A, Foughali L (2014) A new and economic approach to fabricate resistant porous membrane supports using kaolin and CaCO3. J Eur Ceram Soc 34(5):1329–1340. https://doi.org/10.1016/j.jeurceramsoc.2013.11.007
37. Eom JH, Kim YW, Yun SH, Song IH (2014) Low-cost clay-based membranes for oily wastewater treatment. J. Ceram. Soc. Japan 122(1429):788–794. https://doi.org/10.2109/jcersj2.122.788

38. Zhu Z, Wei Z, Sun W, Hou J, He B, Dong Y (2016) Cost-effective utilization of mineral-based raw materials for preparation of porous mullite ceramic membranes via in-situ reaction method. Appl Clay Sci 120:135–141. https://doi.org/10.1016/j.clay.2015.11.020
39. Almandoz MC, Marchese J, Prádanos P, Palacio L, Hernández A (2004) Preparation and characterization of non-supported microfiltration membranes from aluminosilicates. J. Memb. Sci. 241(1):95–103. https://doi.org/10.1016/j.memsci.2004.03.045
40. Kazemimoghadam M, Mohammadi T (2011) Preparation of nano pore hydroxysodalite zeolite membranes using of kaolin clay and chemical sources. Desalination 278(1–3):438–442. https://doi.org/10.1016/j.desal.2011.05.042
41. Gao N et al (2018) Ceramic membranes with mussel-inspired and nanostructured coatings for water-in-oil emulsions separation. Sep Purif Technol 212(2019):737–746. https://doi.org/10.1016/j.seppur.2018.11.084
42. Huang et al (2018) Fabrication of zinc oxide nanostructure coated membranes for efficient oil/water separation. J Membr Sci 566:249–257. https://doi.org/10.1016/j.memsci.2018.09.007
43. Jafari B, Abbasi M, Hashemifard SA (2020) Development of new tubular ceramic microfiltration membranes by employing activated carbon in the structure of membranes for treatment of oily wastewater. J Clean Prod 244:118720. https://doi.org/10.1016/j.jclepro.2019.118720
44. Monfared MA, Kasiri N, Mohammadi T (2016) Microscopic modeling of critical pressure of permeation in oily waste water treatment: Via membrane filtration. RSC Adv 6(75):71744–71756. https://doi.org/10.1039/c6ra12266c
45. Abbasi M, Salahi A, Mirfendereski M, Mohammadi T, Pak A (2010) Dimensional analysis of permeation flux for microfiltration of oily wastewaters using mullite ceramic membranes. Desalination 252(1–3):113–119. https://doi.org/10.1016/j.desal.2009.10.015
46. Voigt I et al (2019) Treatment of oily and salty mining water by ceramic nanofiltration membranes. Chem-Ing-Tech 91(10):1454–1459. https://doi.org/10.1002/cite.201900064
47. Jamaly S, Giwa A, Hasan SW (2015) Recent improvements in oily wastewater treatment: progress, challenges, and future opportunities. J Environ Sci (China) 37:15–30. https://doi.org/10.1016/j.jes.2015.04.011
48. Pendashteh AR, Abdullah LC, Fakhru'L-Razi A, Madaeni SS, Zainal Abidin Z, Awang Biak DR (2012) Evaluation of membrane bioreactor for hypersaline oily wastewater treatment. Process Saf Environ Prot 90(1):45–55. https://doi.org/10.1016/j.psep.2011.07.006
49. Ghosh S, Chakraborty S (2019) Influence of inoculum variation on formation and stability of aerobic granules in oily wastewater treatment. J Environ Manage 248. https://doi.org/10.1016/j.jenvman.2019.07.010
50. Yong M, Zhang Y, Sun S, Liu W (2019) Properties of polyvinyl chloride (PVC) ultrafiltration membrane improved by lignin: hydrophilicity and antifouling. J. Memb. Sci. 575:50–59. https://doi.org/10.1016/j.memsci.2019.01.005
51. Liu L, Yuan W (2018) A hierarchical functionalized biodegradable PLA electrospun nanofibrous membrane with superhydrophobicity and antibacterial properties for oil/water separation. New J Chem 42(21):17615–17624. https://doi.org/10.1039/C8NJ03112F
52. Ao C et al (2017) Superhydrophilic graphene oxide@electrospun cellulose nanofiber hybrid membrane for high-efficiency oil/water separation. Carbohydr Polym 175:216–222. https://doi.org/10.1016/j.carbpol.2017.07.085
53. Zarghami S, Mohammadi T, Sadrzadeh M, Van der Bruggen B (2020) Bio-inspired anchoring of amino-functionalized multi-wall carbon nanotubes (N-MWCNTs) onto PES membrane using polydopamine for oily wastewater treatment. Sci Total Environ 711:134951. https://doi.org/10.1016/j.scitotenv.2019.134951
54. Siwatch P, Sharma K, Manyani N, Kang J, Tripathi SK (2021) Multifunctional Superhydrophobic adsorbents by mixed-dimensional particles assembly for polymorphic and highly efficient oil-water separation. J Alloys Compd, 1–31. https://doi.org/10.1016/j.jallcom.2021.159409
55. Jamalludin MR, Hubadillah SK, Harun Z, Othman MHD, Yunos MZ (2019) Novel superhydrophobic and superoleophilic sugarcane green ceramic hollow fibre membrane as hybrid oil sorbent-separator of real oil and water mixture. Mater Lett 240:136–139. https://doi.org/10.1016/j.matlet.2018.12.111

56. Ma W, Zhang M, Liu Z, Kang M, Huang C, Fu G (2019) Fabrication of highly durable and robust superhydrophobic-superoleophilic nanofibrous membranes based on a fluorine-free system for efficient oil/water separation. J Memb Sci 570–571:303–313. https://doi.org/10.1016/j.memsci.2018.10.035
57. Kamelian FS, Mohammadi T, Naeimpoor F, Sillanpaä¨ M (2020) "One-Step and low-cost designing of two-layered active-layer superhydrophobic silicalite-1/PDMS membrane for simultaneously achieving superior bioethanol pervaporation and fouling/biofouling resistance. ACS Appl Mater Interfaces 12(50):56587–56603. https://doi.org/10.1021/acsami.0c17046
58. Suresh K, Pugazhenthi G (2016) Development of ceramic membranes from low-cost clays for the separation of oil–water emulsion. Desalin Water Treat 57(5):1927–1939. https://doi.org/10.1080/19443994.2014.979445
59. Nandi BK, Uppaluri R, Purkait MK (2009) Treatment of oily waste water using low-cost ceramic membrane: flux decline mechanism and economic feasibility 44(12)
60. Emani S, Uppaluri R, Purkait MK (2014) Microfiltration of oil-water emulsions using low cost ceramic membranes prepared with the uniaxial dry compaction method. Ceram Int 40(1) PART A, pp 1155–1164. https://doi.org/10.1016/j.ceramint.2013.06.117
61. Zhu L, Chen M, Dong Y, Tang CY, Huang A, Li L (2016) A low-cost mullite-titania composite ceramic hollow fiber microfiltration membrane for highly efficient separation of oil-in-water emulsion. Water Res 90:277–285. https://doi.org/10.1016/j.watres.2015.12.035
62. Vasanth D, Pugazhenthi G, Uppaluri R (2013) Performance of low cost ceramic microfiltration membranes for the treatment of oil-in-water emulsions. Sep Sci Technol 48(6):849–858. https://doi.org/10.1080/01496395.2012.712598
63. Rasouli Y, Abbasi M, Hashemifard SA (2017) Oily wastewater treatment by adsorption-membrane filtration hybrid process using powdered activated carbon, natural zeolite powder and low cost ceramic membranes. Water Sci Technol 76(4):895–908. https://doi.org/10.2166/wst.2017.247

Chapter 6
Treatment of Wastewater from Electroplating Industry Using Low-Cost Ceramic Membranes

Raja Ben Amar, Wala Aloulou, Hajer Aloulou, and Afef Attia

1 Introduction

Industrial electroplating activities generated multiple pollutants such as heavy metals and oil which are considered one of the greatest concerns because of their toxicity and their accumulation in the environment (air, water and soil). In particular, heavy metals like cadmium (Cd), chromium (Cr),cobalt (Co), mercury (Hg) and lead (Pb), even in trace amounts, cause harmful effects like respiratory irritation and hepatic cell destruction [12, 13, 31]. In aquatic environments near industrialized coastline areas, these heavy metals can affect marine organisms (fish, mollusks, tuna, etc.) [32] and [57]. In addition, some contaminants such as phenols, petroleum hydrocarbons, and polyaromatic hydrocarbons. are toxic and can inhibit the growth of plants and animals. To human beings, they also bring mutagenic and carcinogenic risks [11] and [27].

The evaluation of the pollutants contained in the wastewater is a critical consideration for the selection of suitable treatment methods for oily and heavy metal removal [20, 52]. Conventional water treatment processes such as centrifugation [19], coagulation [41], adsorption [45], electrocatalytic oxidation [38], Fenton process [18], etc. can be used for treating free-floating oils and dispersed oils. However, most of them are not suitable for treating emulsions with microscopic oil droplets smaller

R. B. Amar (✉) · W. Aloulou · H. Aloulou · A. Attia
Research Unit "Advanced Technologies for Environment and Smart Cities", Faculty of Science of Sfax, University of Sfax, Sfax, Tunisia
e-mail: benamar.raja@yahoo.com

A. Attia
e-mail: afef.attia@fss.usf.tn

H. Aloulou
Department of Chemical, Preparatory Institute for Engineering Studies of Gabes, University of Gabes, Gabes 6029, Tunisia

S. Bandyopadhyay (ed.), *Application of Low Cost Ceramic Membranes in Wastewater Treatment*, Materials Horizons: From Nature to Nanomaterials,
https://doi.org/10.1007/978-981-97-5842-5_6

than 20 μm due to the low treatment efficiency [55]. Membrane separation technology represented by porous ceramic membranes appears to be a highly promising and efficient method for treating oily wastewater because of their higher separation efficiency, excellent mechanical resistance, high chemical and thermal stability in harsh environments, ease of processing, long durability and low maintenance cost [25, 30, 39, 40, 48, 54].

Several researchers have explored numerous treatment technologies to eliminate heavy metals from industrial wastewater. These treatment approaches include adsorption [42, 43], membrane filtration [6, 26], coagulation–flocculation [46], ion exchange [51] and electrochemical treatment technologies [53, 56]. Membrane technology has been widely applied to remove heavy metal ions from contaminated water [28] thanks to its relatively low energy consumption, satisfactory treatment performances, and the possibility of recycling with low co-product generation making this process more efficient and robust [37]. In addition, it was considered as successful method, for the effective removal of emulsified oils (size of drops <5 μm).

Today, various types of porous ceramic separation membranes, such as microfiltration (MF), ultrafiltration (UF) and nanofiltration (NF) are extensively studied owing to their chemical and thermal stability [24], good mechanical strength, pressure resistance, long working time and catalytic properties from their intrinsic nature [22], compared with organic membranes [36]. The commercial membranes from alumina, titania and silica are not suitable for large-scale applications because they are too expensive from a techno-economic point of view [14, 21]. In recent years, ceramic membranes based on natural minerals, such as clay [16], carbon [50], phosphate [34], mineral coal fly ash [29], zeolite [3] and sand [2] have increasingly attracted attention for their low cost. Other authors have prepared supports from natural non-expensive material for coating the MF or UF top layer, such as graphite [10], granitic sand [17] and clay [35, 44].

Against this background, the preparation of different tubular ceramic membranes from natural material was presented here. The evaluation of their performances to treat real wastewater from the electroplating industry was reported.

2 Electroplating Wastewater Characterization

In this study, the wastewater generated from SOPAL electroplating Company located in Sfax, Tunisia was selected taking into account the pollutants content in the wastewater. Two types of wastewater were identified, **EF1** (oily wastewater contaminated with oil and heavy metals) and **EF2** (wastewater contaminated with heavy metals).

Various membranes were applied for the treatment of these effluents contaminated by oil and heavy metals in order to study the influence of the effluent's composition on the membrane separation efficiency. The average effluent composition is given in Table 1.

Table 1 Characteristics of the raw wastewater [2, 3, 4, 6]

Parameters	EF1	EF2
pH	6.6–7,56	2.9
COD (mg/L)	3850–6570	1048
Conductivity (mS/cm)	3.33–4.97	3.25
Turbidity (NTU)	3610–3810	41
Oil (g/L)	40–65	–
Color (Abs. λ max)	–	1.95 (λ 550 nm)
Cr (mg/L)	0.02	79.83
Co (mg/L)	–	2.02
Fe (mg/L)	0.117	5.75
Cu (mg/L)	1.1–2.6	76.4
Cl (mg/L)	–	965
Pb (mg/L)	17–21.5	–
Zn (mg/L)	4.1–10.4	–

Raw and treated wastewater were analyzed for the determination of the physico-chemical characteristics to evaluate the pollutants removal, especially heavy metals and oil content.

The pollutant retention (R %) through the membrane was calculated according to the following equation:

$$R(\%) = \frac{(C_f - C_p)}{C_f} * 100 \quad (1)$$

where $\mathbf{C_f}$ and $\mathbf{C_p}$ represent the concentration of pollutants in the feed and permeate, respectively.

The concentration of heavy metals and oil was determined by Atomic Absorption Spectroscopy (AAS) (Perkin Elmer A Analyst 200) and UV-Spectrophotometer (at $\lambda = 363$ nm), respectively.

3 Low-Cost Ceramic Membrane Preparation and Characterization

The asymmetric membrane preparation requires two principal steps; the elaboration of the tubular ceramic support and then the coating layer deposition.

The fabrication of the tubular support requires the preparation of a plastic paste composed of ceramic powder and organic additives, then extrusion of the paste to produce the tube that can be consolidated by sintering at a given temperature, between 800 and 1200 °C (Fig. 1).

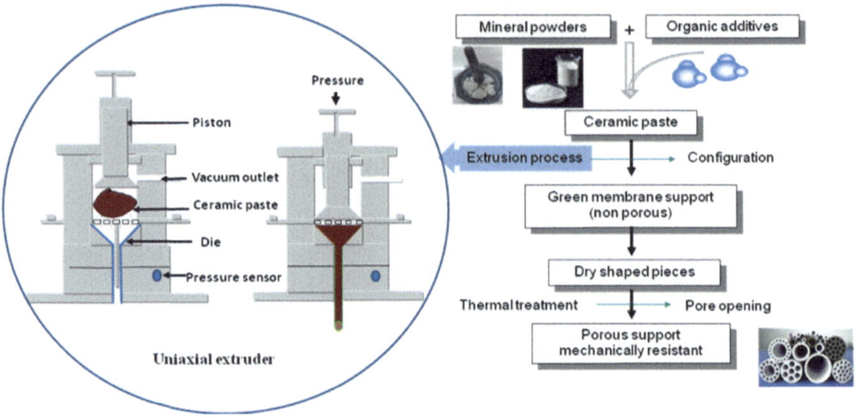

Fig. 1 The main steps of support preparation [2]

The active layer from natural powder or from nanoparticles (NP) can be prepared by different processes (slip casting and layer-by-layer) using a deflocculated slip obtained by mixing fine powder with water and polyvinyl alcohol (12% w/w aqueous solution) [2] (Fig. 2).

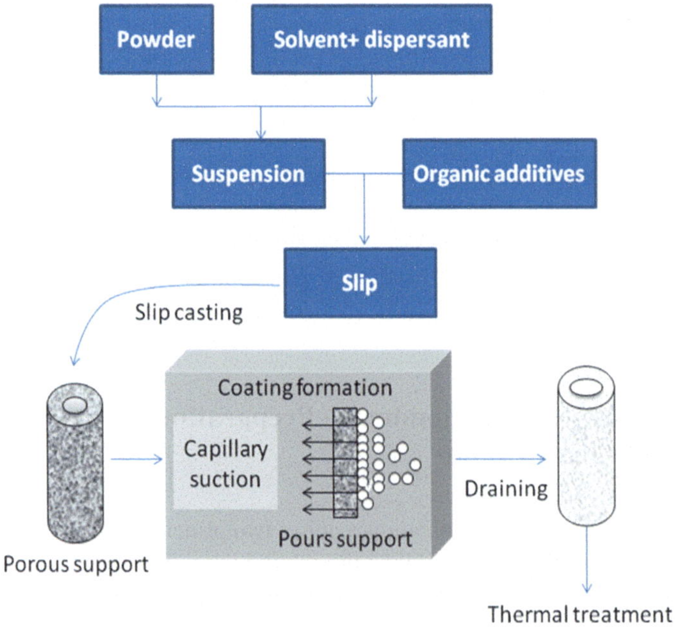

Fig. 2 The main steps of membrane preparation [50]

Table 2 Characteristics of different fabricated ceramic membranes

Membranes	Support	Active layer	Average pores size	Water permeability (L/h m^2 bar)	References
Sm/Z3	Zeolite	Smectite NP	18 nm	80	[9]
Sm/Z6	Zeolite	Smectite NP	3 nm < dp Sm/Z6 < 5 nm	59	[6]
Sm-Ti/Z 3	Zeolite	70% Smectite NP + 30% TiO_2	16 nm	95	[3]
Kaolin/Sand	Sand	Kaolin	4–17 nm	491	[2]
Kaolin/Zeolite	Zeolite	Kaolin	28 nm	182	[2]
Grafted Sand	Sand	Sand	60–160 nm	434	[4]
MFZ	Zeolite	Zeolite	0.18 μm	534	[3]

The characteristics of different ceramic membranes used for the electroplating wastewater treatment are summarized in Table 2.

4 Membrane Separation Performances to Electroplating Wastewater Treatment

4.1 Treatment of Oily Wastewater Contaminated with Heavy Metals (EF1)

The oily wastewater contaminated with heavy metals (**EF1**) was treated by different membrane separation technologies such as microfiltration (MF), ultrafiltration (UF) and membrane distillation via air gap membrane distillation (AGMD) using different types of membranes as classified below:

(i) A hydrophobic sand membrane, modified by grafting with 1H,1H,2H,2H-perfluorodecyltriethoxysilane molecule (C_8), for membrane distillation (MD) [4],
(ii) A zeolite membrane [3] for MF process,
(iii) Two types of nanoparticles-based composite membranes for the UF process: **Sm/Z3** [9] and **Sm-Ti/Z 3** [7]**,** respectively prepared from Smectite nanoparticles and a mixture of 70% Smectite nanoparticles with 30% of commercial TiO_2 over zeolite supports, and also
(iv) Two clay-based composite membranes, **kaolin/sand** and **kaolin/zeolite** successfully prepared by coating the kaolin layer over sand and a zeolite support [5].

In the case of MF and UF, the tests were conducted at 60 °C as the optimal temperature for the treatment of oily wastewater [4] and at transmembrane pressure

(TMP) of 1 bar for MF and 3 bar for UF. Using the air gap membrane distillation (AGMD), filtration tests were conducted at a transmembrane pressure of 1 bar and the temperature of the feed side was fixed at 85 °C while the cooling system temperature at the permeate side was kept constant at 5 °C.

The treatment of the industrial electroplating oily wastewater contaminated with heavy metals (**EF1**) exhibits a high stabilized permeate flux in the range of 100–336 L/h.m^2.bar at the optimal conditions using the various membrane separation technologies. From Table 3, MF, UF and AGMD membranes process showed good efficiency in the treatment of oily wastewater. These membranes have been proven to be successful in removing turbidity (>98%), COD (> 93%) and oil (≈99%). Particularly, ultrafiltration membranes (**Sm/Z3** and **Sm-Ti/Z3**) displayed the best rejection of conductivity of 46% compared to the modified sand membrane (6%) and microfiltration membrane (MFZ) (3%) due to the smallest pore size of these membranes (18 nm and 16 nm) compared to that of others.

To evaluate the decrease in conductivity, the concentration of heavy metals was estimated before and after treatment using **Sm-Ti/Z 3** UF membrane (Table 4) which showed high retention of copper, plumb and zinc, exceeding 96% [7]. This result is similar to that observed by Aloulou et al. [5], who used two composite UF membranes made from Kaolin over sand and zeolite supports, which highlighted the prominence of these new natural kaolin-based membranes (Fig. 3, Table 3).

Table 3 The percentage rejection (R %) of oily wastewater after membrane treatment [4, 5, 7]

	Oil (R %)	COD (R %)	Turbidity (R %)	Conductivity (R %)
AGMD using **Grafted Sand** (T = 85 °C, P = 1 bar)	99.7	98.0	99.9	7.0
MF membrane (**MFZ**) (T = 60 °C, P = 1 bar)	99.1	94.0	98.0	3.0
UF membrane (**Sm/Z3**) (T = 60 °C, P = 3 bar)	99.4	95.0	99.6	46.0
UF membrane (**Sm-Ti/Z 3**) (T = 60 °C, P = 3 bar)	99.9	93	99.9	46.0
UF membrane (**Kaolin/Sand**) (T = 60 °C, P = 1 bar)	99.8	97.9	99.9	–
UF membrane (**Kaolin/Zeolite**) (T = 60 °C, P = 3 bar)	99.8	98.1	99.9	–

Table 4 Retention of high metal after treatment using **Sm-Ti/Z 3** UF membrane [7]

Parameters	Raw effluent (mg/L)	Permeate Sm-Ti/Z 3 membrane (mg/L)	Limits* (mg/L)
Copper	2.631 ± 0.02	0.095 ± 0.002 **(TR: 96%)**	2
Iron	0.117 ± 0.001	0.117 ± 0.001	10
Lead	17.05 ± 0.02	0.029 ± 0.002 **(TR: 99.82%)**	1
Nickel	<0.1 ± 0.001	< 0.1 ± 0.001	1
Zinc	4.115 ± 0.02	0.102 ± 0.002 **(TR: 97.51%)**	5
Chromium	<0.02 ± 0.001	< 0.02 ± 0.001	1.5

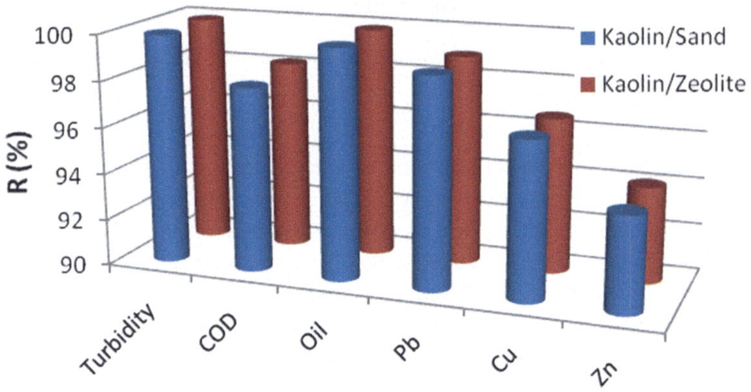

Fig. 3 Retention of pollutants by different membranes at 1 bar for kaolin/sand and 3 bar for kaolin/zeolite [5]

4.2 *Treatment of Wastewater Contaminated with Heavy Metals (EF2)*

The performance of the composite UF membrane **Sm/Z6** based on the sixth layer of smectite nanoparticles **Sm** (previously prepared by sol–gel method) over zeolite support **Z** [6] was also assessed for the treatment of **EF2** at ambient temperature and a transmembrane pressure from 3 to 7 bar in recycling mode (i.e., only the permeate was recovered). During the first 30 min of filtration (Fig. 4), the permeate flux decreased slightly with time., the flux stabilized at a constant value. The permeate flux was about 70 L/h.m^2 at a pressure of 3 bar and reached 175 L/h.m^2 at 7 bar. For a variety of ceramic membrane types, researchers have previously found an increase in the permeate flux with pressure [15, 33, 49]. For all pressure levels, the decrease in flux with time does not exceed 15%, demonstrating the antifouling properties of the UF coating layer formed from composite smectite nanoparticles. This characteristic is critical for long-term applications that require membrane stability.

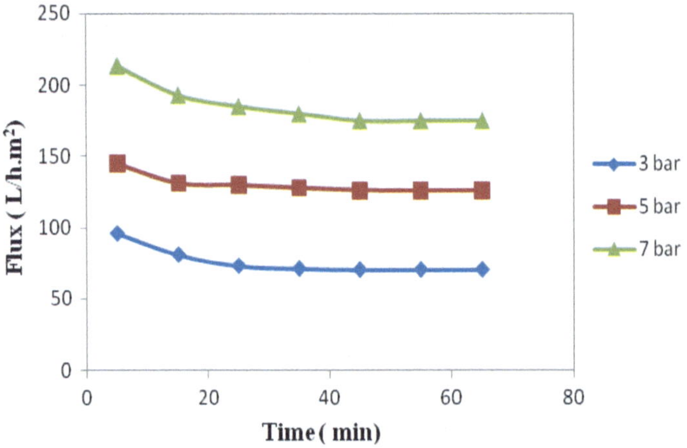

Fig. 4 **a** Evolution of the permeate flux with time at different pressures: 3, 5 and 7 bars for **EF2** using **Sm/Z6** UF membrane [6]

Figure 5 shows a quasi-total retention of the turbidity at different applied pressures. For the rest of the parameters, it can be shown that the rejection of various contaminants reduced slightly when pressure increased from 3 to 7 bars. At 3 bars, the results revealed good performance with strong rejection (>94%) of color, COD, and turbidity, as well as a 30% rejection of conductivity. In addition, the fabricated membrane displayed heavy metals removal efficiency of 90% for Cr (VI) and 60% for Co. Indeed, at pressures exceeding 3 bar, metal rejection decreased due to the passage of contaminants through the membrane's pores. Consequently, increased heavy metal retention rates were linked to pore size and effluent-membrane interaction [6].

The element Cr (VI) can exist in different ionic forms in the aqueous solution ($HCrO_4^-$, CrO_4^{-2}, $Cr_2O_7^{-2}$) depending on Cr (VI) concentration and solution pH

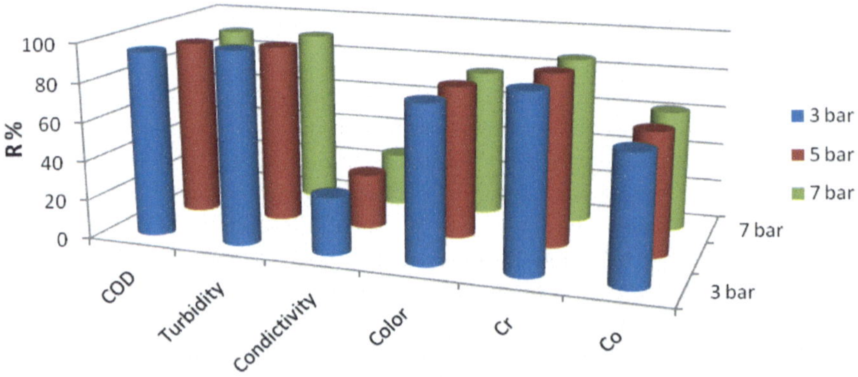

Fig. 5 **b** Percentage of pollutant retention (R %) for **EF2** using **Sm/Z6** UF membrane (P = 3, 5 and 7 bars; T = 25 °C) [6]

[6]. To determine the interaction with the membrane, the point of zero charge (pHpzc) was determined using a zetameter. In this case, the membrane is positively charged for a pH < pHpzc and negatively charged for a pH > pHpzc. The pHpzc value found is around 1.7 or 1.8, therefore **Sm/Z6** has a negative charge in the presence of **EF2** (pH = 2.9) [6]. For a pH of 2.9, the most dominant anionic species for chromium is $HCrO_4^-$ having a radius of 0.375 nm which is lower than the membrane pores (3 to 6 nm). Therefore, the strong repulsion explains the relatively high rejection of the Cr ions (90%) and the retention of Cr (VI) was not significantly impacted by the size exclusion. The high removal via the membrane may be explained by the probability of a bonding between the water molecules and smaller radius ions, which would increase their size [1]. According to Dedečk et al. [23], Co exists as Co2 + ions, which implies that the electrostatic interaction with the membrane can account for the Co rejection. It is important to note that all prepared membranes operate well when treating wastewater from industrial electroplating.

5 Conclusion

Generally, toxic substances such as heavy metals, oil, and other non-biodegradable pollutants from industrial wastewater are considered to be harmful to humans and the environment. Membrane separation technology using ceramic membranes provides permeate quality to meet the standard discharge limit as per environmental regulations. This study underlines the performances of MF, UF and MD technologies for the treatment of electroplating wastewaters including oil and heavy metals employing ceramic membranes of varying origin with respect to selectivity consideration.

The purification of the **EF1** showed a good efficiency of the MF, UF and AGMD processes in terms of Oil and heavy metals removal showing values up to 96% at optimal conditions with a stabilized permeate flux in the range of 100–336 L/h.m^2. The UF of **EF2** at 3 bar exhibits a stabilized permeate flux of 70 L/h.m^2 and heavy metals removal efficiency of 90% for Cr (VI) and 60% for Co.

The acquired results of the current study illustrate that the membrane process can play an essential role in electroplating wastewater treatment, especially for oil and heavy metals removal. The membrane properties in terms of hydrophilicity/ hydrophobicity, are very important to determine the efficiency of the membrane process to be applied in the case of oily wastewater treatment:

- The hydrophilic UF membranes prepared from zeolite, sand, clay and zirconia materials show good oil removal efficiency when applied at lab and pilot scales.
- The use of nanomaterials such as TiO2-Sm for the formation of UF active layer, enhanced the membrane hydrophilicity and consequently the UF separation performances since the coated membranes exhibited better antifouling ability.

Generally, the pollutant content varies in real oily wastewater (the presence of heavy metal is frequently detected) and their concentration usually fluctuates as the industrial processes are altered. A high rate (>95%) of retentions can be achieved with UF and MD which is very encouraging.

This particular study underlines the performances of UF and MD technologies for the treatment of oily wastewater employing low-cost ceramic membranes. However, further work is needed to enhance the throughput, anti-fouling properties and large-scale module design of these membranes.

References

1. Al Rashdi BAMD, Johnson J, Hilal N (2013) Removal of heavy metal ions by nanofiltration. Desalination 315:2–17
2. Aloulou H, Bouhamed H, Ben Amar R, Khemakhem S (2017) New ceramic microfiltration membrane from Tunisian natural sand: application for tangential wastewater treatment. Desal Wat Treat 78:41–48
3. Aloulou H, Bouhamed H, Ghorbel A, Ben Amar R, Khemakhem S (2017) Elaboration and characterization of ceramic microfiltration membranes from natural zeolite: application to the treatment of cuttlefish effluents. Desalin Water Treat 95:9–17
4. Aloulou H, Aloulou W, Daramola MO, Ben Amar R (2021) Silane-grafted sand membrane for the treatment of oily wastewater via air gap membrane distillation: study of the efficiency in comparison with microfiltration and ultrafiltration ceramic membranes. Mater Chem Phys 261:124186
5. Aloulou H, Aloulou W, Duplay J, Baklouti L, Dammak L, Ben Amar R (2022) Development of ultrafiltration kaolin membranes over sand and zeolite supports for the treatment of electroplating wastewater. Membranes 12:1066
6. Aloulou W, Aloulou H, Khemakhem M, Duplay J, Daramola MO, Ben Amar R (2020) Synthesis and characterization of clay-based ultrafiltration membranes supported on natural zeolite for removal of heavy metals from wastewater. Environ Technol Innov 18:100794
7. Aloulou W, Aloulou H, Ben Amar R (2022) Low-cost composite ultrafiltration membrane made from TiO_2 and nanocomposite clay materials over zeolite support for oily wastewater purification and heavy metals removal. Desal Wat Treat 246:166–173
8. Aloulou H, Attia A, Aloulou W, Chakraborty S, Baklouti L, Dammak L, Amar RB (2022) Statistical simulation, a tool for the process optimization of oily wastewater by crossflow ultrafiltration. Membranes 12:676. https://doi.org/10.3390/membranes12070676
9. Aloulou W, Hamza W, Aloulou H, Oun A, Khemakhem S, Jada A, Chakraborty S, Curcio S, Ben Amar R (2018) Developing of titania-smectite nanocomposites UF membrane over zeolite based ceramic support. Appl Clay Sci 155:20–29
10. Ayadi S, Jedidi I, Rivallin M, Gillot F, Lacour S, Cerneaux S, Cretin M, Ben Amar R (2013) Elaboration and characterization of new conductive porous graphite membranes for electrochemical advanced oxidation processes. J Membr Sci 446:42–49
11. Azimi A, Azari A, Rezakazemi M, Ansarpour M (2017) Removal of Heavy Metals from Industrial Wastewaters: a review. Chem Bio Eng Rev 4:37–59
12. Banni M, Jebali J, Daubeze M, Clerandau C, Guerbej H, Narbonne JF (2005) Monitoring pollution in Tunisian coasts: application of a classification scale based on biochemical markers. Biomarkers 10:105–116
13. Barhoumi S, Messaoudi I, Deli T, Saïd K, Kerkeni A (2009) Cadmium bioaccumulation in three benthic fish species, Slaria basilisca, Zosterisessor ophicephalus, Solea vulgaris collected from the Gulf of Gabes in Tunisia. J Environ Sci 21:980
14. Benito M, Sanchez MJ, Pena P, Rodriguez MA (2007) Development of a new high porosity ceramic membrane for the treatment of bilge water. Desalination 214:91–101

15. Bhattacharya P, Roy A, Sarkar S, Ghosh S, Majumdar S, Chakraborty S, Mandal A, Bandyopadhyay S (2013) Combination technology of ceramic microfiltration and reverse osmosis for tannery wastewater recovery. Water Res Ind 3:48–62
16. Bousbih S, Errais E, Darragi F, Duplay J, Trabelsi-Ayadi M, Daramola MO, Ben Amar R (2021) Treatment of textile wastewater using monolayered ultrafiltationceramic membrane fabricated from natural kaolin clay. Environ Technol 42:3348–3359
17. Bouzerara F, Harabi A, Achour S, Larbot A (2006) Porous ceramic supports for membranes prepared from kaolin and doloma mixtures. J Eur Ceram Soc 26:1663–1671
18. Cai QQ, Wu MY, Li R, Deng SH, Lee BCY, Ong SL, Hu JY (2020) Potential of combined advanced oxidation—Biological process for cost-effective organic matters removal in reverse osmosis concentrate produced from industrial wastewater reclamation: Screening of AOP pretreatment technologies. Chem Eng J 389:123419
19. Cambiella A, Benito JM, Pazos C, Coca J (2006) Centrifugal separation efficiency in the treatment of waste emulsified oils. Chem Eng Res Des 84:69–76
20. Carolin CF, Kumar PS, Saravanan A, Joshiba GJ, Naushad M (2017) Efficient techniques for the removal of toxic heavy metals from aquatic environment. Rev J Environ Chem Eng 5:2782–2799
21. Colle RD, Fortulan CA, Fontes SR (2011) Manufacture and characterization of ultra and microfiltration ceramic membranes by isostatic pressing. Ceram Int 37:1161–1168
22. Cui Z, Peng W, Fan Y, Xing W, Xu N (2013) Ceramic membrane filtration as seawater RO pre-treatment: influencing factors on the ceramic membrane flux and quality. Desal Wat Treat 51:2575–2583
23. D˘edečk J, Čapek L, Kaucký D, Zobalíck Z, Wichterlová B (2002) Siting and distribution of the coions in beta zeolite: a UV–VIS-NIR and FTIR study. J Catal 211:198–207
24. Gestel TV, Vandecasteele C, Buekenhoudt, Dotremont AC, Luyten J, Leysen R (2002) Alumina and titania multilayer membranes for nano-filtration: preparation, characterization and chemical stability. J Membr Sci 207:73–89
25. Gohari RJ, Korminouri F, Lau WJ, Ismail AF, Matsuura T, Chowdhury MNK (2015) A novel superhydrophilic PSf/HAO nanocomposite ultrafiltration membrane for efficient separation of oil/water emulsion. Sep Purif Technol 150:13–20
26. Hosseini SM, Alibakhshi H, Jashni E, Parvizian F, Shen JN, Taheri M, Ebrahimi M, Rafiei N (2020) A novel layer-by-layer heterogeneous cation exchange membrane for heavy metal ions removal from water. J Hazard Mater 381:120884
27. Huang S, Ras RHA, Tian X (2018) Antifouling membranes for oily wastewater treatment: interplay between wetting and membrane fouling. Curr Opin Colloid Interface Sci 36:90–109
28. Hubadillah SK, Othman MHD, Harun Z, Ismail AF, Rahman MA, Jaafar J (2017) A novel green ceramic hollow fiber membrane (CHFM) derived from rice husk ash as combined adsorbent-separator for efficient heavy metals removal. Ceram Int 43:4716–4720
29. Jedidi I, Saïdi S, Khemakhem S, Larbot A, Elloumi-Ammar N, Fourati A, Charfi A, Ben Salah A, Ben Amar R (2009) Elaboration of new ceramic microfiltration membranes from mineral coal fly ash applied to waste water treatment. J Hazard Mater 172:152–158
30. Kaner P, Rubakh E, Kim DH, Asatekin A (2017) Zwitterion-containing polymer additives for fouling resistant ultrafiltration membranes. J Membr Sci 533:141–159
31. Kessabi K, Kerkani A, Saïd K, Messaoudi I (2008) Involvement of Cd bioacumulation in spinal deformities occurrence in natural populations of Mediterranean killifish. Biol Trace Elem Res 128:72–81
32. Kessabi K, Navarro A, Saïd K, Messaoudi I, Pifia B (2010) Evaluation of environnemental impact on natural populations of the Mediterranean killifish aphanius fasciatus by quantitative RNA biomarkers. Mar Environ Res 70:327–333
33. Khemakhem M, Khemakhem S, Ben Amar R (2013) Emulsion separation using hydrophobic grafted ceramic membranes by air gap membrane distillation process. Colloids Surf A 436:402–407
34. Khemakhem M, Khemakhem S, Ayedi S, Ben Amar R (2011) Study of ceramic ultra-filtration membrane support based on phosphate industry sub-product: application for the cuttlefish conditioning effluents treatment. Ceram Int 37:3617–3625

35. Khemakhem S, Ben Amar R, Ben Hassen R, Larbot A, Ben Salah A, Cot L (2006) Fabrication of mineral supports of membranes for microfiltration/ultrafiltration from Tunisian clay. Ann Chim Sci Mat 31:169–181
36. Larbot A (2003) Fundamentals on inorganic membranes: present and new developments. Pol J Chem Technol 6:8–13
37. Le NL, Nunes SP (2016) Materials and membrane technologies for water and energy sustainability. Sustain Mater Technol 7:1–28
38. Linares-Hernandez I, Barrera-Diaz C, Bilyeu B, Juarez-Garciarojas P, Campos-Medina E (2010) A combined electrocoagulation– electrooxidation treatment for industrial wastewater. J Hazard Mater 175:688–694
39. Muppalla R, Jewrajka SK, Reddy AVR (2015) Fouling resistant nanofiltration membranes for the separation of oil-water emulsion and micropollutants from water. Sep Purif Technol 143:125–134
40. Mustafa G, Wyns K, Buekenhoudt A, Meynen V (2016) Antifouling grafting of ceramic membranes validated in a variety of challenging wastewaters. Water Res 104:242–253
41. Perdigón-Melón JA, Carbajo JB, Petre AL, Rosal R, García-Calvo E (2010) Coagulation-Fenton coupled treatment for ecotoxicity reduction in highly polluted industrial wastewater. J Hazard Mater 181:127–132
42. Priastomo Y, Setiawan HR, Kurniawan YS, Ohto K (2020) Simultaneous removal of lead(II), chromium(III), and copper(II) heavy metal ions through an adsorption process using C-phenylcalix [4]pyrogallolarene material. J Environ Chem Eng 8:103971
43. Sadeghi MH, Tofighy MA, Mohammadi T (2020) One-dimensional graphene for efficient aqueous heavy metal adsorption: Rapid removal of arsenic and mercury ions by graphene oxide nanoribbons (GONRs). Chemosphere 253:126647
44. Saffaj N, Persin M, Alami Younssi S, Albizane, Cretin AM, Larbot A (2006) Elaboration and characterization of micro-filtration and ultra-filtration membranes deposited on raw support prepared from natural Moroccan clay: application to filtration of solution containing dye sand salts. Appl. Clay Sci. 31:110–119
45. Shi W, Ren H, Huang X, Li M, Tang Y, Guo F (2020) Low cost red mud modified graphitic carbon nitride for the removal of organic pollutants in wastewater by the synergistic effect of adsorption and photocatalysis. Sep Purif Technol 237:116477
46. Sun J, Liu L, Yang F (2020) A WO3/PPy/ACF modified electrode in electrochemical system for simultaneous removal of heavy metal ion Cu2+ and organic acid. J Hazard Mater 394:122534
47. Sun Y, Zhou S, Pan SY, Zhu S, Yu Y, Zheng H (2020) Performance evaluation and optimization of flocculation process for removing heavy metal. Chem Eng J 385:123911
48. Susan LY, Ismail S, Ooi BS, Mustapa H (2017) Surface morphology of pvdf membrane and its fouling phenomenon by crude oil emulsion. J. Water Process. Eng. 15:55–61
49. Tahri N, Jedidi I, Ayadi S, Cerneaux S, Cretin M, Ben Amar R (2016) Preparation of an asymmetric microporouscarbon membrane for ultrafiltration separation: application to the treatment of industrial dyeing effluent. Desalin Water Treat, 1–16
50. Tahri N, Jedidi I, Cerneaux S, Cretin M, Ben Amar R (2013) Development of an asymmetric carbon microfiltration membrane: application to the treatment of industrial textile wastewater. Sep Purif Technol 118:179–187
51. Tavakoli O, Goodarzi V, Saeb MR, Mahmoodi NM, Borja R (2017) Competitive removal of heavy metal ions from squid oil under isothermal condition by CR11 chelate ion exchanger. J Hazard Mater 334:256–266
52. Tomczak W, Gryta M (2021) Application of ultrafiltration ceramic membrane for separation of oily wastewater generated by maritime transportation. Sep Purif Technol 26:1118259
53. Tran TK, Chiu KF, Lin CY, Leu HJ (2017) Electrochemical treatment of wastewater: selectivity of the heavy metals removal process. Int J Hydrogen Energy 42:27741–27748
54. Wei Z, Hou J, Zhu Z (2016) High-aluminum fly ash recycling for fabrication of cost-effective ceramic membrane supports. J Alloy Compd 683:474–480
55. Wu P, Jiang LY, He Z, Song Y (2017) Treatment of metallurgical industry wastewater for organic contaminant removal in China: Status, challenges, and perspectives. Environ Sci Water Res Technol 3:1015–1031

56. Ya V, Martin N, Chou YH, Chen YM, Choo KH, Chen SS, Li CW (2018) Electrochemical treatment for simultaneous removal of heavy metals and organics from surface finishing wastewater using sacrificial iron anode. J Taiwan Inst Chem Eng 83:107–144
57. Zhu WP, Gao J, Sun SP, Zhang S, Chung TS (2015) Poly (amidoamine) dendrimer (PAMAM) grafted on thin film composite (TFC) nanofiltration (NF) hollow fiber membranes for heavy metal removal. J Membr Sci 487:117–126

Part II
Wastewater Treatment for Food Processing and Pharmaceutical Industry

Chapter 7
Application of Low-Cost Ceramic Membranes in the Food Processing Industries Wastewater Treatment

S. Lakshmi Sandhya Rani, G. Pugazhenthi, and Vinoth Kumar Raja

1 Introduction

Water is a valuable resource for human life as it is an essential commodity for numerous purposes, including domestic use, agriculture, industries, etc. However, growing population and industrialization lead to the deterioration of water resources in many countries. For example, according to a study by the Central Pollution Control Board of India, approximately 500 billion m^3 of freshwater is used in industries. In addition, 10 billion m^3 of water is being utilized in processing industries [1].

Food processing industries produce several edible products to meet ever-increasing demand by a growing population. Various food processing industries, including fruit and vegetable processing industries, dairy industry, beverage industry, meat and poultry industry, and edible oil processing industry (Fig. 1), require a considerable amount of freshwater. In addition to edible products and other value-added products, food processing industries generate harmful waste materials that cause pollution to the environment. Specifically, the wastewater generated from these industries may be detrimental to human living conditions due to the presence of perishable constituents. Furthermore, aquatic life also gets seriously affected if wastewater is discharged into the environment without any treatment.

S. Lakshmi Sandhya Rani · V. K. Raja (✉)
Membrane Research Laboratory, Department of Chemical Engineering, National Institute of Technology Andhra Pradesh, Tadepalligudem, Andhra Pradesh 534101, India
e-mail: vinoth@nitandhra.ac.in

S. Lakshmi Sandhya Rani
e-mail: slsandhyarani.sclr@nitandhra.ac.in

G. Pugazhenthi
Department of Chemical Engineering, Indian Institute of Technology Guwahati, Guwahati, Assam 781039, India
e-mail: pugal@iitg.ac.in

S. Bandyopadhyay (ed.), *Application of Low Cost Ceramic Membranes in Wastewater Treatment*, Materials Horizons: From Nature to Nanomaterials,
https://doi.org/10.1007/978-981-97-5842-5_7

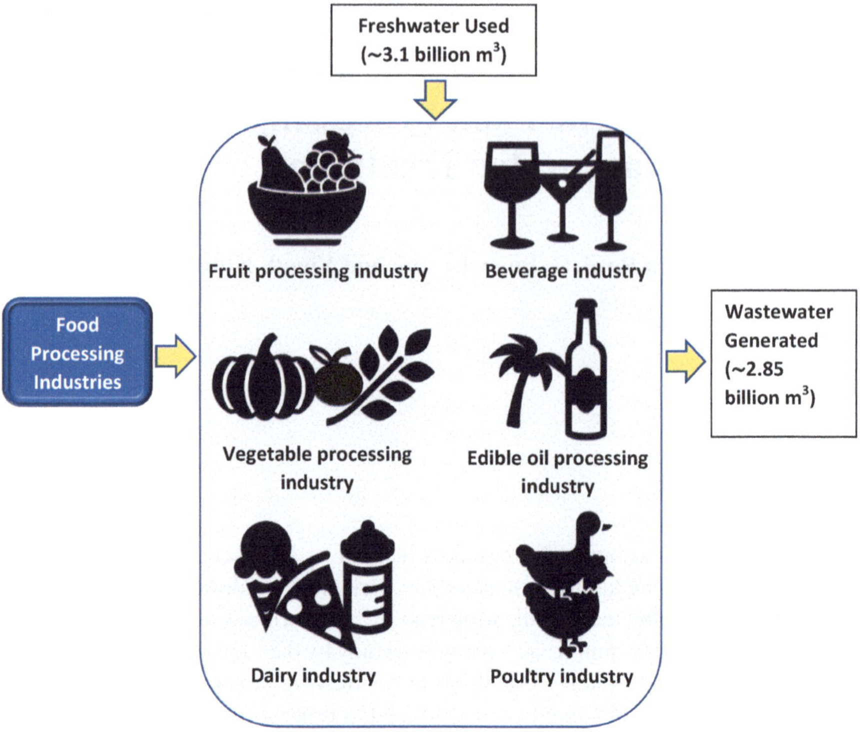

Fig. 1 Various types of food processing industries [2]

Several methods, such as coagulation, centrifugation, adsorption, constructed wetlands, microbial fuel cells, etc. are available to treat wastewater produced by the food processing industries. However, nowadays, researchers are focusing on eco-friendly, inexpensive, and highly efficient methods for wastewater treatment. Membrane technology, incredibly low-cost ceramic membranes, performs well in this aspect due to their robust characteristics such as high stability towards the corrosive environment, fouling resistance, and longer lifetime.

Accordingly, the present chapter discusses wastewater characteristics from different food processing industries, including the dairy industry, edible oil industry, poultry industry, and soft-drink industry. The application of low-cost ceramic membranes in treating these wastewaters is discussed later. In addition, a comparative analysis of the performance of low-cost and commercial membranes is presented in detail.

2 Wastewater from Food Processing Industries

Food processing industries require vast quantities of fresh water for processing during several operations such as washing, production, and cooling. Water can be used as an ingredient, cleaning agent, conveyor, and sanitizing agent. In addition, the wastewater generated from food processing industries is generally biodegradable and often less toxic.

2.1 Sources and General Characteristics

The wastewater characteristics from food processing show significant variation depending on the nature of the raw material processed and the end product. Consequently, these characteristics differ from one industry to another. In general, pollutants present in wastewater are mainly due to the colour, oils, sugars, and preservatives used in manufacturing products. The wastewater generated from food processing industries usually comprises high BOD, COD, TSS, phosphorous, nitrogen, oil, and grease. Sometimes, they may possess acidic/alkaline matter, toxic compounds, and harmful bacteria. Therefore, it is necessary to treat the wastewater before discharging it into the surface water bodies. Wastewater characteristics from different industries are discussed in the following sub-sections.

2.1.1 Dairy Wastewater

Dairy industries generally involve processing the milk into various consumable dairy products such as cheese, butter, yogurt, condensed milk, and ice creams. Therefore, wastewater generated from dairy industries varies and depends on the end product [3]. Moreover, wastewater can be generated from different processing steps for a particular product. For example, Fig. 2 describes wastewater generation from various steps involved in milk processing.

Sanitization wastewater is generated during various sanitization processes such as washing milk containers, equipment, floors, etc. This wastewater consists of organic compounds such as milk proteins and fats and inorganic compounds, including phosphates, chlorine, and nitrogen from cleaning agents. Therefore, it must be adequately treated before discharge. Furthermore, water is used as a coolant in different equipment, including the cooling tower, boiler, air compressor, etc. The outlet water from these types of equipment is often free from any harmful chemicals and can be discharged directly into the water bodies. Generally, this kind of water is recycled in most plants.

Furthermore, whey and buttermilk are typical by-products of milk industries which are often wasted in many sectors. However, these by-products consist of valuable proteins and can be efficiently recovered using membrane technology.

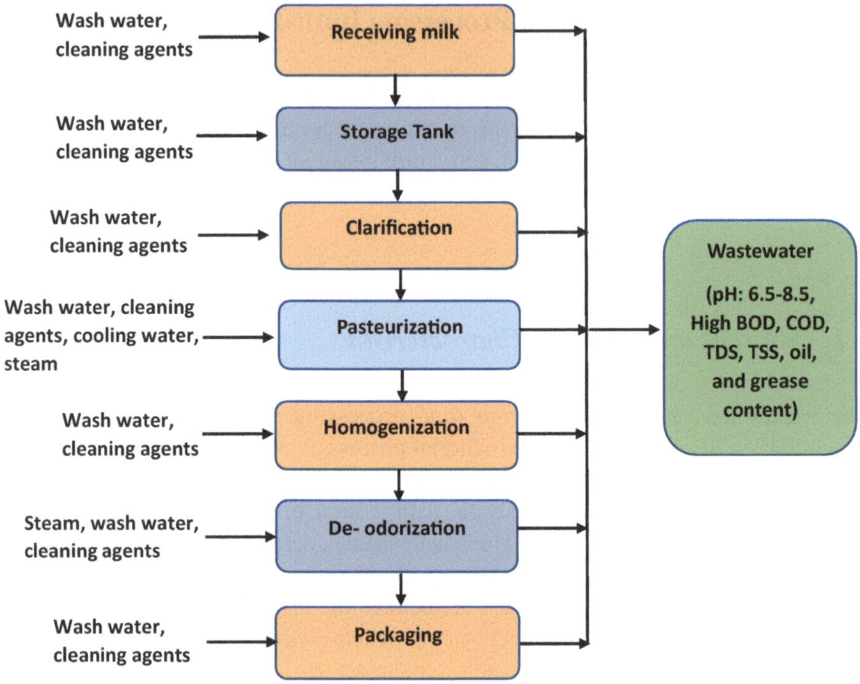

Fig. 2 Sources of effluents from various processing steps of milk processing [4]

Processing losses in dairy industries include entrainments from evaporators, wastage of products during start-up and shutdown operations, loss due to unexpected damage of containers, losses during conveying operations, etc.

Dairy wastewater consists of organic and inorganic constituents, and it is usually white and often turbid. Organic compounds are obtained from waste milk and milk products, cleaning containers, etc. Dairy wastewater's BOD and COD lie in the range of 40–48,000 mg/L and 80–95,000 mg/L, respectively. In addition, the pH of wastewater generally varies from 4.7 to 11. The considerable variation in pH is due to cleaning and sanitizing agents and acidic and alkaline agents in the wastewater.

Furthermore, due to the presence of minute curd, the waste from cheese production is rich in suspended solids. The average yearly temperature of dairy wastewater ranges from 17–25 °C. In summer, the effluent temperature is higher, in the range of 22–25 °C, and during winter, it lies between 17–18 °C [5]. BOD and COD values are low in winter compared to those in summer [6]. Therefore, the wastewater from dairy industries contains fewer inorganic compounds compared to a considerable number of inorganic substances like phosphates used as deflocculants and emulsifiers, chlorine used in cleaning and sanitizing agents, and nitrogen used in wetting agents. Usually, the wastewater from dairy industries does not contain highly toxic compounds. However, it is malodorous and can cause discomfort in the surrounding places [7].

2.1.2 Edible Oil Industry Wastewater

Edible oils are a part of the regular human diet as they provide fatty acids, energy, and vitamins that are soluble in fat. Due to the increased demand for edible oils, oil industries are growing worldwide and thus generating a large amount of wastewater. According to a study, the global oil market is expected to increase from $96 billion in 2019 to $119 billion by 2025 [8]. Palm, sunflower, cottonseed, groundnut, sesame, and olive are a few sources for extracting edible oils. The wastewater from the edible oil industry is primarily generated from the cleaning and washing of process equipment.

Generally, the wastewater is black and sticky liquid and contains high nutrients, BOD, COD, lipids such as fats and oils, and other organic compounds. According to a study, the COD of vegetable oil wastewater ranges from 2,000 to 30,000 mg/L. In general, lipids are significant constituents of edible oil industry wastewater, the concentration of lipids is above 100 mg/L. In some cases, phosphorous content is observed in vegetable oil wastewater as phosphoric acid is used in the degumming process. Furthermore, carbohydrates and phenols are also present in wastewater. However, the variety of oil and processing conditions decide the final characteristics of wastewater. Therefore, the amount and properties of wastewater differ between industries, and in some cases, the characteristics may vary from one batch to another batch of operations.

2.1.3 Poultry Industry Wastewater

Wastewater from poultry industries generally consists of fats, lards, proteins, and blood residues. As a result, the disposal of poultry wastewater without treatment causes eutrophication and damages aquatic life. In addition, contaminated water may cause serious health effects to nearby human beings. Based on the data obtained from the poultry world, the global poultry market increased by 6% in 2019 and is expected to grow continuously [9]. Consequently, the demand for fresh water and the generation of wastewater is also increasing in poultry industries.

In general, similar processing steps are followed in poultry industries where various by-products are generated. For example, in the slaughtering step, for a single animal weighing around 1.7–1.8 kg, nearly 70 g of blood is produced along with other constituents such as feathers, head, feet, etc. Usually, by-products such as feathers, carcass tissues, and blood are removed from liquid effluents and further modified into valued products. In these poultry industries, wastewater is often generated during the cleaning of equipment, carcass washing, and removal of fats, blood, and microorganisms. According to a study, 26.5 L of water is required to process a single bird into meat [10]. Figure 3 describes the average water requirement for various processing steps in the poultry industry.

Slaughterhouse wastewater consists of more organic matter, including suspended solids, oil, fats, grease, nitrogen, and phosphorous. It is also called high-strength wastewater because it possesses heavy metals such as Co and Zn. Organic matter

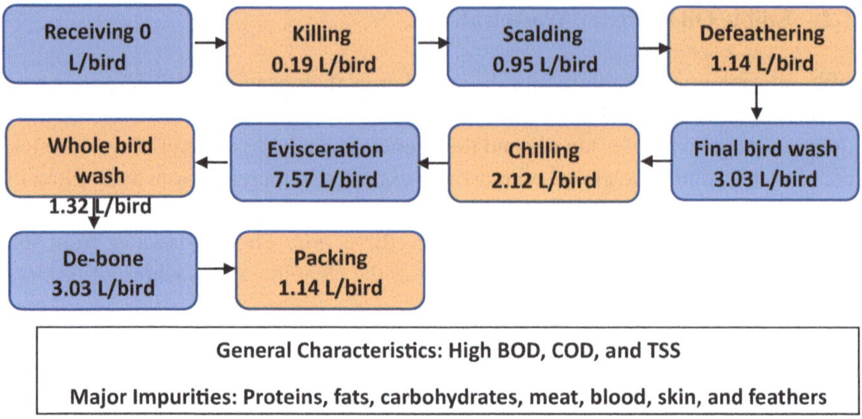

Fig. 3 Average usage of water for different processing steps in the poultry industry [10]

of slaughterhouse wastewater originates from the blood, constituting 90% of the total BOD. In addition, the existence of cleaning agents and sanitizing agents made wastewater more complex and included nearly 18–20% of the entire wastewater. However, the composition varies depending on the type of bird being processed, the instruments, and the process operation [11].

2.1.4 Soft Drink Industry Wastewater

In soft drink industries, nearly 80% of raw material constitutes water. Approximately 1/3rd of water is consumed for backflushing membranes, cleaning equipment, bottles, and washing floors and pipelines. Therefore, these industries generate a tremendous amount of wastewater which should be adequately treated before discharging. Commonly, sugar, colour additives, flavouring agents, and cleaning agents are primary contaminants in soft drink industry wastewater. In addition, it contains more concentration of sodium in the range of 320–397 mg/L and possesses a high pH of about 9 due to the presence of NaOH. Organic loading in the wastewater is mainly due to sugar's existence, leading to high BOD [12].

2.2 Environmental Threats

In every food processing industry, wastewater is generated in various processing steps and considerably affects the environment. As shown in Table 1, different quality and quantities of wastewater are generated based on the type of food industry,

Table 1 Sources, general characteristics, and major problems of wastewater generated from various food processing industries

Types of food processing industry	Source of wastewater generation	General characteristics	Major problems	Refs.
Dairy industry	Milk processing, sanitization, and cleaning	High lactose, TDS, TSS, fats, and nutrients in the form of ammonia and phosphates	Environmental pollution	[13]
Vegetable oil refinery	Degumming, deacidification, deodorization	High COD and turbidity	A dangerous threat to freshwater resources	[14]
Poultry industry	Carcass washing, removal of fats, blood, and microorganisms	Fats, lard, proteins, and blood residues	Eutrophication and threat to aquatic lives	[11]
Fish products processing industry	Fish evisceration	A high content of blood, guts, lipids, and proteins	Oxygen abatement in water bodies and eutrophication	[15]
Soft drink industry	Drink production and washing bottles	Caustic soda and sugar	Pollution of water	[16]

and methods used in processing and handling jeopardize the environment accordingly. The generated effluents are not treated properly, creating severe environmental problems.

2.2.1 Effect of Dairy Wastewater on the Environment

Proteins, lactose, and fat are the major constituents of organic load in dairy wastewater. Based on their biodegradable nature and solubility, they will affect water quality. The presence of highly biodegradable organic compounds enhances the BOD of water. In addition, low molecular compounds such as lactose promote sewage fungus growth. Furthermore, coloured wastewater from dairy industries tends to change the colour of water bodies. The apparent colour of dairy wastewater is due to colloidal and suspended particulates, which reflect the light, whereas the suspended particles give turbidity to the water [17].

In the case of nitrogen, it mainly exists in the form of amino groups from milk proteins. Other nitrogen compounds include urea and uric acid. Bacteria will convert nitrogen into various forms, such as ammonia, nitrate, and nitrite ions. Different nitrogenous compounds formed due to bacterial action can affect the quality of the wastewater/effluent in different ways. For example, nitrate ions are highly toxic and adversely affect human health. In infants, nitrate ions in contaminated drinking water may be converted into nitrite ions and convert hemoglobin into methemoglobin, leading to a condition called methemoglobinemia due to reduced oxygen transport in the bloodstream. Due to these severe environmental and health hazards, wastewater treatment is mandatory.

3 Various Technologies for Wastewater Treatment in the Food Processing Industry

As mentioned in the above section, direct disposal of wastewater into the environment is not recommended. There are various methods for wastewater treatment in food processing industries such as wetland treatment method, coagulation-flocculation followed by sedimentation, membrane separation, electrocoagulation, etc. In addition, biological treatment methods, including activated sludge, anaerobic filters, trickling filters, etc., are also being used [5]. However, every method has its advantages and limitations. For example, the wetland treatment method is conventional, sustainable, and economical. However it requires a larger surface area, may cause risk to surface and groundwater, and there is a possibility of harmful volatile compounds. Likewise, the merits and demerits of different wastewater treatment technologies are mentioned in Table 2.

4 Application of Membrane Technology in the Treatment of Wastewater from Food Processing Industries

The application of membrane technology in treating wastewater from food processing industries offers numerous benefits, such as the feasibility of cleaning, ease of operation, and eco-friendly nature. In addition, membrane separation processes do not use any additional chemicals and thus simplify the entire process. Although polymeric membranes are affordable due to their low cost, inorganic membranes are being developed to overcome the disadvantages of polymeric membranes, such as frequent fouling and low resistance to harsh cleaning conditions. The development of inorganic membranes started in the early 1940s [20]. Ceramic membranes made up of refractory oxides such as alumina, zirconia, and titania are usually highly expensive. However, the higher capital cost can be compensated to a large extent by reducing the operating costs owing to their unique features such as longer life span, higher fouling resistance, and accessibility to use harsh conditions for cleaning instead of replacing the membrane. Nowadays, to reduce the cost of ceramic membranes, they are made up of inexpensive ceramic precursors such as mullite, clays, and some industrial waste. While the cost of commercial ceramic membranes is often more than 500 $\$/m^2$ due to the high price of raw materials, ceramic membranes prepared using very inexpensive raw materials cost less than 50 $\$/m^2$ [21, 22].

Table 2 Merits and demerits of various wastewater treatment methods [18, 19]

S. No.	Method	Merits	Demerits
(i)	Coagulation-flocculation followed by sedimentation	• It helps in the agglomeration of colloidal particles for easy sedimentation • Efficient removal of COD and total suspended solids can be observed	• Requires high quantities of chemicals • Sludge is formed during the process, which needs to be removed again • Difficulty in handling
(ii)	Electrocoagulation	• It can remove pathogens and other organic matter • More than 90% rejection of COD can be obtained	• Requires high energy • It is not a cost-effective method
(iii)	Dissolved air flotation	• 30–90% of COD removal and 70–80% of BOD rejection can be achieved • High nutrient removal can be obtained	• This method requires high energy • TSS cannot be separated efficiently
(iv)	Adsorption	• Simple and adaptable • Efficient process proving fast kinetics • High-quality effluent can be achieved	• High initial investment • Regeneration of material is sometimes expensive • This process is not economical for some industries, such as paper, pulp, and textile industries
(v)	Liquid-liquid extraction	• It can be used for large-scale operations • Extractants can be recycled • Comparatively lower operating costs	• The investment cost is high • Not economical when the concentration of contaminants is less than 0.5 g/L

4.1 Low-Cost Ceramic Membranes for Wastewater Treatment in Food Processing Industries

Numerous types of low-cost raw materials/precursors are identified to fabricate ceramic membranes replacing costly ceramic oxides like zirconia, alumina, etc. Cheaper raw materials can be broadly divided into naturally available minerals and waste materials [23]. A wide variety of natural minerals, including clays, natural zeolites, natural perlite, and natural pozzolan, which contain natural aluminosilicates, are used for ceramic membrane fabrication. Due to their local abundance and low cost, these materials are gaining the interest of the researchers. In general, natural minerals need no additional chemical processing before membrane fabrication. However, they may require some unit operations such as crushing, grinding, and sieving. Various clays, namely kaolin, sepiolite, bentonite, and ball clay, have been extensively utilized to elaborate ceramic membranes [24]. In addition, few studies focused on waste materials in fabricating ceramic membranes [25]. Some industries, including the sugar industry, rice industry, and palm oil industry, generate an enormous amount of solid waste materials in addition to liquid effluents. If they are not appropriately handled, it creates environmental pollution. Efficient utilization of these solid waste materials is one of the best alternatives to avoid disposal issues. Consequently, several researchers used various waste materials, such as rice husk ash, coal fly ash, etc. for the preparation of ceramic membrane.

Ceramic membranes prepared from low-cost alternative materials perform well, similar to conventional ceramic membranes. Accordingly, they have also been used for several applications, including the separation of fruit juice clarification, the concentration of biomass, etc., in addition to the treatment of food processing effluents [26, 27]. For example, kaolin and bentonite were used in a study to prepare low-cost ceramic membranes by a simple pressing route followed by sintering at 1000 °C for oily wastewater treatment. As a result, they have achieved ceramic membranes with a high mechanical strength of 58 MPa and a porosity of 34% [28]. Furthermore, ceramic membranes fabricated from Moroccan perlite with a porosity of nearly 42% were utilized to remove turbidity from the baking powder solution and achieved more than 95% rejection [29]. Similarly, a few other ceramic membranes prepared from various low-cost raw materials with high porosity and flexural strength comparable to commercial ceramic membranes are shown in Table 3.

Table 3 Properties of ceramic membranes prepared from various low-cost raw materials

S. No.	Major raw material	Porosity (%)	Pore size (μm)	Flexural strength (MPa)	References
1.	Kaolin clay	34.5	2.28	7.1	[30]
2.	Coal fly ash	50	–	69.8	[31]
3.	Talcum fly ash	37.8	5	25	[32]
4.	Moroccan bentonite	32.12	1.7	22	[33]
5.	Potter's clay	52.51	0.49	–	[21]
6.	Natural Moroccan perlite	52.11	1.7	21.68	[34]
7.	Natural Indian clay	42	4.58	11.55	[35]
8.	Natural phosphate	25.6	0.26	19.7	[36]
9.	Indian natural white clay	41	0.52	–	[37]
10.	Alumina	39.8	2.96	87	[38]
11.	α-Alumina	35	0.8	–	[39]

4.1.1 Dairy Wastewater Treatment

Dairy wastewater is generally characterized by high BOD, COD, and total suspended solids. Several treatment technologies are available, including adsorption, anaerobic digestion, etc. However, membrane separation can be considered a promising technology to treat dairy wastewater as it possesses several advantages, including low maintenance requirements and being environmentally safe. The permissible discharge limits of pH, suspended solids, oil, and grease are given in Table 4.

Generally, dairy industries' wastewater is treated in three stages: first, intermediate, and final stages. The initial stage involves physiochemical methods, whereas the intermediate stage consists of biological treatment. The final stage is disinfecting the water, after which it can be recycled into the dairy industry. Membrane separation processes can eliminate suspended solids, oils, and fats in the first stage. In most cases, initially, water is pre-treated by different methods, such as coagulation, flotation, and sedimentation, before the membrane separation process. In the case of membrane separation, primarily ceramic micro and ultrafiltration membranes are used, which dramatically reduces the BOD and COD of wastewater (Fig. 4). However, in all stages, ceramic membranes can be used. For example, using membrane bioreactors, ceramic membranes can immobilize particular microorganisms in secondary treatment [41].

As dairy wastewater does not contain toxic compounds, several researchers have shown interest in recycling the wastewater to minimize freshwater consumption. In addition, the scarcity of water resources and strict environmental regulations have driven industries toward the recovery and reuse/recycling of wastewater. Though dairy wastewater may not contain harmful chemicals, it has high organic compounds such as proteins, lactose, and fats. Therefore, the BOD, COD, and TSS should be within acceptable limits before recycling.

A few researchers studied the performance of low-cost ceramic membranes in dairy wastewater treatment. A low-cost ceramic membrane was fabricated from local clay materials using the extrusion method, and this membrane with 0.31 μm pore size and 53% porosity was applied for dairy wastewater treatment to reduce the COD to the permissible limit. As a result, a COD rejection of 91% was achieved with a permeate flux of 2.59×10^{-6} m^3/m^2 s [13].

Table 4 Environmental standards for effluents of the dairy industry (Central Pollution Control Board, India [40])

S. No.	Parameter	Standard
1.	pH	6.5–8.5
2.	Suspended solids	150 mg/l
3.	Oil and grease	10 mg/l
4.	BOD (3 days at 27 °C)	100 mg/l
5.	Generation of wastewater	3 m^3/1000 l of milk

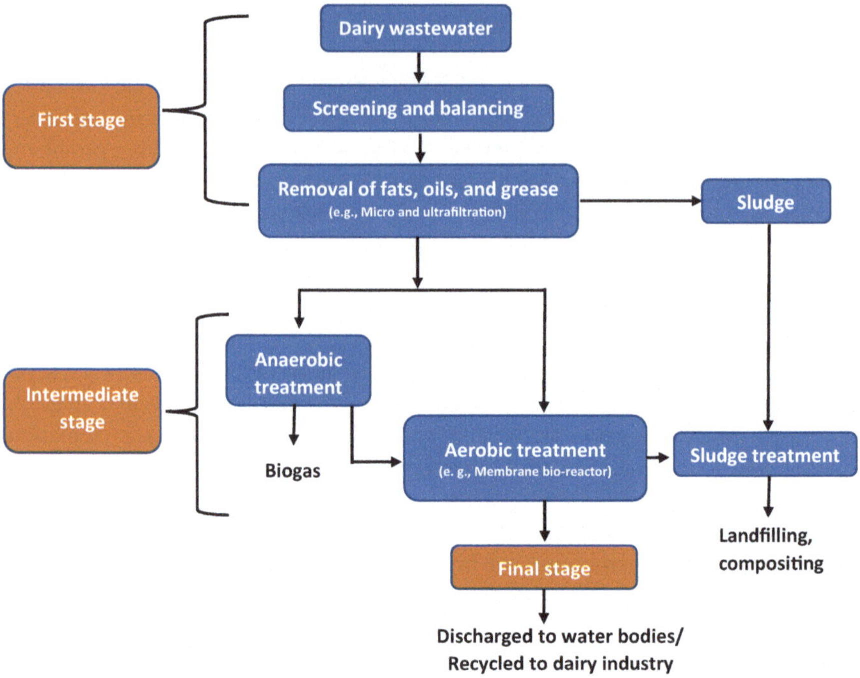

Fig. 4 An overview of wastewater treatment in the dairy industry [42]

Furthermore, a dry compaction method elaborated for the preparation of a low-cost ceramic MF membrane using 50% natural clay and 50% pyrrhotite ash exhibited optimum membrane characteristics such as 34% porosity, 27.42 MPa flexural strength, and low corrosion resistance. This membrane was utilized to treat dairy wastewater and achieved 94% rejection of oil and fats with complete removal of turbidity [43].

4.1.2 Poultry Wastewater

In a study, a low-cost ceramic membrane with a pore size of 0.133 μm was prepared from fly ash by extrusion method. The fabricated membrane was applied for the treatment of poultry slaughterhouse wastewater. The membrane has shown almost complete removal of turbidity, COD, and total suspended solids from the wastewater [44]. Furthermore, Almandoz et al. prepared inexpensive composite ceramic membranes from locally available natural clays. In this study, tubular support was fabricated by extrusion, whereas the active layer was deposited by slip casting technique. They utilized the membrane to treat slaughterhouse wastewater and efficiently removed 99.4% of bacteria and 91% of COD [45].

4.1.3 Edible Oil Industry Wastewater

In a study, tubular ceramic membranes with three different pore sizes, namely 0.2 μm, 450, and 200 Da, were used to treat the palm oil industry effluent. It was reported that all three membranes reduced more than 99% of turbidity, whereas COD rejection was ineffective [46]. Furthermore, wastewater treatment from the edible oil industry with a ceramic membrane having 200 nm pore size resulted in 60% COD rejection and a maximum permeate flux of 160 L/m^{2}h. In another study, a hollow fiber ceramic membrane prepared from silica sand using a phase inversion/sintering technique having a high mechanical strength of 210 MPa, was applied to treat Carotino red palm oil-water emulsion with an oil rejection of 99.7% [47]. In another study by Dionisio et al., sunflower oil-water emulsion was treated with a composite membrane, fabricated using α-alumina support and a polymeric selective layer of polyamide 66 to achieve an oil-retention retention up to 99.5% [48].

4.1.4 Soft Drink Industry Wastewater

Generally, soft drink industries use membrane separation processes to treat feed water before bottling, as membranes can produce water with sufficiently low TDS. Similarly, membrane technology also treats wastewater from soft drink industries [49] for removal of suspended solids usually by micro and ultrafiltration processes to recycle water in beverage industries. As ceramic membranes can be operated over a wide range of pH (1–14), they are gaining interest in wastewater treatment from soft drink industries. In a study, a TiO_2 ultrafiltration ceramic membrane having a pore size of 5 nm was used to treat beverage wastewater to demonstrate the influence of crossflow velocity and transmembrane pressure on the performance of the membrane [50]. Kumar et al. fabricated a nanocomposite membrane with polyvinylidene fluoride (PVDF) and titanium dioxide nanofibers by hot-pressing ceramic nanofibers onto a flat sheet of polymeric and conducted experiments to reduce the glucose content and COD [51].

As shown in Table 5, ceramic membranes were widely used in wastewater treatment from various types of food processing industries which have shown great performance in removing COD, TSS, and turbidity from wastewater.

5 Membrane Fouling, Cleaning and Regeneration with Recovery and Reuse

Ceramic membranes offer a longer lifetime compared to conventional polymeric membranes. The lifetime of ceramic membranes can be estimated as nearly 20 years, whereas the lifetime of polymeric membranes lies between 7–10 years [53]. However, after a long time of operation, membrane fouling also occurs in the case of ceramic

Table 5 Treatment of various food processing industries' wastewater using ceramic membranes

Types of food processing industry	Characteristics of wastewater treated	Characteristics of membrane used	Membrane performance			Refs.
			Operating conditions	Flux	Separation	
Dairy	COD: 1462 mg/L	Tubular membrane with 0.309 μm pore size, 53% porosity	Pressure: 207 kPa Crossflow rate: 3.996 Lph	2.59×10^{-6} m^3/m^2s	91% COD removal	[13]
Poultry	COD: 271.95 mg/L TSS: 414.5 mg/L	Tubular membrane with 0.133 μm pore size, 40.17% porosity	Pressure: 207 kPa Crossflow rate: 40 Lph	18.92 L/m^2h	100% COD removal 100% TSS removal	[44]
Edible oil	Oil concentration: 500 mg/L	Hollow fiber membrane with 0.857 μm pore size, 17.23% porosity	Pressure: 2.5 bar	–	99.7% Rejection of oil	[47]
Soft drink	COD: 3750 mg/L Turbidity: 58 NTU	5 nm tubular membrane	Crossflow velocity: 3.1 m/s Pressure: 300 kPa	–	–	[50]
Whey processing	BSA, lactoferrin, and immunoglobulins	300 kDa tubular membrane	pH: 9 Pressure: 1 bar Crossflow velocity: 3.5 m/s	91–87 L/m^2h	60% Purity	[52]

membranes. This fouling phenomenon adversely affects the overall lifetime of the membranes. Therefore, to increase the lifetime of the membrane along with recovery of initial permeate flux along with the consistent performance of the membrane, an analysis of fouling behaviour is needed. In addition, appropriate methods must be followed to clean the fouled membrane based on the type of fouling. In addition to inherent membrane fouling, other resistances exist, such as gel layer resistance, concentration polarization resistance, pore blockage, etc. The fouling mechanism is usually analysed by four types of Hermia's fouling models, namely cake filtration, complete, standard, and intermediate pore blocking models [13].

5.1 Fouling Mechanism in Dairy Wastewater and Cleaning Aspects

Membrane fouling is a primary challenging factor in the case of dairy wastewater treatment applications. It results in a substantial decline of permeate flux and a notable change in the membrane performance [54]. This phenomenon occurs due to the physical and chemical interactions between milk components and the active separating layer of the membrane. Fouling may be either external, in which solute particles are stuck on the surface of the membrane, or internal fouling, in which particles constrict membrane pores. In dairy industries, fouling is mainly caused by milk proteins. Fouling on the surface is usually caused by macromolecules; conversely, blocking pores is primarily due to smaller molecules.

A few studies analysed the fouling mechanism in dairy wastewater treatment and found the cake layer filtration model as the best fit [13, 43], with a correlation coefficient between 0.98 and 0.99. Therefore, the fouling phenomenon that occurs while treating dairy industry wastewater is primarily reversible, and the membrane can be easily regenerated after proper cleaning. As membrane fouling is unavoidable, it must be cleaned regularly using proper cleaning methods. Either physical or chemical cleaning methods can clean fouling. Physical cleaning methods include backwashing, mechanical scraping, negative pressure cleaning, etc. Usually, physical cleaning cannot completely clean the blocked pores. In such cases, chemical cleaning is done using appropriate cleaning agents. The selection of a suitable cleaning agent based on the composition of the feed and deposited cake layer is the primary step in the chemical cleaning method. The cleaning agent should remove the pollutant by dissolving it without causing any damage to the membrane. In most cases, the cleaning agent is selected by a trial-and-error approach. HCl, HNO_3, H_2SO_4, NaOH, and EDTA are a few examples of commonly used chemical cleaning agents.

5.2 Recovery of Valuable Products from Dairy Industry Processing Wastewater

Whey is a significant by-product of the dairy industry, usually obtained during milk product production, such as cheese, casein, and paneer. In India, paneer is one of the most important dairy products. In many places, whey is simply drained off, which creates a loss of nutritious compounds and environmental pollution. Separation of nutrients from whey by traditional methods is time and energy-consuming. But membrane technology can efficiently fractionate/concentrate valuable whey products, including α-lactalbumin, β-lactoglobulin, etc. In addition, we can get whey protein concentrate (WPC) and whey protein isolate (WPI) by whey processing. These whey proteins are very useful in food industries due to their inherent properties. Whey proteins are highly nutritious as they contain amino acids. They are used as ingredients in food due to their special characteristics, such as more solubility, absorption tendency towards the water, and gelatinization capacity.

In a study, ceramic support was fabricated from alumina and coated with zirconia top layer. The resulting composite membrane was used to concentrate the proteins of whey [55]. Slukova et al. have used a commercial ceramic membrane having a pore size of 50 kDa to concentrate proteins from cheese whey [56]. Using a tubular ceramic UF membrane, they can retain 80.9% of proteins and nitrogen compounds. Furthermore, a tubular ceramic membrane with a cut-off of 300 kDa was used to fractionate proteins from clarified whey. They intended to retain BSA, lactoferrin, and IgG proteins on the retentate side and α-lactalbumin and β-lactoglobulin on the permeate side [52]. In a study, the recovery of proteins from whey was analysed for ceramic membrane (ZrO_2) and organo-mineral (polymer-ceramic composite) membrane (ZrO_2/PSf). The performance of those membranes was compared with a traditional polymeric membrane (PSf/PVP). They observed that the permeability of the ZrO_2/PSf membrane was three-fold higher than the other two membranes due to high surface porosity [57]. As low-cost ceramic membranes exhibit excellent properties similar to conventional ceramic membranes (Table 3), they may be utilized to recover valuable products from whey efficiently.

6 A Comparative Study of Low-Cost Ceramic Membranes Over Conventional Polymeric And Ceramic Membranes

In wastewater treatment applications, polymeric and ceramic membranes are significantly used. However, most of the membranes are made with polymeric membranes due to their high flexibility, lower cost, and easier fabrication. But ceramic membranes provide durability and thermal and mechanical stability despite their high cost. For example, Guerra et al. studied the performance of ceramic membranes and polymeric membranes for the treatment of water-containing bentonite. They have performed experiments for both membranes under the same hydrodynamic conditions to analyse the influence of membrane material alone. It was observed that ceramic membranes showed higher production costs. However, it was concluded that ceramic membranes are cost-competitive because they can be operated under harsh conditions, avoiding the need for frequent membrane replacement and reducing labour costs [58].

Therefore, the application of low-cost ceramic membranes in wastewater treatment would be a promising option for obtaining a more cost-effective separation process. In a study, Khemakem et al. treated the cuttlefish effluent using commercial alumina membrane and natural Tunisian clay membrane and compared the performance of both membranes by measuring the flux and quality of the permeate. They confirmed that clay membranes treated the wastewater more efficiently than commercial ceramic membranes [59]. In addition, the fabrication of polymer-low-cost ceramic composite membranes also might provide more effective separation performance in wastewater treatment applications as they retain the advantages of both polymeric and ceramic membranes [60]. Mittal et al. prepared a polymer-ceramic composite membrane for oily wastewater treatment. They have prepared ceramic support from low-cost natural clays by the paste casting technique. In contrast, a selective polymeric layer was prepared from cellulose acetate, and it was coated by the dip-coating method. As a result, they achieved an oil rejection of 93% by using a 56% porous composite membrane with a pore size of 28 nm [61].

7 Scope for Further Work

The use of low-cost materials can effectively enable the application of ceramic membranes for wastewater treatment in food processing industries on an industrial scale. To further widen their application, various combinations of low-cost raw materials and pore formers can be performed to get much better permeate flux. The cost of fabrication can be further reduced if sintering-free materials are used.

8 Summary and Conclusions

Food processing industries consume a large amount of fresh water and generate considerable quantities of wastewater which has the potential to recycle the treated water with recovery of valuable by-products. This chapter mainly discussed the sources and characteristics of wastewater from various food processing industries, including the dairy industry, poultry industry, soft drink industry, and edible oil industry. Treatment of these industrial effluents is necessary to minimize freshwater consumption, avoid water scarcity, and protect natural water resources. Consequently, the present chapter discussed the scope of using various low-cost ceramic membranes for application in wastewater treatment from the food processing industries. Several membranes prepared from different raw materials, including kaolin clay, fly ash, and other natural, low-cost raw materials, showed high performance similar to commercial ceramic membranes. Most of the low-cost ceramic membranes can be used for removal of the undesirable constituents in the wastewater, including turbidity and COD of the industrial wastewater, according to the permissible discharge limits. As fouling is an inevitable phenomenon in membrane filtration processes, a few researchers analysed the fouling mechanism using Hermia's model and concluded that the cake layer filtration model is the best fit for fouling in dairy wastewater treatment.

Though polymeric membranes are more cost-effective than commercial ceramic membranes, low-cost ceramic membranes can efficiently replace them with their excellent inherent characteristics. However, the large-scale application of low-cost ceramic membranes is still under development. As the cost of the low-cost ceramic membranes is much lower than the commercial ceramic membranes, industrial application after optimization of the process parameters would be economically beneficial.

Acknowledgment This work was supported by the Science and Engineering Research Board, Department of Science and Technology, Government of India [File No: EEQ/2018/001432].

References

1. "www.environmentalpollution.in." [Online]. https://www.environmentalpollution.in/essay/water-consumption-in-industrial-sector-in-india/392
2. Katsuyama AM (1979) A guide for waste management in the food processing industry. The Food Processors Institute, Washington, DC
3. Sharma P (2008) Performance evaluation of wastewater treatment plant for milk based food industry. Master's Thesis, Department of Environmental Science & Technology, Thapar University, Patiala, India
4. Shete MBS (2013) Comparative study of various treatments for dairy industry wastewater. IOSR J Eng 03(08):42–47. https://doi.org/10.9790/3021-03844247
5. Ahmad T et al (2019) Treatment and utilization of dairy industrial waste : a review. Trends Food Sci Technol 88:361–372. https://doi.org/10.1016/j.tifs.2019.04.003
6. Kolhe AS, Pawar VP (2011) Physico-chemical analysis of effluents from dairy industry. Recent Res Sci Technol 3(5):29–32
7. Sarkar B, Chakrabarti PP, Vijaykumar A, Kale V (2006) Wastewater treatment in dairy industries—possibility of reuse. Desalination 195(1–3):141–152. https://doi.org/10.1016/j.desal.2005.11.015
8. global-edible-oil-market-forecasts-from-2020-to @ www.researchandmarkets.com. [Online]. https://www.researchandmarkets.com/reports/5138736/global-edible-oil-market-forecasts-from-2020-to
9. fe3e9e8fd6a99a23a23dd59bd0f16da5fd406c28 @ www.poultryworld.net. [Online]. https://www.poultryworld.net/Meat/Articles/2020/7/Popularity-of-poultry-continues-globally-615517E/
10. Avula RY, Nelson HM, Singh RK (2009) Recycling of poultry process wastewater by ultrafiltration. Innov Food Sci Emerg Technol 10(1):1–8. https://doi.org/10.1016/j.ifset.2008.08.005
11. Hilares RT, Atoche-Garay DF, Pagaza DAP, Ahmed MA, Andrade GJC, Santos JC (2021) Promising physicochemical technologies for poultry slaughterhouse wastewater treatment: a critical review. J Environ Chem Eng 9(2). https://doi.org/10.1016/j.jece.2021.105174
12. Remya N, Swain A (2019) Soft drink industry wastewater treatment in microwave photocatalytic system—exploration of removal efficiency and degradation mechanism. Sep Purif Technol 210:600–607. https://doi.org/10.1016/j.seppur.2018.08.051
13. Kumar RV, Goswami L, Pakshirajan K, Pugazhenthi G (2016) Dairy wastewater treatment using a novel low cost tubular ceramic membrane and membrane fouling mechanism using pore blocking models. J Water Process Eng 13:168–175. https://doi.org/10.1016/j.jwpe.2016.08.012
14. Šereš Z et al (2016) Treatment of vegetable oil refinery wastewater using alumina ceramic membrane: optimization using response surface methodology. J Clean Prod 112:3132–3137. https://doi.org/10.1016/j.jclepro.2015.10.070
15. Picos-Benítez AR, Peralta-Hernández JM, López-Hincapié JD, Rodríguez-García A (2019) Biogas production from saline wastewater of the evisceration process of the fish processing industry. J Water Process Eng 32:100933. https://doi.org/10.1016/j.jwpe.2019.100933

16. Haroon H, Waseem A, Mahmood Q (2013) Treatment and reuse of wastewater from beverage industry. J Chem Soc Pak 35(1):5–10
17. Shete B, Shinkar NP (2013) Dairy industry wastewater sources, characteristics & its effects on environment. Int J Curr Eng Technol 3(5):1611–1615. ISSN 2277-4106. http://inpressco.com/category/ijcet
18. Musa MA, Idrus S (2021) Physical and biological treatment technologies of slaughterhouse wastewater: a review. Sustainability 13(9):1–20. https://doi.org/10.3390/su13094656
19. Crini G, Lichtfouse E (2019) Advantages and disadvantages of techniques used for wastewater treatment. Environ Chem Lett 17(1):145–155. https://doi.org/10.1007/s10311-018-0785-9
20. (1996) Chapter 2 Historical development and commercialization of inorganic membranes. Membr Sci Technol 3(C):15–22. https://doi.org/10.1016/S0927-5193(96)80021-0
21. Saikia J, Sarmah S, Bora JJ, Das B, Goswamee RL (2019) Preparation and characterization of low cost flat ceramic membranes from easily available potters' clay for dye separation. Bull Mater Sci 42(3):1–13. https://doi.org/10.1007/s12034-019-1767-7
22. Singh G, Bulasara VK (2015) Preparation of low-cost microfiltration membranes from fly ash. Desalin Water Treat 53(5):1204–1212. https://doi.org/10.1080/19443994.2013.855677
23. Abdullayev A, Bekheet MF, Hanaor DAH, Gurlo A (2019) Materials and applications for low-cost ceramic membranes. Membranes (Basel) 9(9). https://doi.org/10.3390/membranes9090105
24. Vinoth Kumar R, Kumar Ghoshal A, Pugazhenthi G (2015) Elaboration of novel tubular ceramic membrane from inexpensive raw materials by extrusion method and its performance in microfiltration of synthetic oily wastewater treatment. J Membr Sci 490:92–102. https://doi.org/10.1016/j.memsci.2015.04.066
25. Rawat M, Bulasara VK (2018) Synthesis and characterization of low-cost ceramic membranes from fly ash and kaolin for humic acid separation. Korean J Chem Eng 35(3):725–733. https://doi.org/10.1007/s11814-017-0316-6
26. Nandi BK, Das B, Uppaluri R, Purkait MK (2009) Microfiltration of mosambi juice using low cost ceramic membrane. J Food Eng 95(4):597–605. https://doi.org/10.1016/j.jfoodeng.2009.06.024
27. Issaoui M, Limousy L, Lebeau B, Bouaziz J, Fourati M (2017) Manufacture and optimization of low-cost tubular ceramic supports for membrane filtration: application to algal solution concentration. Environ Sci Pollut Res 24(11):9914–9926. https://doi.org/10.1007/s11356-016-8285-6
28. Eom JH, Kim YW, Yun SH, Song IH (2014) Low-cost clay-based membranes for oily wastewater treatment. J Ceram Soc Jpn 122(1429):788–794. https://doi.org/10.2109/jcersj2.122.788
29. Majouli A, Younssi SA, Tahiri S, Albizane A, Loukili H, Belhaj M (2011) Characterization of flat membrane support elaborated from local Moroccan Perlite. Desalination 277(1–3):61–66. https://doi.org/10.1016/j.desal.2011.04.003
30. Agarwalla A, Mohanty K (2022) Comprehensive characterization, development, and application of natural/Assam Kaolin-based ceramic micro filtration membrane. Mater Today Chem 23:100649. https://doi.org/10.1016/j.mtchem.2021.100649
31. Cao J, Dong X, Li L, Dong Y, Hampshire S (2014) Recycling of waste fly ash for production of porous mullite ceramic membrane supports with increased porosity. J Eur Ceram Soc 34(13):3181–3194. https://doi.org/10.1016/j.jeurceramsoc.2014.04.011
32. Cheng C, Fu H, Wu J, Zhang H, Chen H (2020) Study on the preparation and properties of talcum-fly ash based ceramic membrane supports. Membranes (Basel) 10(9):1–20. https://doi.org/10.3390/membranes10090207
33. Bouazizi A et al (2016) Elaboration and characterization of a new flat ceramic MF membrane made from natural Moroccan bentonite. Application to treatment of industrial wastewater. Appl Clay Sci 132–133:33–40. https://doi.org/10.1016/j.clay.2016.05.009
34. Saja S et al (2018) Elaboration and characterization of low-cost ceramic membrane made from natural Moroccan perlite for treatment of industrial wastewater. J Environ Chem Eng 6(1):451–458. https://doi.org/10.1016/j.jece.2017.12.004

35. Jana S, Purkait MK, Mohanty K (2010) Preparation and characterization of low-cost ceramic microfiltration membranes for the removal of chromate from aqueous solutions. Appl Clay Sci 47(3–4):317–324. https://doi.org/10.1016/j.clay.2009.11.036
36. Belgada A, Achiou B, Alami S, Zohra F (2020) Low-cost ceramic microfiltration membrane made from natural phosphate for pretreatment of raw seawater for desalination. J Eur Ceram Soc. https://doi.org/10.1016/j.jeurceramsoc.2020.09.064
37. Meghnani R, Kumar M, Pugazhenthi G, Dhakshinamoorthy V (2020) Synthesis of ceramic membrane using inexpensive precursors and evaluation of its biocompatibility for hemofiltration application. Sep Purif Technol 256:117814. https://doi.org/10.1016/j.seppur.2020.117814
38. Dong Y et al (2011) Corrosion resistance characterization of porous alumina membrane supports. Mater Char 62(4):409–418. https://doi.org/10.1016/j.matchar.2011.01.012
39. Mueller J, Cen Y, Davis RH (1997) Crossflow microfiltration of oily water. J Membr Sci 129(2):221–235. https://doi.org/10.1016/S0376-7388(96)00344-4
40. The Environment (Protection) Rules, 1986 449 (1996) 475(1):1996
41. Purkait MK, Singh R (2018) Membrane technology in separation science, 1st edn. Taylor & Francis. https://doi.org/10.1201/9781315229263
42. Ryan MP, Walsh G (2016) The characterisation of dairy waste and the potential of whey for industrial fermentation. EPA Research Report, Environmental Protection Agency, Ireland
43. Hatimi B et al (2020) Low cost pyrrhotite ash/clay-based inorganic membrane for industrial wastewaters treatment. J Environ Chem Eng 8(1):103646. https://doi.org/10.1016/j.jece.2019.103646
44. Goswami KP, Pugazhenthi G (2020) Treatment of poultry slaughterhouse wastewater using tubular microfiltration membrane with fly ash as key precursor. J Water Process Eng 37:101361. https://doi.org/10.1016/j.jwpe.2020.101361
45. Almandoz MC, Pagliero CL, Ochoa NA, Marchese J (2015) Composite ceramic membranes from natural aluminosilicates for microfiltration applications. Ceram Int 41(4):5621–5633. https://doi.org/10.1016/j.ceramint.2014.12.144
46. Ang WL, Balakrishnan T, Veeraiya T, Haiqal M, Elham M, Faidzal MS (2020) Performance evaluation of tubular ceramic membrane for palm oil mill effluent treatment. J Eng 32(3):513–521. https://doi.org/10.17576/jkukm-2020-32(3)-16
47. Alftessi SA et al (2021) Novel silica sand hollow fibre ceramic membrane for oily wastewater treatment. J Environ Chem Eng 9(1):104975. https://doi.org/10.1016/j.jece.2020.104975
48. Da Silva Biron D, Zeni M, Bergmann CP, Dos Santos V (2017) Analysis of composite membranes in the separation of emulsions sunflower oil/water. Mater Res 20(3):843–852. https://doi.org/10.1590/1980-5373-MR-2016-0732
49. Applications of membrane technologies in food and beverage (2017) American Membrane Technology Association @ https://www.amtaorg.com
50. Agana BA, Reeve D, Orbell JD (2013) Performance optimization of a 5 nm TiO_2 ceramic membrane with respect to beverage production wastewater. Desalination 311:162–172. https://doi.org/10.1016/j.desal.2012.11.027
51. Kumar M, Jaafar J (2018) Preparation and characterization of TiO_2 nanofiber coated Pvdf membrane for softdrink wastewater treatment. Environ Ecosyst Sci 2(2):35–38. https://doi.org/10.26480/ees.02.2018.35.38
52. Almécija MC, Ibáñez R, Guadix A, Guadix EM (2007) Effect of pH on the fractionation of whey proteins with a ceramic ultrafiltration membrane. J Membr Sci 288(1–2):28–35. https://doi.org/10.1016/j.memsci.2006.10.021
53. Zuo X, Zhang S, Ma G, Lv Y, Li P (2020) Analysis of ceramic membrane fouling behavior and cleaning technology. In: E3S web of conferences, vol 194, pp 1–4. https://doi.org/10.1051/e3sconf/202019404048
54. Tan TJ, Wang D, Moraru CI (2014) A physicochemical investigation of membrane fouling in cold microfiltration of skim milk. J Dairy Sci 97(8):4759–4771. https://doi.org/10.3168/jds.2014-7957

55. Erdem I, Çiftçioğlu M, Harsa Ş (2006) Separation of whey components by using ceramic composite membranes. Desalination 189(1-3 SPEC ISS):87–91. https://doi.org/10.1016/j.desal.2005.06.016
56. Sluková M et al (2016) Cheese whey treated by membrane separation as a valuable ingredient for barley sourdough preparation. J Food Eng 172:38–47. https://doi.org/10.1016/j.jfoodeng.2015.03.023
57. Doyen W, Adriansens W, Molenberghs B, Leysen R (1996) A comparison between polysulfone, zirconia and organo-mineral membranes for use in ultrafiltration. J Membr Sci 113(2):247–258. https://doi.org/10.1016/0376-7388(95)00124-7
58. Guerra K, Pellegrino J (2013) Development of a techno-economic model to compare ceramic and polymeric membranes. Sep Sci Technol 48(1):51–65. https://doi.org/10.1080/01496395.2012.690808
59. Khemakhem S, Ben Amar R (2012) Purification of industrial effluent by microfiltration and ultrafiltration ceramic membranes: comparative study between commercial and elaborated Tunisian clay membranes. Desalin Water Treat 39(1–3):182–189. https://doi.org/10.1080/19443994.2012.669173
60. Kotobuki M, Gu Q, Zhang L, Wang J (2021) Ceramic-polymer composite membranes for water and wastewater treatment: bridging the big gap between ceramics and polymers. Molecules 26(11). https://doi.org/10.3390/molecules26113331
61. Mittal P, Jana S, Mohanty K (2011) Synthesis of low-cost hydrophilic ceramic-polymeric composite membrane for treatment of oily wastewater. Desalination 282:54–62. https://doi.org/10.1016/j.desal.2011.06.071

Chapter 8
Novel Composite Ceramic Membranes in Bioelectrochemical Systems for Distillery Wastewater

Shraddha Yadav, Akash Tripathi, and M. M. Ghangrekar

1 Introduction

The inflating energy demand emerges as a massive challenge for the current society as it gradually depletes the available natural resources/fossil fuels. Therefore, a shift towards renewable and sustainable energy sources is considered as the top-most priority. In that direction bioenergy (in the form of biofuels or bioelectricity) could satisfy the future needs of renewability, sustainability, and even carbon neutrality [1]. Nowadays, the most attractive source for deriving bioenergy is either wastewater or solid waste, which solves the multiple problems of waste remediation, water crisis, and energy production. The recovery of value-added products and bioenergy from wastewater is also in tandem with the United Nation's sustainable development goals (SDG) of providing clean water (i.e., SDG 6) and clean renewable energy (i.e., SDG 7) [2].

The bioenergy can be harvested from distillery wastewater by employing different technologies, such as anaerobic digestion, dark fermentation, and bioelectrochemical systems (BESs). Among them, the BES is the most prominent technology as it provides the opportunity to harness other value-added products such as biohydrogen, acetate, and ethanol along with bioelectricity.

Mohanakrishna et al. (2010) reported the simultaneous reduction/removal of multiple pollutants, salinity, total dissolved solids (23.69%), COD (72.84%), and colorants (31.67%) from the distillery effluent using BESs [3]. The performance of BESs greatly depends on the membrane incorporated, which mainly separates the

S. Yadav
School of Environmental Science and Engineering, Indian Institute of Technology Kharagpur, Kharagpur, India

A. Tripathi · M. M. Ghangrekar (✉)
Department of Civil Engineering, Indian Institute of Technology Kharagpur, Kharagpur, India
e-mail: ghangrekar@civil.iitkgp.ac.in

S. Bandyopadhyay (ed.), *Application of Low Cost Ceramic Membranes in Wastewater Treatment*, Materials Horizons: From Nature to Nanomaterials,
https://doi.org/10.1007/978-981-97-5842-5_8

anodic and cathodic compartments in the system. Despite the success of BES on the lab-scale, its field-scale applicability is still limited due to the high capital cost of the membrane. Nafion is the conventional membrane material and it is widely applied in the BESs. However, apart from high economic investment, it also faces challenges, such as substrate loss, internal resistance, back diffusion of oxygen, poor proton permeability, and biofouling [4]. Therefore, low-cost and more efficient membrane materials, such as Polyvinyl alcohol (PVA)-Nafion composites, Ceramic, Ceramic composites, etc., have been developed [1]. Implementation of these membranes lowers the overall cost and improves the performance of the BESs, which is discussed in this chapter. Additionally, the remediation of endocrine-disrupting chemicals (EDCs), a class of emerging contaminants, from distillery effluent and the future perspective of the low-cost membrane are also highlighted in the present chapter.

2 Membrane-Based BESs

Membrane-based technologies have been extensively used for the removal of odours, taste, colour, organic matter, dissolved solids, suspended solids, bacteria, and viruses from distillery effluent. Additionally, the membrane-based BESs technologies were reported to efficiently treat distillery effluent. Further advancement in membrane production and module design reduces the operating and capital cost of BESs for large-scale applications, which are discussed in this section.

2.1 Basic Mechanism of Membrane-Based BESs

The bio- or microbial-electrochemical systems are emerging platforms for energy-efficient and sustainable wastewater treatment, biomass remediation, desalination, biofuels (biohydrogen) generation, and biochemicals recovery (such as ethanol, methanol, and hydrogen peroxide) [5]. Compared to conventional fuel cells, the BES proves to offer a low-cost technological solution as it uses microbes as a catalyst for the biochemical reactions in the ambient temperature range (15–45 °C) via utilization of organic matter [6]. The BESs preliminary consists of an anode, cathode, external electric circuit, and an ion exchange membrane/separator, which separates anodic and cathodic chambers. The flow of electrons through an external circuit can be directly used to harvest bioelectricity (in the microbial fuel cell, MFC) and biohydrogen or other value-added substances (hydrogen peroxide, bio-hydrogen, acetate, etc.) using microbial electrolysis cell (MEC) and/or microbial electrosynthesis (MES) [7].

At first, the MFCs are potential options to be used as bi-functional devices to recover water and energy together and therefore have gained incredible research interest and importance. The MFC can extract bioelectricity by anaerobic oxidation of the organic substrate (acetate, glucose, lactate, ethanol, municipal and industrial wastewater, etc.) in the anodic compartment during which proton, electron, and carbon dioxide (CO_2) are generated [8]. The released electrons from the anodic chamber are transferred to the cathodic chamber via an external circuit and in a process, electricity is generated; while the protons are transferred to the cathodic chamber through a proton exchange membrane (PEM). Oxygen is the most commonly used electron acceptor in the cathodic chamber due to its high potential for oxidation, easy availability, and capability for the generation of the clean reduced product (water or hydrogen peroxide (H_2O_2)) (Eqs. 1–3). [9].

$$O_2 + 4H^+ + 4e^- \rightarrow 2H_2O \tag{1}$$

$$O_2 + 2H^+ + 2e^- \rightarrow H_2O_2 \tag{2}$$

$$H_2O_2 + 2H^+ + 2e^- \rightarrow 2H_2O \tag{3}$$

In MFC, the membrane plays a vital role in restricting the transfer of cathodic oxygen towards the anode as the passage of oxygen can adversely affect the anaerobes (e.g., *Geobacter*) in the anodic chamber. Moreover, oxygen diffusion enhances the growth of aerobic bacteria, which reduces the availability of substrate for electricity generation [4, 10].

The second analogue of MFC is MEC, which utilizes exo-electrogenic microorganisms for the oxidation of organic matter present in wastewater. Further, the released electrons and protons are transferred to the cathode side via an external circuit and separating membrane, respectively. Then, the protons get chemically reduced to the hydrogen (H_2) gas at the cathode with the help of additional external voltage input (to overcome thermodynamic limitations). In MEC, the optimum membrane selection is very necessary for maintaining the purity of produced H_2 gas. Otherwise, the H_2 gas will substantially mix with the CO_2, methane (CH_4), and hydrogen sulphide (H_2S) produced at the anode. Therefore, membranes in BESs should be highly efficient in blocking the transfer of oxygen, H_2 gas, and H_2O_2.

The third modification of MFC with one extra desalination chamber is microbial desalination cell (MDC). The MDC can be explained as a completely membrane-based technology as it requires both anion exchange membrane (AEM) and cation exchange membrane (CEM). The desalination chamber lies between the anodic and cathodic chambers, and the anolyte and catholyte are, respectively, separated by the AEM and CEM (*discussed in* Sect. 3.2). To maintain the electroneutrality condition, the negative ions (such as Cl^-, NO^{3-}, $(SO_4)^{2-}$) from the desalination cell migrate towards the anodic chamber, while the positive ions (such as Na^+, Ca^{2+}, Mg^{2+}) migrate towards the cathodic chamber. Hence, MDC simultaneously desalinates and generates electricity while oxidizing organic matter from wastewater in an anodic chamber. A recent modification in MDC configuration is stacking with multiple desalination chambers to enhance the overall efficiency of the system. These stacked MDCs further require the alternate placing of suitable AEM and CEM. Hence, the membrane selectivity and stability significantly influence the performance of BESs.

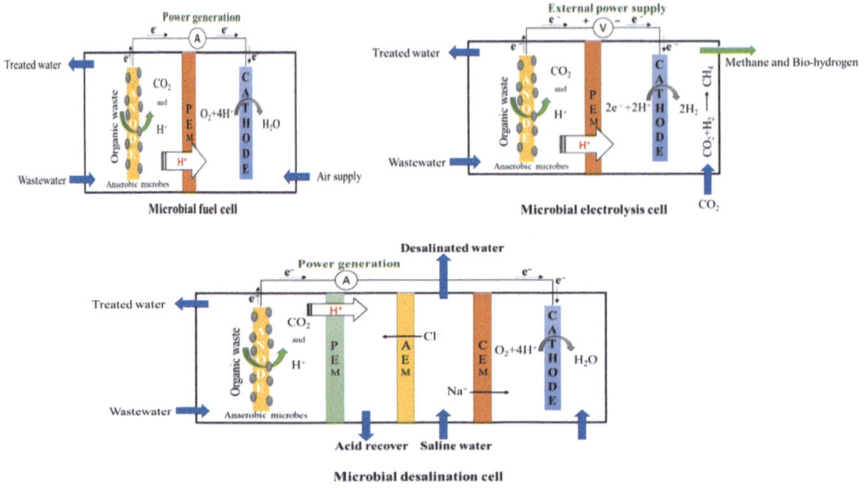

Fig. 1 Stipulation of mechanism and membrane function involved in the bio-electrochemical systems: MFC, MEC, and MDC

The basic mechanism of the above three BESs (i.e., MFC, MEC, and MDC) and the role of different membranes (i.e., PEM, CEM, AEM) are represented in Fig. 1.

2.2 *Conventional Membranes and Their Classifications*

Based on the requirement of the BESs, the different membranes have been used for smooth operation and high energy output of the system. As discussed, the main motive of these membranes is to reduce the interactions between the cathodic and anodic chambers. In addition, acclimating the bio-electrodes, preventing the loss of desirable products, and transferability of the ions through the membrane are also desirable properties. Hence, the membrane used should have properties like good ionic conductivity, mechanical strength, biocompatibility, and chemical inertness [11]. Traditionally, different ionic-based separators, such as CEM and AEM, have been used based on the anticipated functions in the BES reactors as discussed in the following section.

2.2.1 Cation Exchange Membranes

A semipermeable membrane that allows the transport of only positive ions (cations) through them can be considered as CEM. Some of the conventionally used CEM in MFCs are Nafion, Hyflons, Zirfon, and Ultrex CMI 7000, which allow transport of the proton generated in the anodic chamber towards the cathode. For the first time, in the 1960s, Dupont developed a perfluorinated ion exchange membrane "Nafion" during work on fuel cells [12]. Nafion is the most popularly used CEM in the MFC/MDC due to its high permeability towards electro-active species and lower ohmic resistance. Nafion being a perfluoro sulfonic acid (PFSA), is a copolymer of poly-tetrafluorethylene (PTFE) and sulfonic acid group (SO_3H) [13]. The PTFE polymer is the backbone of the Nafion and imparts a hydrophobic nature to the membrane. While the negatively charged hydrophilic SO_3H group is responsible for the proton conductivity through the membrane. Several different variant of Nafion, i.e., Nafion 112, Nafion 117, Nafion 115, and Nafion 212 are available depending on their thickness and surface area. Amongst them, the Nafion 117 has been found effective in MFC due to its lower oxygen cross-over and high mechanical strength [14].

Another commonly used membrane in MFC is Ultrex CMI 7000, which has improved mechanical durability compared to Nafion. The Ultrex CMI 7000 is a firm acid polymer that contains numerous groups of SO_3H and has excellent cation exchange properties. However, its high thickness (0.046 cm) compared to the Nafion 117 (0.019 cm) leads to elevated ohmic resistance. The higher membrane resistance adds up to the total internal resistance and has an inversely proportional relationship with the maximum power output of MFC [15]. Also, the experiments performed by Harnisch et al. found that the ohmic resistance of CMI 7000 was 45 $\Omega\ m^2$, which was five times higher than the Nafion 117 (9.2 $\Omega\ m^2$) [16].

Zirfon is an organo-mineral membrane made of polysulfone and ZrO_2 and it is the economically viable membrane for the BES reactors. Comparing Zirfon as a CEM to Nafion 117, a significant reduction in ohmic resistance is achieved [17]. The limitation of Zirfon is its high oxygen mass transfer capacity with the oxygen mass transfer coefficient of 1.9×10^{-3} cm/s [18]. Hayflon, a perfluoro sulfonyl fluoride vinyl ester with reduced ohmic resistance, is one of the other conventional membranes that is less frequently utilized. Additionally, its applicability in lab-scale MFCs is constrained by the low chemical stability [19]. The cost of the fluorinated membrane (Nafion) is one of the most significant contributors to the BES's total fabrication cost, hence the non-fluorinated membrane (such as polyether-ketone, Victrex Poly) have also been explored. Usually, non-fluorinated membranes have aromatic or hydrocarbon bases with sulfonated or similar pendant groups attached. These membranes have better mechanical and chemical stability, higher ohmic losses, and oxygen mass transfer capacity. The requirement of the appropriate membrane with low thickness, low oxygen mass transfer capacity, high proton transfer ability, and economical synthesis process is still a future research interest [20].

2.2.2 Anion Exchange Membranes

After the CEM, the AEM is the widely used membrane in the BESs, allowing the negatively charged ions to move through it. In addition to the anions, small-sized cations like protons can also be transported through the membrane by forming a hydrogen bond with water molecules [21]. The AEM is fabricated by developing the covalent structure between the positive cations group ($NR_x\ H_x^+$, phosphonium, pyridinium) and polymer base. Recent investigations have shown that AEM is more effective in the operation of MFC than CEM for higher power density generation. The AEM is effective in reducing the pH splitting between anolyte (pH: 3–4) and catholyte (pH: 8–9) in the MFC. As the proton transfer in AEM does not compete with the other cations (Ca^{+2}, Mg^{+2}, Na^{+}) in comparison to CEM, thereby maintaining the neutral pH in the anodic chamber. While the catholyte preserves its pH by transporting the excess hydroxide and other negatively charged ions to the anodic chamber [22]. Moreover, recent investigations have proved that hydroxide ion transfer from the cathodic section is critical for smooth oxygen reduction reaction (ORR) in MFC [23]. These are the reasons that the AEM is gaining more interest in MFC operations. For instance, Rossi et al. operated the MFC using AEM and recovered the power density of 7.1 W/m^2 with an internal resistance of 7.2 $m\Omega\ m^2$ [24]. Similarly, Kook et al. also reported that the power density of AEM-based MFC was five times higher than that of CEM-based MFC [25].

Similar to CEM, the AEM is also commercially available with different commercial names such as AFM, ACS, AM-1, AEM-7000, and AEM-700. In a desalination cell, the AEM facilitates the removal of the ions from wastewater and simultaneous chemical recovery. While, during the MES operation, the AEM connected to the cathodic chamber allows the acetate ions to pass through it. The high synthesis cost of the AEM is also a significant limitation in the upgradation of the technology. Also, low mechanical strength, chemical stability, and high substrate cross-over reduce the efficacy of the AEM-based reactor. The low stability causes a deformation in the membrane, increases biofouling, and limits the application for long-term operations [26]. More investigation to determine these drawbacks and their suitable prevention technologies can be a way forward to the large-scale operation of the BES system for wastewater treatment.

3 Advancement in Low-Cost Membranes

The field-scale demonstration of BESs technologies is hindered by the high cost of commercial polymer membranes as it alone comprises 38% of total capital cost considering Nafion 117 as PEM [1, 27]. However, to achieve the future goal of low-cost and environment-friendly BESs, it is imperative to use low-cost membrane materials such as ceramic, PVA-borosilicates, nylon mesh, J-cloth, and polypropylene and their composites [28]. The composite membrane is effective in altering the membrane properties like hydrophilic/hydrophobic nature. The composite membrane also effectively reduces the fouling properties in the wastewater treatment process. In this section, some emerging low-cost separators are discussed in detail for investigating their technological and economic feasibility.

3.1 PVA-Nafion Borosilicate Membrane

Nafion has been the most adopted PEM owing to its high proton conductivity via electrostatic interaction due to the presence of negatively charged sulphonated groups. It also serves the advantages of high mechanical strength, chemical stability, and high electrical conductivity. However, the Nafion-based membrane possesses certain drawbacks i.e., high-cost, cations (Na^+ and K^+) permeability, which obstructs the proton transfer, requires several activation steps, increases the overall internal resistance due to fouling, and also allows the back diffusion of oxygen to the anodic chamber [29]. The efficacies of Nafion membrane can be enhanced by minimizing the oxygen transfer coefficient and acetate diffusion coefficient, or by the development of less expensive composites. Modification using titanium dioxide (TiO_2) proved to counter the above limitations and provided higher power density when incorporated in the polymer-based sulfonated polyether ether ketone (SPEEK) membrane.

Nearly, 7.5% doping of TiO_2 in Nafion was reported to be optimum for best output, and above that, the blocking effect was observed. The blocking effect refers to the blockage in the movement of long-chain polymers, which helps to transport protons in the exchange membranes. The uniform distribution of TiO_2 in the Nafion enhances the amorphous structure and ultimately leads to an increase in porosity. The increment in porosity results in less pressure drop, less fouling problem in the membrane, and higher sustainability in long-run operation. However, TiO_2 doping increases the overall cost of MFC. Hence, low-cost alternative materials for Nafion composites were developed such as polyvinylpyrrolidone (PVP), polypropylene (PP), and sulfonated polystyrene ethylene butylene polystyrene (SEEBS) [30]. Recently, the hydrophilic polymer PVA-based Nafion borosilicate membrane emerged as an ideal candidate for low-cost PEM in comparison to the perfluoro sulfonic acids-based membranes (Nafion, Acipex, Flemions). The PVA provides good chemical stability, and mechanical resistance, offers less oxygen permeation, decent film formation ability, and forms cross-linkages with –OH functional groups [31].

In MFCs, the oxygen permeation coefficient is a significant factor as oxygen is toxic for obligate anaerobes in the anodic chamber of MFC; hence, a lesser oxygen mass transfer coefficient (K_{O2}) is required for membranes. For PVA membranes the $K_{O2} = 1.70 \times 10^{-4}$ cm/s, which is comparatively lower than bare Nafion $K_{O2} = 2.71 \times 10^{-4}$ cm/s [32]. The PVA-Nafion complex has more uniformity in structure due to hydrophilic interaction and hydrogen bonding (both fluorinated and non-fluorinated). Hence, PVA and Nafion composite performed better in MFCs when incorporated with a borosilicate backbone. Tiwari et al. obtained a COD removal efficiency of 76–80% in PVA-Nafion borosilicate membrane-based MFC, when operated on synthetic wastewater having COD around 3000 mg/L [29]. However, relatively lower COD removal efficiency (in the range of 54–65%) was obtained from the same configuration MFC operated on the distillery wastewater with three different organic loading of 2300, 4200, and 6300 mg/L. Furthermore, the maximum power density of 4.3 W/m^3 was obtained with the use of 11-fold lower cost membrane (PVA-Nafion borosilicate) material compared to Nafion 117 [33].

3.2 Ceramic Composite Membrane

The application of ceramic membrane in MFC started back in 2010, by adopting a cylindrical membrane made up of locally available clay [34]. The MFCs are often operated in the ambient environment as the mesophilic microorganisms thrive at less extreme temperatures. Compared to other low-cost separators, the ceramic membrane is an ideal material for such BES application due to its capability for microbial colonization, chemical properties, high structural strength, low substrate cross-over, and ease of availability [35]. Moreover, the ceramic membrane can better withstand external resistance and current variation compared to conventional membranes [36].

Usually, the clay-derived ceramic membrane consists of TiO_2, SiO_2, aluminium oxide (Al_2O_3), Al_2SiO_5, and other prominent ionic groups. The net negative charge on the mineral surface of the clay imparts better ion exchange capability to the ceramic membranes. The protons (positive ions) generated in the anodic chamber get attracted towards the negatively charged mineral surface of the membrane to neutralize it. Additionally, these protons are also transferred from one side to another through the intermolecular hopping of the loosely bound cations in the membrane. At lower pH, the protons replace other metal cations (Ca^{2+}, Mg^{2+}, Na^+, K^+), and thus ion exchange through ceramic is highly influenced by the cationic sites and membrane composition [37]. For instance, Behera and Ghangrekar performed the initial investigations regarding the utilization of the ceramic membrane in MFC and achieved a maximum power density of 24.32 mW/m^2. Also, Behera et al. examined the importance of membrane thickness in proton diffusion and found a positive correlation between the coulombic recovery and the membrane thickness. They optimized the membrane thickness around 3 mm for 2.5 mW power recovery and noticed the reduction in power output with further reduction in thickness. This might be due to an increase in brittleness and oxygen cross-over across the membrane, which decreased the efficiency of MFC [38].

However, certain limitations like the high membrane thickness and low active cationic sites of bare ceramic membranes resulted in incomplete proton consumption for the cathodic reduction reaction. In MFC, the incomplete cathodic reduction reaction leads to low energy conversion and the excess accumulation of the proton in the anodic chamber, which further suppresses the oxidation activity of the microbes [39]. In addition to it, the porous structure of the ceramic membrane also causes substrate and oxygen to cross-over through the chambers, which results in lower power generation. The oxygen cross-over was found to have a toxic effect on the biofilm of anaerobic exoelectrogens such as *Geobacter* sp. Additionally, long-term exposure to the oxygen in an anodic chamber can cause the formation of the aerobic biofilm with the dominance of the non-electrochemical bacteria such as *Acinetobacter* sp., and *Comamonas* sp. This further reduces the coulombic efficiency of MFC and limits the commercialization of technology [40, 41].

Further, the ceramic composite membranes were invented to improve the properties of the ceramic membrane for an efficient BES. The composite membrane amends the membrane properties such as pore size, surface roughness, and hydrophilic/

hydrophobic nature. These alternations in the membrane properties effectively reduce membrane fouling and increase the economic viability of the membrane system. Addition of the compounds like lime into the ceramic matrix or pre-treatment of clayey soil with a basic solution also enhances the proton transfer efficiency. The pre-treatment of clay with an alkaline solution induces the OH^- radicals, which augment the proton transfer through the membrane. As per the investigations, the pre-treatment of the clay with sodium hydroxide resulted in a 4 times higher proton transfer capacity and 1.5 times higher power density with the reduction in the oxygen cross-over compared to the untreated ceramic membrane [42].

The adsorption capacity of the clay precursor also has a significant impact on the proton transfer efficiency, which in turn depends upon the firing temperature. As the rapid firing of the clay results in more water adsorption, ionic conductivity, and porous structure, than the slow-firing clays, therefore, the power output of the rapidly fired clay was reported to be enhanced by 65% [41]. Also, the composition of clay is an essential factor in the long-term application of the BES. For example, adding hygroscopic silica oxides (SiO_2) compounds in the ceramic composition erects the Grotthuss mechanism pathway for proton diffusion and results in better power output compared to conventional Nafion 117 [43]. Hence, adding SiO_2-rich waste material like rice husk ash (which contains around 90% SiO_2) with clay can be an economical and effective way to increase the proton exchange capacity of the membrane. The investigation shows that the 10% mixing of rice husk with clay enhances the proton transfer capacity by 3 times as the rice husk ash aids the proton transfer through the membrane. The rice husk ash increases the number of proton exchange sites by retaining the bounded water in the membrane matrix [36]. Similar investigations regarding the addition of iron oxide in the ceramic matrix resulted in better structural porosity and low pore size. The MFC was also found to be long-term operational with high power output by adding iron oxide [44]. Thus, it can be concluded that adding the ionic-rich compounds or their substitutes improves the proton transfer mechanism with high energy production.

3.2.1 Ceramic-Ceramic Composite

Ceramic-ceramic composite refers to the mixing of two or more different types of clay, such as natural clay, terracotta, pyrophyllite, montmorillonite, and kaolinite. The addition of montmorillonite in the ceramic has positively affected the proton transfer ability and oxygen diffusion. The high alumina-silicate composition of the montmorillonite has weak van der wall and electrostatic interactions between its different layers, which consequently result in a high adsorption capacity compared to other clays (Fig. 2a) [45]. The addition of the montmorillonite and vermiculite with red clay-based ceramic membrane was reported to have a 5-times higher power density (84.3 mW/m^3) compared to the Nafion 117, and the average operating power of 82 mW in MFC [46, 47].

Another experiment with the goethite (iron ore)—clay composite resulted in a power density of 115 mW/m^2 with a lower mass transfer capacity (1.95×10^{-5} cm/

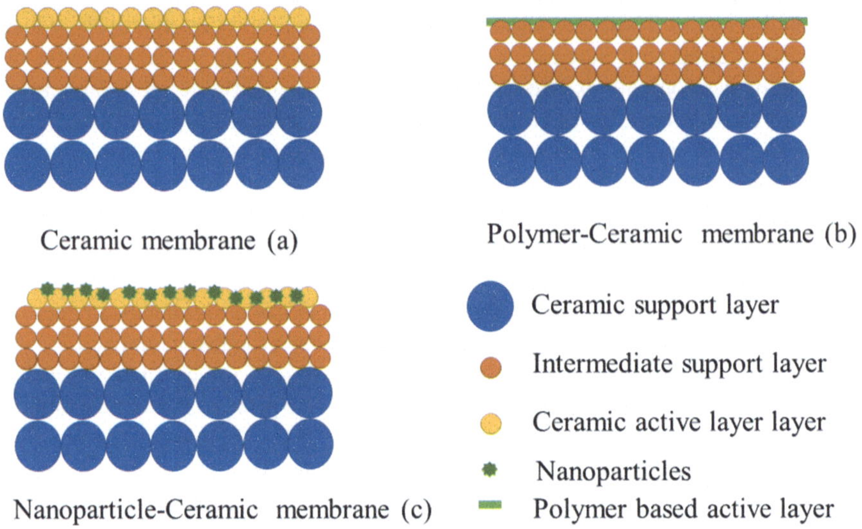

Fig. 2 Layer distribution in the **a** ceramic membrane, **b** polymer ceramic composite membrane, **c** nanoparticle ceramic composite membrane

s) and proton conductivity (78×10^{-3} S/cm) than the clay membrane [48]. The addition of other non-ceramic materials like polymers and carbon-based materials can also influence membrane performance. For instance, the addition of low-cost biopolymers like chitosan was found to be effective in the proton exchange capacity of the membrane [49]. Similar findings of polymer-modified ceramic materials have been reported to have better performance in the BESs compared to single-phase ceramic material. Like, the MFC separator synthesized using semolina and clay as the low-cost ceramic membrane showed 80 mW/m^2 of power density with 75% lower oxygen mass transfer capacity (3.8×10^{-4} cm/s) as compared to Nafion 117 (16×10^{-4} cm/s) [50]. A similar experiment with the starch and clay-based MFC separator has shown a power density of 2250 mW/m^2 with high coulombic efficiency (44%) and low internal resistance [51]. The investigation trend suggests an excellent future outcome of the ceramic separator in the BES for stable renewable energy production from wastewater.

3.2.2 Polymer-Ceramic Composite

The polymer-ceramic composite membrane consists of the ceramic substrate as a support/intermediate layer with the thin active layer of the polymers considered as the polymer-ceramic membrane (Fig. 2b). The ceramic support enhances the chemical and thermal stability of the composite membrane. At the same time, the active polymer layer maintains the pore distribution and pore size of the composite membrane [52]. Prominent polymers like polyamide, PVA, PVP, and chitosan have been used with ceramic supports like clay, TiO_2, ZrO_2, SiO_2, and Al_2O_3. Using natural materials such as clay and kaolin has also been found effective in reducing the cost of the polymer-ceramic composite. For instance, Jana et al. (2011) fabricated a low-cost polymer-ceramic membrane using clay support and chitosan as the dip-coated polymer active layer. The chitosan-clay membrane (pore size: 13 nm) was applied as an ultrafiltration membrane for the removal of the heavy metal, and 100% removal was observed [53].

The polymer-ceramic membrane is more effective in removing the deposited foulant particles during the back-washing compared to the ceramic membrane. Moreover, the low cost of waste-derived polymers provides the viability of removing the active layer after membrane clogging and replacing it with a new active membrane layer [54]. A similar investigation was performed with reusable Al_2O_3 support and methyl-ammonium chloride-based active polymer layer [55]. Apart from the fouling resistance, the polymer-ceramic membrane can withstand high temperatures as the ceramic support limits the swelling of the active layer [56]. Recently, the waste-derived non-woven propylene-ceramic composite membrane was examined for the long-run performance of air–cathode MFC. Two techniques, i.e., sandwiching the polymer layer directly in contact with the ceramic and other including placing the ceramic membrane in a polymer (dissolved in xylene) bath, were adopted for the development of a propylene-ceramic composite. However, the layer-deposited composite resulted in 92% higher power than the unmodified ceramic membrane. The outstanding performance is attributed to the strong antifouling effect of propylene-ceramic composite. Also, the non-woven propylene possesses a fibrous structure that develops the microscale access paths for the passage of ions and maintains porosity, even in the fouling condition [57, 58].

Another researcher evaluated the suitability of the polymer-derived ceramic montmorillonite composite membrane for MFC and reported the ion exchange capacity of 0.6072 meq/gm (67% of Nafion) and cation exchange number of 0.6988 (72% of Nafion). Additionally, the significant factor for MFC is the minimal oxygen transfer coefficient, which is found to be around 5.62×10^{-4} cm/s (similar to the commercially available Nafion) [59]. Hence, the fabricated polymer-ceramic composite membranes satisfy all the essential requirements for their use in MFC and overcome the limitations of polymeric membranes. Hence, the application of the polymer-ceramic membrane provides a solution to be used in the long-term application with low membrane fouling. However, the challenges of achieving a strong interfacial connection between ceramic and polymer still require further research.

3.2.3 Ceramic-Nanoparticle Composite

The need for a membrane with suitable microstructure and surface properties has led the research towards nanoparticles-doped ceramic membranes (Fig. 2c). The nanoparticles of different metals and non-metals are doped on the active layer. The incorporation of nanoparticles enhances the membrane efficiency by augmentation of non-ceramic functions like antifouling, high abrasion strength, and photocatalytic oxidation. The most commonly used nanoparticles in ceramic membranes are TiO_2, Fe_2O_3, SiO_2, silver, carbon nanotube, and graphene-based compounds. Each nanoparticle has been used to attain different properties in the membrane system [60]. The application of the activated carbon with the clay-based separator in the MFC resulted in lower oxygen diffusion coefficient (36×10^{-6} cm^2/s) with 290 mV of average operating potential (2 times of Nafion 117) [61].

The Zirconia-modified clay ceramic membrane was also evaluated as a separator in MFC and the decrease in the internal resistance from 1000 to 230 Ω was observed compared to the single-phase ceramic membrane; hence the current density of the MFC also increased from 102 to 440 mA/m^2 [62]. Apart from the clay-based ceramic, other ceramic materials like Al_2O_3, zirconia, and carbon fibre with suitable modifications were tested in MFC. For instance, the graphene oxide-modified carbon fibre membrane showed high adsorption and lower internal resistance compared to the clay membrane, and the 826 mW/m^2 of power density was observed [63]. The commonly used ultrafiltration ceramic membrane (Zirconia) was modified for application in the MFC for long-term use, and the fouling was found to be significantly low compared to the Nafion 117. Moreover, the modified zirconia membrane was effective in proton conductivity (0.08 S/cm) and power generation (1600 mW/m^2) [64].

These modifications have been helpful to improve the operation of MFC and subsequently reducing the cost and also limiting the drawbacks of the operation. Moreover, the large-scale applications of ceramic membrane-based MFC were found in the range of 1 to 1500 L; which gave the current density between 0.17 and 33 A/m^3. A tubular MFC setup of 2 L volume produced a power density of 7 W/m^3 during the HRT of 14 h and an internal resistance of 100 Ω [65]. Ghangrekar et al. operated a major large-scale ceramic-based MFC with an anodic chamber working volume of 750 L, which derives electricity from the toilet waste, and approximately 8 mW/m^2 of power density was reported. Similarly, the performance of other MFC based on the ceramic membrane with a working volume of 1500 L was also evaluated [34]. Considering the cost aspect, the ceramic membrane and their composites were reported to have expenses in the range of 6–607 $\$/m^2$. Similarly, the cost of PVA-Nafion borosilicate was around 151 $\$/m^2$, which is much less compared to the conventional Nafion membrane (1659 $\$/m^2$) [34]. Thus, the conclusion can be drawn that the ceramic membrane and its composite can be a sustainable solution for the scale-up application compared to other commercially available membranes.

4 BESs for Distillery Wastewater Treatment

The distillery industry is the most crucial segment of the global economy and asseverates the production of potable liquors, spirits, ethyl alcohol, and grain alcohol. The bioethanol (biofuel) serves as an important substitute for fossil fuels in the transportation sector and it is primarily produced in the distillery (fermentation) industries.

Additionally, the alcohol produced is also used in the pharmaceutical, personal care products, beverage, food, and perfume industries [66]. Distillery industries are one of the major agro-processing industries, which generate ample quantities of wastewater with high organic matter concentration [67]. The discharge of untreated distillery effluent into inland waterbodies or onto land causes serious environmental threats, such as eco-toxicity, eutrophication, water pollution, and requires proper handling, treatment, and disposal [68].

The enormous demand and hike in production amplify the issue of water pollution due to the distilleries, as nearly 88% of raw substrates end up in the wastewater discharge. High chemical oxygen demand (COD) (90,000–190,000 mg/L), high biochemical oxygen demand (BOD) (40,000–60,000 mg/L), total solids (60,000–115,000 ppm), high ash content, low pH, dark brown colour, polysaccharides, proteins, lignin, waxes, melanoidins, and unpleasant odour characterize the high strength liquid effluent (spent wash). Disposal of such chemically diverse effluent into water bodies has carcinogenic and mutagenic effects on aquatic ecosystems and pollutes surface water as well as groundwater. Therefore, the treatment of wastewater using suitable remediation technology, either physiochemical, biological or coupled is preferable. Amongst them, membrane-based bio-electrochemical systems (BESs) is the most adapted for distillery wastewater treatment.

4.1 Distillery Efflux and Its Adverse Impacts

Distilleries are one of the major agro-based polluting industries in developing countries like India. In addition, it also consumes a high amount of freshwater as well as natural feedstock (sugarcane molasses), which in turn produces dark-brown coloured wastewater, also known as spent wash. The effluent from distilleries is characterized by a high amount of organic content, which includes spent malt grain wash, yeast sludge, and spent wash. Generally, 12–15 L of spent wash is produced from one litre of ethanol. The composition of spent wash greatly depends on the raw material utilized in the distillery process of distillation, fermentation, and sugar extraction. The high COD, BOD, and dark brown colour effluent are the most obvious indicator of water and soil pollution and thus require adequate treatment and a safe disposal strategy.

Apart, from the above characteristics, the spent wash also comprises of high amount of lignocellulosic content, phenolic compounds, heavy metals, resins, polysaccharides, organic acids, tannins, fatty acids, nutrients, sulphate, lipids, hemicelluloses, and natural colorants [69]. Furthermore, the organic load is also due to the presence of melanoidins, which is a product of the Maillard reaction between the amino acids, reducing sugars, and caramel at elevated temperatures [70]. The natural colorant melanoidins have high molecular weight, polymeric acid, nitrogenous structure, antioxidant, antihypertensive, antimicrobial, and are recalcitrant to biological wastewater treatment (due to toxicity to microorganisms). Melanoidins also form organometallic complexes with the heavy metals (Pb^{2+}, Zn^{2+}, Cu^{2+}, Cd^{2+},

Co^{2+}, Fe^{2+}), and aromatic halogenated disinfection by-products with the polyphenols during disinfection. Besides melanoidins, other toxic chemicals present in spent wash are furan carboxylic acid, tricarballylic acid, phthalate, benzene acetic acid, methyl ester, benzoic acid (polyphenols), TMS ester, 3,4 dihydroxy mandelic acid, and butanedioic acid [71]. These toxic chemicals are considered as EDCs, which cause hormonal imbalance as well as physiological, metabolic, and reproductive disorders in humans and animals. The typical concentration range of physical, chemical, and biological parameters of distillery spent wash and the subsequent adverse impacts are highlighted in Table 1 and Fig. 3.

Table 1 Concentration range of diverse parameters (physical, chemical, and biological) and heavy metal content in the distillery effluent

Physical characteristics	Concentration range
Colour	Dark brown
Odour	Pungent
pH	3–4
Electrical conductivity (μS/cm)	80,000
Alkalinity (mg/L)	1441
Hardness (mg/L)	475
Total solids (mg/L)	3263–83,084
Total suspended solids (mg/L)	233–550.25
Total dissolved solids (mg/L)	3029–77,776
Total volatile solids (mg/L)	80,000–120,000
Biochemical characteristics	
BOD (mg/L)	40,000–60,000
COD (mg/L)	90,000–190,000
Total nitrogen (mg/L)	9000–10,000
Total organic carbon (mg/L)	25,368
Cellulose (w/w)	31.6%
Reducing sugar (mg/L)	16,170–16,500
Phosphate (mg/L)	5–8766
Sulphate (mg/L)	5760–9000
Heavy metals	
Lead (mg/L)	Up to 1.24
Copper (mg/L)	Up to 0.337
Cadmium (mg/L)	Up to 0.025
Iron (mg/L)	163–1018
Zinc (mg/L)	1–2.5
Chromium (mg/L)	Up to −0.172
Manganese (mg/L)	4–5

Note The above data is adapted from the published articles by [72–77]

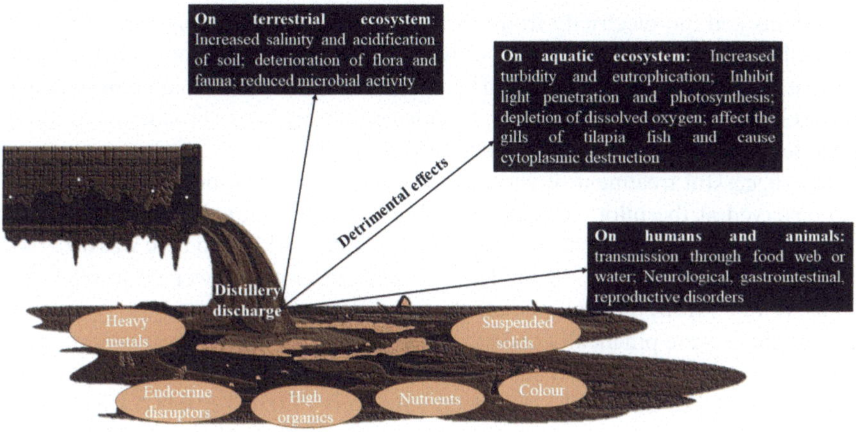

Fig. 3 Schematic diagram of contamination of terrestrial and aquatic ecosystems by distillery wastewater and its adverse impact on the environment

4.2 *Case Studies on the Membrane-Based BESs for Distillery Wastewater Treatment*

Classical anaerobic digestion successfully recovered energy and removed COD from distillery wastewater in the mesophilic and thermophilic temperature range at the industrial scale. However, the microbial process is hindered at low temperatures (lower than 13 °C) with low-strength wastewater, which negatively affects the efficiency of anaerobic digestion. In this direction, the BES is an alternative wastewater treatment system that serves the additional advantages of treating low-strength wastewater, low cost, and can be sustained in temperate climates as well. From the viewpoint of recovering energy, the three best BESs i.e., MFC, MEC, and MDC. Amongst them, MFC has relatively simpler reactor configuration, requires no additional energy input, and produces electricity directly from wastewater.

Comparing the performance efficiency of anaerobic digestion and MFC, the former achieved 95% COD removal efficiency in a significantly shorter retention time. For instance, an investigation performed by Mersinkova et al. revealed an average COD removal rate of 376.2 mg/dm^3/day in anaerobic digestion and 118.3 mg/dm^3/day in MFC. [78]. However, the MFC offers a potential opportunity to recovery energy from wastewater directly in the form of electricity and scientists are exerting to improve the performance of MFC further. On the other hand, anaerobic digestion requires additional technological stages of combustion of biogas to harvest energy. Furthermore, the carbon foot print of MFC is very less due to the formation of only CO_2 as an end product instead of CH_4, and CO_2 is much less potent GHG compared to CH_4. Another reported advantage of MFC is the much lower generation of excess sludge (~0.09 g/g COD) than anaerobic digestion (~0.159 g/g COD) [78]. Recently, the MFC is also considered as prominent technology for the simultaneous recovery of nutrients and bio-electricity from wastewater as it eliminates the need for chemical addition for the precipitation of struvite, an ammonium phosphate-based fertilizer that precipitates at alkaline pH. The introduction of air in the cathodic chamber increases the oxygen availability thereby enhances the hydroxyl production resulting in alkaline pH and struvite precipitation [79].

The successful treatment of wastewater and energy generation using BESs was also observed at the pilot- or field-scale. For example, A modularized pilot-scale stacked MFC consisting of 50 modules and 1000 L of working volume was observed to remove 70–90% (below 50 mg/L) COD and generated power up to 60 W/m^3 (~3.64 W m^{-2}) from the real municipal wastewater, when operated for 1 year. The MFC modules were practically operated under 2 different water flow connections in 2 municipal wastewater treatment plants to test for both low COD (i.e., 80 mg/L) and high COD (i.e., 250 mg/L). The energy recovery of 0.033 ± 0.005 kWh per m^3 of municipal wastewater was achieved, with a hydraulic retention time (HRT) of 2 h [80]. Another field trial of MFC was done in 2015 at the Frenchay Campus, University of the West of England, which transformed urine into power, water, and minerals. The power generated was then utilized for internal LED lighting and mobile charging purposes. The technical specification includes: 8 modules of MFC (nearly 288 MFC in total) of dimension 70 cm × 30 cm × 16 cm connected using plastic elbow connectors and pipes. After the successful 3-month field trial, which processed an average of 2.5–5 L urine daily for the HRT 2–3 weeks, the stacked MFC was reported to treat a maximum of 98% of COD, 50% of total nitrogen, and generate maximum power of 0.4 W for 75 h for LED lighting in urinals [81].

Later, in India, Ghangrekar et al., investigated the performance of a hexagonal centrally chambered MFC (720 L total working volume and consisting of 6 MFCs) having low-cost clayware separators for the onsite treatment of sanitary wastewater. They utilized $Co_{0.5}Zn_{0.5}Fe_2O_4$ and Sn_5Cu_{84} as cathode catalyst, goethite as anode catalyst, and carbon felt as electrode. The scaled MFC significantly removed COD $78.45\% \pm 19.12\%$ in 18 h, which increased to $87.29\% \pm 7.28\%$ with an increase in HRT to 36 h and also generated power of 61 mW that was used to illuminate the toilet having low power LED bulbs [82].

However, relatively fewer field-scale investigations were reported for the treatment of distillery wastewater by MFC or other BESs. Previously, a pilot scale MEC of 1000 L continuous flow MEC reactor was operated for H_2 production and COD removal from winery wastewater. The net energy obtained from the system was 2.11 kWh/m^3.day with a cathodic surface area of 18.1 m^2/m^3. They encountered certain difficulties including low volumetric H_2 production, the presence of methane along with H_2 and a highly polluted cathode [83]. Later, in 2015, a 90 L stackable pilot-scale MFC (consisting of 5 modules) was designed for the remediation of brewery wastewater and simultaneous electricity generation. The designed MFC consisted of 2 rolling cathodes made of activated carbon and PTFE, 2 rows of carbon brush anode, and 2 layers of textile separators. The system successfully demonstrated COD removal efficiency of 87.6% and produced energy of 0.097 kWh/m^3 to power the pumping system of 0.027 kWh/m^3. The net energy harvested by the system was 0.034 kWh/m^3 while treating the real wastewater with zero energy input

[84]. However, the singular large reactor unit approach is comparatively less efficient than the stacked multiple small reactors together as it can be maintained separately. Lu et al. [85] and Zhuang et al. [86] worked in the former direction, with stackable multiple MFCs of smaller volume, for treatment of distillery wastewater. Additional investigations related to the deployment of MFC, MEC, and MDC for the removal of COD, TDS, and other pollutants with simultaneous resource recovery from distillery efflux are highlighted in Table 2.

4.3 Emerging Contaminants Removal from Distillery Efflux

At present, the EDCs (melanoidins, polyphenols, phthalate, esters, etc.) present in the distillery wastewater are considered emerging contaminants and need to be tackled before discharge, along with organic load remediation. Generally, advanced oxidation processes, electrochemical processes, photocatalysis, and adsorption have been adopted for the removal of EDCs from wastewater. However, the bioremediation strategy has invariably provided an inexpensive, sustainable, and emerging treatment approach for wastewater. The highly recalcitrant organic contaminants, such as natural colourants, have been biotreated using many thermotolerant yeasts (*Candida tropicalis*) and anaerobic bacteria. Most recently, melanoidins were effectively removed by using a laccase-producing mixed consortium of *Enterococcus faecalis, Escherichia fergusonil, Citrobacter werkmanni* integrated with an enzyme-mediated MFC. Nearly, 86.02 $\pm$ 0.05% (after 7 days of operation) of melanoidins removal was achieved along with the current density of 0.540 A/m^2 and power density of 0.080 W/m^2. The mechanism pertains to the specific activity of lactic acid bacteria (*Lactobacillus plantarum, Lactobacillus casei,* and *Pediococcus parvulus*), *Enterobacter aerogens*, and *Bacillus* species in the degradation of melanoidins.

The acclimation of melanoidins degrading consortium in the anodic chamber served the dual purpose of exoelectrogens for MFC and biological catalyst for electrochemical oxidation of melanoidins [95]. Previously, simultaneous removal of melanoidins and bioelectricity generation from distillery effluent were also explored by Mohamed et al., however, more investigations are necessary as there is very limited number of studies to draw inferences [89]. The biocatalyst used in the degradation process was reported to be effectively recovered by immobilization on water-insoluble solid matrices. In this direction, the silica and alumina were utilized as immobilization support for laccase, which was later used for bioelectrolysis of synthetic distillate. The bioelectrolytic oxidation of melanoidins in the presence of yeast cells and enzymes (laccase and peroxidase) results in 70% removal after 6 h operational time and a final current of 12 μA was obtained parallelly [96].

Furthermore, technologies, such as advanced oxidation process, electrocoagulation, adsorption, and coagulation-flocculation, have been investigated for the removal of melanoidins from distillery wastewater. For instance, the batch electrocoagulation investigation using aluminium electrodes resulted in 98% decolourization efficiency (initial concentration of melanoidins = 100 mg/L) after 60 min of operating time

Table 2 Membrane-based BESs performance for treatment and resource recovery from distillery wastewater

BESs	Membranes	Electrodes (anode/cathode)	Operating conditions	Maximum output and removal efficiency	References
Conventional systems					
MFC	Nafion NE 1135 CEM (89 μm)	Carbon yarn/ open-air	Initial COD = 2303 mg/L pH = 7.8; Rint = 6 Ω	CD = 1815 ± 189 mA/m^2 PD = 364 ± 11 mW/m^2	[87]
H-shaped MFC	Nafion-117	Graphite rod	Initial COD = 33,075 mg/L pH = 6.0	PD = 836.81 mW/m^2 RE_{COD} = 78.66% RE_{TDS} = 39.66%	[88]
MFC	Nafion117	Graphite plate/ graphite plate	Initial COD = 4000 ± 20 mg/ L	CD = 580 mA/m^2 PD = 168 mW/m^2 RE_{COD} = 68.2%	[89]
MEC	Nafion 117 PEM	Plain carbon cloth/Pt mesh	Initial COD = 1500 ± 50 mg/ L	CD = 811.7 mA/m^2 HPR = 0.52 ± 0.03 mmol/Ld RE_{COD} = 75.6%	[90]
MDC	Nafion 117 PEM	Graphite plate/ open-air	SLR = 16.20 kg COD/ m^3 pH = 6.0	CD = 354.29 mA/m^2 PD = 124.35 mW/m^2 RE_{COD} = 72.84% RE_{TDS} = 23.96% (i.e., 1150 mg TDS/L)	[3]
Low-cost systems					
MFC	0.16 mm thickness 60 × 50 mm dimensions Biobag (1); Ceramic (2)	Carbon veil/ Carbon veil	Initial COD = 1200 ± 95 mg/ L	CD 1 = 32 A/m^3 CD 2 = 60 A/m^3	[91]
MFC	PVA/SSA crosslinked polymer membranes	Carbon cloth/ carbon cloth (surface-attached, PVA/PAC beads)	Initial COD = 800–3600 mg/ L	PD = 50.6–67.1 mW/m^2 RE_{COD} = 98–62%	[92]

(continued)

Table 2 (continued)

BESs	Membranes	Electrodes (anode/cathode)	Operating conditions	Maximum output and removal efficiency	References
MFC	SPEEK	Plain carbon paper/carbon paper coated with Pt	–	PD = 58.19 mW/m^2 RE_{COD} = 94.8%	[93]
MFC	Clay ware ceramic	Carbon felt/carbon felt	Initial COD = 5920 mg/L pH = 7.5	CD = 246 mA/m^2 RE_{COD} = 92.5%	[94]

Note SSA = Sulfosuccinic acids; PD = Power density; CD = Current density; OR = Ohmic resistance; HPR = Hydrogen production rate; SLR = Substrate loading rate; RE_{COD} = COD Removal efficiency; RE_{TDS} = TDS Removal efficiency

[97]. Another major EDC found in distillery wastewater is the phthalate compounds and several advanced oxidation processes (such as O_3/H_2O_2, O_3/TiO_2, O_3/activated carbon, O_3/Al_2O_3, $O_3/Fe^{2+}/H_2O_2$, and UV/TiO_2) have been investigated for its treatment. Among these, the heterogeneous catalytic ozonation with aluminium oxide was reported to have maximum diethyl phthalate (nearly 100%) removal efficiency after 15 min of operational time for the initial concentration of 200 mg/L [98]. Membrane-based technologies, such as ceramic membrane-based anaerobic bioreactor, were applied for the phenolic compounds from the wastewater and resulted in 99% removal efficiency [99]. Therefore, the hybrid system of BES with advanced oxidation or the modification of existing BES with the EDC degrading consortium and suitable low-cost membrane could be a potent solution for distillery wastewater treatment.

5 Membrane Fouling

The membrane-based technologies facilitating wastewater treatment face the drawback of membrane fouling, which results in poor effluent quality and high maintenance costs. Membrane fouling occurs due to the accumulation of suspended solids, colloidal, and other particles into membrane pores and surfaces. The membrane fouling can be classified into organic, inorganic, and biofouling based on the type of foulant [100]. Moreover, the membrane fouling process occurs in three different phases (Fig. 4). The first phase is pore clogging, in which the particles smaller than the membrane pore size get settled or adsorbed into the pores. The second phase starts with the accumulation of bacteria, biopolymers, and other inorganic substances. This accumulation developed layer by layer on the membrane surface is called the cake layer formation [101]. The final stage is pore blocking, where the pore gets blocked by larger particles, and membrane maintenance becomes a necessity for the continuous operation of the technology [102].

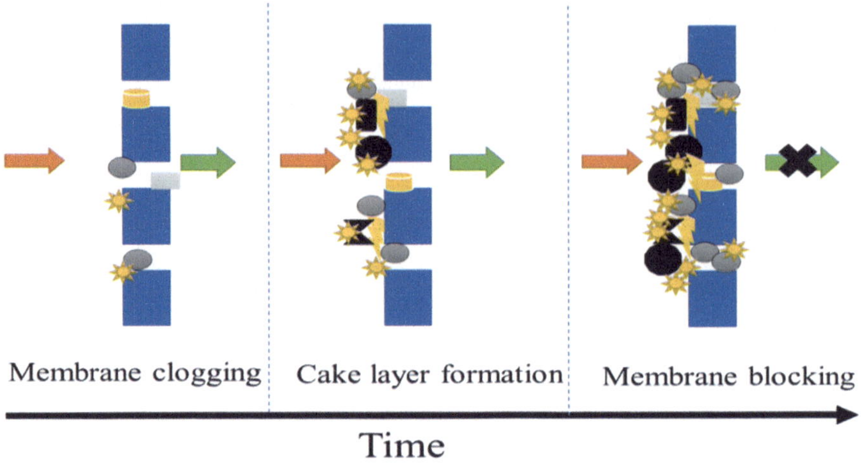

Fig. 4 Three-phase interpretation of membrane fouling

Detailed investigation of the membrane fouling shows that the organic foulant and cake layer formation are the major reasons for membrane fouling [103]. Further, the analysis of the organic foulant illustrates that the extracellular polymer substrate (EPS) contributes significantly to organic fouling. The EPS is usually secreted by the microbial community, and it is an unmetabolized substrate component made up of proteins and polysaccharides. In addition, the EPS also provides a hydrated gel matrix that acts as a binder or adhesive for the pollutants [100].

Apart from the biological parameters, wastewater characteristics like pH, salinity, and alkalinity have also been found to affect membrane fouling. For instance, high salinity enhances the EPS production and distresses the floc formation, zeta potential, and other factors leading to a high fouling rate. Similarly, the investigation regarding the effect of pH/alkalinity on fouling has revealed that lower pH (<5.5) has an increasing impact on fouling [102]. At the same time, alkaline pH (>9) also increases the inorganic precipitation leading to membrane clogging. In addition to these factors, membrane characteristics like material, pore size, morphology, zeta potential, and water affinity influence membrane fouling [104].

The surface properties of membrane materials play an important role in fouling due to interaction with the wastewater since hydrophobic materials have shown more association and sorption capacity for foulant compared to hydrophilic materials. Fouling prevention and mitigation is the primary concern in membrane-based technologies as the fouling increases the maintenance cost of the system and also limits the membrane performance. Numerous fouling preventive strategies have been evaluated in different membrane-based methods, and these can be broadly categorized as: (a) feed wastewater treatment, (b) control of operating conditions, (c) membrane cleaning, (d) adjustment of sludge suspension properties, (e) membrane surface modification, and (f) other advance methods [105]. Further discussion on control strategy is beyond the scope of the present chapter.

5.1 *Fouling Susceptibility of Polymer-Based Versus Ceramic Membrane*

The polymeric membranes are more susceptible to fouling due to their hydrophobic surface nature. In addition, the low chemical stability of polymeric membranes against chlorine and ozone-like oxidants also lowers their usability in wastewater treatment [106]. On the other hand, the ceramic membrane is chemically inert, and its hydrophilic surface nature limits membrane fouling [107]. For instance, Hofs et al. investigated the performance of four different ceramic membranes (alumina, zirconia, titania, and silicon) and compared the result with a polymeric membrane. The finding shows that all the ceramic membranes had low reversible fouling, which might be due to a higher volume/area ratio and less fine structure compared to polymeric membranes. Also, the ceramic membranes exhibit higher chemical stability (in the order $TiO_2 > ZrO_2 > Al_2O_3 > SiO_2$) compared to polymeric membranes; whereas, the hydrothermal stability follow the sequence of $ZrO_2 > Al_2O_3 > TiO_2 > SiO_2$ [108]. A similar outcome was also observed by Lee and Kim, where the fouling of the polymeric membrane was found to be 76% higher compared to the ceramic membrane. The extension of the same investigation shows that physical and chemical cleaning is more effective for ceramic membranes [104]. Overall, it can be concluded that the ceramic membrane has additional advantages over the polymeric membrane in terms of fouling prevention and management.

5.2 *Fouling of BESs for Distillery Wastewater Treatment*

Membranes, a crucial part of the BES system always face fouling especially with real wastewater and seawater treatment [109]. However, due to the large part of the BES-based research still being in the lab phase, the fouling on the AEM and CEM is not addressed compared to other parameters such as reactor configuration, electrode distance, and different ion exchange membranes. In contrast, membrane fouling was found to significantly affect the coulombic efficiency and current density in long-term operations. The membrane fouling in BES was found to follow a general trend in which the membrane separating the anolyte has high biofouling, while the membrane facing the catholyte was prominently covered with the inorganic scaling of calcium and magnesium compounds. For instance, Luo et al. operated the MDC reactor for around eight months for the treatment of domestic wastewater with AMI 7001 and CMI 7000 as membrane separators and the continuous decrease in the current density (47% decline), desalination efficiency (27% decline) and coulombic efficiency (46% decline) was observed due to the biofouling on the AMI 7001 [110].

A similar experiment of the desalination of the seawater by MDC was performed through an eight-month-long operation using AMI 7001 and a combination of CMI 7000 or Nafion 117 as the membrane separator [26]. The investigation observed that the MDC reactor with the Nafion 117 (52%) has shown more decline in the current generation compared to the CMI 7000 (43%). The continuous decline in the current generation is associated with membrane resistance, which increases with the membrane fouling along with the MDC operation. The investigation shows that the internal resistance of both AEM and CEM contributed to the membrane resistance; however, the biofouling due to anolyte and replacement of the proton-binding sites (sulfonate groups) in CEM membranes by other cations contributed most to the internal resistance. The reduction in the pores also leads to pH splitting and negatively affects the exoelectrogens in the MFC, which further deteriorates the MFC performance.

In addition, it was also observed in the MFC for long-term operation, and the SEM result revealed that the fouling on the membrane was promptly the protein-based compounds with high C–N, H–N and C=O bonds [111]. The biofouling in the MFC also caused a decline in the power generation during long-term operation; such as the 180 operations of the MFC with Nafion 117 membrane separator, and a 40% decline in the power generation was observed [112]. As the ceramic membrane has superior antifouling properties compared to the polymer and has been proven to be effective in long-term operation due to its hydrophilic nature compared to the polymer membrane. For instance, the comparative analysis of the zirconia-based ceramic and Nafion membrane for an eight-month-long operation concluded the ceramic membrane had not shown a decline in the current generation throughout the process, whereas the Nafion membrane showed a 38% decline in the current generation [64].

Alike experiment was performed by Ghadge et al. on a pilot scale MFC (45 L) using a ceramic separator. During the whole year of operation, the decline in the current generation was not observed, however, after the year of the operation, the fouling was observed with the decrease in the current generation [113]. The literature shows that the ceramic-based separator is adequate to compare with the Nafion for long-term operation. However, there is a lacuna in the field of distillery wastewater using the ceramic membrane with the BES technology. Hence, further investigation should focus on filling the research gap regarding the distillery wastewater treatment.

6 Summary and Future Perspective

The microbial electrochemical systems or BESs prove to be an efficient technology for the treatment as well as resource recovery from the distillery wastewater. The field-scale implementation of BESs requires a reduction in the overall cost of reactor fabrication, which directly depends on the cost and recyclability of the membrane. Conventional polymeric membranes such as Nafion limit the scalability of BES due to their high cost and therefore, are replaced by low-cost alternatives. Further, modified polymeric membranes like PVA-Nafion, PVA-Nafion borosilicate, ceramic membrane, and composites are attractive options to replace costly polymeric membranes. Such alternatives performed equally better at the field-scale and overcame the drawbacks, such as high oxygen diffusion rate, high cost, and biofouling when incorporated in the BESs for the treatment of wastewater.

The issues related to the ceramic membrane i.e., optimization of the thickness to enhance the water absorption, better quality catholyte, and simultaneous high-power output in MFCs, were overcome by the ceramic composite membranes. Hence, from the future perspective, these low-cost membranes based BESs can be a potent option for the large-scale treatment of industrial or distillery wastewater. As discussed, the generation of bioelectricity through MFC and MDC, biohydrogen, and other value-added products through MEC, could make the technology self-sustainable and carbon neutral. Additionally, the implementation of a less expensive membrane could decrease the overall energy requirement in comparison to industrial chemical-based synthetic membranes. Also, these BESs could potentially remediate the emerging contaminant EDCs present in distillery effluent in their hybrid or modified mode. However, the investigations related to the removal of EDCs and deploying low-cost membranes in BESs are still very limited and need to be explored further.

Abbreviations

AEM	Anion exchange membrane
BESs	Bio-electrochemical systems
BOD	Biochemical oxygen demand
CEM	Cation exchange membrane
CH_4	Methane
CO_2	Carbon dioxide
COD	Chemical oxygen demand
EDCs	Endocrine disrupting chemicals
EPS	Extracellular polymer substrate
H_2	Hydrogen
H_2O_2	Hydrogen peroxide
H_2S	Hydrogen sulphide
K_{O2}	Oxygen mass transfer coefficient
MDC	Microbial desalination cell
MEC	Microbial electrolysis cell
MES	Microbial electrosynthesis
MFC	Microbial fuel cell
OLR	Organic loading rate
ORR	Oxygen reduction reactions
PA	Polyamide
PEM	Proton exchange membrane
PFSA	Perfluoro sulfonic acid
PP	Polypropylene
PTFE	Polytetrafluorethylene

PVA	Polyvinyl amide
PVP	Polyvinylpyrrolidone
SDG	Sustainable development goals
SEEBS	Sulfonated polystyrene ethylene butylene polystyrene
SiO_2	Silica oxide
SO_3H	Sulfonic acid group
SPEEK	Sulfonated polyether ether ketone
TiO_2	Titanium dioxide
ZrO_2	Zirconium oxide

References

1. Hernández-Flores G, Poggi-Varaldo HM, Solorza-Feria O (2016) Comparison of alternative membranes to replace high cost Nafion ones in microbial fuel cells. Int J Hydrog Energy 41:23354–23362. https://doi.org/10.1016/j.ijhydene.2016.08.206
2. Kuddus M (2022) Organic waste to biohydrogen. Clean energy and production technologies. Springer Nature Singapore Pte Ltd. https://doi.org/10.1007/978-981-19-1995-4
3. Mohanakrishna G, Venkata Mohan S, Sarma PN (2010) Bio-electrochemical treatment of distillery wastewater in microbial fuel cell facilitating decolorization and desalination along with power generation. J Hazard Mater 177:487–494. https://doi.org/10.1016/j.jhazmat.2009.12.059
4. Dhar BR, Lee HS (2013) Membranes for bioelectrochemical systems: challenges and research advances. Environ Technol (United Kingdom) 34:1751–1764. https://doi.org/10.1080/09593330.2013.822007
5. Zheng T, Li J, Ji Y, Zhang W, Fang Y, Xin F, Dong W, Wei P, Ma J, Jiang M (2020) Progress and prospects of bioelectrochemical systems: electron transfer and its applications in the microbial metabolism. Front Bioeng Biotechnol 8:1–10. https://doi.org/10.3389/fbioe.2020.00010
6. Santoro C, Arbizzani C, Erable B, Ieropoulos I (2017) Microbial fuel cells: from fundamentals to applications. A review. J Power Sources 356:225–244. https://doi.org/10.1016/j.jpowsour.2017.03.109
7. Escapa A, Mateos R, Martínez EJ, Blanes J (2016) Microbial electrolysis cells: an emerging technology for wastewater treatment and energy recovery. From laboratory to pilot plant and beyond. Renew Sustain Energy Rev 55:942–956. https://doi.org/10.1016/j.rser.2015.11.029
8. Logan BE, Hamelers B, Rozendal R, Shroder U, Keller J, Freguia S, Aelterman P, Verstraete W (2006) Critical review microbial fuel cells: methodology and technology. Environ Sci Technol 40:5181–5192
9. Wang Z, Cao C, Zheng Y, Chen S, Zhao F (2014) Abiotic oxygen reduction reaction catalysts used in microbial fuel cells. ChemElectroChem 1:1813–1821. https://doi.org/10.1002/celc.201402093
10. Antonopoulou G, Stamatelatou K, Bebelis S, Lyberatos G (2010) Electricity generation from synthetic substrates and cheese whey using a two chamber microbial fuel cell. Biochem Eng J 50:10–15. https://doi.org/10.1016/j.bej.2010.02.008
11. Haile SM (2003) Fuel cell materials and components. Acta Mater 51:5981–6000. https://doi.org/10.1016/j.actamat.2003.08.004
12. Dharmalingam S, Kugarajah V, Sugumar M (2018) Membranes for microbial fuel cells. https://doi.org/10.1016/B978-0-444-64052-9.00007-8

13. Rahimnejad M, Adhami A, Darvari S, Zirepour A, Oh SE (2015) Microbial fuel cell as new technology for bioelectricity generation: a review. Alexandria Eng J 54:745–756. https://doi.org/10.1016/j.aej.2015.03.031
14. Dharmalingam S (2022) PEM Fuel Cells. Candice Janco. Elsevier Inc. https://doi.org/10.1016/C2020-0-00143-X
15. Manohar AK, Mansfeld F (2009) The internal resistance of a microbial fuel cell and its dependence on cell design and operating conditions. Electrochim Acta 54:1664–1670. https://doi.org/10.1016/j.electacta.2008.06.047
16. Harnisch F, Schröder U, Scholz F (2008) The suitability of monopolar and bipolar ion exchange membranes as separators for biological fuel cells. Environ Sci Technol 42:1740–1746. https://doi.org/10.1021/es702224a
17. Aerts P, Greenberg AR, Leysen R, Krantz WB, Reinsch VE, Jacobs PA (2001) The influence of filler concentration on the compaction and filtration properties of Zirfon®-composite ultrafiltration membranes. Sep Purif Technol 22–23:663–669. https://doi.org/10.1016/S1383-5866(00)00165-9
18. Pant D, Van Bogaert G, De Smet M, Diels L, Vanbroekhoven K (2010) Use of novel permeable membrane and air cathodes in acetate microbial fuel cells. Electrochim Acta 55:7710–7716. https://doi.org/10.1016/j.electacta.2009.11.086
19. Tofighi A, Rahimnejad M (2020) Synthetic polymer-based membranes for microbial fuel cells. Elsevier Inc. https://doi.org/10.1016/B978-0-12-818485-1.00014-9
20. Scott K (2016) Membranes and separators for microbial fuel cells. Elsevier Ltd. https://doi.org/10.1016/B978-1-78242-375-1.00005-8
21. Afsar NU, Erigene B, Irfan M, Wu B, Xu T, Ji W, Emmanuel K, Ge L, Xu T (2018) High performance anion exchange membrane with proton transport pathways for diffusion dialysis. Sep Purif Technol 193:11–20. https://doi.org/10.1016/j.seppur.2017.10.062
22. Kim JR, Cheng S (2007) Membranes in microbial fuel cells. Environ Sci Technol 41:1004–1009. https://doi.org/10.1021/es062202m
23. Rossi R, Logan BE (2021) Using an anion exchange membrane for effective hydroxide ion transport enables high power densities in microbial fuel cells. Chem Eng J 422:1–10. https://doi.org/10.1016/j.cej.2021.130150
24. Rossi R, Wang X, Logan BE (2020) High performance flow through microbial fuel cells with anion exchange membrane. J Power Sources 475:228633. https://doi.org/10.1016/j.jpowsour.2020.228633
25. Koók L, Le QED, Bakonyi P, Zitka J, Trably E, Tóth G, Pavlovec L, Pientka Z, Bernet N, Bélafi-Bakó K, Nemestóthy N (2019) Behavior of two-chamber microbial electrochemical systems started-up with different ion-exchange membrane separators. Bioresour Technol 278:279–286. https://doi.org/10.1016/j.biortech.2019.01.097
26. Ping Q, Cohen B, Dosoretz C, He Z (2013) Long-term investigation of fouling of cation and anion exchange membranes in microbial desalination cells. Desalination 325:48–55. https://doi.org/10.1016/j.desal.2013.06.025
27. Chatterjee P, Ghangrekar MM (2014) Design of clayware separator-electrode assembly for treatment of wastewater in microbial fuel cells. Appl Biochem Biotechnol 173:378–390. https://doi.org/10.1007/s12010-014-0846-x
28. Kondaveeti S, Lee J, Kakarla R, Kim HS, Min B (2014) Low-cost separators for enhanced power production and field application of microbial fuel cells (MFCs). Electrochim Acta 132:434–440. https://doi.org/10.1016/j.electacta.2014.03.046
29. Tiwari BR, Noori MT, Ghangrekar MM (2016) A novel low cost polyvinyl alcohol-Nafion-borosilicate membrane separator for microbial fuel cell. Mater Chem Phys 182:86–93. https://doi.org/10.1016/j.matchemphys.2016.07.008
30. Banerjee A, Calay RK, Eregno FE (2022) Role and important properties of a membrane with its recent advancement in a microbial fuel cell. Energies 15:444. https://doi.org/10.3390/en15020444
31. Bhowmick GD, Dhar D, Ghangrekar MM, Banerjee R (2020) TiO_2-Si- or $SrTiO_3$-Si-impregnated PVA-based low-cost proton exchange membranes for application in microbial fuel cell. Ionics (Kiel) 26:6195–6205. https://doi.org/10.1007/s11581-020-03779-z

32. González-Pabón MJ, Figueredo F, Martínez-Casillas DC, Cortón E (2019) Characterization of a new composite membrane for point of need paper-based micro-scale microbial fuel cell analytical devices. PLoS One 14:1–22. https://doi.org/10.1371/journal.pone.0222538
33. Tiwari BR, Ghangrekar MM (2018) Electricity production during distillery wastewater treatment in a microbial fuel cell equipped with low cost PVA-nafion-borosilicate membrane. J Clean Energy Technol 6:155–158. https://doi.org/10.18178/jocet.2018.6.2.452
34. Jadhav DA, Park S, Eisa T, Mungray AK, Celik E, Olabi A, Ali M (2022) Current outlook towards feasibility and sustainability of ceramic membranes for practical scalable applications of microbial fuel cells. Renew Sustain Energy Rev 167:112769. https://doi.org/10.1016/j.rser.2022.112769
35. Winfield J, Gajda I, Greenman J, Ieropoulos I (2016) A review into the use of ceramics in microbial fuel cells. Bioresour Technol 215:296–303. https://doi.org/10.1016/j.biortech.2016.03.135
36. Raychaudhuri A, Behera M (2020) Ceramic membrane modified with rice husk ash for application in microbial fuel cells. Electrochim Acta 363:137261. https://doi.org/10.1016/j.electacta.2020.137261
37. Merino Jimenez I, Greenman J, Ieropoulos I (2017) Electricity and catholyte production from ceramic MFCs treating urine. Int J Hydrog Energy 42:1791–1799. https://doi.org/10.1016/j.ijhydene.2016.09.163
38. Satyam BSR, Behera M, Ghangrekar MM (2011) Performance and economics of low cost clay cylinder microbial fuel cell for wastewater treatment. In: Proceedings of the World Renew Energy Congress—Sweden, 8–13 May, 2011 Linköping, Sweden, vol 57, pp 1189–1196. https://doi.org/10.3384/ecp110571189
39. Rismani-Yazdi H, Carver SM, Christy AD, Tuovinen OH (2008) Cathodic limitations in microbial fuel cells: an overview. J Power Sources 180:683–694. https://doi.org/10.1016/j.jpowsour.2008.02.074
40. De Oliveira MAC, D'Epifanio A, Ohnuki H, Mecheri B (2020) Platinum group metal-free catalysts for oxygen reduction reaction: applications in microbial fuel cells. Catalysts 10:1–22. https://doi.org/10.3390/catal10050475
41. Merino-Jimenez I, Gonzalez-Juarez F, Greenman J, Ieropoulos I (2019) Effect of the ceramic membrane properties on the microbial fuel cell power output and catholyte generation. J Power Sources 429:30–37. https://doi.org/10.1016/j.jpowsour.2019.04.043
42. Das I, Das S, Sharma S, Ghangrekar MM (2020) Ameliorated performance of a microbial fuel cell operated with an alkali pre-treated clayware ceramic membrane. Int J Hydrog Energy 45:16787–16798. https://doi.org/10.1016/j.ijhydene.2020.04.157
43. Neethu B, Bhowmick GD, Ghangrekar MM (2018) Enhancement of bioelectricity generation and algal productivity in microbial carbon-capture cell using low cost coconut shell as membrane separator. Biochem Eng J 133:205–213. https://doi.org/10.1016/j.bej.2018.02.014
44. Salar-García MJ, Walter XA, Gurauskis J, de Ramón Fernández A, Ieropoulos I (2021) Effect of iron oxide content and microstructural porosity on the performance of ceramic membranes as microbial fuel cell separators. Electrochim Acta 367. https://doi.org/10.1016/j.electacta.2020.137385
45. Ghadge AN, Ghangrekar MM (2015) Development of low cost ceramic separator using mineral cation exchanger to enhance performance of microbial fuel cells. Electrochim Acta 166:320–328. https://doi.org/10.1016/j.electacta.2015.03.105
46. Suransh J, Tiwari AK, Mungray AK (2020) Modification of clayware ceramic membrane for enhancing the performance of microbial fuel cell. Environ Prog Sustain Energy 39:1–13. https://doi.org/10.1002/ep.13427
47. Cheraghipoor M, Mohebbi-Kalhori D, Noroozifar M, Maghsoodlou MT (2019) Comparative study of bioelectricity generation in a microbial fuel cell using ceramic membranes made of ceramic powder, Kalporgan's soil, and acid leached Kalporgan's soil. Energy 178:368–377. https://doi.org/10.1016/j.energy.2019.04.124
48. Das I, Das S, Dixit R, Ghangrekar MM (2020) Goethite supplemented natural clay ceramic as an alternative proton exchange membrane and its application in microbial fuel cell. Ionics (Kiel) 26:3061–3072. https://doi.org/10.1007/s11581-020-03472-1

49. Yousefi V, Mohebbi-Kalhori D, Samimi A (2020) Start-up investigation of the self-assembled chitosan/montmorillonite nanocomposite over the ceramic support as a low-cost membrane for microbial fuel cell application. Int J Hydrog Energy 45:4804–4820. https://doi.org/10.1016/j.ijhydene.2019.11.216
50. Patel BR, Noroozifar M, Mohebbi-kalhori D, Kerman K (2022) Durum wheat semolina-modified ceramic membranes as novel porous separators for enhanced power generation and wastewater remediation using microbial fuel cell. Bioresour Technol 361:127752. https://doi.org/10.1016/j.biortech.2022.127752
51. Daud SM, Daud WRW, Bakar MHA, Kim BH, Somalu MR, Muchtar A, Jahim JM, Muhammed Ali SA (2020) Low-cost novel clay earthenware as separator in microbial electrochemical technology for power output improvement. Bioprocess Biosyst Eng 43:1369–1379. https://doi.org/10.1007/s00449-020-02331-7
52. He Z, Lyu Z, Gu Q, Zhang L, Wang J (2019) Ceramic-based membranes for water and wastewater treatment. Colloids Surf A Physicochem Eng Asp 578:123513. https://doi.org/10.1016/j.colsurfa.2019.05.074
53. Jana S, Saikia A, Purkait MK, Mohanty K (2011) Chitosan based ceramic ultrafiltration membrane: preparation, characterization and application to remove Hg(II) and As(III) using polymer enhanced ultrafiltration. Chem Eng J 170:209–219. https://doi.org/10.1016/j.cej.2011.03.056
54. Kotobuki M, Gu Q, Zhang L, Wang J (2021) Ceramic-polymer composite membranes for water and wastewater treatment: bridging the big gap between ceramics and polymers Molecules 26. https://doi.org/10.3390/molecules26113331
55. Amirilargani M, Merlet RB, Nijmeijer A, Winnubst L, de Smet LCPM, Sudhölter EJR (2018) Poly (maleic anhydride-alt-1-alkenes) directly grafted to Γ-alumina for high-performance organic solvent nanofiltration membranes. J Membr Sci 564:259–266. https://doi.org/10.1016/j.memsci.2018.07.042
56. Chong JY, Wang R (2019) From micro to nano: polyamide thin film on microfiltration ceramic tubular membranes for nanofiltration. J Membr Sci 587. https://doi.org/10.1016/j.memsci.2019.06.001
57. Pasternak G, Ormeno-Cano N, Rutkowski P (2021) Recycled waste polypropylene composite ceramic membranes for extended lifetime of microbial fuel cells. Chem Eng J 425:130707. https://doi.org/10.1016/j.cej.2021.130707
58. Szustakiewicz K, Cichy B, Gazińska M, Pigłowski J (2013) Comparative study on flame, thermal, and mechanical properties of HDPE/clay nanocomposites with MPP or APP. J Reinf Plast Compos 32:1005–1017. https://doi.org/10.1177/0731684413481508
59. Ahilan V, Wilhelm M, Rezwan K (2018) Porous polymer derived ceramic (PDC)-montmorillonite-$H_3PMo_{12}O_{40}/SiO_2$ composite membranes for microbial fuel cell (MFC) application. Ceram Int 44:19191–19199. https://doi.org/10.1016/j.ceramint.2018.07.138
60. Balkenov A, Anuarbek A, Satayeva A, Kim J, Inglezakis V, Arkhangelsky E (2020) Complex organic fouling and effect of silver nanoparticles on aquaporin forward osmosis membranes. J Water Process Eng 34. https://doi.org/10.1016/j.jwpe.2020.101177
61. Neethu B, Bhowmick GD, Ghangrekar MM (2019) A novel proton exchange membrane developed from clay and activated carbon derived from coconut shell for application in microbial fuel cell. Biochem Eng J 148:170–177. https://doi.org/10.1016/j.bej.2019.05.011
62. Cheraghipoor M, Mohebbi-kalhori D, Noroozifar M (2022) Enhancing the efficiency of ceramic native soil membrane using Zircon in a continuous microbial fuel cell for wastewater treatment and sustainable energy. J Environ Chem Eng 10:108255. https://doi.org/10.1016/j.jece.2022.108255
63. Xu M, Wu L, Zhu M, Wang Z, Huang ZH, Wang MX (2022) Self-supporting nitrogen-doped reduced graphene oxide@carbon nanofiber hybrid membranes as high-performance integrated air cathodes in microbial fuel cells. Carbon NY 193:242–257. https://doi.org/10.1016/j.carbon.2022.03.024
64. Daud SM, Wan Daud WR, Abu Bakar MH, Kim BH, Somalu MR, Jahim JM, Muchtar A, Ghasemi M (2019) A comparison of long-term fouling performance by zirconia ceramic filter

and cation exchange in microbial fuel cells. Int Biodeterior Biodegrad 136:63–70. https://doi.org/10.1016/j.ibiod.2018.11.001
65. Thanh CND, Nitisoravut R (2015) Earthen membrane microbial fuel cell in septage treatment. Elsevier B.V. https://doi.org/10.1016/j.egypro.2015.11.489
66. Mikucka W, Zieli M (2020) Mikucka-Zielińska2020_Article_DistilleryStillageCharacterist.pdf. Appl Biochem Biotechnol 770–793
67. Kharayat Y (2012) Distillery wastewater: bioremediation approaches. J Integr Environ Sci 9:69–91. https://doi.org/10.1080/1943815X.2012.688056
68. Wani A, Bhasarkar JB, Gaikwad RW (2022) Photocatalytic degradation of sugar and distillery industry effluent. J Inst Eng Ser E 103:79–86. https://doi.org/10.1007/s40034-020-00190-1
69. Al-Khalid T, Surkatti R, El-Naas MH (2020) Organic contaminants in industrial wastewater: prospects of waste management by integrated approaches.https://doi.org/10.1007/978-981-15-0497-6_10
70. Rani A, Pal N (2013) Biodegradation of melanoidin from distillery effluent: role of microbes and their potential enzymes. Biodegrad Hazard Spec Prod. https://doi.org/10.5772/56252
71. Chowdhary P, Raj A, Bharagava RN (2018) Environmental pollution and health hazards from distillery wastewater and treatment approaches to combat the environmental threats: a review. Chemosphere 194:229–246. https://doi.org/10.1016/j.chemosphere.2017.11.163
72. Arora PK (2019) Microbial metabolism of xenobiotic compounds.https://doi.org/10.1007/978-981-13-7462-3
73. Naresh Bharagava R (2018) Distillery wastewater: it's impact on environment and remedies. Environ Anal Ecol Stud 1:4–7. https://doi.org/10.31031/eaes.2018.01.000507
74. Ghosh Ray S, Ghangrekar MM (2019) Comprehensive review on treatment of high-strength distillery wastewater in advanced physico-chemical and biological degradation pathways. Int J Environ Sci Technol 16:527–546. https://doi.org/10.1007/s13762-018-1786-8
75. Bustamante MA, Paredes C, Moral R, Moreno-Caselles J, Pérez-Espinosa A, Pérez-Murcia MD (2005) Uses of winery and distillery effluents in agriculture: characterisation of nutrient and hazardous components. Water Sci Technol 51:145–151. https://doi.org/10.2166/wst.2005.0018
76. Sun W, Xu M, Xia C, Li A, Sun G (2013) Enhanced production of laccase by Coriolus hirsutus using molasses distillery wastewater. Front Environ Sci Eng 7:200–210. https://doi.org/10.1007/s11783-012-0457-0
77. Amare E, Kebede F, Mulat W (2017) Analysis of heavy metals, physicochemical parameters and effect of blending on treatability of wastewaters in Northern Ethiopia. Int J Environ Sci Technol 14:1679–1688. https://doi.org/10.1007/s13762-017-1270-x
78. Mersinkova Y, Nenov V, Yemendzhiev H (2021) Comparative analysis of conventional anaerobic digestion and bio-electrochemical systems in waste organics utilization. J Agric Chem Environ 10:428–435. https://doi.org/10.4236/jacen.2021.104028
79. Baby MG, Ahammed MM (2022) Nutrient removal and recovery from wastewater by microbial fuel cell-based systems—a review. Water Sci Technol 86:29–55. https://doi.org/10.2166/wst.2022.196
80. Liang P, Duan R, Jiang Y, Zhang X, Qiu Y, Huang X (2018) One-year operation of 1000-L modularized microbial fuel cell for municipal wastewater treatment. Water Res 141:1–8. https://doi.org/10.1016/j.watres.2018.04.066
81. Ieropoulos IA, Stinchcombe A, Gajda I, Forbes S, Merino-Jimenez I, Pasternak G, Sanchez-Herranz D, Greenman J (2016) Pee power urinal-microbial fuel cell technology field trials in the context of sanitation. Environ Sci Water Res Technol 2:336–343. https://doi.org/10.1039/c5ew00270b
82. Das I, Ghangrekar MM, Satyakam R, Srivastava P, Khan S, Pandey HN (2020) On-site sanitary wastewater treatment system using 720-L stacked microbial fuel cell: case study. J Hazard Toxic Radioact Waste 24:1–7. https://doi.org/10.1061/(asce)hz.2153-5515.0000518
83. Cusick RD, Bryan B, Parker DS, Merrill MD, Mehanna M, Kiely PD, Liu G, Logan BE (2011) Performance of a pilot-scale continuous flow microbial electrolysis cell fed winery wastewater. Appl Microbiol Biotechnol 89:2053–2063. https://doi.org/10.1007/s00253-011-3130-9

84. Dong Y, Qu Y, He W, Du Y, Liu J, Han X, Feng Y (2015) A 90-liter stackable baffled microbial fuel cell for brewery wastewater treatment based on energy self-sufficient mode. Bioresour Technol 195:66–72. https://doi.org/10.1016/j.biortech.2015.06.026
85. Lu M, Chen S, Babanova S, Phadke S, Salvacion M, Mirhosseini A, Chan S, Carpenter K, Cortese R, Bretschger O (2017) Long-term performance of a 20-L continuous flow microbial fuel cell for treatment of brewery wastewater. J Power Sources 356:274–287. https://doi.org/10.1016/j.jpowsour.2017.03.132
86. Zhuang L, Yuan Y, Wang Y, Zhou S (2012) Long-term evaluation of a 10-liter serpentine-type microbial fuel cell stack treating brewery wastewater. Bioresour Technol 123:406–412. https://doi.org/10.1016/j.biortech.2012.07.038
87. Sonawane JM, Marsili E, Chandra Ghosh P (2014) Treatment of domestic and distillery wastewater in high surface microbial fuel cells. Int J Hydrog Energy 39:21819–21827. https://doi.org/10.1016/j.ijhydene.2014.07.085
88. Nayak JK, Amit GUK (2018) An innovative mixotrophic approach of distillery spent wash with sewage wastewater for biodegradation and bioelectricity generation using microbial fuel cell. J Water Process Eng 23:306–313. https://doi.org/10.1016/j.jwpe.2018.04.003
89. Mohamed SN, Thota Karunakaran R, Manickam M (2018) Enhancement of bioelectricity generation from treatment of distillery wastewater using microbial fuel cell. Environ Prog Sustain Energy 37:663–668. https://doi.org/10.1002/ep.12734
90. Samsudeen N, Spurgeon J, Matheswaran M, Satyavolu J (2020) Simultaneous biohydrogen production with distillery wastewater treatment using modified microbial electrolysis cell. Int J Hydrog Energy 45:18266–18274. https://doi.org/10.1016/j.ijhydene.2019.06.134
91. Winfield J, Chambers LD, Rossiter J, Ieropoulos I (2013) Comparing the short and long term stability of biodegradable, ceramic and cation exchange membranes in microbial fuel cells. Bioresour Technol 148:480–486. https://doi.org/10.1016/j.biortech.2013.08.163
92. Lin CW, Wu CH, Huang WT, Tsai SL (2015) Evaluation of different cell-immobilization strategies for simultaneous distillery wastewater treatment and electricity generation in microbial fuel cells. Fuel 144:1–8. https://doi.org/10.1016/j.fuel.2014.12.009
93. Sedighi M, Aljlil SA, Alsubei MD, Ghasemi M, Mohammadi M (2018) Performance optimisation of microbial fuel cell for wastewater treatment and sustainable clean energy generation using response surface methodology. Alexandria Eng J 57:4243–4253. https://doi.org/10.1016/j.aej.2018.02.012
94. Ghosh Ray S, Ghangrekar MM (2015) Enhancing organic matter removal, biopolymer recovery and electricity generation from distillery wastewater by combining fungal fermentation and microbial fuel cell. Bioresour Technol 176:8–14. https://doi.org/10.1016/j.biortech.2014.10.158
95. Thipraksa J, Chaijak P, Michu P, Lertworapreecha M (2022) Biocatalysis and agricultural biotechnology biodegradation and bioelectricity generation of melanoidin in palm oil mill effluent (POME) by laccase-producing bacterial consortium integrated with microbial fuel cell. Biocatal Agric Biotechnol 43:102444. https://doi.org/10.1016/j.bcab.2022.102444
96. Tsiakiri EP, Sompatzi E, Voukia MP, Sotiropoulos S, Pantazaki AA (2020) Biocatalytic and bioelectrolytic decolorization of simulated melanoidin wastewaters by Saccharomyces cerevisiae cells suspended and conjugated on silica and alumina. J Environ Chem Eng 8:104078. https://doi.org/10.1016/j.jece.2020.104078
97. Kobya M, Gengec E (2012) Decolourization of melanoidins by a electrocoagulation process using aluminium electrodes. Environ Technol (United Kingdom) 33:2429–2438. https://doi.org/10.1080/09593330.2012.671371
98. Mansouri L, Tizaoui C, Geissen SU, Bousselmi L (2019) A comparative study on ozone, hydrogen peroxide and UV based advanced oxidation processes for efficient removal of diethyl phthalate in water. J Hazard Mater 363:401–411. https://doi.org/10.1016/j.jhazmat.2018.10.003
99. Wang W, Wang S, Ren X, Hu Z, Yuan S (2017) Rapid establishment of phenol- and quinoline-degrading consortia driven by the scoured cake layer in an anaerobic baffled ceramic membrane bioreactor. Environ Sci Pollut Res 24:26125–26135. https://doi.org/10.1007/s11356-017-0284-8

100. Asif MB, Zhang Z (2021) Ceramic membrane technology for water and wastewater treatment: a critical review of performance, full-scale applications, membrane fouling and prospects. Chem Eng J 418:129481. https://doi.org/10.1016/j.cej.2021.129481
101. Li C, Sun W, Lu Z, Ao X, Li S (2020) Ceramic nanocomposite membranes and membrane fouling: a review. Water Res 175:115674. https://doi.org/10.1016/j.watres.2020.115674
102. Guo W, Ngo HH, Li J (2012) A mini-review on membrane fouling. Bioresour Technol 122:27–34. https://doi.org/10.1016/j.biortech.2012.04.089
103. Wang H, Ding A, Gan Z, Qu F, Cheng X, Bai L, Guo S, Li G, Liang H (2017) Fluorescent natural organic matter responsible for ultrafiltration membrane fouling: fate, contributions and fouling mechanisms. Chemosphere 182:183–193. https://doi.org/10.1016/j.chemosphere.2017.04.148
104. Lee SJ, Kim JH (2014) Differential natural organic matter fouling of ceramic versus polymeric ultrafiltration membranes. Water Res 48:43–51. https://doi.org/10.1016/j.watres.2013.08.038
105. Xia L, Vemuri B, Saptoka S, Shrestha N, Chilkoor G, Kilduff J, Gadhamshetty V (2018) Antifouling membranes for bioelectrochemistry applications. Elsevier B.V.
106. Do VT, Tang CY, Reinhard M, Leckie JO (2012) Effects of chlorine exposure conditions on physiochemical properties and performance of a polyamide membrane-mechanisms and implications. Environ Sci Technol 46:13184–13192. https://doi.org/10.1021/es302867f
107. Zhao Yy, Wang Xm, Yang Hw, Xie YfF (2018) Effects of organic fouling and cleaning on the retention of pharmaceutically active compounds by ceramic nanofiltration membranes. J Membr Sci 563:734–742. https://doi.org/10.1016/j.memsci.2018.06.047
108. Hofs B, Ogier J, Vries D, Beerendonk EF, Cornelissen ER (2011) Comparison of ceramic and polymeric membrane permeability and fouling using surface water. Sep Purif Technol 79:365–374. https://doi.org/10.1016/j.seppur.2011.03.025
109. Noori MT, Ghangrekar MM, Mukherjee CK, Min B (2019) Biofouling effects on the performance of microbial fuel cells and recent advances in biotechnological and chemical strategies for mitigation. Biotechnol Adv 37:107420. https://doi.org/10.1016/j.biotechadv.2019.107420
110. Luo H, Xu P, Ren Z (2012) Long-term performance and characterization of microbial desalination cells in treating domestic wastewater. Bioresour Technol 120:187–193. https://doi.org/10.1016/j.biortech.2012.06.054
111. Xu J, Sheng GP, Luo HW, Li WW, Wang LF, Yu HQ (2012) Fouling of proton exchange membrane (PEM) deteriorates the performance of microbial fuel cell. Water Res 46:1817–1824. https://doi.org/10.1016/j.watres.2011.12.060
112. Flimban SGA, Hassan SHA, Rahman MM, Oh SE (2020) The effect of Nafion membrane fouling on the power generation of a microbial fuel cell. Int J Hydrog Energy 45:13643–13651. https://doi.org/10.1016/j.ijhydene.2018.02.097
113. Ghadge AN, Jadhav DA, Ghangrekar MM (2016) Wastewater treatment in pilot-scale microbial fuel cell using multielectrode assembly with ceramic separator suitable for field applications. Environ Prog Sustain Energy 35:676–680. https://doi.org/10.1002/ep.12403

Chapter 9
Performances of Ceramic Membranes for Cuttlefish Conditioning Wastewater Treatment

Raja Ben Amar, Afef Attia, Hajer Aloulou, and Wala Aloulou

1 Introduction

Recently, the East Mediterranean Sea's marine resources and ecosystems have come under growing strain, driven not only by demographic and economic expansion but also by diversification and the intensity of marine and maritime activities. Fishing around the Mediterranean Sea is of great importance to its countries, where various ways of exploitation of this resource have evolved since antiquity. The common cuttlefish (*Sepia officinalis*) lives in a broad variety of water temperatures from South Norway and the North Atlantic Ocean to the shores of Mauritanian and Senegal, as well as from Eastern Atlantic seas distributed to the East Mediterranean Sea through the Aegean Sea, Black Sea, Adriatic Sea, and Marmara Sea (FAO 2016) [10, 23]. Waste generated by industries processing marine species, as well as the concentration of conditioning activity on the shores, are becoming a big issue in nations where fishing is a significant sector.

The industry of conservation of sea products is very important in Tunisia. However, it produces a great amount of wastewater which is generally rejected in the littoral and then is responsible for an enormous pollution, by supporting the eutrophication phenomenon [21]. Before freezing, the cuttlefish must be washed to eliminate the black color caused by the ink contained in the animal bag, resulting in highly colored wastewater [21].

In brief, the cuttlefish processing industry produces a lot of solid and liquid waste, particularly; the presence of the ink in the animal produces colored rinsing

R. B. Amar · A. Attia · H. Aloulou · W. Aloulou (✉)
Research Unit "Advanced Technologies for Environment and Smart Cities", Faculty of Science of Sfax, University of Sfax, Sfax, Tunisia
e-mail: walaaloulou6@gmail.com

A. Attia
e-mail: afef.attia@fss.usf.tn

S. Bandyopadhyay (ed.), *Application of Low Cost Ceramic Membranes in Wastewater Treatment*, Materials Horizons: From Nature to Nanomaterials,
https://doi.org/10.1007/978-981-97-5842-5_9

water, which can have a serious negative impact on the environment. Wastewaters contain high concentrations of nitrogenous impurities, organic compounds, and salt constituents [17, 18, 27]. Because of the organisms' plasmolysis, biological treatment of these saline wastewaters with conventional microorganisms results in low chemical oxygen demand (COD) removal efficiency [16]. In this regard, membrane technology, employed in the treatment and exploitation of effluents from companies that handle sea products, is particularly appreciated. Commercially available ceramic membranes are suitable for use in harsh conditions and aggressive environments despite these advantages their use is still limited [9, 25]. The reason for such a situation is generally attributed to their high manufacturing cost due to the use of expensive synthetic mineral oxides which require high sintering temperatures in membrane manufacturing. A major cost reduction can be expected by replacing the synthetic precursor materials with natural minerals or industrial waste in manufacturing ceramic membranes. In recent years, many attempts have been made to develop porous ceramic membranes using alternative naturally sourced raw materials such as clay, kaolin, zeolite, bauxite, etc.

Several papers on this issue were discovered in the literature, and some of them have been mentioned and explained in this work. A brief summary of treatment technologies has been presented, followed by an in-depth analysis discussion of the advantages of using membrane technology, along with a detailed review highlighting the various technologies for cuttlefish effluent treatment. Specifically, the effectiveness of ceramic membranes on the removal of color, carbon oxygen demand (COD) and turbidity was monitored and discussed. Recommendation is given at the end of this chapter for efficiently handling such waste.

2 Direct Treatment Technology

2.1 Coagulation-Flocculation

In order to reduce the turbidity and carbon oxygen demand (COD), Coagulation-flocculation was used to treat cuttlefish effluent, with sulfate alumina (SA) as the coagulant, then a flocculation step with starch, SiO, MgO, and after that poly-dimethyl ammonium chloride. (PDMAC) [11]. Kinetic features, adsorption equilibrium, pH, and conductivity were investigated to find the optimal coagulation conditions and SA amounts. The inclusion of organic polymer as a flocculent outperformed the use of salt alone, reducing SA usage by 50%. The inclusion of organic polymer as a flocculent outperformed the use of salt alone, reducing SA usage by 50%. The combination of 165.5 mg l^{-1} sulfate alumina, 750 mg l^{-1} MgO, and 35 mg l^{-1} PDMAC 400,000 produced the greatest results, with a 90% elimination rate in chemical oxygen demand (COD) and a turbidity value of 7 NTU.

2.2 *Membrane Technology*

Interest in separation by the use of membrane processes has rapidly increased recently. The membrane technology is utilized in various fields: chemistry, food, biotechnology and wastewater treatment [6]. For economic reasons, much research has been conducted on the creation of new types of inorganic membranes that have demonstrated better resistance to heat, chemicals, and corrosion. Rapid development and innovation have already occurred in this field [15]. Clay minerals have well-known structural adsorption, rheological and thermal properties [22]. Research on clay membranes has primarily focused on pillared clays [28]. Studies of membranes based entirely on clay have been largely investigated [1, 5, 7, 13, 20].

2.2.1 MF Zeolite Membrane

Slip casting was used to develop ceramic microfiltration membranes on tubular microporous Turkish zeolite supports [3]. The main steps of the support preparation and microfiltration layer deposition have been detailed as follows:

Plastic paste has been prepared from zeolite powder (84 wt% of zeolite powder) mixed with organic additives working as porosity agent (8%), binder (4 wt%) and plasticizer (4 wt%). The tubular Zeolite support sintered at 850 °C for 3 h demonstrated appropriate mechanical resistance, porosity, and chemical stability. The water permeability values of the support and the membrane were 1218 and 534 L/h m^2 bar, respectively. A comparison of the results showed that the study using Turkish zeolite [3] was different compared with others made from zeolite from China [8] and zeolite from Bulgaria [14]. In fact, Dong et al. [8] fabricated zeolite multilayer MF membranes on tubular porous zeolite supports by dip-coating using natural zeolite from China as the starting material. The low-cost zeolite membranes could be used in the purification of wastewater due to their good permeability and pore structure. On the other hand, Hristov et al. [14] fabricated low-cost multilayer zeolite membranes on porous zeolite support by dip-coating process using natural zeolite from Bulgaria as the starting material. The good adhesion between the support and the layers allows the use of these zeolite membranes in several MF applications (Fig. 1).

The application of the microfiltration membrane to the treatment of cuttlefish effluents (Table 1) exhibited good performances in terms of permeate flux and efficiency (good retention of turbidity and chemical oxygen demand and total color removal).

2.2.2 MF Clay Membrane

Khemakhem prepared tubular microfiltration clay membranes using natural powder derived from Tunisian natural material [21]. The ceramic support was formed by the extrusion of a ceramic paste of silty malts followed by sintering at 1190 °C and

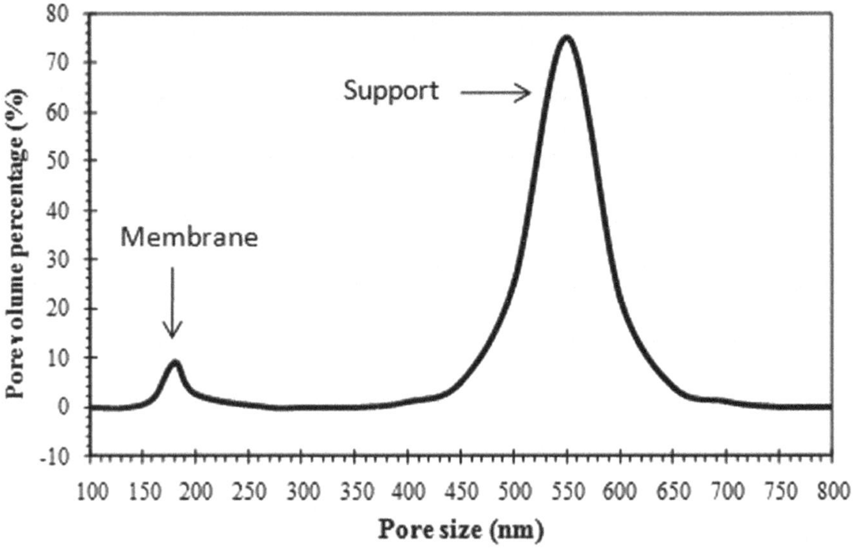

Fig.1 Pore size distribution of zeolite membrane sintered at 850 °C [3]

Table 1 Main characteristics of the effluent before and after filtration using MF membrane fabricated using zeolite at 850 °C [3]

	Raw effluents	Permeate ($P = 1$ bar)
pH	7.14	7.28
Conductivity (mS/cm)	1.3	1.31
Turbidity (NTU)	105	3.7
COD (mg/L)	425	180

characterization by mercury porosimetry showed mean pore diameter and porosity of about 9.2 μm and 49%, respectively (Fig. 2).

The microfiltration layer based on clay was achieved by deposition onto the support, by the slip-casting method. An excellent compatibility between the support and the microfiltration layer was obtained. The support pore sizes could be carefully controlled by the choice of the particle size of calcination of natural materials (silty marls). The desired mean pores size of the microfiltration layer (<1 μm) was achieved by optimization of the calcination temperature. Microfiltration membranes from Tunisian clay with pore sizes as small as 0.18 μm could be then produced (Fig. 3).

This membrane has a water permeability of 870 l h^{-1} m^{-2} bar^{-1}.The application was performed for the treatment of wastewater collected from a sea product-freezing factory located in Sfax (Tunisia). The dark color in this type of effluent was due to the presence of dark particles in suspension (sepia ink) having a size range between 56 and 161 nm (Table 2 and Fig. 4) [21].

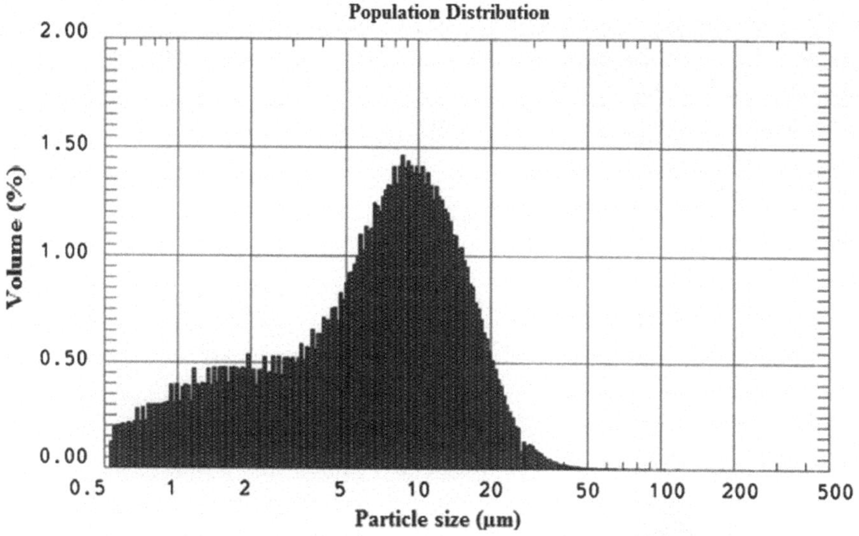

Fig. 2 Siltymarls particle size distribution [21]

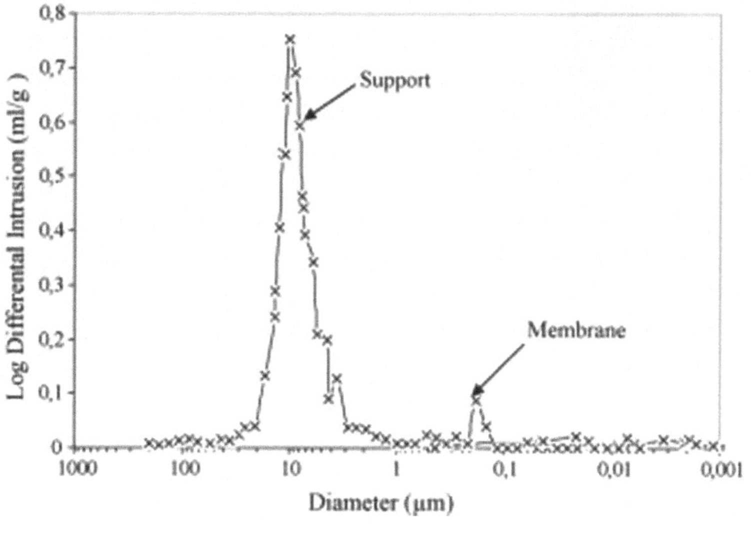

Fig. 3 Pore diameters of the clay membrane (support was sintered at 1190 °C, and the microfiltration layer base on clay was obtained by the slip-casting method and sintered at 900 °C) [21]

Table 2 Characteristics of the effluent before and after filtration

Sample	Conductivity (mS/cm)	Turbidity (NTU)	COD (mg/L)	Flux L/h.m^2
Raw effluents	80	>2000	6000–6500	140 (at $P = 1$ bar)/ 180 (at $P = 2$ bar)
Filtrate ($P = 1$ bar)	72	0.89	2460	70
Filtrate ($P = 2$ bar)	70.5	0.95	1930	120

Fig.4 A photo of cuttlefish effluent before and after MF treatment (clay support and microfiltration layer based on clay and sintered at 900 °C)

2.2.3 UF Natural Kaolin Membrane

The preparation of the porous membranes involved the formulation of a plastic and homogeneous paste for the clay ceramic powder mixed with organic additive and water.

The optimal formulation of the ceramic precursor paste was carried out in an empirical way. The selected composition for the plastic paste preparation was 90 wt% Kaolin (K) and 10 wt.% starch as a plasticizer under sintering temperature between 1100 and 1250 °C with a soaking time of 1 h and heating rate of 5 °C min^{-1} [26]. Tubular support was prepared with an external/internal diameter of 16/11 mm and a length of 150 mm.

Table 3 Characteristics of the effluent before and after filtration on kaolin membranes [26]

	TMP (bar)	Turbidity (NTU)	COD (mg L^{-1})	Conductivity (ms cm^{-1})
Raw effluents	–	335	2615	204
Permeate	2	0.62	350	140
	4	0.95	460	146
	6	1.36	530	148

Several ultrafiltration membranes with different layer thicknesses (between 3 and 30 μm) were prepared. The casting time was fixed to 6 min. The achieved membrane has an average pore size diameter of about 11 nm and a water permeability of 78 l/h m^2 bar and can be used in water treatment in the ultrafiltration range. The developed UF Kaolin membranes were applied to the treatment of cuttlefish effluents. The main physicochemical parameters determined for the permeate are listed in Table 3. The turbidity value varied as 0.62, 0.95, and 1.36 NTU with the increase in transmembrane pressure (TMP) of 2, 4 and 6 bar, respectively while COD was found to increase between 350 and 530 mg $L^{-1}$with an increase in TMP from 2 to 6 bar. The conductivity values remained almost constant in the range of 140–148 mS cm^{-1}.

Both turbidity and COD reduction were found to increase as TMP decreased, which could be attributed to higher rejection at lower TMP. As pressure increases, more melanin permeates the membrane, allowing most solutes to pass through the pores of the membrane and, as a result, decreasing rejection. When UF treatment was performed at a TMP of 2 bar, turbidity of the permeate was reduced by 99.8%, but COD was reduced by only 87%. On the other hand, the conductivity decreased slightly in the permeate and was pressure independent.

2.2.4 UF Membrane from Phosphate Industry Sub-Product

A new support for ultrafiltration membranes using the mud of phosphate hydrocyclone has been developed by Khemakhem et al. [19]. This material was chosen considering its abundance in Tunisia. The determination of the composition reveals that silica (SiO_2: 38.89%) and calcium oxide (CaO: 19.98%) were the two main components of this mud along with P_2O_5 (8.40%), Al_2O_3 (6.75%). With the exception of the relatively high proportion of P_2O_5, the phase composition is similar to that of clay [21].

The porous tubular support has been characterized by an average pore diameter and a porosity of about 1.05 mm and 39%, respectively.

A low-cost ceramic membrane was fabricated with an active separating layer of zirconia deposited over the support based on the mud of phosphate by extrusion, heating at 900 °C followed with the top layer coating by slip-casting technique. The average pore size of the membrane after heating at 700 °C was around 5 nm with a water permeability of 86 l/h m^2 bar (Fig. 5).

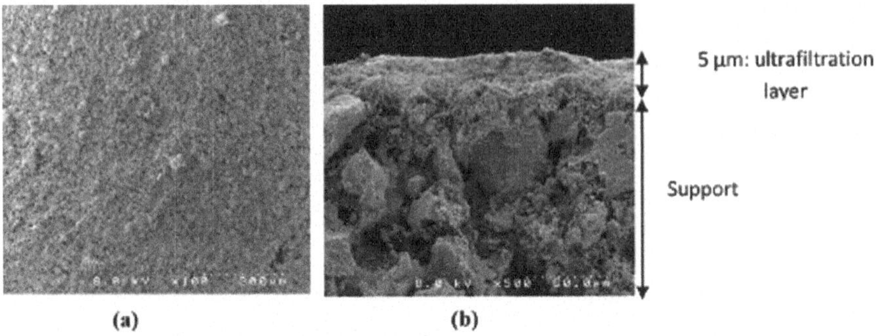

Fig. 5 SEM of zirconia ultrafiltration layer after heating at 700 °C over mud of phosphate support **a** general view of the surface; **b** cross-sectional view [19]

The fabricated zirconia membrane over mud of phosphate-based support was used for the treatment of the wastewater produced from the sea product (Sepia) freezing and conditioning [19].

Permeate flux increased linearly with transmembrane pressure up to 7 bar before becoming pressure independent. This behavior can be explained by the creation of a polarization layer, which stabilized the flux at 66 l/h m^2.

Table 4 illustrates the key features of raw and processed effluent. The permeate's turbidity can be reduced to 1.5 NTU, and the COD retention determined at 3 bar exceeds 60%. However, the conductivity in the permeate reduces little and remains pressure independent. These results validate the efficacy of low-cost UF membranes for cuttlefish conditioning effluent treatment. The results indicate that UF performances were comparable to those of alumina-based commercial UF membranes with a 300 KDa weight cut-off.

Table 4 Characteristics of the effluent before and after treatment using zirconia membrane over mud of phosphate-based support [19]

		Conductivity (mS/cm)	Turbidity (NTU)	TRCOD (%)
	Raw effluents	81.5	1300	–
Elaborated low-cost zirconia UF membrane	Permeate ($P =$ 3 bar)	67.6	1.25	60
	Permeate ($P =$ 5 bar)	69.3	1.38	65
	Permeate ($P =$ 7 bar)	70.1	1.40	67
Alumina-based commercial UF membrane	Permeate ($P =$ 3 bar)	72	1.46	70

3 Hybrid Treatment

3.1 MF in Conjunction with UF

Aloulou [4] presented results related to the treatment and exploitation of wastewater from cuttlefish conditioning using centrifugation, microfiltration and ultrafiltration.

The distribution of the effluent between the black residue and the clear supernatant could be determined by centrifugation, while pigment concentration and wastewater pollution reduction were achieved using membrane processes. Cuttlefish samples that were salinized similarly to the original effluent were used to reconstitute the synthetic wastewater in ratios of 1–50 for centrifugation and 1–100 for membrane filtration. The main characteristics of the membranes used are illustrated in Table 5.

The operating parameters were transmembrane pressure of 1.5 bar, circulation velocity of 1.5 ms^{-1}, and temperature of 25 °C. Membrane separation was performed using two different procedures: either with a constant feed composition to find the ideal operating conditions or with increasing effluent concentration and monitoring of the Volumetric Reduction Factor.

At a constant concentration, the ultrafiltration (UF) and microfiltration (MF) procedures operated differently. The MF process showed a quick decline in flux in the first few minutes, followed by stabilization at 30 L h^{-1} m^{-2} after 30 min, whereas the UF process showed a rapid stabilization at around 25 L h^{-1} m^{-2}.

The decline in flux at the start of the MF process could be attributed to partial fouling of the membrane pores by melanin particles ranging in size from 55 to 160 nm, which is on the same order of magnitude as the MF membrane pore size of 100 nm. In contrast, the slight drop in flux seen during ultrafiltration was mostly due to the creation of a polarization layer and probable interactions between the membrane material and the solution. Batch concentration was carried out for 4 and 5 h using MF and UF, respectively.

For a VRF of 2.64 (final concentration), the values of the permeate flux were found to be 2.8 L h^{-1} m^{-2} and 15.2 L h^{-1} m^{-2} for MF and UF respectively. Despite,

Table 5 Main characteristics of the membranes used [4]

Membrane	Ultrafiltration	Microfiltration
Reference	IRIS 3026	IRIS 3065
Pore size/weight cut-off	10000Daltons	0.1 μm
Material	PES	PVDF
Geometry	Flat	Flat
Membrane area	0.0106 m^2	0.0106 m^2
Flux (25 °C, 2 bars)	60–2000 L h^{-1} m^{-2}	1500–15,000 L h^{-1} m^{-2}
Chemical resistance	Good with chlorine	Very good
Thermal resistance	Good: 75–80 °C	Very good

the significant difference in pure water flux of MF and UF membranes (2690 against 34 L h^{-1} m^{-2} bar^{-1}), lower permeate flux values are a consequence of the internal pore fouling of the MF membranes.

Pollutant elimination was slightly higher for UF than for MF, with around 65% for COD, 98% for TOC, and 95% for nitrogen content (NTK) for both processes. In terms of remediation of pollution and permeate COD levels, the membrane separation efficiency does not appear to be effective. Membrane regeneration was accomplished following ultrafiltration via a simple rinse with water. On the other hand, rinsing, rinsing was inefficient for the microfiltration (PVDF) membrane and chemical regeneration was necessary to restore the membrane's initial permeability. Therefore, cuttlefish-washing wastewater is more suitable for treatment through the ultrafiltration process. However, given the degree of residual COD in the ultrafiltration permeate, more effective post-treatment approaches must be developed.

3.2 MF/Coagulation-Flocculation in Conjunction with UF

Ellouz [12] found that the membrane processes offer a very efficient method for treating cuttlefish effluents. In fact, to test the performance of this procedure in reducing COD and proteins (Npr) in the wastewater produced by the cuttlefish conditioning and freezing process, cross-flow microfiltration was used alone or in combination with ultrafiltration or coagulation-flocculation. The influence of operational conditions (velocity and pressure) on ceramic microfiltration membrane fouling (0.1 μm pore size made of Al_2O_3–TiO_2) using real effluent showed that the maximum permeate flux of about 120 1/h m^2 was obtained at 5.6 m/s and 2 bar. For the same initial COD, when the conductivity decreased from 50 to 0.6 mS/cm and the pH decreased from 7 to 4, the final value of flux increased by 28% and 40% respectively. Nevertheless, the adjustment of these parameters (0.8 NTU, 55% and 60%, respectively) had no effect on the turbidity of the permeate or the retention of COD and nitrogen proteins. Additionally, the combination of alumina salt and MgO coagulation-microfiltation (SA-MgO-MF) and microfiltration-ultrafiltration (MF-UF) was used to study the flux enhancement. Almost similar performance was obtained in the case of combined SA-MgO-MF and MF-UF wherein the impact of fouling was less prominant. However, the irreversible resistance appeared to be effective in all cases. Maximum steady state flux was found to be 140 1/h m^2 under the optimum conditions as mentioned previously. The comparable COD rejection rate was higher for SA-MgO-MF treatment (75% vs 65%). However, the retention rate of nitrogen proteins was higher with MF-UF treatment (78% vs 57%).

Based on permeate flux performances, the best operating conditions applied to raw effluents using ceramic MF membrane were: 2 bars transmembrane pressures and 5.6 m/s cross-flow velocity. The variation in feed effluent quality was also studied using reconstituted cuttlefish effluent samples under varying COD, pH and conductivity. It was found that permeate flux was also improved compared to the real effluent conditions by 25% and 26% when the model effluent conductivity and pH decreased

Table 6 Effect of MF alone and MF combined with ultrafiltration or coagulation on conductivity, turbidity and retention of COD and nitrogen protein using real cuttlefish effluents [12]

Processes used	MF alone (Al_2O_3–TiO_2)	SA + MF	SA + MgO + MF	MF + UF (Al_2O_3–TiO_2)
Turbidity, NTU	1	0.75	0.6	0.8
R COD%	55	62	75	65
R Npr %	65	41	57	78
Conductivity, mS/cm	50	38	30	46
Water flux L/h m^2	220	190	300	340
Permeate flux l/h m^2	100	100	140	140
Total resistance m^{-1}, 10^{11}	42.35	42.35	30.25	30.25
Irreversible resistance m^{-1}, 10^{11}	32.72	37.89	24	21.17
Reversible resistance m^{-1}, 10^{11}	7.62	2.46	4.25	7.08

from 50 to 0.6 mS/cm and from 7 to 4 respectively. The variation of these parameters was observed to be relatively ineffective in COD and nitrogen protein retention rate, which were about 55% and 60%, respectively.

With a view to obtain higher permeate flux and improved treated effluent quality, MF was used in conjunction with UF (Al_2O_3–TiO_2 with 15 kDa MWCO) or chemical coagulation. The use of SA-induced coagulation prior to microfiltration did not affect the stabilized permeate flux. However, the regenerated water flux after washing improved significantly when the treatment was accomplished using the combined (MF/UF) treatment system. Thus, it appears that the flux decrease in the MF/UF and SA-MgO/MF are affected by the membrane system and the nature of pretreatment. Hence membrane processes with a suitable selection of membrane cut-off and pretreatment steps (Table 6) could be used for pollution abatement with treated water recycling into the process and protein recovery.

3.3 *Performance Comparison*

The performance of different types of membranes used and process conditions applied for the treatment of cuttlefish conditioning was summarized in Table 7 which showed the prospect of low-cost ceramic membrane in providing a low-cost solution for cuttlefish wastewater treatment and reuse.

Table 7 Performance of different processes for cuttlefish effluent treatment

Process	Characterization	Results	Ref
MF tubular membrane from natural zeolite T_f = 850 °C	Porosity 43.69% average pore size 0.55 μm Mechanical resistance (12.65 MPa) Water permeate flux 1.218 L/h m^2 bar) Optimum firing temperature (850 °C) 1 layer; Thickness (2.2 μm) Average pore size (0.18 μm) Water permeate flux (534L/h m^2bar)	– Decrease of turbidity (96%) – Retention of COD (58%) – Total color removal	[3]
UF tubular membranes from natural kaolin T_f = 650 °C	*Support*: – Average pore size 1 μm – Pore volume was 44%, T_f = 1150 °C – Flexural strength of about 15 MPa *Active layer* (slip casting) Average pore diameter 11 nm Thickness was around 9 μm. Water permeability shows a value of 78 l/h m^2 bar	– Decrease of turbidity, inferior to 1.5 NTU – Retention rate of 87% for COD	[26]
UF tubular membrane from mud of the hydrocyclone laundries of phosphate T_f = 700 °C	Average pore diameter 1.05 μm porosity 39%, support sintered at 900 °C The deposition of one layer from zirconium by slip-casting (average pore size of 5) Water permeability 86 l/h m^2 bar	Decrease of turbidity, inferior to 1.5 NTU COD retention rate of about 60%	[19]
MF membrane from Tunisian clay T_f = 900 °C	Two layers: a support (SI2) with a pore diameter centered near 6.3 μm and porosity of about 48.5%, and a microfiltration layer (JM18) with a mean pore size of 0.18 μm	Turbidity1 NTU retention of COD (65%)	[21]
UF tubular membrane from apatite T_f = 600 °C	UF membranes based on hydroxyapatite slip-casting process	– UF reduced totally the turbidity (>99%), the number of total germs (100%) – 97% retention of COD – ~ 85% reduction of – conductivity pH remains stable	[24]
UF tubular membrane from apatite T_f = 600 °C	– Membrane thickness of about 32.5 μm; – Mean pores diameter of 0.5 μm – Porosity of 48% – Filtration flux of 125 l/h m^2 – The optimum composition of slip: PVA: 40%, apatite powder: 15% and dispersant solution: 48%	– Retention of turbidity (99%) – Color and COD (92%)	[24]

(continued)

Table 7 (continued)

Process	Characterization	Results	Ref
Flat MF (PVDF) membrane (0.1 μm) and Flat UF (PES) membrane (10 000 Daltons)	– MF process flux 28 L·h^{-2}·m^{-2} – UF process 15.2 L·h^{-1}·m^{-2} – VRF = 2.64	Retention of COD (65%) Retention of TOC (98%) Retention of NTK. (95%)	[4]
MF (Al_2O_3–TiO_2) in conjunction with UF (Al_2O_3–TiO_2)	– MF membrane of 0.1 μm pore size and UF membrane of 15 kDaMWCO	Turbidity, 0.8 NTU Retention of COD (65%) Retention of NTK (78%) Conductivity, 46 μS/cm	[12]
Coagulation and flocculation	The combination using: −165.5 mg l^{-1} of sulphate aluminina, 750 mg l^{-1} of MgO and 35 me l^{-1} of PDMAC 400,000	Turbidity 7 NTU Retention of COD (90%)	[11]

4 Membrane Regeneration

The application of membrane separation is frequently limited by the unavoidable phenomena of fouling, which reduces membrane performance. In the case of cuttlefish conditioning effluent treatment, the fouling process is induced by hydrophilic-hydrophobic interactions between the membrane surface and the effluent when ultrafiltration membranes are used. The absence of hydrogen bonding at the membrane interface causes hydrophobic membrane fouling [2]. Several approaches for regenerating membranes have been performed and results demonstrated that it was necessary to clean the membrane to eliminate foulants by chemical solutions [2, 19, 26].

In general, the regeneration of the membrane was carried out as follows: firstly, the membrane was exposed to an acid-basic washing sequence (Table 8), and secondly, by rinsing the membrane with distilled water till neutral pH. The used protocol appears sufficient to regenerate the initial permeability of the membranes.

Masmoudi [24] proposed two methods for the cleaning of the membrane: three successive back-flushing procedures for 15 min and thermal treatment at 450 °C for 3 h.

Aloulou et al. [2] used a cleaning method based on distilled water followed by a basic solution of NaOH (pH ≈ 8.3), and an acidic solution of HNO3 (pH ≈3.5) for three UF zeolite membranes. The efficiency of these cleaning methods (evaluated based on the recovery of the flux) is depicted in Table 8, which shows the water permeate flux as a function of the transmembrane pressure (TMP) for the unused and the regenerated membranes [2]. In all of the regeneration experiments conducted, the permeability values of the unused and the regenerated membranes were very similar,

Table 8 Membrane regeneration protocol [26]

Sequence	Agent	Concentration	T (°C)	Pressure (bar)	Time (min)
Rinsing then evacuation	Water		Room temperature	2	
Basic washing	NaOH	5–10 g L^{-1}	60–65	2	30
Rinsing until neutrality	Water		Room temperature	2	
Acidic washing	HNO_3	5 ml L^{-1}	40–50	2	30
Rinsing until neutrality	Water		Room temperature	2	

once again confirming the high quality of the prepared membranes and the ease of membrane regeneration after fouling.

5 Conclusion and Future Scope

In this chapter, different types of commercial and low-cost ceramic membranes were used for the treatment of cuttlefish conditioning effluent treatment. This chapter also discussed certain aspects for low-cost membrane fabrication used for this purpose based on naturally abundant materials (clay and apatite) and industrial sub-product (mud from phosphate industry transformation), and their characterizations, especially in terms of pore size. Indeed, tailoring the pore size of the membrane is very important to determine the membrane performances in terms of permeate flux and retention of undesirable constituents.

Effluent from the cuttlefish conditioning and freezing industry containing a high concentration of soluble organic waste, in particular ink from the animal and high concentration of nitrogen as well as salt was treated with MF, UF and low cut-off UF membranes. Almost total retention of color, turbidity and high retention of COD exceeding 60% and NTK exceeding 80% were obtained by UF membranes. It was demonstrated also that membranes with a large pore size such as MF necessitate a pretreatment to increase the treatment performances in terms of pollutant retention and limitation of fouling. The UF membranes could be utilized in post-treatment to MF or other separation processes such as coagulation/flocculation.

This study, therefore, is contemplated as a significant attempt towards the preparation of various low-cost ceramic membranes from various types of natural materials and their subsequent performance for application in the treatment of cuttlefish conditioning effluent.

Based on the findings of the current work, the following future studies can be performed:

- Development of relevant process modeling to predict performance parameters for pilot plant study along with cost estimation of the entire system.
- Development of membranes having super hydrophilicity and antibacterial properties to mitigate membrane fouling for improving membrane longevity.
- Development of UF membrane process coupling the system with water backflushing/pulsing or air spurge to contribute in solving/minimizing the membrane fouling problem.

References

1. Aloulou H, Attia A, Aloulou W, Chakraborty S, Baklouti L, Dammak L, Amar R (2022) Statistical simulation, a tool for the process optimization of oily wastewater by cross flow ultrafiltration. Membranes 12(juin):676. https://doi.org/10.3390/membranes12070676
2. Aloulou H, Bouhamed H, Daramola MO, Khemakhem S, Amar RB (2020) Fabrication of asymmetric ultrafiltration membranes from natural zeolite and their application in industrial wastewater treatment. Euro-Mediterr J Environ Integr 5(2):36. https://doi.org/10.1007/s41207-020-0150-9
3. Aloulou H, Bouhamed H, Ghorbel A, Amar RB, Khemakhem S (2017) Elaboration and characterization of ceramic microfiltration membranes from natural zeolite: application to the treatment of cuttlefish effluents. Desalination and Water Treat
4. Aloulou I, Walha K, Amar R, Quemeneur F, Jaouen P (2006) Étude préliminaire du traitement d'effluentscontenant de l'encre de seiche par centrifugation et procédés à membranes. Revue des sciences de l'eau/J Water Sci 19(4)383–392. https://doi.org/10.7202/014422ar
5. Aloulou W, Aloulou H, Attia A, Chakraborty S, Amar R (2022) Treatment of tuna cooking juice via ceramic ultrafiltration membrane: optimization using response surface methodology. Membranes 12(août):813. https://doi.org/10.3390/membranes12080813
6. Asif MB, Zhang Z (2021) Ceramic membrane technology for water and wastewater treatment: a critical review of performance, full-scale applications, membrane fouling and prospects. Chem Eng J 418:129481. https://doi.org/10.1016/j.cej.2021.129481
7. Bousbih S, Errais E, Amar R, Duplay J, Trabelsi-ayadi M, Darragi F (2019) Elaboration and characterization of new ceramic ultrafiltration membranes from natural clay: application of treatment of textile wastewater. 195–198. https://doi.org/10.1007/978-3-030-01575-6_47
8. Dong Y, Chen S, Zhang X, Yang J, Liu X, Meng G (2006) Fabrication and characterization of low cost tubular mineral-based ceramic membranes for micro-filtration from natural zeolite. J Membr Sci 281(septembre):592–599. https://doi.org/10.1016/j.memsci.2006.04.029
9. Dong Y, Wu H, Yang F, Gray S (2022) Cost and efficiency perspectives of ceramic membranes for water treatment. Water Res 220:118629. https://doi.org/10.1016/j.watres.2022.118629
10. El-Dakar A, Aly S, El-Aiatt A, Al-Beak A (2020) Status of Sepia officinalis and 19 main fisheries species caught from Southeast Mediterranean Sea, North Sinai, Egypt. Egypt J Aquat Biol Fish 24(septembre):393–404. https://doi.org/10.21608/ejabf.2020.116577
11. Ellouz E, Amar R, Abdelhamid BS (2003) Coagulation-flocculation performances for cuttlefish effluents treatment. Environ Technol 24(décembre):1357–1366. https://doi.org/10.1080/09593330309385680
12. Ellouz E, Amar R, Abdelhamid BS (2005) Cross-flow microfiltration using ceramic membranes applied to the cuttlefish effluents treatment: effect of operating parameters and the addition of pre or post-treatment. Desalination 177(juin):229–240. https://doi.org/10.1016/j.desal.2004.12.010
13. Hamden MB, Bouaziz J (2021) Preparation and characterization of tubular cermet membrane for microfiltration separation: application to the treatment of textile wastewater. Comptes Rendus Chimie 24(1):135–146.https://doi.org/10.5802/crchim.69

14. Hristov P, Yoleva A, Djambazov S, Chukovska I, Dimitrov D (2012) Preparation and characterization of porous ceramic membranes for micro-filtration from natural zeolite. 47:476–480
15. Hubadillah SK, Jamalludin MR, Othman MHD, Iwamoto Y (2022) Recent progress on low-cost ceramic membrane for water and wastewater treatment. Ceram Int 48(17):24157–24191. https://doi.org/10.1016/j.ceramint.2022.05.255
16. Kapdan IK, Erten B (2007) Anaerobic treatment of saline wastewater by *Halanaerobium lacusrosei*. Process Biochem 42(3):449–453. https://doi.org/10.1016/j.procbio.2006.09.001
17. Kechaou E, Bergé J-P, Jaouen P, Amar R (2015) Optimization of common cuttlefish (*Sepia officinalis*) protein hydrolysate using pepsin by response surface methodology. J Aquat Food Prod Technol 24(avril):270–282. https://doi.org/10.1080/10498850.2013.773116
18. Kechaou ES, Dumay J, Donnay-Moreno C, Jaouen P, Gouygou J-P, Bergé J-P, Amar RB (2009) Enzymatic hydrolysis of cuttlefish (*Sepia officinalis*) and sardine (*Sardina pilchardus*) viscera using commercial proteases: effects on lipid distribution and amino acid composition. J Biosci Bioeng 107(2):158–164. https://doi.org/10.1016/j.jbiosc.2008.10.018
19. Khemakhem M, Khemakhem S, Ayedi S, Amar RB (2011) Study of ceramic ultrafiltration membrane support based on phosphate industry subproduct: application for the cuttlefish conditioning effluents treatment. Ceram Int 37(8):3617–3625. https://doi.org/10.1016/j.ceramint.2011.06.020
20. Khemakhem S, Ben Amar R, Larbot A (2007) Synthesis and characterization of a new inorganic ultrafiltration membrane composed entirely of Tunisian natural illite clay. Desalination 206(1):210–214. https://doi.org/10.1016/j.desal.2006.03.567
21. Khemakhem S, Larbot A, Amar RB (2009) New ceramic microfiltration membranes from Tunisian natural materials: application for the cuttlefish effluents treatment. Ceram Int 35:55–61
22. López-Galindo A, Viseras C, Cerezo P (2007) Compositional, technical and safety specifications of clays to be used as pharmaceutical and cosmetic products. Appl Clay Sci 36(1):51–63. https://doi.org/10.1016/j.clay.2006.06.016
23. Aly SM, El-Dakar AY, El-Aiatt AAO, AL-Beak AM (2020) Feeding habits and reproduction period of sepia officinalis (Linnaeus, 1758) captured from East Mediterranean Sea. Egypt J Aquat Biol Fish 24(7-Special issue):171–182. https://doi.org/10.21608/ejabf.2020.119760
24. Masmoudi S, Ben Amar R, Larbot A, El Feki H, Ben Salah A, Cot L (2005) Elaboration of inorganic microfiltration membranes with hydroxyapatite applied to the treatment of wastewater from sea product industry. J Membr Sci 247(1):1–9. https://doi.org/10.1016/j.memsci.2004.03.047
25. Mestre S, Gozalbo A, Lorente-Ayza MM, Sánchez E (2019) Low-cost ceramic membranes: a research opportunity for industrial application. J Eur Ceram Soc 39(12):3392–3407. https://doi.org/10.1016/j.jeurceramsoc.2019.03.054
26. Rekik SB, Deratani A, Bouaziz J, Baklouti S (2016) Development of an asymmetric ultrafiltration membrane from naturally occurring kaolin clays: application for the cuttlefish effluents treatment. J Membr Sci Technol 6(3):1–12. https://doi.org/10.4172/2155-9589.1000159
27. Souissi N, Ellouz-Triki Y, Bougatef A, Blibech M, Nasri M (2008) Preparation and use of media for protease-producing bacterial strains based on by-products from Cuttlefish (*Sepia officinalis*) and wastewaters from marine-products processing factories. Microbiol Res 163(4)473–480. https://doi.org/10.1016/j.micres.2006.07.013
28. Zereffa EA, Desalegn T (2019) Preparation and characterization of sintered clay ceramic membranes water filters. Open Mater Sci 5(décembre):24–33. https://doi.org/10.1515/oms-2019-0005

Chapter 10
Pharmaceutical Wastewater Treatment Using Nanoparticle-Coated Low-Cost UF Membranes

Priyankari Bhattacharya, Sathi Banerjee, and Sibdas Bandyopadhyay

1 Introduction

Emerging contaminants (ECs) are basically synthetic organic chemicals found recently in water bodies (Fig. 1) [1]. They include human and veterinary pharmaceuticals and personal care products (PPCPs), steroid hormones, agrochemicals, endocrine disruptor compounds (EDCs), antibiotics, perfluorinated compounds, water disinfection byproducts, gasoline additives, etc. They have deleterious impacts on aquatic as well as human life and have thus created concern among scientists; government agencies and society [2, 3]. They negatively impact ecology and have drawbacks to human health and aquatic organisms [4]. ECs are generally classified into pharmaceutical (Ps) and personal care products (PCPs) commonly known as PPCPs. These are detected in municipal sewage, household products, wastewater, hospitals, pharmaceutical production plants, natural water bodies, etc. Their concentration varies from n/L to g/L as depicted in Table 1 [5]. At this high concentration, ECs impose a threat to life as they are known to cause endocrine disruption of living organisms, microbial resistance in soil, etc. [6]. ECs are present in a variety of daily use products like medicines including analgesics, lipid regulators, antibiotics, diuretics, non-steroid anti-inflammatory drugs (NSAIDs), etc., cosmetics, pesticides, antiseptics, food supplements, deodorants, etc. [7, 8]. ECs bioaccumulate in the environment and cause damage to reproductive organs, reduce fertility, endometriosis, and affect hormonal balance and antibiotic resistance [9]. Some ECs noted as major pollutants

P. Bhattacharya (✉) · S. Banerjee
Metallurgical and Materials Engineering Department, Jadavpur University, Kolkata 700032, India
e-mail: priyankari2004@yahoo.co.in

S. Bandyopadhyay
Department of Environmental Studies (Ex-Emeritus Scientist CSIR), Visva-Bharati, Santiniketan 731235, India

S. Bandyopadhyay (ed.), *Application of Low Cost Ceramic Membranes in Wastewater Treatment*, Materials Horizons: From Nature to Nanomaterials,
https://doi.org/10.1007/978-981-97-5842-5_10

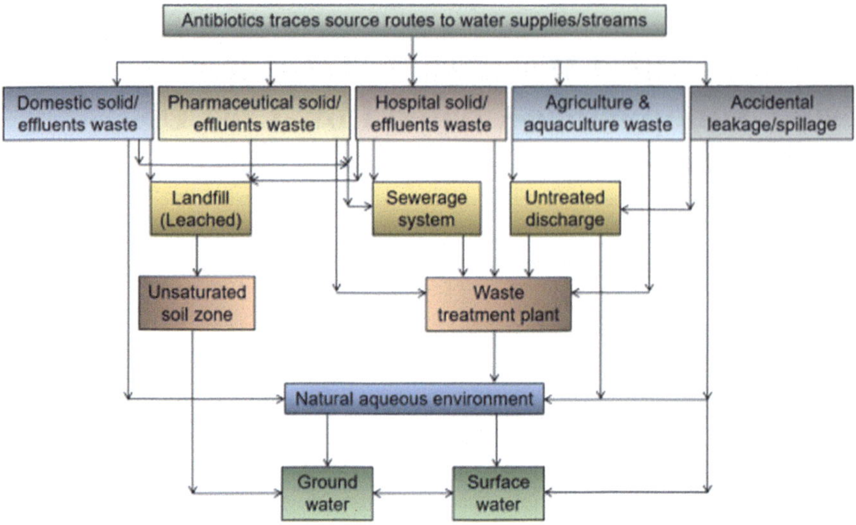

Fig. 1 Routes of ECs from source to water bodies (*Source* Reprinted from [4], Copyright (2023) with permission from Elsevier)

include polychlorinated biphenyls (PCBs), chloroacetanilides, triazines, organophosphates, dichloro diphenyl trichloroethane (DDT), carbamates, etc. [1, 10]. Since ECs are known to cause damage even in low concentrations, it is necessary to treat and effectively remove ECs from water and wastewater. Various treatment technologies exist for the removal of ECS like biological, chemical and hybrid treatment [11, 12]. Biological treatment includes activated sludge process, microalgal bioreactor, fungal bioreactor membrane bioreactor, aerobic bioreactor, anaerobic bioreactor, biosorption, etc. The chemical process includes ozonation, coagulation, photocatalysis, oxidation, Fenton process, etc. Membrane-based treatment using ultrafiltration, nanofiltration, reverse osmosis, etc. is associated with several drawbacks like operational cost, sludge generation, etc. [13–23]. Thus hybrid or integrated processes are now being undertaken for the removal of ECs. This book chapter, deals with the scope of using ceramic ultrafiltration membranes coated with metal oxide nanoparticles for treatment of ECs.

2 Pharmaceutical Manufacturing Operations

Pharmaceutical manufacturing processes include transforming the active drug ingredient followed by inserting it into a suitable carrier to a particular dosage form. Solid dose form accounts for more than 80% of active dosage forms. These forms of pharmaceutical operations include tablets, capsules, powders, suppositories, etc. Preparation of these forms requires blending followed by granulation and then sieving. For

Table 1 Concentrations of different ECs detected in water bodies of different geographic regions

Category of ECs	ECs	Max. concentration (ng/L)	Source/country
Pharmaceutical products	Paracetamol	106,970	Surface water/Kenya
	Paraxanthine	116	Surface water/China
	Lincomycin	57.2	Surface water/China
	Roxithromycin	27.4	Surface water/China
	Nalidixic acid	2.7	Groundwater/China
	Naproxen	2.62	Groundwater and surface water/India
	Acetaminophen	4.17	Groundwater and surface water/India
	Sulfapyridine	3.5	Drinking water/United States
	Sulfapyridine	10.2	Groundwater and surface water/India
	Triclosan	8.6	Surface water/Brazil
	Triclosan	1.05	Drinking water/Italy
	Ciprofloxacin	28.8	Groundwater and surface water/India
	Atenolol	1.3	Groundwater and surface water/India
	Clofibric acid	91.7	Drinking water/United States
	Hydrochlorothiazide	4.07	Groundwater and surface water/India
	Meprobamate	14.18	Drinking water/United States
	Sulfamethoxazole	161	Source water/United States
	Sulfamethoxazole	8.2	Drinking water/United States
	Chloramphenicol	660	Surface water/Kenya
	Chloramphenicol	17,440	Surface water/Kenya
	Paracetamol	630	Surface water/Kenya
	Ibuprofen	49.4	Groundwater and surface water/India
	Ibuprofen	106,970	Surface water/Kenya
	Carbamazepine	586	Source and drinking water/United States
	Carbamazepine	660	Surface water/Kenya
	Amitriptyline	430	Groundwater and surface water/Kenya

(continued)

Table 1 (continued)

Category of ECs	ECs	Max. concentration (ng/L)	Source/country
	Amitriptyline	10.3	Groundwater (for drinking)/Italy
Pesticides	Metolachlor	0.11	Sea water/Western Mediterranean, Italy
	Diuron	23.05	Surface water/Spain
	Fipronil	29.2	Surface water/Brazil
	Atrazine	130	Source and drinking water/United States
	Atrazine	171.3	Surface water/Brazil
	Tebuthiuron	1,000	Surface water/Australia
	Hexazinone	44.1	Surface water/Brazil
	Hexazinone	100	Surface water/Australia
	Piperonyl butoxide	10	Surface water/Australia
	2-hydroxy atrazine	129.5	Surface water/Brazil
	Malathion	50.4	Surface water/Brazil
	Chlorpyrifos	67.6	Surface water/Malaysia
	Deisopropylatrazine	18.74	Surface water/Spain
	Buprofezin	9.28	Surface water/Spain
	Quinalphos	64	Surface water/Malaysia
	Simazine	100	Surface water/Australia
Personal care products	Benzophenone-3	5.7	Drinking water/Italy
	Diethyltoluamide	98	Drinking water/United States
	Diethyltoluamide	1.21	Sea water/Western Mediterranean, Italy
	Galaxolide	110	Drinking water/United States
	1,4-Dioxane	180	Surface water/United States
	1,4-Dioxane	10.2	Surface water/China
	1,4-Dioxane	22.3	Groundwater and surface water/India
	Propylparaben	3.61	Surface water/Spain
	Methylparaben	3.86	Surface water/Spain

(continued)

Table 1 (continued)

Category of ECs	ECs	Max. concentration (ng/L)	Source/country
Surfactants and plasticizers	Bisphenol A	4,800	Surface water/Spain
	Bisphenol A	683	Groundwater (for drinking)/Italy
	Bis(2-ethylhexyl) phthalate	2,050	Surface water/Taiwan
	4-n-nonylphenol (4-n-NP)	1,030	Surface water/Spain
	Tri(2-butoxyethyl) phosphate	470	Source water/United States
	Dibutyl phthalate	163	Surface water/Taiwan
Anthropogenic markers	Bromoform	3,300	Source water/United States
	Nicotine	28.3	Groundwater (for drinking)/Italy
	Benzotriazole methyl-1H	1,200	Source water/United States
Hormones	17α-ethinyl estradiol	38.5	Surface water/Brazil
	17α-ethinyl estradiol	1.08	Groundwater/China
	Estrone	0.29	Drinking water/United States
	Estriol	11.9	Surface water/Brazil
Flame retardants	Tris(2-chloroisopropyl) phosphate	128	Surface water/Spain
	Tris(2-butoxyethyl) phosphate	60	Surface water/Spain
	Tri(2-chloroethyl) phosphate	65	Drinking water/United States
Fluorinated compounds	Perfluorobutanoic acid	725	Surface water/Spain
	Perfluorobutanoic acid	104	Drinking water/United States
	Perfluorooctane sulfonate	48.3	Source water/United States
	Perfluorooctane sulfonate	2	Surface water/Australia
	Perfluorooctanoic acid	11,200	Source water/United States
	Sulfluramid	2	Surface water/Australia

(*Source* Reprinted from [7], Copyright (2023) with permission from Elsevier) tablet formation, five methods followed include wet granulation, dry granulation, direct compression, fluid bed granulation and spray-dried granulation. The granules are mixed with lubricants and compressed into respective forms like tablets, gelcaps, etc. Tablets may also need colouring, coating or printing followed by packaging. The semisolid dosage form of tablets is generally hydrocarbon base, absorption base, water removal base or water-soluble base. They contain additives like antimicrobial or colours. Liquid dosage forms are prepared by mixing active ingredients like preservatives, antioxidants, pH adjusters, colours, flavors, alcohol, water, etc. Parental dosage forms include injections or vials composed of ampicillin, sodium, penicillin, etc. Further antibiotics also are included in parental dosages that are aseptically manufactured. Pharmaceuticals also deal with biotechnology for developing vaccines that focus on recombinant DNA technology (rDNA) [24–26].

2.1 Pharmaceutically Active Compounds (PhACs)

Contaminants arising from pharmaceuticals and personal care products are termed emerging contaminants (ECs) and are classified into pharmaceuticals and personal care products (PCPs) [27]. They are organic synthetic chemicals and are found in sewage water, aquatic environments, wastewater, landfills, etc. from the ng/L-g/L range. They impose deleterious effect on plants and animals. As they are not removed by conventional water treatment processes, they ought to remain in the environment, potentially affecting water quality, human health and ecosystem. These emerging contaminants interfere with the endocrine system of high organisms [28]. ECs or pharmaceutically active compounds (PhACs) include antibiotics, beta-blockers, antiseptics, cosmetics, antimicrobials, fragrances, etc. They bioaccumulate disturbing normal hormonal balance, causing cancers, decreases fecundity. Endocrine-disrupting compounds can modify the natural function of endocrine systems thereby affecting a progeny. Some visible changes associated with it are broken eggs of birds, fish or turtles. Pesticides, surfactants and other personal care products exhibit immuno-depressive behavior in mammals, fish, and affect human growth hormone and other endocrine activities [29–32].

2.2 Typical Composition of Wastewaters Generated in Different Stages

Manufacturing of PPCPs includes multistage processes that generate effluent of different compositions at different stages. The five major manufacturing units of pharmaceutical industries involve: (1) fermentation plants (2) synthesized organic chemicals plants (3) fermentation/synthesized organic chemicals plants (4) natural/biological product extractions for antibiotics/vitamins/enzymes, etc. (5) drug mixing, formulation, and preparation plants for tablets, capsules, and solutions, etc. [33]. Drugs are produced by the use of inorganic and organic chemicals in the chemical synthesis process. The process involves many stages generating intermediates and by-products along with the generation of complex wastewater that includes toxic residual products like acids, bases, cyanides, nitrates, sulfates, etc. along with the use of huge quantities of fresh water. Wastewater released after the purification stage is most toxic due to the presence of solvents, spills, finished products, etc. and thus needs immediate treatment rather than release in water treatment plants. The typical composition of pharmaceutical wastewater includes high biological oxygen demand (BOD), chemical oxygen demand (COD), total suspended solids (TSS) levels, pH ranging from 7–8, nitrates, phosphates, metals, phenols, detergents, etc. Waste generated after the fermentation process includes fermentation broth, phenols, dead cell waste, etc. For extraction of biologically active ingredients from natural origin, a biological extraction process is carried out which generates effluent containing solvents, metals, phenols, aromatic hydrocarbons, etc. In last or final stage of the drug manufacturing process includes the compounding process where processes like milling, mixing, grinding, compression, packaging, etc. are undertaken thereby generating effluent containing complex chemicals like fillers, binders, flavoring agents, preservatives, and antioxidants. In a nutshell, typical composition of wastewater from pharmaceutical processing is very complex and its removal is not easy with conventional treatment processes [33–35] (Table 2).

2.3 Range of Organic and Inorganic Compounds and Impurities and Discharge Limit.

Impurities in pharmaceuticals usually arise during primary packaging or end product. Contact between processing materials like storage bags, tubing materials, filters, etc. gives rise to impurities. During storage also impurities come from labels, ink, boxes, etc. Genotoxic impurities as called arise during the synthesis process. Impurities are detected by isolating and characterizing; Column chromatography; Gas chromatography; Flash chromatography; TLC; HPLC; Gas chromatography; Capillary electrophoresis; HPTLC etc.

Impurities in pharmaceuticals consist of organic, inorganic or residual components that develop during formulation or aging or synthesis process and remain within the active pharmaceutical ingredients known as APIs. These impurities can be related to products that come from chemical synthesis or biosynthetic routes. During the decomposition of drug, during storage or during exposure to extreme conditions, impurities may be formed in the final product. Volatile or non-volatile, arising throughout the synthesis process or storage that may come from raw materials or developed intermediate and not removed during synthesis is termed organic impurities. These impurities may develop in the bulk manufacturing process. Inorganic impurities are developed during manufacturing processes and usually include

Table 2 Chemical composition of pharmaceutical wastewater with ECs content collected from different parts of the world

pH	BOD (mg/L)	COD (mg/L)	TOC(mg/L)	TDS (mg/L)	TSS (mg/L)	Main compound and conc. (mg/L)	Analysis method	Source/Location
6.8	8600	16,240	NA	NA	400	Ampicillin (3.5)	HPLC	Fuzhou Pharmaceutical Company, China
6.6	350	2300	NA	NA	30	Aureomycin (4.6)	HPLC	Fuzhou Pharmaceutical Company, China
6.9	140	540	NA	1760	680	NA	BOD/COD analysis	Pharmaceutical industry in Dewas, India
5.8–7.8	20–620	128–960	NA	650–1250	230–830	NA	BOD/COD analysis	ETP plant, India
NA	10,184 ± 2574	16,547 ± 1827	8083 ± 578	24,899 ± 1758	285 ± 175	Triethylamine (9872 ± 2142)	GC	Pharmaceutical company in Singapore
7.18	NA	2840 ± 135	634 ± 42	846 ± 108	311 ± 0.83	Ampicillin (12.34 ± 1.32) Ceftriaxone (2.61 ± 0.22) Amoxicillin (18.68 ± 0.97) Cefazolin (8.74 ± 0.75)	HPLC	Pharmaceutical factory, Lanzhou, China
6.8	466	3500	NA	600	360	Paracetamol (48)	HPLC	Pharmaceutical industry

(continued)

Table 2 (continued)

pH	BOD (mg/L)	COD (mg/L)	TOC(mg/L)	TDS (mg/L)	TSS (mg/L)	Main compound and conc. (mg/L)	Analysis method	Source/Location
12.8	21,560 ± 160	37,410 ± 225	8250 ± 145	21,340 ± 450	6780 ± 180	NA	BOD/COD analysis	Pharmaceutical manufacturing unit, Derabassi, Punjab, India
13–14	NA	34,348 ± 1425	17,215 ± 1784	64,383 ± 2113	433 ± 120	Antibiotics (penicillin family)	COD analysis	Pharmaceutical factory, Singapore

NA-Not applicable
(*Source* Reprinted from [36], Copyright (2023) with permission from Elsevier)

reagents, ligands, catalysts, heavy or residual metals, inorganic salts, filter aids, or charcoals. Residual solvents on the other hand are volatile organic chemicals that are unwanted. Even traces of these substances might interfere with the efficiency of drugs. Their presence affects the nature of pharmaceuticals like crystallinity, dissolution, colour, etc. Impurities above 0.1% must be identified and reduced to an acceptable level. Restricted use of solvents like benzene (<2 ppm), 1,2 Dichloroethane (<5 ppm), 1,1 Dichloroethene (<8 ppm),1,1,1 Trichloroethane (<1500 ppm), and Carbon tetrachloride (4 ppm) are denoted as class I impurities. Class II impurities include solvents whose use must be limited because of their toxicity. Class III impurities are solvents that have lesser toxicity potential and use should be limited to 50 mg per day.

Resistance to biodegradation may be beneficial to patients by increasing drug benefits but causes toxic effects on animals and humans as they have increased half-life. For this, there is a need to set certain standards for discharge into the environment. The government must set rules for effluent discharge from manufacturing locations, hospitals, pharmaceutical treatment plants, etc. In-situ processes for reducing pharmaceutical effluxes must be considered to decrease discharge. Country-wise certain approaches are being undertaken by the government but International standards must be set. The European Union and USFDA have set some guidelines for discharge concentrations of pharmaceuticals i.e. 1 and 0.01 μg/L. The USA, Canada, and Spain have issued rules for discarding unused and expired drugs. The Netherlands, Switzerland and Germany followed the incineration of municipal solid waste and thus spread of pharmaceuticals from municipal solid waste to agricultural lands or landfills decreased. Advanced oxidation, ultrafiltration, adsorption also provide 95% removal of pharmaceuticals. The Stockholm Country Council of Sweden classified pharmaceuticals on the basis of persistence, bioaccumulation, and toxicity index from 0–3 where 0 corresponds to lesser toxic and 3 highest toxic and encourage physicians to prescribe drugs with low PBT index value [35, 37]. The Central Pollution Control Board of India has set discharge limits for bulk drugs and formulation which is shown in Table 3.

2.4 Limitations of Conventional Wastewater Treatment Systems

Various treatment methods are employed for the removal of emerging contaminants includes physical, chemical and biological methods Table 4 [39]. Activated sludge process; aerobic, anaerobic, microalgal, fungal and membrane bioreactors; trickling filter; constructed wetland; enzyme treatment; biosorption, etc. Most of the emerging contaminants are not treated by biological processes. Biological treatment works when ECs have a large molecular weight that can be easily broken down by microorganisms and employed for their cell development [40]. But as most ECs are themselves toxic or their byproducts are non-suitable for the growth of microorganisms and degradation is blocked. Then there comes the need for substrate for the growth of

Table 3 Discharge limit of pharmaceutical Industry [38]

Parameter (effluent standards)	Standards Limiting concentration in mg/l, except for pH
Compulsory parameters	
pH	6.0–8.5
Oil and grease	10
BOD (3 days at 27 °C)	100[a]
COD	250[a]
Total suspended solids (TSS)	100
Total dissolved solids (TDS)**	–
Bioassay test	90% survival after 96 h in 100% effluent[b]
Additional parameter*	
Mercury	0.01
Arsenic	0.2
Chromium (hexavalent)	0.1
Lead	0.1
Cyanide	0.1
Phenolics	1.0
Sulphides	2.0
Phosphate	5.0

(*Source* Rana et al. 2017)

microorganisms as in aerobic, anaerobic or membrane bioreactors. Biodegradation experiences drawbacks when certain ECs like polar pharmaceuticals or beta blockers are present in wastewater. Some pharmaceuticals like ibuprofen, diclofenac, ketoprofen, sulfamethoxazole, etc. are removed by activated sludge process with removal efficiency varying from 40%-100% respectively. Carbamazepine on the other hand is not removed by the activated sludge process [41–47]. The type of wastewater as well as season may play an important role in removal efficiency. A study conducted by Kosma et al. at WWTP in Greece revealed that PPCP removal in summer was more effective than in winter [48]. This is due to microbial activity being slowed down in winter than in summer [49]. In this situation, chemical treatment may show some positive results.

Chemical treatment methods like ozonation; Fenton's oxidation; coagulation and sedimentation; flotation; advanced oxidation process; photocatalytic oxidation; ultrasound oxidation; electrochemical; oxidation; ozonation etc. may remediate these contaminants. UV radiation, gamma rays, oxidizing agents, etc. may convert the ECs to lesser toxic products. Chlorine, hydrogen peroxide, metal-based catalysts may reorganize the charge on ECs or their structure which can easily be degraded by biological methods [50, 51]. Advanced oxidation process includes ozonation, gamma radiolysis, sonolysis, etc. which might remove PPCPs [52]. Removal of PPCPs by

Table 4 ECs removal by conventional process, membrane process and other integrated process

Treatment	ECs	Initial conc. (μg/L)	Operational conditions	Removal %
Conventional WWTPS	Bisphenol A	0.16	F/M ratio 0.07–0.10 kg BOD_5/kg VSS/d	60
	Diuron	0.3	F/M ratio 0.07–0.10 kg BOD_5/kg VSS/d	18
	Estrone—3-sulfate	100	Anaerobic/anoxic/ oxic biological treatment	83
	Estradiol-3-glucuronide	0.05	Step-feed anoxic/ oxic process	100
	Amoxicillin	0.17	HRT: 30 h; MLSS: 3.8 g/L; F/M: 0.12	50
	Triclosan	0.78	Denitrification/ anaerobic/anoxic/ oxic; HRT: 10.3 h; SRT 18.2 d	80
	Estriol	0.80	Single-stage aerobic and anoxic process; HRT: 7.9 h; SRT 20 d	100
	Atenolol	0.479	Activated sludge process; HRT: 24 h	23
	Bisoprolol	0.925	Activated sludge process; HRT: 24 h	20
	Carbamazepine	3.78	Activated sludge process; HRT: 12 h; SRT 1.5 d	11
	Naproxen	8.50	Activated sludge process; HRT: 12 h; SRT 1.5 d	43
	Ketoprofen	5.52	Activated sludge process; HRT: 12 h; SRT 1.5 d	52
Enzymatic membrane reactor	Benzophenone	100	HF membrane (6 kDa, 0.19 m^2 surface area); Laccase dose: 46 mg/L d	100

(continued)

Table 4 (continued)

Treatment	ECs	Initial conc. (μg/L)	Operational conditions	Removal %
	Estriol	100	HF membrane (6 kDa, 0.19 m^2 surface area); Laccase dose: 46 mg/L d	90
	Triclosan	100	HF membrane (6 kDa, 0.19 m^2 surface area); Laccase dose: 46 mg/L d	95
	17β-Estradiol	100	HF membrane (6 kDa, 0.19 m^2 surface area); Laccase dose: 46 mg/L d	99
	17β-Estradiol	5	Pressure: 10 bar; Crossflow velocity: 40.2 cm/s; 25 °C; HRT: 16 h	65
	Ibuprofen	50	HF membrane (6 kDa, 0.19 m^2 surface area); Laccase dose: 46 mg/L d	80
	Atrazine	50	HF membrane (6 kDa, 0.19 m^2 surface area); Laccase dose: 46 mg/L d	70
	Bisphenol A	5	Pressure: 10 bar; Crossflow velocity: 40.2 cm/s; 25 °C; HRT: 16 h	99
Membrane bioreactor	Androsterone	1.20	Permeate flux: 10 LMH; VS/TS: 0.80; 6 h HRT; 30d SRT; 7.5 pH	98
	Testosterone	1.85	Permeate flux: 10 LMH; VS/TS: 0.80; 6 h HRT; 30d SRT; 7.5 pH	99
	Lorazepan	84.8	MF membrane (0.10 μm); TSS: 5 g/L; HRT: 19 h	100

(continued)

Table 4 (continued)

Treatment	ECs	Initial conc. (μg/L)	Operational conditions	Removal %
	17α-ethinylestradiol	50	pH: 4.5; fungus concentration: 0.4 g/L; incubation time: 24 h	93
	17β-estradiol	50	pH: 4.5; fungus concentration: 0.4 g/L; incubation time: 24 h	99
	Bisphenol A	50	pH: 4.5; fungus concentration: 0.4 g/L; incubation time: 24 h	94
	Ranitidine	0.485	8 m^2 flat sheet membrane (0.4 μm pore); 16.1 °C; 45d SRT; 12.5 h HRT	90
	Erythromycin	0.049	8 m^2 flat sheet membrane (0.4 μm pore); 16.1 °C; 45d SRT; 12.5 h HRT	80
	Hydrocodone	0.046	8 m^2 flat sheet membrane (0.4 μm pore); 16.1 °C; 45d SRT; 12.5 h HRT	95
	Acetaminophen	21.95	HRT: 2–6 h; MLSS: 4,000–13,000 mg/L; MBR flux: 15–22 gfd	96
	Carbamazepine	0.299	HRT: 2–6 h; MLSS: 4,000–13,000 mg/L; MBR flux: 15–22 gfd	96
	Ibuprofen	70.35	HRT: 2–6 h; MLSS: 4,000–13,000 mg/L; MBR flux: 15–22 gfd	99
Anaerobic membrane bioreactor	Naproxen	5.0	MLSS: 15.4 g/L; SRT: 250d; HRT: 48 h; flux 0.94 LMH; MF membrane	60

(continued)

Table 4 (continued)

Treatment	ECs	Initial conc. (μg/L)	Operational conditions	Removal %
	Ibuprofen	5.0	MLSS: 15.4 g/L; SRT: 250d; HRT: 48 h; flux 0.94 LMH; MF membrane	30
	Triclosan	5.0	MLSS: 15.4 g/L; SRT: 250d; HRT: 48 h; flux 0.94 LMH; MF membrane	60
	Diclofenac	0.05	PVDF membrane (0.04 m^2 surface area, <0.14 μm pore); MLSS 29 g/L	78
	Erythromycin	0.31	PVDF membrane (0.04 m^2 surface area, <0.14 μm pore); MLSS 29 g/L	86
	17β-estradiol	1.70	Permeate flux: 10 LMH; VS/TS: 0.80; 6 h HRT; 30d SRT; 7.5 pH	15
	Estriol	1.75	Permeate flux: 10 LMH; VS/TS: 0.80; 6 h HRT; 30d SRT; 7.5 pH	90
	Estrone	2.12	Permeate flux: 10 LMH; VS/TS: 0.80; 6 h HRT; 30d SRT; 7.5 pH	35
	Bisphenol A	5.0	Permeate flux: 1.8 LMH; 1.3 gCOD/L d; HRT: 4 h; SRT 180 d; pH: 7	99
Membrane bioreactor–reverse osmosis	Ranitidine	0.932	Polyamide RO membrane (Flow: 179.35 ± 1.28 L/h; TMP: 10 ± 1 bar)	100
	Atenolol	2.44	Polyamide RO membrane (Flow: 179.35 ± 1.28 L/h; TMP: 10 ± 1 bar)	100

(continued)

Table 4 (continued)

Treatment	ECs	Initial conc. (μg/L)	Operational conditions	Removal %
	Carbamazepine	0.099	Polyamide RO membrane (Flow: 179.35 ± 1.28 L/h; TMP: 10 ± 1 bar)	100
	Diazepam	0.017	Polyamide RO membrane (Flow: 179.35 ± 1.28 L/h; TMP: 10 ± 1 bar)	100
	Estriol	0.226	Polyamide RO membrane (2:1 array, 85 ft^2 surface area, 12 gdf flux)	99
	Estrone	0.10	Polyamide RO membrane (2:1 array, 85 ft^2 surface area, 12 gdf flux)	99
	Diclofenac	0.10	Polyamide RO membrane (2:1 array, 85 ft^2 surface area, 12 gdf flux)	99
	Naproxen	21.0	Polyamide RO membrane (2:1 array, 85 ft^2 surface area, 12 gdf flux)	99
Nanofiltration	Carbamazepine	17	MWCO: 200 Da.; flux: 4.32 LMH; cross-flow velocity: 0.50 cm/s	91
	Amoxicillin	25,000	MWCO: 300 Da; Surface area: 0.37 m^2; pH: 10; Pressure: 0.5 MPa	99
	Atrazine	17	MWCO: 200 Da.; flux: 4.32 LMH; cross-flow velocity: 0.50 cm/s	97
	Acetaminophen	17	MWCO: 200 Da.; flux: 4.32 LMH; cross-flow velocity: 0.50 cm/s	67

(continued)

Table 4 (continued)

Treatment	ECs	Initial conc. (μg/L)	Operational conditions	Removal %
	Diclofenac	>0.80	Flux: 2.2 LMH; TMP: 0.3 bar; permeability: Operating pressure: 14 bar	60
	Naproxen	>0.95	Flux: 2.2 LMH; TMP: 0.3 bar; permeability: Operating pressure: 14 bar	60
	Norfloxacin	5,000	MWCO: 200 Da.; surface area: 0.036 m^2; flowrate: 480 L/h; TMP: 8 bar	97
	Sulfamethoxazole	25,000	MWCO: 200 Da.; surface area: 0.036 m^2; flowrate: 480 L/h; TMP: 8 bar	97
Membrane bioreactor–Nanofiltration	Carbamazepine	15	HRT 12 h; UF (30 nm pore, 310 cm^2 area, 6 LMH); NF (200 Da.,10 bar)	90
	Atrazine	10	HRT 12 h; UF (30 nm pore, 310 cm^2 area, 6 LMH); NF (200 Da.,10 bar)	90
	17β-estradiol	0.08	MWCO: 400 Da.; surface area: 7.6 m^2; flowrate: 9.5 m^3/h; TMP: 8 bar	84
	Testosterone	0.28	MWCO: 400 Da.; surface area: 7.6 m^2; flowrate: 9.5 m^3/h; TMP: 8 bar	100
Electrocoagulation	Amoxicillin	50,000	Aluminum electrode plates; current: 0.7 A; pH: 3; EC time: 60 min	52
	Estrone	200	Iron electrode (60 cm^2 surface area); pH: 7; 2 h EC time; 16.7 mA cm^{-2}	81

(continued)

Table 4 (continued)

Treatment	ECs	Initial conc. (μg/L)	Operational conditions	Removal %
	Estriol	200	Iron electrode (60 cm^2 surface area); pH: 7; 2 h EC time; 16.7 mA cm^{-2}	85
	17β-estradiol	200	Iron electrode (60 cm^2 surface area); pH: 7; 2 h EC time; 16.7 mA cm^{-2}	87
	17α-ethynylestradiol	200	Iron electrode (60 cm^2 surface area); pH: 7; 2 h EC time; 16.7 mA cm^{-2}	30
	Diclofenac	1.0	Al anode; stainless steel cathode; EC time: 3 h; current: 0.3 mA cm^{-2}	46
Ultrafiltration	Amoxicillin	10,000	Hollow fibre membrane (100 kDa, 6.6 cm^2); permeate flux: 150 LMH	38
	Diclofenac	10,000	Hollow fibre membrane (100 kDa, 6.6 cm^2); permeate flux: 150 LMH	25
	Carbamazepine	10,000	Hollow fibre membrane (100 kDa, 6.6 cm^2); permeate flux: 150 LMH	20
	Estradiol	100	Flat sheet membrane (100 kDa); pH: 7.5; NOM conc.: 0–40 mg C/L	25
Wetland	Estrone	0.01	Influent: 8100 m^3/d; surface area: 0.15 m^2 (0.5 × 0.3 m); depth: 60 cm	68

(continued)

Table 4 (continued)

Treatment	ECs	Initial conc. (μg/L)	Operational conditions	Removal %
	Estrone	0.058	5–20 L/h inflow; size (80 × 60 × 80 cm); initial density; 25 plants/m^2	60
	17β-estradiol	0.009	Influent: 8100 m^3/d; surface area: 0.15 m^2 (0.5 × 0.3 m); depth: 60 cm	84
	17β-estradiol	0.023	5–20 L/h inflow; size: 80 × 60 × 80 cm; initial density; 25 plants/m^2	15
	17β-estradiol	0.12	3.6 m^3/s base flow; size: 82.8 × 8.9 m; 0.5 m water level depth; 10d HRT	36
	17α-ethynylestradiol	0.006	Influent: 8100 m^3/d; surface area: 0.15 m^2 (0.5 × 0.3 m); depth: 60 cm	75
	17α-ethynylestradiol	0.12	3.6 m^3/s base flow; size: 82.8 × 8.9 m; 0.5 m water level depth; 10d HRT	41

(*Source* Reprinted from [7], Copyright (2023) with permission from Elsevier)

ozonation, Fenton's oxidation depends on the formation of free radicals that oxidize PPCPs and this limits their large-scale applications.

Removal of PPCPs by activated carbon has been reported. Removal depends on hydrophobicity and charge. The higher the hydrophobicity and positive charge more the removal by adsorption [53]. Adsorption by carbon nanotubes of different PPCPs like ketoprofen, carbamazepine, sulfamethoxazole, etc. proved to be highly effective. However the cost of graphene or its oxides is a major drawback [54]. Membranes on the other hand remove PPCPs based on their size, electrostatic repulsion and adsorption. Factors like charge, size, hydrophobicity of PPCPs and membrane properties like pore size, molecular weight cutoff, zeta potential determine the removal of PPCPs by membrane-based processes. Membrane-based treatment methods like reverse osmosis; nano-filtration; ultra-filtration; or hybrid treatment schemes under physical treatment of ECs are found to be promising [55–57].

In a hybrid system, ozonation followed by biologically activated carbon effectively removed pesticides, beta-blockers and other pharmaceuticals. Ozonation combined with ultrasound completely eliminated ibuprofen, naproxen, acetaminophen, ulfamethazine-like pharmaceuticals [41]. Although these hybrid treatment schemes or membrane-based treatment processes may have high operational costs employing integrated or hybrid treatment processes can contribute to lesser cost and higher efficiency [58]. Another approach can be the application of ceramic membranes which have initial installation costs but have higher shelf life than their polymeric counterparts and also can be operated at high temperatures and harsh chemical atmospheres. Metal oxide nanoparticles may be employed for coating ceramic membranes to achieve desirable pore size for removal of ECs [59]. Further discussions on these membranes will be detailed in this book chapter.

3 Membrane-Based Treatment of Pharmaceutical Compounds

Membranes are generally of two types i.e. polymeric and ceramic. Commercial membranes are mostly polymeric membranes developed from synthetic polymers like polysulfone, polyacrylonitrile, polypropylene, cellulose acetate, polyacrylonitrile, etc. Reverse osmosis (RO) is generally composed of polyamide where, polysulfone, poluamide, ceramics are used for the development of nanofiltration (NF) membranes. Polyvinylidene fluorides are used for the development of ultrafiltration (UF) membranes embedded with metal oxide nanoparticles for enhanced performance [60]. For water purification, organic, inorganic and organic–inorganic hybrid materials have been studied. Polymeric membranes exhibit chemical resistance, diverse mechanical properties, cost-effective and processing is flexible. Inorganic ceramic membranes developed from zeolites, montmorillonite, boehmite, etc. show high rejection efficiency but the cost of production is high and complex. The incorporation of nanomaterials within a polymer matrix or coated on ceramics provides excellent rejection along with an increase in hydrophilicity, anti-fouling properties, strength and permeation. Electrospun nanofibers are now studied for effective removal along with cost-effectiveness and ease in the synthesis process [61, 62].

It is well-known fact that membranes remove PPCPs through size exclusion, surface charge, electrostatic repulsion and adsorption. Microfiltration and ultrafiltration membranes remove suspended solids owing to their large pore size. PPCPs are thus removed by nanofiltration and reverse osmosis membranes. The use of nanofiltration (NF) for the removal of amoxicillin from aqueous solution is widely conducted. NF is used for the removal as well as recovery of amoxicillin from pharmaceutical wastewater. About 97% rejection was obtained with 47% removal of COD and permeation flux of 1.5L/min/m^2. Most studies carried out in RO and NF membranes use synthetic solutions and pure water. Few case studies on real wastewater were carried out using combined UF and RO pilot plants [63, 64]. Raw municipal wastewater and secondary effluent containing 12 PPCPs were treated in UF and RO membranes in northern Spain where <20% removal was obtained with only UF membranes and >99% removal obtained for UF and RO membranes. Another study was conducted in Sydney on MF-treated wastewater containing 10 types of PPCPs using an NF membrane. PPCPs had a molecular weight of 119–296 g/mol and were negatively charged. NF90 membrane was used showing removal rates of 35–98% removal rates. The author observed that 7 out of 10 PPCPs like ibuprofen, diclofenac, naproxen, gemfibrozil, triclosan, etc. were removed by NF membrane with >90% efficiency. PPCPs also get removed by adsorption on an organic substrate that might develop on the membrane surface during filtration study causing fouling. A membrane bioreactor was studied for the removal of the etodolac drug. Several fragrances, pharmaceuticals and ECs were subjected to pilot-scale MBR treatment and compared with the conventional activated sludge process [65]. It was observed that 90% removal of ibuprofen was achieved but carbamazepine removal was hindered even at high SRT. In another study, two different types of pilot-scale MBRs were used for the removal of pharmaceuticals. One MBR was a hollow fiber UF membrane and the other was an MF flat sheet membrane. The authors observed maximum removal for the MBR process due to adsorption on the surface. These might result in fouling which needs to be regularized and also the pH of the system to be maintained for proper growth of microorganisms. Few pharmaceuticals like diclofenac are removed by adsorption whereas 17β-estradiol is biodegraded [66, 67]. Application of two-stage MBR for removal of several pharmaceuticals was observed. Granular activated carbon was used as a medium for the growth of microorganisms as well as for adsorption. The combined process resulted in 90% removal efficiency for sulfadiazine, naproxen, ciprofloxacin, carbamazepine, etc. along with a 95% reduction of COD [66, 68]. Enzyme-mediated degradation of pharmaceuticals was also studied. Application of laccases for enzymatic degradation of pharmaceuticals from concentrated wastewater was observed. Laccases depleted estrogen with 1 h contact time and antibiotics like tetracycline, doxycycline, and chlortetracycline received a similar fate with 16–48% removal within 4 h reaction time [69]. Laccases from *Trametes versicolor* removed 25% diclofenac and 95% mefenamic within 20 h of reaction time [70]. Other membrane-based treatments of pharmaceutical compounds have been explored where Hassan et al. used ZnO-based coagulant followed by microfiltration for removal of various pharmaceutical compounds from human urine for its safe disposal and reuse [71]. Nanofiltration was employed for the removal of pharmaceutical and estrogenic components from urine [72]. MF and RO were employed in sequence for the removal of 20 micropollutants by Garcia et al., where the removal of ibuprofen and nonylphenol was not satisfactory (~75%) which were further removed (~99%) by employing a membrane bioreactor system [73]. Removal of ibuprofen and ketoprofen was carried out using an emulsion liquid membrane with 99.3% removal efficiency from distilled water [74]. Nanofiltration membrane in cross-flow filtration equipment was investigated for the removal of carbamazepine diclofenac and ibuprofen from drinking water [75]. The application of nanofiltration membrane and reverse osmosis for the removal of pharmaceuticals

Table 5 Removal of some pharmaceuticals by membrane process and MBR process [78]

Contaminant	Percent removal				
	MF (3)	UF (5)	UF/MBR	NF (3)	RO (9)
Acetaminophen	<20	<20	>80	20–50	>80
Caffeine	<20	<20	>80	50–80	>80
Carbamazepine	<20	<20	20–50	50–80	>80
Diazepam	ND	20–50	<20	50–80	>80
Diclofenac	<20	<20	<20	50–80	>80
Dilantin	<20	<20	<20	50–80	>80
Erythromycin	<20	20–50	20–50	>80	>80
Gemfibrozil	<20	<20	20–50	50–80	>80
Hydrocodone	<20	<20	20–50	50–80	>80
Ibuprofen	<20	<20	50–80	50–80	>80
Iopromide	<20	<20	<20	>80	>80
Meprobamate	<20	<20	20–50	50–80	>80
Naproxen	<20	<20	>80	20–50	>80
Pentoxifylline	<20	<20	>80	50–80	>80
Sulfamethoxazole	<20	20–50	20–50	50–80	>80
Triclosan	20–50	>80	50–80	>80	>80
Trimethoprim	<20	<20	20–50	50–80	>80

from groundwater was studied. Residues of drugs like ketoprofen, diclofenac, carbamazepine, metoprolol, etc. were detected in groundwater. The study showed better results in removing most of the drugs with greater efficiency for use in drinking water purposes [76]. Modified cellulose acetate nanofiltration membranes were studied for the removal of traces of pharmaceuticals and personal care products from two water treatment plants with good removal of ibuprofen and poor removal of carbamazepine, respectively [77]. Table 5 depicts the removal of some pharmaceuticals by membrane process and membrane bioreactor process [78].

3.1 Hybrid Process

Remediation of emerging contaminants using a hybrid process was observed. It involved the use of a combination of physical, chemical or biological process coupled with membranes or adsorbents coupled with membranes. Adsorption can successfully remove ECs. Adsorbents like activated carbon, biochar, composite adsorbent, nanoadsorbent can remove ECs. Surface-modified adsorbents are more effective for the removal of ECs. Agricultural waste materials can be used as valuable biosorbents for ECs removal. Surface modifications in terms of porosity, surface area, functional group activation by carbonization, activation, milling, sieving, grafting, acid/alkali treatment, etc. showed enhanced performance. About 56.3% and 59.7% removal of ciprofloxacin were achieved using corn cob and rice husk, respectively. Phenol removal by fly ash was reported with the highest removal of 95%. Apart from agricultural wastes, other biosorbents prepared from biological origin were investigated. Removal of ibuprofen, diclofenac, celecoxib was studied using wood rot fungi. Diclofenac from municipal wastewater was removed by various fungal strains with >99.9% efficiency. The application of nanoadsorbent for ECs removal was demonstrated. Although nanoadsorbent effectively removed ECs at lose dose, lesser time and low ECs concentration but feasibility of its wide-scale application is restricted. Pesticide lindane removal was observed using modified aluminium hydroxide nanomaterial as adsorbent with 93.68% efficiency at an adsorbent dose of only 25 mg/L and 60 min contact time. Membrane-based treatment of ECs is already an established one. But all processes bear their own limitations for which hybrid treatment technologies are now being proposed to overcome these challenges. Adsorption/membrane filtration for EC removal is increasing considerably. The hybrid system depicts some advantages such as better separation, low-energy footprint, resuability, rapid kinetic, and less fouling. Hybrid treatment methods can be of different combinations like adsorption/activated sludge process or adsorption/MBR process. Also, adsorption combined with chemical treatment and physical treatment might be another option. Hybrid treatment of steroids was attempted using activated carbon + UF/NF. Adsorption + UF showed 90% removal efficiency with adsoption + NF showing no significant removal. A granular activated carbon and ozonation hybrid treatment system was applied for the removal of various ECs. Granular activated carbon and Fento removed >99% carbamazepine from raw wastewater. Caffeine, ibuprofen and acetaminophen removal by a hybrid system of activated sludge and activated carbon resulted in maximum removal than either single process. Hybrid system of photocatalytic oxidation and membrane filtration removed some ECs like nicotine, ranitidine hydrochloride, sulfamethoxazole, etc. Combination of activated carbon, NF and coagulation as pretreatment for the removal of diclofenac, caffeine and octyl phenol showed better efficiency than a single system. Ceramic membranes coated with MnO2-Co3O4 nanoparticles were used in crossflow mode for the removal of benzophenone—(BP-3) with reduced toxicity (Fig. 2). Antibiotic removal was also studied using adsorption + membrane [36, 79–81].

3.2 Mechanism of Removal

Removal of contaminants is dependent on the pore size of membrane. For separation of contaminants via membrane it requires driving force, pressure and concentration difference. Physicochemical characteristics of contaminants determine their transport through membrane pores. Also, membrane morphology determines the retention

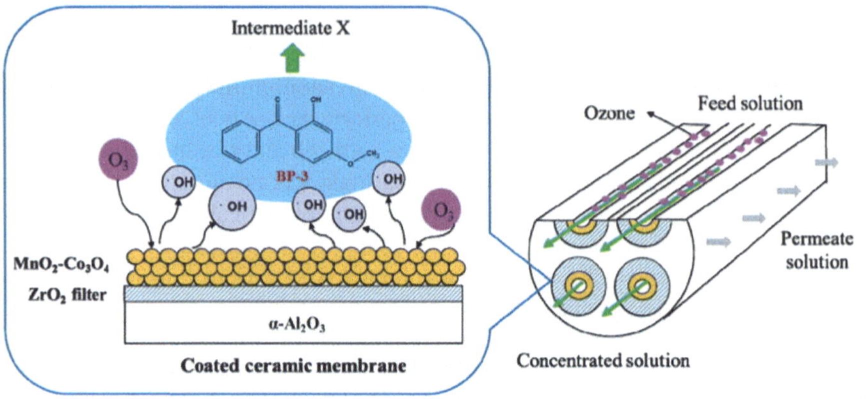

Fig. 2 Removal of benzophenone-3 (BP-3) by catalytic ceramic membrane fabricated from MnO_2 CO_3O_4 particles ([78] Copyright (2023) with permission from Elsevier)

of contaminants. The mechanism of removal of EC depends on size exclusion, electrostatic and hydrophobic interaction. The mechanism of size exclusion depends on sieving where larger molecules get trapped and smaller molecules pass out. Electrostatic interaction involves electrostatic attraction or repulsion between contaminants and membrane surfaces. Negatively charged EC will be rejected through a negatively charged membrane and vice versa. The hydrophobicity of contaminants is another factor that determines its removal. Hydrophobic ECs can be removed by a hydrophobic membrane. It gets absorbed on the membrane surface when present in feed water in immiscible conditions. Physicochemical characteristics of ECs and membrane properties determine the removal of contaminants. Exclusion of ECs through the membrane depends on size exclusion, hydrophobic interaction and electrostatic force of attraction. The size exclusion mechanism depends on the sieving of molecules through membrane pores where smaller molecules pass and larger molecules are retained. The electrostatic force of attraction depends on the surface charge of ECs where negatively charged ECs get attracted by the positively charged membrane surface and vice versa. Neutral ECs find it difficult to separate via surface charge. ECs that are hydrophobic in nature are separated and concentrated on hydrophobic membrane surfaces as these are immiscible in water. UF and MF membranes can separate pharmaceuticals when used in a hybrid process. UF generally removes pathogens and MF-suspended materials. RO and NF can remove pesticides, pharmaceuticals with or without pre-treatment. Feed foulant properties can cause membrane fouling and decrease permeability. Also, permeability depends on trans-membrane pressure where an increase in pH increases permeation.

In the case of NF membrane, removal depends largely on molecular size, charge, hydrophobic nature, solubility and diffusivity of contaminants and surface charge, hydrophobicity, pore size and permeability of the membrane. The membrane process needs force like electrical, pressure, concentration difference, etc. to allow the separation of pollutants or contaminants from solution. Therefore pore size of membranes determines the retention of pollutants. Separation of contaminants through membranes with smaller pores requires high pressure. Figure 3 depicts the interaction between pharmaceutical molecules and membrane in the removal of ciprofloxacin. Figure 3a shows pharmaceutical molecules without charge. In forward osmosis molecular radius is the main factor where molecules without charge are retained. Here number of molecules affects the collision between pharmaceutical molecules and membrane. Figure 3b shows the contact between pharmaceutical molecules and membrane surfaces having the same charge. The negative charge of both molecules and membranes hamper removal and thus molecules cannot pass the membrane surface affecting forward osmosis. Figure 3c depicts the opposite charge of pharmaceutical molecules and membrane surface. This reduces electrostatic repulsion and the retention rate of negatively charged molecules is decreased in the presence of positively charged pharmaceuticals. Attraction between positively charged and negatively charged pharmaceutical molecules is shown in Fig. 3d resulting in the formation of molecules with larger molecular weight and increased retention of negatively charged molecules [79, 82–84].

3.3 Treatment of Some Selected Emerging Contaminants Including Pesticides, Drugs and Personal Care Products

Previously in this book chapter, it is discussed that adsorption and membrane-based hybrid process is the most viable option for treatment of ECs. Here further discussion will be done based on some case studies where hybrid treatment of a few selected ECs is conducted. Ibuprofen adsorption by surface-modified activated carbon was investigated. Adsorption occurred at pH 3 and followed Langmuir isotherm [85]. Iron particles were loaded onto mesoporous silica SBA-15 by arc discharge treatment and were used for the removal of ibuprofen [86]. R- and S-enantiometers were studied for the treatment of Ibuprofen and it was observed that about 23% EC was adsorbed on flat-sheet stainless steel equipment [87]. Graphene oxide and reduced graphene oxide were used for modification of polysulphone NF membrane and efficiency was observed against the removal of ibuprofen where coated membrane performed better than non-coated membrane [88]. Removal of ibuprofen and atenolol was reported by Bhattacharya et al. (2020), where zinc oxide nanoparticles were used as a coating material for the development of the UF membrane. Removal efficiency of about 96% and 99% respectively was achieved for atenolol and ibuprofen [22]. Removal of atenolol from solutions using ozone-modified multi-walled carbon nanotubes were studied with a removal efficiency of about 75.79% [89]. An aerobic sequencing batch reactor was employed for the removal of atenolol, gliclazide and prazosin. 83% atenolol was biodegraded in the reactor with 41% of gliclazide and prazosin degradation [90]. The authors observed that *Mycobacterium* sp. was responsible for degradation. Ceramic composite membrane was employed for the removal of ciprofloxacin from synthetic solution with 99.3% rejection. In this

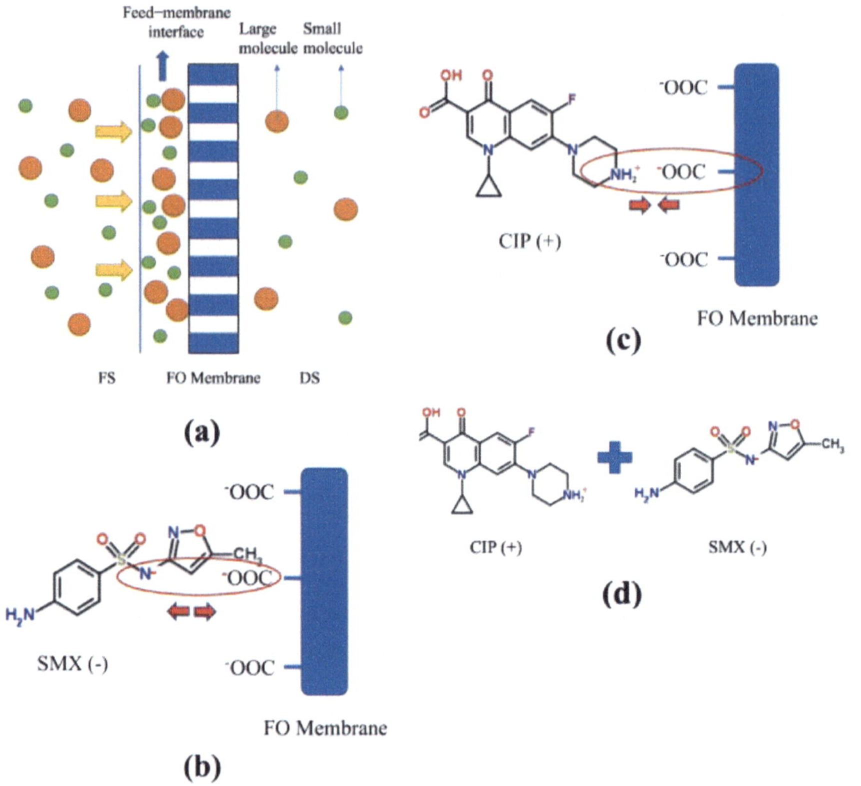

Fig. 3 Mechanism of forward osmosis in removal of pharmaceuticals **a** interaction between pharmaceutical molecules without any surface charge **b** interaction between pharmaceutical molecules and membrane surface with similar charge **c** interaction between pharmaceutical molecules and membrane with opposite charge **d** interaction between pharmaceuticals with opposite charge (Reprinted from [84], Copyright (2020) with permission from Elsevier.) CIP: ciprofloxacin and SMX: sulfamethoxazole

study, the authors used cross-linked polydimethylsiloxane (PDMS) to coat over a hydrophilic MF ceramic membrane. The membrane was thus made hydrophobic for the remediation of hydrophobic ECs [91]. CuO/TiO2 coated ceramic UF membrane was studied for removal of ciprofloxacin with 99% rejection efficiency [92]. Bhattacharya et al. (2021) applied the CuO/TiO2 coated ceramic membrane for the removal of well-known ECs phthalates and parabens. Removal efficiency of >99% was observed [24]. Atrazine removal was studied by surface modification of MF ceramic membrane. The authors performed two types of modifications. First, they immobilized chromium from tannery waste in single channel ceramic membrane and extruded. This chromium caused mullite phase formation that further increased the mechanical and chemical stability of membranes. Next PDMS coating was applied for increasing hydrophobicity. The modifications resulted in >95% removal of atrazine from solution [93]. Magnetic carbon nanocomposite was prepared from pineapple as a precursor for the removal of ciprofloxacin. The adsorbent was used in single and hybrid processes along with the NF membrane. 100% removal efficiency was obtained in the hybrid process [94].

3.3.1 Membrane Morphology

Membrane morphology plays a major role in the removal of contaminants. Surface modification of membranes with target compound can improve separation efficiency. Apart from membrane morphology, experimental conditions like pH, composition of materials, temperature, pressure, etc. also play an important role in removal. It was observed that hydrophobic compounds were adsorbed and saturated on a hydrophilic membrane thus increasing hydrophobicity and permeability. Acero et al. used UF membranes with a molecular weight cut-off of 2000 kDa and studied the removal of hydroxybiphenyl whose molecular weight is 170 g/mol. According to the size exclusion phenomenon, it should have passed through membrane but hydroxybiphenyl was surface adsorbed on the membrane [95, 96]. Dense membranes exhibit high selectivity compared to their porous counterparts. Dense RO membranes remove ECs depending on their pore size. Coating nanoparticles onto ceramic membranes can offer pore size in the nano range. Increased catalytic properties in membranes fabricated with catalytic nanoparticles were observed. Coating ceramic membranes takes place in layers of different pore sizes. First a support layer and then a thin top layer. For coating substrates are mixed with metal oxide powders, binders and plasticizers and then extrusion, casting or pressing followed by sintering. Thus membrane morphology is dependent not only on coating materials but also on coating method and sintering conditions. To achieve tailor-made pore size, size of nanoparticles and fabrication techniques must be controlled [97]. Bhattacharya et al. have used green synthesized copper oxide nanoparticles along with titanium dioxide to produce ceramic ultrafiltration membranes with high flux for the removal of ciprofloxacin. The coating was conducted over ceramic support having a porosity of 37% and pore size of 1.14 μm. The dip coating technique was followed for coating with nanoparticles over the inside wall of ceramic support. The crystallite size of CuO was determined to be 16.4 nm and TiO2 to be 23 nm but that of the unsupported membrane was 7.02 nm. Incorporating CuO decreased the transition temperature from anatase to rutile. Unsupported membrane exhibited BET surface area of about 37 m^2/g, pore diameter of 35 nm and molecular weight cut off of about 31 KDa (Fig. 4a). Surface roughness and morphology were observed by AFM (Fig. 4b). This membrane was used for the removal of ciprofloxacin [92].

In another study, Mukherjee et al. fabricated a membrane by incorporating dried tannery sludge along with α-Al2O3 and clay. Starch, methocel was used as binder and PEG was used as plasticizer and was added and mixed. Single channel hollow membranes were extruded having 8.5 mm/ 6.5 mm (OD/ID). Membranes were sintered at different temperatures viz. 1300 °C, 1350 °C and 1400 °C at a slow heating rate to ensure densification of membranes. Tannery sludge was added at

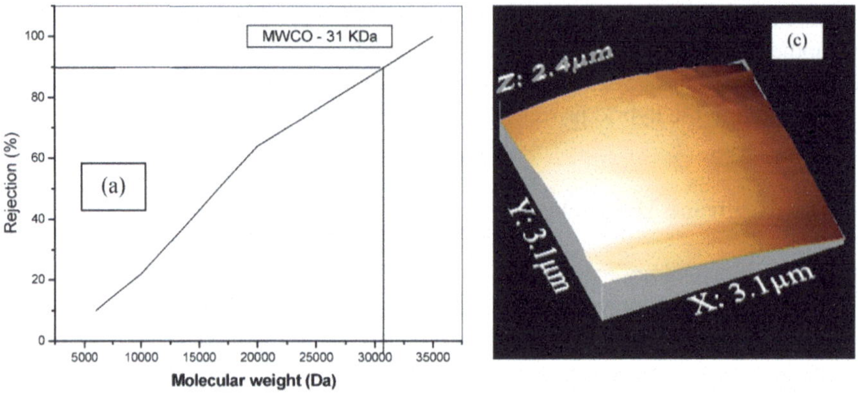

Fig. 4 **a** Graph depicting molecular weight cut-off of the membrane **b** AFM images of the coated membrane (Reprinted from [92], Copyright (2020) with permission from Elsevier.)

various wt% i.e. 5–15 and results were compared to control membranes sans the addition of tannery sludge. The MF membrane was coated with PDMS cross-linked with TEOS and DBTDL in the (1:0.1:0.008, vol%) and made hydrophobic by dip coating. Membrane was studied for chromium leaching. As observed from the microstructure analysis of the membrane (Fig. 5), a glassy amorphous phase was observed for samples sintered at 1300 °C, i.e., M9 and M10 but increasing the sintering temperature to 1400 °C, alumina grains started to form with increase in grain size. Mullite crystals were formed in M1, M2, M5 and M6 samples. The chemical stability of membranes is discussed in Table 6 where weight loss of ceramic membranes in the presence of acid, base is shown along with chromium leaching. Here chromium was immobilized into the membrane by replacing Al + 3 ions with Cr + 3 ions. This membrane was used for the removal of atrazine [93].

Membrane morphology of ceramic MF was observed after coating with CuO and TiO2 nanoparticles for the removal of phthalates and parabens. Microhardness increased greatly after coating where 90.0 VH increased to 115.7 VH. The results indicated uniform dispersion of nanoparticles over support that increased mechanical strength. FESEM micrographs also show densification of membrane pores (Fig. 6). FTIR micrographs (Fig. 7) show stretching vibrations at 1635 cm^{-1} that corresponded to C=O and N–H groups of amide that are found in algal proteins acting as capping agents. Stretching vibrations that are found at 451 cm^{-1} and 709 cm^{-1} were due to the OTi-O bond of TiO2-anatase form. The presence of the Cu–O bond of CuO NPs along (101) direction was observed at 608 cm^{-1} and 516 cm^{-1} respectively. After filtration with phthalate solution vibrations at 1637 cm^{-1} corresponded to the C-O bond of the pthalic ester. The presence of paraben molecules on the membrane surface was detected in Fig. 7c where peaks at 1650 cm^{-1} corresponded to O–H and C=O vibration of paraben molecules [92].

The hydrophilic behavior of the membrane was increased by the addition of ZnO NPs for the development of the UF membrane. The membrane was used for the

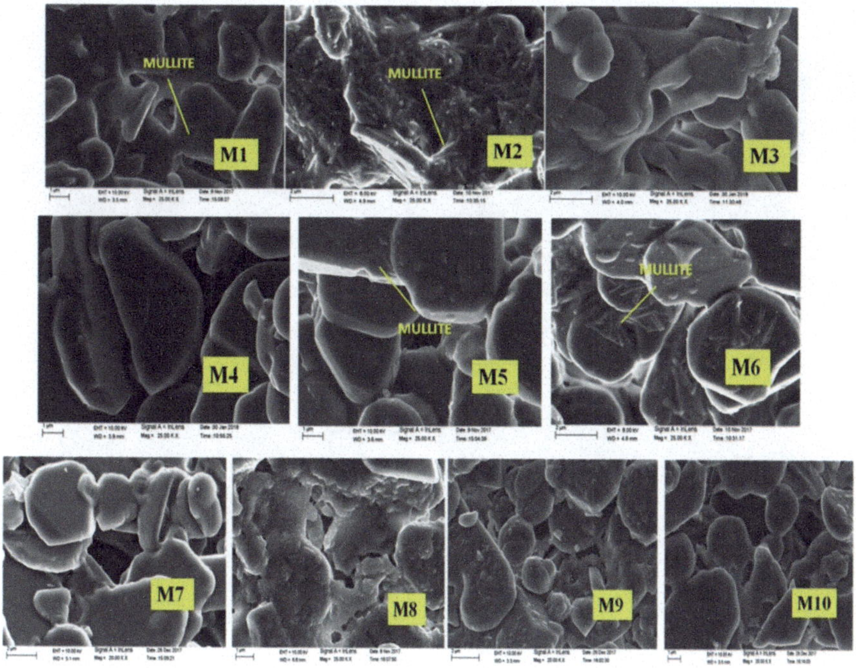

Fig. 5 Microstructure analysis of fabricated MF membrane (Reprinted from [93], Copyright (2020) with permission from Elsevier.)

Table 6 Leaching test and stability test of fabricated MF membrane

Properties	Membrane									
	M1	M2	M3	M4	M5	M6	M7	M8	M9	M10
Chemical stability:										
Acidic medium weight loss (%)	0	0	2	3	3	0	3	5	4	7
Basic medium weight loss (%)	0	0	2	3	2	0	5	6	5	7
Chromium leaching (mg/L)	0	0.01	0.53	0	0	0.02	0.58	0	0.07	0.27
Mechanical Strength										
Flexural strength (MPa)	85	92	64	34	79	86	58	29	47	59

(Reprinted from [93], Copyright (2020) with permission from Elsevier.)

removal of atenolol and ibuprofen. FESEM images show reduced pores in the coated membrane as compared to support (Fig. 8). Pure phase ZnO is observed from EDX analysis where peaks of Zn and O are observed [22].

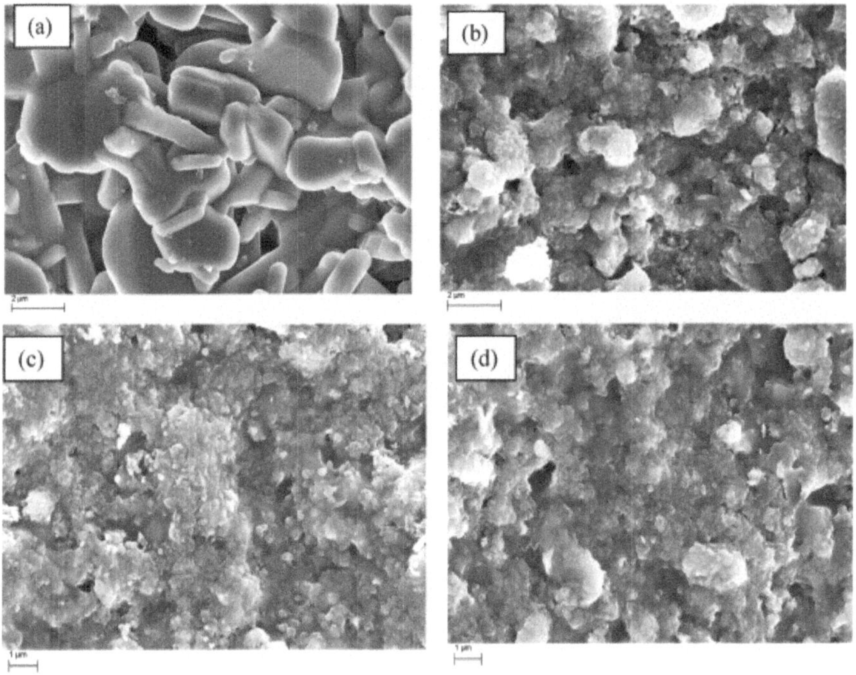

Fig. 6 FESEM micrographs of **a** support **b** After coating with nanoparticles **c** after filtration with phthalate **d** after filtration with paraben (Reprinted from [24], Copyright (2020) with permission from Elsevier.)

3.3.2 Permeability

The affinity of membrane material towards permeating components and vice versa is defined as permeability.

$$Pi = Di \times Si \tag{1}$$

where D is the diffusivity of the component in the membrane and S is the sorption coefficient of the component in the membrane

Membrane permeability of NF and RO membranes in thin film composite configuration favors the separation process as well as exhibits good water permeability as compared to dense membranes [63]. Permeability decreases when coating promotes rejection. As in the case of rejection of NF270 where unfavorable interaction between PA and MPs enhanced sorption of MPs within PA developing a hydrophobic structure that enhanced rejection but reduced permeability by 4–5 times[98]. Baransi-Karkaby et al. modified the surface of ESPA1 which increased the removal of BPA and acetaminophen but reduced permeability [99]. It is observed that the longer and denser the coating, the lesser the permeability of the membrane. This might be due to

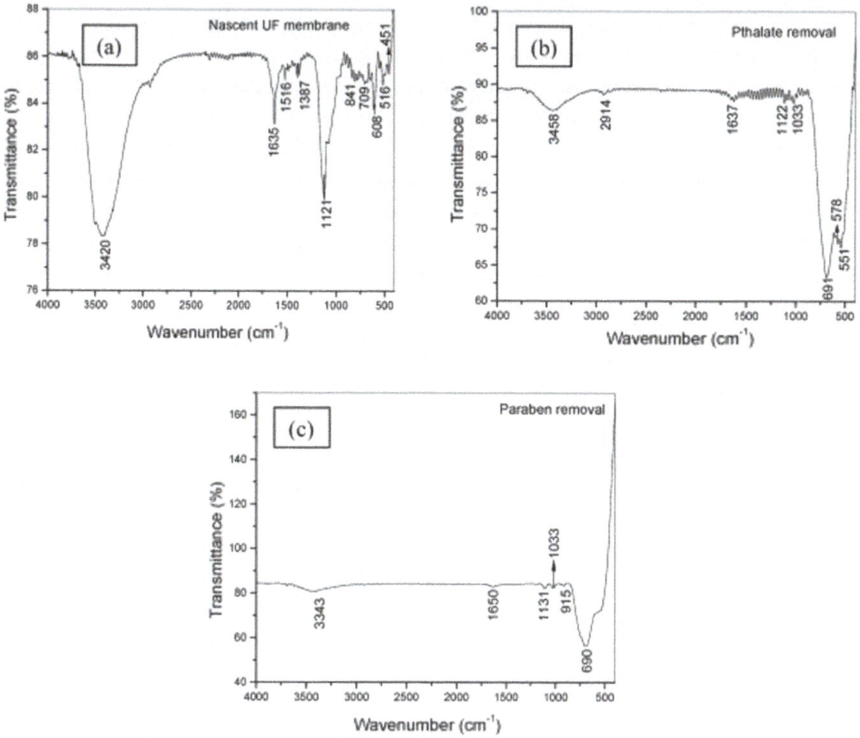

Fig. 7 FTIR images **a** naïve membrane **b** after filtration with phthalate (c) after filtration with paraben (Reprinted from [24], Copyright (2020) with permission from Elsevier.)

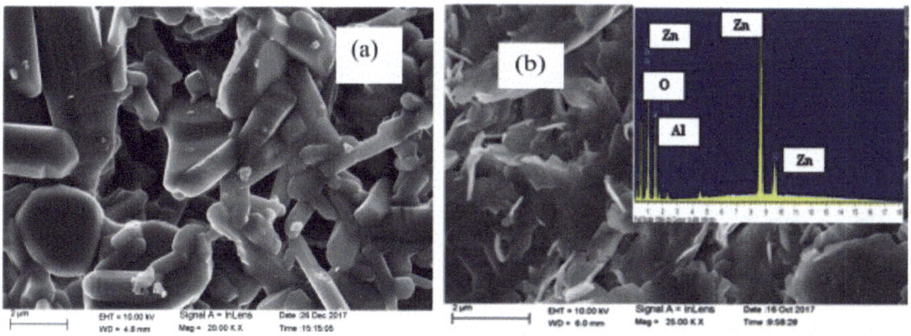

Fig. 8 FESEM micrographs of **a** support **b** After coating with ZnO NPs (Reprinted from [22], Copyright (2020) with permission from Elsevier.)

the denser layer acting as an inhibitor for water transport and thus reduced flow [63]. It was also observed that membranes exhibiting higher flux show decreased removal as in the case of TFN membranes where coating enhanced permeation but decreased removal efficiency [100]. Green synthesized ZnO NPs were applied for the development of a UF membrane for the removal of atenolol and ibuprofen. Figure 9a–d shows the effect of time on the permeation and removal of ECs was shown. The initial decrease in flux was due to surface adsorption on the membrane surface which almost reached steady state flux. But removal increased as compared to permeation value. About 95% removal was achieved whereas flux of about 233.7 $Lm^{-2}\ h^{-1}\ bar^{-1}$ was obtained [22].

UF membrane developed from green synthesized CuO NPs and TiO2 was used for the removal of ciprofloxacin. When clean water permeability was compared with support and coated membrane it was observed that permeability reduced for coated membrane from 216 $Lm^{-2}\ h^{-1}\ bar^{-1}$ for support to 150 $Lm^{-2}\ h^{-1}\ bar^{-1}$ for membrane as shown in Fig. 9 [92].

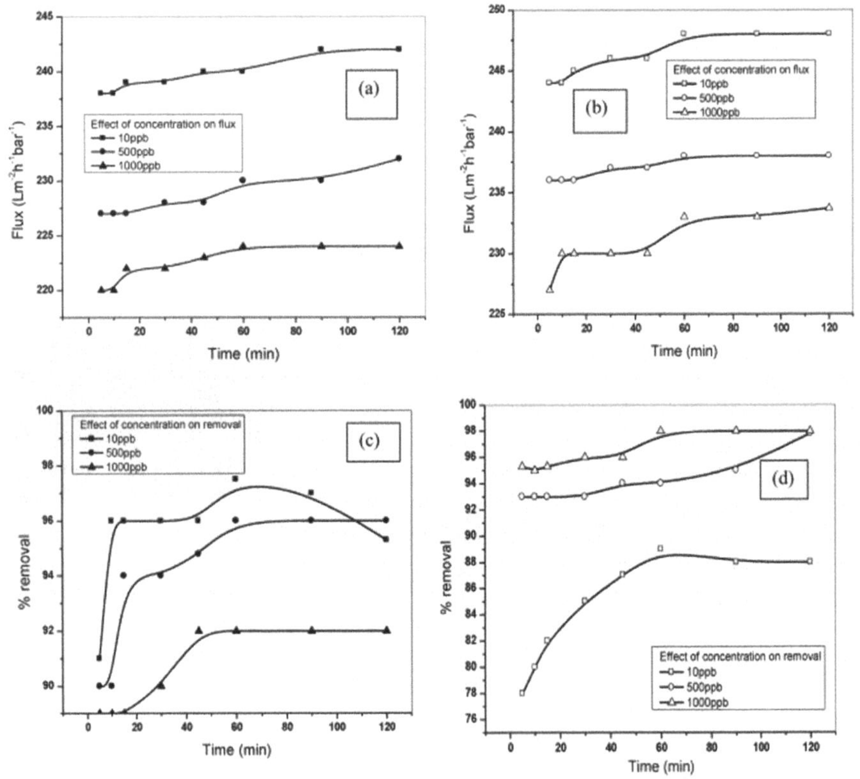

Fig. 9 Flux profile for atenolol and ibuprofen **a, b** effect of concentration on flux with time effect of concentration on removal with time **c, d** atenolol and ibuprofen (Reprinted from [22], Copyright (2020) with permission from Elsevier.)

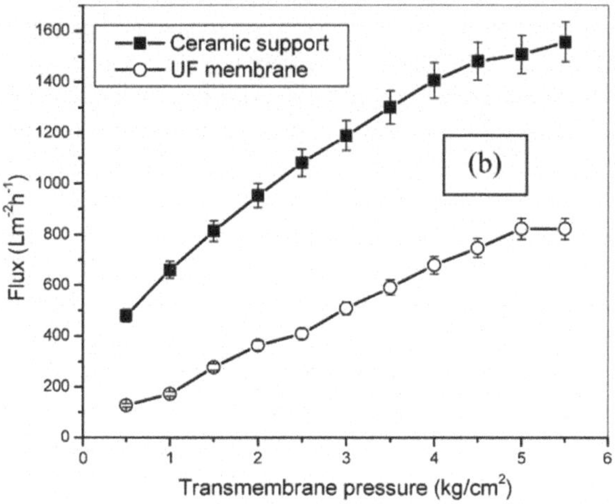

Fig. 10 Clean water permeability of ceramic support and ultra-filtration (UF) membrane (Reprinted from [92], Copyright (2020) with permission from Elsevier.)

Table 7 Various conditions optimized from RSM

Number	TMP (bar)	CFV (LPM)	Time (min)	Flux (L m^{-2} h^{-1} bar^{-1})	Removal (%)	Desirability
Paraben						
1	5.0	3.05	91.19	540.94	99.32	0.977
2	5.0	3.06	91.76	541.0	99.33	0.977
Phthalate						
1	4.99	3.57	110.72	544.0	99.95	1.0
2	5.0	3.21	114.69	540.39	99.83	1.0

(Reprinted from [24], Copyright (2020) with permission from Elsevier.)

CuO and TiO2 coated UF membrane was studied for removal of phthalate and parabens. Conditions optimized from RSM suggest that TMP and CFV have a positive effect on flux. Flux attains stability after 60 min of operation and remains steady for 120 min. As time increased, contaminants deposited on the membrane surface resulting in reduced flux (Fig. 10 and Table 7) [92].

3.3.3 Separation Characteristics

Different removal technologies are used for the separation of ECs. The advanced oxidation process generally targets degrading micropollutants by producing reactive oxygen species and microbes present in water. Surface adsorption of ECs is another method for the separation of pollutants. However different adsorbents have different removal behavior towards a single compound. Like removal of BPA by using sulfonic acid was about 236.80 mg/g but using *Ulva prolifera* based biochar was able to absorb only 9.68 mg/g of BPA [101, 102]. Membrane bioreactor combines the process of biodegradation via conventional activated sludge process and membrane separation. As compared to MBR, application of MF, UF, NF and RO offer more removal efficiency and are able to remove pollutants like diclofenac, carbamazepine, fenoprop, metronidazole and trimethoprim that are not removed by MBR. Still, further small molecules of pollutants that are not separated by MF or UF are removed by RO/NF. Coating on the membrane surface alters membrane properties and ECs are separated by size exclusion phenomenon or adsorption. Commercial NF membrane when surface modified by coating resulted in >95% BPA rejection, >99% ibuprofen rejection. Incorporation of nanoparticles like zeolite, silver in TFC membranes, increased the rejection of methylparaben, ethylparaben, propylparaben and benzyl paraben to about 90% [63]. CuO NPs and TiO2-coated UF membrane showed rejection of phthalate and paraben to about 99.7% and ciprofloxacin to about 99.5% [23, 92]. ZnO-coated UF membrane successfully separated atenolol and ibuprofen to more than 95% by surface adsorption mechanism [22]. Hydrophobic membrane developed from PDMS deposition on chromium-laden ceramic support for removal of atrazine to >95% [93]. Table 8 depicts the separation efficiency of different current technologies like photocatalytic membrane reactors, Forward osmotic membrane bioreactor, Catalytic membrane reactor, Microfiltration membrane, Metal–organic frameworks with ultrafiltration hybrid systems (MOF-UF), etc.

3.3.4 Role of Nanoparticle Coating in Membranes

Surface modification of polymeric and ceramic membranes has been suggested to overcome fouling issues and wetting. Carbon nanomaterials or metal oxide nanoparticles are generally used for coating of MF, UF, NF or RO membranes. Coating increases stabilization and performance. Surface modifications with graphene oxide, carbon nanotubes are in operation for desalination [103, 104]. Nanoparticle coating in membranes requires control of the desired particle size, nature and crystallinity of synthesized nanoparticles. Then nanoparticles' addition in porous membranes must be controlled so that pores are not blocked nor membrane integrity is lost [105]. Vapor deposition, electrophoretic coating, dip coating, etc. are techniques that are followed during metal oxide nanoparticle coating. In general hybrid mixtures of organic polymers with metal oxide nanoparticles are used to enhance membrane properties. SiO_2 was used for the development of a thin-film nanocomposite RO membrane causing an increase in hydrophilicity [106]. The application of TiO_2 as a coating material is widely used as TiO_2 has photocatalytic properties, is highly stable in nature, is non-toxic and cost-effective. CA/PEG membranes were coated with TiO_2 of various loadings to get RO membrane having high salt rejection capacity. Further modifications with TiO_2 were made to get membranes having antifouling properties, photocatalytic properties, superhydrophobic nature, etc. [107, 108]. Another metal oxide that is used for surface modification of membranes is ZnO. ZnO is a low-cost material with high mechanical, chemical, physical, thermal stability along with high surface area and anti-microbial, anti-corrosion properties. Anti-microbial

Table 8 Removal efficiency of various new technologies for separation of ECs

Technology	Target compound	Degradation %	Experimental conditions
Photocatalytic membrane reactor	Diclofenac 0.5 mg/L	56–100% 52% removal of TOC	pH = 7.5–8, Vtot = 25 L, Vphoto = 15 L, Vmem = 10 L, TMP = 8–34 kPa, HRT = 30–60 min
Forward osmotic membrane bioreactor	COD, NH4 + -N, Carbamazepine	94.77–97.45%, 93.56–99.28%, 88.20–94.45%,	DO = 5 mg/L, HRT = 25.25 and 157.89 h, SRT = 30 days
Catalytic membrane reactor	Diclofenac	65%	pH 3, 42.9 mg/L H2O2
Microfiltration membrane	Triclosan, Acetaminophen Ibuprofen	10–95%	Flow rate of 1 mL/min for 120 min
Metal–organic frameworks with ultrafiltration hybrid systems (MOF-UF)	Ibuprofen and 17α-ethinyl estradiol	53.2%	ΔP = 520 kPa (75 psi); Stirring speed = 200 rpm; MOF = 20 mg/L; Initial selected PhAC concentration = 10 μM; Conductivity = 300 μS/cm; Pre-contact time with MOF = 2 h

properties of the RO membrane were enhanced by the incorporation of ZnO nanoparticles in the CA membrane [109]. The incorporation of carbon nanotubes, graphene oxide, and a mixture of CNTs and GO for the development of a modified RO membrane was studied. The membranes exhibited properties like hydrophobicity, porosity, anti-fouling, etc. [110].

Metal oxide nanoparticles are also used for coating ceramic membranes to enhance surface properties. A mixture of CuO NPs and TiO2 NPs was used to surface coat ceramic membranes to develop surface-modified UF membranes for the removal of personal care products and pharmaceuticals. CuO NPs were synthesized from extracts of algae [93]. Coating with iron oxide nanoparticles green synthesized from *Aloe vera* extracts and cross-linked with chitosan to develop UF membrane. The efficiency of the membrane was observed on atrazine removal [111]. Chromium from tannery waste was used as coating material over alumina support to develop the MF membrane. The incorporation of chromium initiated the mullite phase and thus increased the mechanical, thermal and chemical stability of membrane. The membrane was further modified by incorporating PDMS that increased hydrophobicity. Pesticide removal was studied using this membrane [94]. Green-synthesized

ZnO NPs were used as coating material for the development of UF membrane to remove drugs like ibuprofen and atenolol [88].

Another innovation was conducted by Heredia et al. (2022) where photodegradation of pollutants like diclofenac, iopamidol, methylene blue and metoprolol was observed using a photocatalytic ceramic membrane. The photocatalytic membrane was synthesized by incorporation of TiO2 in alumina matrix within the PMMA holder inside the cupboard to prevent a reaction of pollutants with ambient light. The membrane resulted in 97% removal of diclofenac and metoprolol, 86% methylene blue and 85% iopamidol, respectively [112].

4 Biotechnological Application

The biotechnology process is dependent on membranes. Processes like media fermentation, buffer purification or protein separation are part of membrane separation. MF, UF or RO plays an important role in different biotechnology processes.

4.1 Polymeric Membranes in Biotechnological Applications

Non-solvent phase inversion technique was used to develop blend membranes. These membranes exhibited enhanced biocompatibility, longer clotting time, reduced platelet adhesion and protein adsorption and lesser activation of complement. They also showed an increase in permeability of cytochrome-c of middle molecular weight suggesting possible application for hemodialysis [113]. Functionalized graphene oxide (f-GO) was developed by immobilizing heparin through polydopamine anchor on GO nanosheets to produce hemocompatible membranes. The membrane morphology, roughness, wet ability, porosity and surface chemistry were altered by this modification. f-GO reduced protein adsorption from platelet-poor plasma and decreased adhesion of platelets and improved blood clotting time. Cell proliferation assay showed the non-toxicity of f-GO membranes. Thus the membranes can effectively used for hemodialysis [114]. Another membrane developed from GO-polyvinylpyrrolidone was found to be effective in removing uremic toxins and thus a potential candidate for hemodialysis [115]. Polyethylene glycol and polydopamine were used for the modification of poly (vinylidene fluoride) (PVDF) ultrafiltration (UF) membranes. These membranes showed excellent rejections of 90.5% and 95.4%, respectively, for humic acid and bovine serum albumin with high fluxes of 96.4 and 81.4 Lm-2 h-1 [116].

4.2 Ceramic Membrane for Biotechnological Application

Separation of *Kocuria rhizophila* from fermentation broth was carried out using a ceramic membrane. Membranes were developed from China clay, quartz and calcium carbonate. At an optimum pressure and rotational speed of 345 kPa and 250 rpm, respectively, flux of 176.8 L m^{-2} h^{-1} was obtained with the bacterial rejection of 83% [117]. A combination of reaction and separation is possible using ceramic membranes. The term biocatalytic membrane reactors is used for such membrane units. An asymmetric ceramic hollow fibre membrane was used in a two-separate-phase reactor. Membranes were synthesized using phase inversion and sintering techniques. These fibers were used as support for the immobilization of lipase covalent for the development of two separate phases. Density of reactive hydroxyl groups on the membrane surface was increased by the functionalization method followed by amino activation and treatment with cross-linker. Results depicted the possibility of lipase immobilization on ceramic membrane surface without changing its catalytic performance [118].

5 Ceramic Membranes Versus Polymeric Membranes

With demand for several PPCPs, there is a surge in the development of the pharmaceutical industry causing more and more serious environmental problems. A variety of pharmaceutical products in different industries, their production scale and processes create different pharmaceutical loads in the environment. Therefore, there is a requirement for robust and high-quality treatment system in the pharmaceutical industry. In this regard comparison might be made between polymeric and ceramic membranes for a more efficient treatment process. Since application in pharmaceutical processes requires repeated sterilization, ceramic membranes might find their application as these membranes are more robust than their polymeric counterparts. They are highly stable under harsh conditions of chemical, mechanical and thermal cleaning. Fouling can be reduced by using ceramic membranes as they are assumed to be highly porous, hydrophilic and exhibit narrow pore size distribution [119, 120]. Moreover, polymeric membranes are sensitive to most solvents, chemicals and temperature.

Polymeric membranes are produced from cheap polymers that are easy to fabricate. This advantage is letting back the use of ceramic membranes. Moreover, polymeric membranes are easy to surface modify for targeted use. But fouling is one issue that affects its stability and one-time installment cost increases. To overcome these difficulties, ceramic membranes developed from cheap raw materials are being considered. Natural minerals like silica, natural zeolites, clay, apatites, etc. are used for the development of ceramic membranes. Kaoline, bentonite, fly ash, sugarcane bagasse, cement, etc. are also used as raw materials for the synthesis of ceramic membranes [121]. Surface modifications of ceramic membranes are done by sol–gel, doping, grafting, hydrothermal, chemical vapor deposition, dip-coating, atomic layer coating, etc. [122]. But, the poor selectivity of ceramic membranes towards another aspect of pharmaceuticals i.e. vitamins, enzymes, ions, antibiotics, etc. creates another drawback for ceramics [123].

6 Scope of Future Research in This Sector

Membrane technology although widely used for water treatment, bears certain disadvantages like inefficiency in the removal of ions or very small molecules. Removal of these contaminants requires membranes with very small pore diameters which reduces flux. Hybrid technology involves ceramic membranes with polyelectrolyte layers that are efficient in removing micropollutants without hampering flux. Polyelectrolytes are polymers with dissociating groups that include polycations, polyanions or zwitterionic polymers. Coating of polymeric membranes with polyelectrolytes is well available but polyelectrolytes on ceramic membranes are rarely studied. Multilayer structure production is based on the electrostatic force of attraction and thus coating layer can be controlled to get the desired layers. Charge of layer, hydrophobicity and density of layer determine the performance of membranes. Charge of membrane and separating solute must be the same otherwise repulsion will occur. Ceramic substrate used for coating must be charged or have functional groups for covalent bonding of polyelectrolytes. Uncharged coating is also possible but in that case, hydrophobic polyelectrolyte must be used. Effect of binding hydrophobic substances on the hydrophobic membrane was studied for the removal of hazardous substances that are not removed by conventional methods [124, 125]. Ceramic-ceramics composite membranes are also considered for applications in the removal of ECs. Ceramics embedded with metal oxide nanoparticles are also studied for removal of ECs. Although most ceramics are made of costly material these are used in pharmaceutical industries because of their robustness, superior performance, high chemical and thermal stability, hydrophilicity etc. Polymeric membranes although cheap in the manufacturing process, their life span gets half because of fouling problems and low water flux. Therefore, ceramic-polymer composites would serve the problem in terms of cost and flux. Ceramic-polymer composites were used in gas separation where ceramic fillers were mixed into cellulose acetate. The fillers increase membrane rigidity. The different types of polymer composite membranes include ceramic fillers in polymeric composite, thin film nanocomposite and ceramic-supported polymeric membranes [126, 127]. The addition of ceramics into the polymeric matrix will overcome the fouling problem and increase flux keeping costs lower. Glucose flux and rejection was increased by incorporating Mg Fe2O4 on the PSf membrane [128]. The addition of Ag, Cu in polymeric membranes increases their antibacterial properties. GO-added membranes also show antibacterial properties. The incorporation of TiO2 or ZnO in polymeric matrix can increase its photocatalytic properties. Similarly, hydrophilicity, roughness, porosity, pore size, mechanical and thermal properties can be enhanced by addition of inorganic components. The addition of MCM-41 to poly-vinylidene chloride membranes increased the strength of the membrane as compared to such polymeric membrane [129].

But like every new technology, composites also bear certain disadvantages like compatibility issues leading to aggregation of inorganic particles and separation of components and affecting stability. Great progress in the field of finding compatible inorganic with polymers is reported. Manufacturing techniques might cost a bit more for composites. Polymeric membranes are fabricated by phase inversion or casting method which can be applied for nanoparticle coating on polymeric matrix. But depositing a polymeric layer on a ceramic matrix requires the preparation of ceramic support by sintering and then dip coating or spin-coating to deposit the polymeric layer on ceramic support thus increasing fabrication cost [130].

Another composite membrane is developed where a thin nanocomposite membrane is supported on polymeric support and ceramic nanoparticles are deposited on the surface without affecting the properties of polymeric support. These membranes are known as ceramic-polymer nanocomposite membranes. They are developed via phase inversion or casting method in flat sheet or hollow fiber configuration. The pore size and pore distribution are tunable and flux profile and life of membranes are increased. Polymeric membranes with deposition of ceramic nanoparticles result in the development of another type of nanocomposite membranes where surface properties like hydrophilicity, roughness, pore size are influenced and also the membranes exhibit additional properties like photocatalytic properties, antibacterial properties, etc. These types of membranes are fabricated by casting and phase inversion with a polymeric solution containing ceramic nanoparticles. Ceramic fillers like SiO_2, Al_2O_3, TiO_2, Fe_3O_4, etc. are used. For development of membrane kaolin, cloisite, montmorillonite are used [131]. The addition of TiO_2 enhances photocatalytic activity and antibacterial activity thereby reducing fouling. The addition of SiO_2 in the PES UF membrane increases porosity thereby increasing flux. Thin film nanocomposite membranes are composed of thin nanocomposite membranes supported on polymeric substrate. These membranes are applied for the removal of salt, heavy metals, pharmaceutically active compounds, micropollutants, etc. Thin polymeric membranes supported on the ceramic substrate are also developed via the dip coating technique. These types of membranes have high chemical, mechanical and thermal stabilities. They are widely used for gas separation, pervaporation, etc. PDMS/Al_2O_3-ZrO_2 exhibits higher pervaporation than pristine polymeric membranes. Polymer-supported and ceramic-supported polymeric membranes show different swelling behavior. Swelling in polymer support occurs in the top and support layer which affects membrane performance but ceramic support maintains its pore structure and swelling is reduced on the top polymeric layer [132]. Also, ceramic-supported polymeric membranes can be cleaned by backwashing which is not possible in polymeric support. Moreover, high-temperature operations are not possible with polymeric support. A study showed how reusable Al_2O_3 was removed after separating NaCl by sodium hypochlorite and again coated with the same Al_2O_3 monolith without hampering the pure water permeability of the membrane. These types of membranes can reduce overall cost and material requirements with high performance. Further research in this field is required to optimize the ceramic-polymeric interface to tailor made performance of the composite membrane. Moreover, since thermodynamically both ceramic and polymeric materials are non-compatible fabrication of these types of membranes requires strategies like modifying the properties of both the polymeric and ceramic materials to increase their compatibility and efficiency. For the incorporation of ceramic nanoparticles in polymeric support, modifications on nanoparticles are done for compatibility. Modifications on nanoparticles are done by surface modification, surface functionalization and grafting. Polymers can also be modified to increase compatibility between ceramic and polymers. Introducing functional groups like sulfonic and carboxyl groups increases interaction with ceramic components as well as increasing hydrophilicity [124, 133, 134].

In a nutshell, different types of fabrication techniques are undertaken to develop tune membrane structure and properties and functioning in terms of flux, antifouling, photocatalytic, thermal, chemical, mechanical stabilities, etc. Increased compatibility between ceramic and polymeric composite membranes can thus be a suitable candidate for solving future problems of ECs removal with great efficiency.

7 Conclusion

Certain understandings of the removal mechanism of ECs are still unclear. As observed ceramic-polymeric membranes with nanoparticle coating play a major role in the effective removal of ECs but the influence of NPs on membrane structures and performance is yet to be understood. Proper materials are yet to be fabricated to synthesize a ceramic-polymeric composite membrane that bears all the desirable characteristics. Although several studies are being reported on the influence of ceramic NPs on polymeric matrixes the origin of the influence has not been detailed researched. So detailed studies on the association between surface properties of ceramic NPs, structures and performance of the membrane are required to develop new composite membranes.

Proper structural analysis apart from basic analysis like SEM, EDS, FTIR and XPS is required for a proper understanding of the type of membrane synthesized from composite type membrane. To determine the actual stability and performance of composite membranes; long-term field study is needed which is lacking. This type of study will provide enough data to understand long-term compatibility, change in membrane properties, environmental impact, harsh conditions, difficulties in operation, discharge of nanoparticles layer, etc. can be sorted. It is particularly necessary for photocatalytic membranes where catalytic performance, TiO2 coating stability will be observed. Any damage to membranes must be characterized via leaching tests, scanning electron micrographs, of the membrane surface and cross-section, roughness and FTIR with attenuated total reflectance (ATR) scans of used membranes, etc.

Cost analysis of ceramic-polymer composite membranes for the removal of ECs is at a nascent stage. Studies on large-scale production and industrial application in terms of cost must be conducted. Comparison between installation cost and durability must be analyzed for proper cost estimation. Techniques proposed for ceramic-polymer composites increase fabrication costs as compared to ceramic membranes. Ceramic membranes in hybrid technology or in combination with nanoparticles might be viable option for the remediation of pharmaceuticals.

References

1. Ahmeda MB, Zhou JL, Ngoa HH, Wenshan G, Thomaidis NS, Xu J (2017) Progress in the biological and chemical treatment technologies for emerging contaminant removal from wastewater: a critical review. J Hazar Mater 323:274–298
2. De la Cruz N, Giménez J, Esplugas S, Grandjean D, De AL, Pulgarin C (2012) Degradation of 32 emergent contaminants by UV and neutralphoto-fenton in domestic wastewater effluent previously treated by activated sludge. Water Res 46:1947–1957
3. Chen K, Zhou JL (2014) Occurrence and behaviour of antibiotics in water and sediments from the Huangpu River, Shanghai, China. Chemosphere 95:604–612
4. Muhammad B, Syed SA, Damiá B, Iqbal HMN (2019) Review Biocatalytic degradation/redefining "removal" fate of pharmaceutically active compounds and antibiotics in the aquatic environment. Sci Total Environ 691:1190–1211
5. Klamerth N, Malato S, Aguera A, Fernandez-Alba A, Mailhot G (2012) Treatment of municipal wastewater treatment plant effluents with modified photo-Fenton as a tertiary treatment for the degradation of micro pollutants and disinfection. Environ Sci Technol 46:2885–2892
6. Lister AL, Van Der Kraak GJ (2001) Endocrine disruption: why is it so complicated? Water Qual. Res J Can 36:175–190
7. Shahid MK, Kashif A, Fuwad A, Choi Y (2021) Current advances in treatment technologies for removal of emerging contaminants from water—a critical review. Coord Chem Rev 442:213993
8. Ghernaout D, Elboughdiri N (2019) Water reuse: emerging contaminants elimination—progress and trends 3. OALib 6:e5981
9. Esplugas S, Bila DM, Krause LGT, Dezotti M (2007) Ozonation and advanced oxidation technologies to remove endocrine disrupting chemicals (EDCs) and pharmaceuticals and personal care products (PPCPs) in water effluents. J Hazard Mater 149:631–642
10. Sarkar B, Mandal S, Tsang YF, Vithanage M, Biswas JK, Yi H, Dou X, Ok YS (2019) Sustainable sludge management by removing emerging contaminants from urban wastewater using carbon nanotubes. In: Prasad MNV, de Campos Favas PJ, Vithanage M, SVBT-I, Mohan MS (Eds), Ind. Munic. Sludge, Butterworth-Heinemann, pp 553–571
11. Zhang Y, Zhou JL (2008) Occurrence and removal of endocrine disruptingchemicals in wastewater. Chemosphere 73:848–853
12. Grover DP, Zhou J, Frickers P, Readman J (2011) Improved removal of estrogenicand pharmaceutical compounds in sewage effluent by full scale granularactivated carbon: impact on receiving river water. J Hazard Mater 185:1005–1011
13. Garcia-Rodríguez A, Matamoros V, Fontàs C, Salvadó V (2014) The ability ofbiologically based wastewater treatment systems to remove emerging organic contaminants—a review. Environ Sci Pollut Res 21:11708–11728
14. Deegan A, Shaik B, Nolan K, Urell K, Oelgemöller M, Tobin J, Morrissey A (2011) Treatment options for wastewater effluents from pharmaceutical companies. Int J Environ Sci Technol 8:649–666
15. Matamoros V, Gutiérrez R, Ferrer I, García J, Bayona JM (2015) Capability of microalgae-based wastewater treatment systems to remove emergingorganic contaminants: a pilot-scale study. J Hazard Mater 288:34–42

16. Raj DSS, Anjaneyulu Y (2005) Evaluation of biokinetic parameters forpharmaceutical wastewaters using aerobic oxidation integrated with chemical treatment. Process Biochem 40:165–175
17. Sreekanth D, Sivaramakrishna D, Himabindu V, Anjaneyulu Y (2009) Thermophilic treatment of bulk drug pharmaceutical industrialwastewaters by using hybrid up flow anaerobic sludge blanket reactor. Bioresour Technol 100:2534–2539
18. Töre GY, Meric S¸ Lofrano G, De Feo G (2012) Removal of trace pollutants fromwstewater in constructed wetlands. In: Lofrano G (Ed), emerging compounds removal from wastewater, Springer, pp 39–58
19. Yang W, Cicek N, Ilg J (2006) State-of-the-art of membrane bioreactors:worldwide research and commercial applications in North America. J. Membrane Sci. 270:201–211
20. Arana J, Melián JH, Rodrıguez JD, Dıaz OG, Viera A, Pena JP, Sosa PM, Jiménez VE (2002) TiO_2-photocatalysis as a tertiary treatment of naturally treatedwastewater. Catal Today 76:279–289
21. Malato S, Maldonado MI, Oller I, Zapata A (2012) Removal of pesticides fromwater and wastewater by solar-driven photocatalysis. In: Lofrano G (ed) Emerging Compounds Removal from Wastewater. Springer, pp 59–76
22. Bhattacharya P, Mukherjee D, Deb N, Swarnakar S, Banerjee S (2020) Application of green synthesized ZnO nanoparticle coated ceramic ultrafiltration membrane for remediation of pharmaceutical components from synthetic water: Reusability assay of treated water on seed germination. J Environ Chem Eng 8:103803
23. Snyder SA, Adham S, Redding AM, Cannon FS, DeCarolis J, Oppenheimer J, Wert EC, Yoon Y (2007) Role of membranes and activated carbonin the removal of endocrine disruptors and pharmaceuticals. Desalination 202:156–181
24. Bhattacharya P, Mukherjee D, Deb N, Swarnakar S, Banerjee S (2021) Indigenously developed CuO/TiO2 coated ceramic ultrafiltration membrane for removal of emerging contaminants like phthalates and parabens: Toxicity evaluation in PA-1 cell line. Mater Chem Phys 258:123920
25. Sarantopoulos PD, Altiok T, Elsayed AE (1995) Trends and perspectives manufacturing in the pharmaceutical industry. J Manuf Syst 14:452–467
26. Pabby AK (2015) Handbook of membrane separations chemical, pharmaceutical, food and biotechnological applications, Second Edition, 2nd ed..ed., Hoboken, pp 878
27. Duscher S (2013) Ceramic membranes in chemical and pharmaceutical applications. Ceramic Applications 1:19–23
28. Majumder A, Gupta B, Gupta AK (2019) Pharmaceutically active compounds in aqueous environment: A status, toxicity and insights of remediation. Environ Res 176:108542
29. Ahmed MB, Zhoua JL, Ngoa HH, Guoa W, Thomaidis NS, Xu J (2017) Progress in the biological and chemical treatment technologies for emerging contaminant removal from wastewater: a critical review. J Hazard Mater 323:274–298
30. Luo Y, Guo W, Ngo HH, Nghiem LD, Hai FI, Zhang J, Liang S, Wang XC (2014) A review on the occurrence of micropollutants in the aquatic environment and their fate and removal during wastewater treatment. Sci Total Environ 473:619–641
31. Esplugas S, Bila DM, Krause LGT, Dezotti M (2007) Ozonation and advancedoxidation technologies to remove endocrine disrupting chemicals (EDCs) and pharmaceuticals and personal care products (PPCPs) in water effluents. J Hazard Mater 149:631–642
32. Rivera-Utrilla J, Sánchez-Polo M, Ferro-García MÁ, Prados-Joya G, Ocampo-Pérez R (2013) Pharmaceuticals as emerging contaminants and theirremoval from water: a review. Chemosphere 93:1268–1287
33. Tijani JO, Fatoba OO, Petrik LF (2013) A review of pharmaceuticals andendocrine-disrupting compounds: sources, effects, removal, and detections. Water Air Soil Pollut 224:1–29
34. Gadipelly C, Perez-Gonzalez A, Yadav GD, Ortiz I, Ibaez R, Rathod VK, Marathe KV (2014) Pharmaceutical industry wastewater: review of the technologies for water treatment and reuse.l Ind. Eng Chem Res 53:11571–11592
35. Browner CM, Fox JC, Frace S, Rubin MB, Hund F (1998) Development document for final effluent limitations guidelines and standards for the pharmaceutical manufacturing point

source category, EPA 821-B-98–009; Engineering and Analysis Division, U.S. Environmental Protection Agency (EPA): Washington, DC, USA
36. Eniola JO, Kumar R, Barakat MA, Rashid J (2022) A review on conventional and advanced hybrid technologies for pharmaceutical wastewater treatment. J Clean Prod 356:131826
37. Patel M, Kumar R, Kishor K, Mlsna T, Pittman CU Jr, Mohan D (2019) Pharmaceuticals of emerging concern in aquatic systems: chemistry, occurrence, effects, and removal methods. Chem Rev 119:3510–3673
38. Rana RS, Singh P, Kandari V, Singh R, Dobhal R, Gupta S (2017) A review on characterization and bioremediation of pharmaceutical industries' wastewater: an Indian perspective. Appl Water Sci 7:1–12
39. Saibaba SV, Sathish KM, Ramu B (2016) Pharmaceutical impurities and their characterization: a review. European J Pharm Med Res 3:190–196
40. Samal K, Mahapatra S, Al MdH (2022) Pharmaceutical wastewater as Emerging Contaminants (EC): Treatment technologies, impact on environment and human health. Energy Nexus 6:100076
41. Ahmed SF, Mofijur M, Nuzhat S, Chowdhury AT, Rafa N, Uddin MdA, Inayat A, Mahlia TMI, Ong HC, Chia WY, Show PL (2021) Recent developments in physical, biological, chemical, and hybrid treatment techniques for removing emerging contaminants from wastewater. J Hazard Mater 416:125912
42. Loganathan P, Vigneswaran S, Kandasamy J, Cuprys AK, Maletskyi Z, Ratnaweera H (2023) Treatment trends and combined methods in removing pharmaceuticals and personal care products from wastewater—a review. Membranes 13:1–19
43. Li Q, Wang P, Chen L, Gao H, Wu L (2016) Acute toxicity and histopathological effects of naproxen in zebrafish (Danio rerio) early life stages. Environ Sci Pollut Res 23:18832–18841
44. Blair B, Nikolaus A, Hedman C, Klaper R, Grundl T (2015) Evaluating the degradation, sorption, and negative mass balances of pharmaceuticals and personal care products during wastewater treatment. Chemosphere 134:395–401
45. Wu M, Xiang J, Que C, Chen F, Xu G (2015) Occurrence and fate of psychiatric pharmaceuticals in the urban water system of Shanghai, China. Chemosphere 138:486–493
46. Carballa M, Omil F, Lema JM, Llompart MA, García-Jares C, Rodríguez I, Gomez M, Ternes T (2004) Behavior of pharmaceuticals, cosmetics and hormones in a sewage treatment plant. Water Res 38:2918–2926
47. Lindqvist N, Tuhkanen T, Kronberg L (2005) Occurrence of acidic pharmaceuticals in raw and treated sewages and in receiving waters. Water Res 39:2219–2228
48. Kasprzyk-Hordern B, Dinsdale RM, Guwy AJ (2009) The removal of pharmaceuticals, personal care products, endocrine disruptors and illicit drugs during wastewater treatment and its impact on the quality of receiving waters. Water Res 43:363–380
49. Kosma CI, Lambropoulou DA, Albanis TA (2014) Investigation of PPCPs in wastewater treatment plants in Greece: occurrence, removal and environmental risk assessment. Sci Total Environ 466:421–438
50. Quednow K, Püttmann W (2008) Organophosphates and synthetic musk fragrances in freshwater streams in Hessen/Germany. Clean: Soil, Air, Water 36:70–77
51. Diamond J, Altenburger R, Coors A, Dyer SD, Focazio M, Kidd K, Koelmans AA, Leung KM, Servos MR, Snape J (2018) Use of prospective and retrospective risk assessment methods that simplify chemical mixtures associated with treated domestic wastewater discharges. Environ Toxicol Chem 37:690–702
52. Samer M (2015) Biological and chemical wastewater treatment processes. Wastewater Treat Eng 150:212
53. Wang JL, Xu LJ (2012) Advanced oxidation processes for wastewater treatment: formation of hydroxyl radical and application. Crit Rev Environ Sci Technol 42:251–325
54. Jamil S, Loganathan P, Listowski A, Kandasamy J, Khourshed C, Vigneswaran S (2019) Simultaneous removal of natural organic matter and micro-organic pollutants from reverse osmosis concentrate using granular activated carbon. Water Res 155:106–114

55. Hao J, Zhang Q, Chen P, Zheng X, Wu Y, Ma D, Wei D, Liu H, Liu G, Lv W (2019) Removal of pharmaceuticals and personal care products (PPCPs) from water and wastewater using novel sulfonic acid (–SO 3 H) functionalized covalent organic frameworks. Environ Sci Nano 6:3374–3387
56. Taheran M, Brar SK, Verma M, Surampalli RY, Zhang TC, Valero JR (2016) Membrane processes for removal of pharmaceutically active compounds (PhACs) from water and wastewaters. Sci Total Environ 547:60–77
57. Parida VK, Saidulub D, Majumder A, Srivastava A, Gupta P, Gupta AK (2021) Emerging contaminants in wastewater: A critical review on occurrence, existing legislations, risk assessment, and sustainable treatment alternatives. J Environ Chem Eng 9:105966
58. Urtiaga AM, Pérez G, Ibáñez R, Ortiz I (2013) Removal of pharmaceuticals from a WWTP secondary effluent by ultrafiltration/reverse osmosis followed by electrochemical oxidation of the RO concentrate. Desalination 331:26–34
59. Neoh CH, Noor ZZ, Mutamim NSA, Lim CK (2016) Green technology in wastewater treatment technologies: Integration of membrane bioreactor with various wastewater treatment systems. Chem Eng J 283:582–594
60. Wu L-G, Zhang X-Y, Wang T, Du H, Yang C-H (2019) Enhanced performance of polyvinylidene fluoride ultrafiltration membranes by incorporating TiO2/graphene oxide. Chem Eng Res Des 141:492–501
61. Xua X, Yanga Y, Liua T, Chu B (2022) Cost-effective polymer-based membranes for drinking water purification Giant 10:100099
62. Boffa V, Marino E (2020) Inorganic materials for upcoming water purification membranes, in: Current Trends and Future Developments on (Bio-) Membranes, INC, 2020, pp 117–140, emerging and advanced membrane technology for wastewater treatment—advanced emergence processing
63. Wen Y, Yuan J, Ma X, Wang S, Liu Y (2019) Polymeric nanocomposite membranes for water treatment : a review. Environ Chem Lett 17:1539–1551
64. Khoo YS, Goh PS, Lau WY, Ismail AF , Abdullah MS ,Ghazali NHM , Khan N, Yahaya EM , Hashim N, Othman AR, Mohammed A, Devi NA/P. Kerisnan (Kerishnan) ,Yusoff MAM, Hashim NHF, Karim J , Abdullah NS (2022) Removal of emerging organic micropollutants via modified-reverse osmosis/nanofiltration membranes: a review. Chemosphere 305:135151
65. Radjenovic J, Petrovic M, Barceló D (2007) Analysis of pharmaceuticals in wastewater and removal using membrane bioreactor. Anal Bioanal Chem 387:1365–1377
66. Aguilera Soriano G, Erb M, Garel C, Audic JM (2003) A comparative pilot-scale study of the performance of conventional activated sludge and membrane bioreactors under limiting operating conditions. Water Environ 75:225–231
67. Cazes Md, Abejón R, Belleville M-P, Sanchez-Marcano J (2014) Membrane bioprocesses for pharmaceutical micropollutant removal from waters membranes, 4:692–729
68. Radjenovic J, Petrovic M, Barcelo D (2009) Fate and distribution of pharmaceuticals in wastewater and sewage sludge of the conventional activated sludge (cas) and advanced membrane bioreactor (mbr) treatment. Water Res 43:831–841
69. Fan H, Li J, Zhang L, Feng L (2014) Contribution of sludge adsorption and biodegradation to the removal of five pharmaceuticals in a submerged membrane bioreactor. Biochem Eng J 88:101–107
70. Tamagawa Y, Yamaki R, Hirai H, Kawai S, Nishida T (2006) Removal of estrogenic activity of natural steroidal hormone estrone by ligninolytic enzymes from white rot fungi. Chemosphere 65:97–101
71. Margot J, Maillard J, Rossi L, Barry DA, Holliger C (2013) Influence of treatment conditions on the oxidation of micropollutants by trametes versicolor laccase. New Biotechnol 30:803–813
72. Hassan SSM, Abdel-Shafy HI, Mansour MSM (2019) Removal of pharmaceutical compounds from urine via chemical coagulation by green synthesized ZnO nanoparticles followed by microfiltration for safe reuse. Arab J Chem 12:4074–4083

73. Abdel-Shafy HI, Mohamed-Mansour MS (2014) Membrane bioreactor for the removal of pharmaceutical compounds from urine. AWWA/AMTA 2014 membrane technology conference and exposition, Las Vegas, NV, USA, 03/2014
74. Daas A, Hamdaoui O (2014) Removal of non-steroidal anti-inflammatory drugs uprofen and ketoprofen from water by emulsion liquid membrane. Environ Sci Pollut Res 21:2154–2164
75. Vergili I (2013) Application of nanofiltration for the removal of carbamazepine, diclofenac and ibuprofen from drinking water sources J Environ Manage 127 C, 177–187
76. Radjenovic J, Petrovic´M, Venturac F, Barcelo D (2008) Rejection of pharmaceuticals in nanofiltration and reverse osmosis membrane drinking water treatment. Water Res 42:3601–3610
77. Narbaitz RM, Rana D, Dang HT, Morrissette J, Matsuura T, Jasim SY, Tabe S, Yang P (2013) Pharmaceutical and personal care products removal from drinking water by modified cellulose acetate membrane: Field testing. Chem Eng J 225:848–856
78. Rosman N, Salleh WNW, Mohamed MA, Jaafar J, Ismail AF, Harun Z (2018) Hybrid membrane filtration-advanced oxidation processes for removal of pharmaceutical residue. J Colloid Interface Sci 532:236–260
79. Dewi R, Shamsuddin N, Abu Bakar MS, Santos JH, Bilad MR, Lim LH (2021) Progress in emerging contaminants removal by adsorption/membrane filtration-based technologies: a review. Indones J Sci Technol 6:577–618
80. Kim S, Muñoz-Senmache JC, Jun BM, Park CM, Jang A, Yu M, Hernández-Maldonado AJ, Yoon Y (2020) A metal organic framework-ultrafiltration hybrid system for removing selected pharmaceuticals and natural organic matter. Chem Eng J 382:122920
81. Snydera SA, Adhamb S, Redding AM, Cannonc FS, DeCarolis J, Oppenheimer J, Werta EC, Yoon Y (2007) Role of membranes and activated carbon in the removal of endocrine disruptors and pharmaceuticals. Desalination 202:156–181
82. Wintgens T, Melin T, Schäfer A, Khan S, Muston M, Bixio D, Thoeye C (2005) The role of membrane processes in municipal wastewater reclamation and reuse. Desalination 178:1–11
83. Hammami A, Charcosset C, Ben Amar RR (2017) Performances of continuous adsorption-ultrafiltration hybrid process for AO7 dye removal from aqueous solution and real textile wastewater treatment. J Membr Sci Technol 07:1–8
84. Cao D-Q, Yang X-X, Yang W-Y, Wang Q-H, Hao X-D (2020) Separation of trace pharmaceuticals individually and in combination via forward osmosis. Sci Total Environ 718:137366
85. Bolong N, Ismail AF, Salim MR, Matsuura T (2009) A review of the effects of emerging contaminants in wastewater and options for their removal. Desalination 239:229–246
86. Guedidi H, Reinert L, Soneda Y, Belakhal N, Duclaux L (2017) Adsorptionof ibuprofen from aqueous solution on chemically surface- modified activated carbon cloth. Arab J Chem 10:S3584–S3594
87. Higgins CJ, Duranceau SJ (2020) Removal of enantiomeric ibuprofen in a nanofiltration membrane process. Membranes 10:383
88. Hidalgo AM, Gómez M, Murcia MD, León G, Miguel B, Gago I, Martínez PM (2022) Ibuprofen removal by graphene oxide and reduced graphene oxide coated polysulfone nanofiltration membranes. Membranes 12:562
89. Dehdashti B, Amin MM, Pourzamani H, Rafati L, Mokhtari M (2018) Removal of atenolol from aqueous solutions by multiwalled carbon nanotubes modified with ozone: kinetic and equilibrium study. Water Sci Technol 2017:636–649
90. Zaini YMM, Purba LDA, Abdullah N, Yuzir A, Iwamoto K, Mohamad SE (2022) Removals of atenolol, gliclazide and prazosin using sequencing batch Reactor. Mater Today Proc 65:3007–3014
91. Mukherjee D, Banerjee S, Ghosh S, Majumdar S (2020) PDMS/ceramic composite membrane synthesis and evaluation of ciprofloxacin removal efficiency. Korean J Chem Eng 37:1985–1998
92. Bhattacharya P, Mukherjee D, Dey S, Ghosh S, Banerjee S (2019) Development and performance evaluation of a novel CuO/TiO_2 ceramic ultrafiltration membrane for ciprofloxacin removal. Mater Chem Phys 229:106–116

93. Mukherjee D, Kar S, Mandal A, Ghosh S, Majumdar S (2019) Immobilization of tannery industrial sludge in ceramic membrane preparation and hydrophobic surface modification for application in atrazine remediation from water. J Eur Ceram Soc 39:3235–3246
94. Ullah A, Zahoor M, Alam S (2019) Removal of ciprofloxacin from water through magnetic nanocomposite/membrane hybrid processes. Desalin Water Treat 137:260–272
95. Kárászová M, Bourassi M, Gaálová J (2020) Membrane removal of emerging contaminants from water: which kind of membranes should we use? Membranes 10:305
96. Acero JL, Benitez FJ, Teva F, Leal AI (2010) Retention of emerging micropollutants from UP water and a municipal secondary effluent by ultrafiltration and nanofiltration. Chem Eng J 163:264–272
97. Qalyoubi L, Al-Othman A, Al-Asheh S (2021) Recent progress and challenges on adsorptive membranes for the removal of pollutants from wastewater. Part I: Fundamentals and classification of membranes. Case studies in chemical and environmental engineering 3: 100086
98. Verliefde ARD, Cornelissen ER, Heijman SGJ, Hoek EMV, Amy GL, Van Der Bruggen B, Van Dijk JC (2009) Influence of solute-membrane affinity on rejection of uncharged organic solutes by nanofiltration membranes. Environ Sci Technol 43:2400–2406
99. Baransi-Karkaby K, Bass M, FregerV (2019) In situ modification of reverse osmosis membrane elements for enhanced removal of multiple micropollutants. Membrane 9:1–14
100. Cacho-Bailo F, Caro G, Etxeberría-Benavides M, Karvan O, Télleza O, Coronas J (2015) High selectivity ZIF-93 hollow fiber membranes for gas separation. Chem Commun 51:11283–11285
101. Ahsan A, Islam T, Hernandez C, Kim H, Lin Y, Curry ML, Gardea-torresdey J, Noveron JC (2018) Adsorptive removal of sulfamethoxazole and bisphenol A from contaminated water using functionalized carbonaceous material derived from tea leaves. J Environ Chem Eng 6:4215–4225
102. Lu J, Zhang C, Wu J, Luo Y (2017) Adsorptive removal of bisphenol A using N-doped biochar made of Ulva prolifera. Water Air Soil Pollut 228:1–9
103. Khan AA, Maitlo HA, Khan IA, Lim D, Zhang M, Kim K-H, Lee J, Kim J-O (2021) Metal oxide and carbon nanomaterial based membranes for reverse osmosis and membrane distillation: a comparative review. Environ Res 202:111716
104. Hofs B, Schurer R, Harmsen DJH, Ceccarelli C, Beerendonk EF, Cornelissen ER (2013) Characterization and performance of a commercial thin film nanocomposite seawater reverse osmosis membrane and comparison with a thin film composite. J Membr Sci 446:68–78
105. Chang Q, Zhou J, Wang Y, Liang J, Zhang X, Cerneaux S, Wang X, Zhu Z, Dong Y (2014) Application of ceramic microfiltration membrane modified by nano-TiO_2 coating in separation of a stable oil-in-water emulsion. J Membr Sci 456:128–133
106. Yin J, Kim E-S, Yang J, Deng B (2012) Fabrication of a novel thin-film nanocomposite (TFN) membrane containing MCM-41 silica nanoparticles (NPs) for water purification. J Membr Sci 423:238–246
107. Lee E-J, Kyoungjin An A, He T, Woo YC, Shon HK (2016) Electrospun nanofiber membranes incorporating fluorosilane-coated TiO2 nanocomposite for direct contact membrane distillation. J Membr Sci 520:145–154
108. Shafiq M, Sabir A, Islam A, Khan SM, Gull N, Hussain SN, Butt MTZ (2018) Cellulaose acetate based thin film nanocomposite reverse osmosis membrane incorporated with TiO_2 nanoparticles for improved performance. Carbohydr Polym 186:367–376
109. Mahmoudi E, Ng LY, Mohammad AW, Ba-Abbad MM, Razza Z (2018) Enhancement of polysulfone membrane with integrated ZnO nanoparticles for the clarification of sweetwater. Int J Environ Sci Technol 15:561–570
110. Zhao H, Qiu S, Wu L, Zhang L, Chen H, Gao C (2014) Improving the performance of polyamide reverse osmosis membrane by incorporation of modified multi-walled carbon nanotubes. J Membr Sci 450:249–256
111. Mukherjee D, Bhattacharya P, Jana A, Bhattacharya S, Sarkar S, Ghosh S, Majumdar S, Swarnakar S (2018) Synthesis of ceramic ultrafiltration membrane and application in membrane

bioreactor process for pesticide remediation from wastewater. Process Saf Environ Prot 116:22–33
112. Heredia Deba SA, Wols BA, Yntema DR, Lammertink RGH (2022) Effects of the water matrix on the degradation of micropollutants by a photocatalytic ceramic Membrane. Membranes 12
113. Kaleekkal NJ, Thanigaivelan A, Tarun M, Mohan D (2015) A functional PES membrane for hemodialysis—preparation, characterization and biocompatibility. Chinese J Chem Eng 23:1236–1244
114. Kaleekkal NJ (2021) Heparin immobilized graphene oxide in polyetherimide membranes for hemodialysis with enhanced hemocompatibility and removal of uremic toxins. J Membr Sci 623:119068
115. Kaleekkal NJ, Thanigaivelan A, Durga M, Girish R, Rana D, Soundararajan P, Mohan D (2015) Graphene oxide nanocomposite incorporated poly(ether imide) mixed matrix membranes for in vitro evaluation of its efficacy in blood purification applications. Ind Eng Chem Res 54:7899–7913
116. Saraswathi MSA, Kausalya R, Kaleekkal NJ, Rana D, Nagendran A (2017) BSA and humic acid separation from aqueous stream using polydopamine coated PVDF ultrafiltration membranes. J Environ Chem Eng 5:2937–2943
117. Dahiya D, Pugazhenthi G, Kumar M, Vasant D (2021) Separation of bacteria Kocuria rhizophila BR-1 from its broth during synthesis of gold nanoparticles using ceramic membrane by shear-enhanced filtration process. Chemosphere 281:130761
118. Ranieri G, Mazzei R, Wu Z, Li K, Giorno L (2016) Use of a ceramic membrane to improve the performance of two-separate-phase biocatalytic membrane reactor. Molecules 21:1–14
119. Wu H, Sun C, Huang C, Zheng X, Zhao M, Gray S, Dong Y (2011) Treatment of oily wastewaters by highly porous whisker-constructed ceramic membranes: separation performance and fouling models. Water Res 2011:118042
120. Abdullayev A, Bekheet MF, Hanaor DAH, Gurlo A (2019) Materials and applications for low-cost ceramic membranes. Membranes 9:105
121. Chen M, Heijman SGJ, Rietveld LC (2021) State-of-the-art ceramic membranes for oily wastewater treatment: modification and application. Membranes 11:1–23
122. Pendergast MM, Hoek EM (2011) A review of water treatment membrane nanotechnologies. Energy Environ Sci 4:1946–1971
123. Radeva J, Roth AG, Göbbert C, Niestroj-Pahl R, Dähne L, Wolfram A, Wiese J (2021) Hybrid ceramic membranes for the removal of pharmaceuticals from aqueous solutions. Membranes 11:1–18
124. Kotobuki M, Gu Q, Zhang L, Wang J (2021) Ceramic-polymer composite membranes forwater and wastewater treatment: bridging the big gap between ceramics and polymers. Molecules 26:3331
125. Ng LY, Mohammad AW, Leo CP, Hilal N (2013) Polymeric membranes incorporated with metal/metal oxide nanoparticles: a comprehensive review. Desalination 308:15–33
126. Saljoughi E, Mousavi SM (2012) Preparation and characterization of novel polysulfone nanofiltration membranes for removal of cadmium from contaminated water. Sep Purif Tech 90:22–30
127. Nambikkattu J, Kaleekkal NJ, Jacob JP (2021) Metal ferrite incorporated polysulfone thin-film nanocomposite membranes for wastewater treatment. Environ Sci Poll Res 28:11915–11927
128. Hoek EMV, Ghosh AK, Huang X, Liong M, Zink JI (2011) Physical-chemical properties, separation performance, and fouling resistance of mixed-matrix ultrafiltration membranes. Desalination 283:89–99
129. Hamm JBS, Ambrosi A, Pollo LD, Marcillo NR, Tessaro IG (2021) Thin polymer layer-covered porous alumina tubular membranes prepared via a dip-coating/phase-inversion process. Mater Chem Phys 265:124511
130. Goossens I, Van Haute A (1976) The influence of mineral fillers on the membrane properties of high flux asymmetric cellulose acetate reverse osmosis membranes. Desalination 18:203–214
131. Wei W, Xia S, Liu G, Dong X, Jin W, Xu N (2011) Effects of polydimethylsiloxane (PDMS) molecular weight on performance of PDMS/ceramic composite membranes. J Membr Sci 375:334–344

132. Wang J-W, Li X-Z, Fan M, Gu J-Q, Hao L-Y, Xu X, Chen C-S, Wang C-M, Hao Y-Z, Agathopoulos S (2017) Porous sialon planar membrane with a robust polymer-derived hydrophobic ceramic surface. J Membr Sci 535:63–69
133. Zhang Y, Wang Y, Cao X, Xue J, Zhang Q, Tian J, Li X, Qiu X, Pan B, Gu AZ, Zheng X (2020) Effect of carboxyl and hydroxyl groups on adsorptive polysaccharide fouling: a comparative study based on PVDF and graphene oxide (GO) modified PVDF surfaces J Membr Sci 595:117514
134. Madhura L, Kanchi S, Sabela MI, Singh S, Inamuddin BK (2018) Membrane technology for water purification. Environm Chem Lett 16:343–365

Part III
Emerging Issues, Challenges and Reusability Aspects

Chapter 11
Separation of Toxic Ions Using Low-Cost Ceramic UF Membrane

P. Roy Choudhury, A. Roy Choudhury, and Sibdas Bandyopadhyay

1 Introduction

The increasing rate of population and the demand for freshwater are two major concerns in today's society. Fresh water is polluted by urbanization and industrialization if the wastewater is disposed of in to the natural water resources making it unfit for utilization. Therefore, water resource management via proper planning, distribution, and treatment of wastewater generated is necessary to restrict contamination for its safe usage. Sewage contains various pollutants like grit, suspended solids, chemical oxygen demand (COD), biochemical oxygen demand (BOD), organic pollutant, etc., whereas, industrial wastewater contains different organic and inorganic pollutants among which heavy metals like arsenic (As), lead (Pb), chromium (Cr), mercury (Hg), etc., are known for their extreme toxicity, persistence characteristics and bio-magnification potential. Heavy metals are also found in natural water sources. Among these toxic contaminants Pb and Cr are often observed in considerable amounts in wastewater from industries like semiconductor, metal plating, textile, chemical, tannery, paint, etc. All these toxic ions are contaminating natural water bodies causing serious threats to the environment. The discharge of heavy metal-containing waste in an inappropriate manner on land leads to leaching in soil

P. R. Choudhury (✉)
Department of Rubber and Plastics Technology, Anna University MIT Campus, Chennai, Tamil Nadu, India
e-mail: piyali.hit@gmail.com; piyaliroychoudhury@mitindia.edu

A. R. Choudhury
BGR Energy Systems Limited, Chennai, Tamil Nadu, India

S. Bandyopadhyay
Department of Environmental Studies, Siksha-Bhavana, Visva-Bharati, Santiniketan, West Bengal, India

S. Bandyopadhyay (ed.), *Application of Low Cost Ceramic Membranes in Wastewater Treatment*, Materials Horizons: From Nature to Nanomaterials,
https://doi.org/10.1007/978-981-97-5842-5_11

and surface water bodies. Thus, it creates hazards for human health and aquatic environments. The consumption of heavy metals creates long-lasting diseases that may affect the brain, lungs, kidneys, etc.

Various methods are available to separate heavy metals from wastewater through chemical and physical processes. Membrane technology is considered as a promising technology due to its high efficiency and universal applicability. Membrane technology is continuously gaining attraction from researchers to industry professionals [1–5]. Consistent output quality with lower energy, minimum chemical and less space requirements makes it more attractive. However, the deposition of solutes on membranes causes fouling shortening its lifespan.

For the past few decades, ceramic membrane have been considered a potential separation tool due to their outstanding mechanical strength, thermal and chemical stability [6, 7]. Antifouling nature of ceramic membranes helps to maintain fluxes and reduces cleaning frequencies, hence increasing the lifespan of the membrane. Many researchers are now introducing nanoparticles (NPs) to enhance surface properties of the membrane to address the common fouling problems [8, 9].

Ultrafiltration (UF) is serving as an advanced technology handling a wide range of input water quality. This chapter mainly focuses on ceramic UF membrane, its development and trends for metal ion separation from wastewater especially through nanoparticle-enhanced ceramic membrane. NPs like copper oxide (CuO) and hydroxyapatite-based low-cost ceramic membrane have been described here in a comprehensive way including synthesis of membranes, their characterization and evaluation of heavy metal ion removal efficiencies [10, 11, 12]. This chapter aims to highlight the scope of nanoparticles embedded low-cost ceramic UF membrane separation process for removal of the metal ions from wastewater.

2 Wastewater Treatment by Membrane Technology

2.1 Heavy Metal Ion Contaminated Wastewater Treatment

Removal of undesirable constituents is the main objective of wastewater treatment. Vo et al., summarized the advantages and disadvantages of adsorptive membranes including nano-enhanced membranes (NEMs) which exhibited high selectivity and adsorption capacity for heavy metal ions [13]. Different methodologies are available to separate toxic metal ions from wastewater as listed in Table 1.

Table 1 Heavy metal removal methods from wastewater with description [14]

Removal methods	Description	Heavy metal
Coagulation and flocculation	Charge neutralization and agglomeration of destabilized particles occurs by coagulation followed by flocculation. Simple operation but incomplete removal	Pb(II), Cd(II), Cr(III), Cr(VI)
Chemical precipitation	Soluble compounds transform insoluble compounds in the presence of chemicals. Inexpensive method but generates sludge	Cd(II), Cu(II), Pb(II), Cr(III), Cr(VI)
Adsorption	Uses physical or chemical interactions of adsorptive type materials. Effective and cost efficient but separation of nanoadsorbent from solution is difficult after treatment	Pb(II), Cr(III), Cr(VI), Fe(III) Zn(II), Cu(II)
Ion exchange	Refers to the exchange of ions with resign and wastewater. Gives high purity but effective for lower metal ion concentration	Cd(II), Cr(III), Cr(VI), Zn(III)
Membrane filtration	Membrane acts as a semi-permeable barrier and separates contaminant by using a pressure gradient between two different concentrated fluid. High-purity product and energy efficient but fouling problem may arise	Cu(II), Cr(III), Pb(II), Cr(VI), Zn(II), Fe(III)
Flotation	Surfactant is added to promote non-surface active material to surface active material. Simple and low-cost operation but effectiveness varies with ion strength	Pb(II), Cd(II), Cu(II), Zn(II)
Complexation	Complex ion or chelating agent is formed with metal ions and thus removed. Not effective for large scale	Cu(II), Cr(III), Cr(VI), Zn(II), Fe(III)
Electrochemical operation	Using of direct current, electrolytic metals are recovered and removed. Enhanced mass transfer but high cost operation	Cu(II), Zn(II), Ni(II), Pb(II)
Distillation	Evaporation, condensation and rinsing are involved in this process. It is used mainly for product recovery but includes complex operation with higher operational cost	Zn(II), Cr(III), Cr(VI), Cu(II)

2.2 Membrane Technology for Heavy Metal Ion Removal

Membrane technology is considered as a high-potential, energy-efficient and low-footprint separation method in wastewater treatment which offers selective separation of constituents. Since 1960, numerous efforts have been involved to develop advanced membrane technology for significant industrial applications [13]. Extensive research is also associated with developing cost-effective commercial applications [13]. However, recently it became the most accepted technology for wastewater treatment to produce high-quality output in less space and with minimum chemical consumption.

Based on pore size, membranes can be classified as microfiltration (MF), ultrafiltration (UF), nanofiltration (NF) and reverse osmosis (RO) (Fig. 1). These membranes are the most common and widely used separation techniques with respect to wastewater characteristics and desired outlet water quality. The same is described here in brief. Moreover, different integrated membrane systems are approaching to mitigate hazardous waste, considering growing environmental concern. Apart from these, forward osmosis (FO), electrodialysis (EDI) and liquid membranes (LM) are now in practice to treat wastewater.

Microfiltration: To enhance the life span of reverse osmosis membranes, MF is generally used as pretreatment to eliminate turbidity. MF is typically used to retain the total suspended solid, colloidal particles, macromolecules and microorganisms with pore size more than >0.1 μm. MF is very useful in treating municipal wastewater

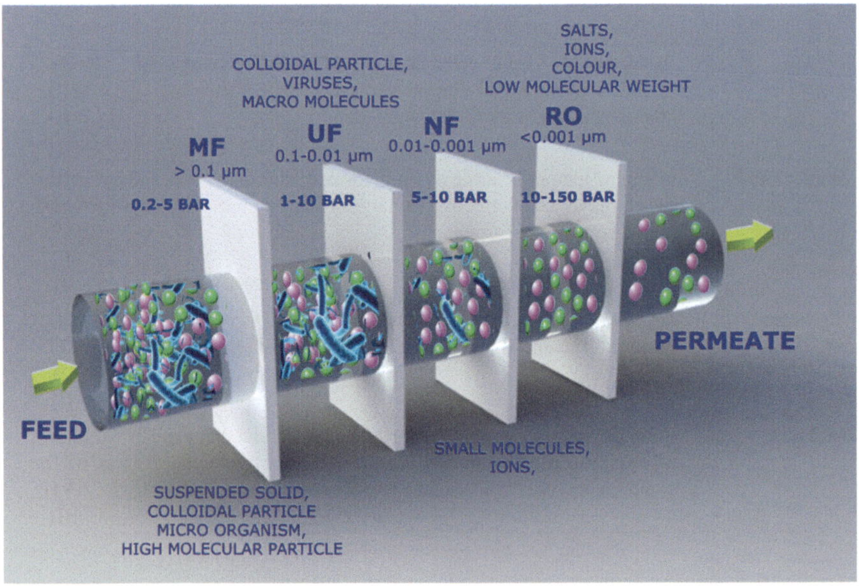

Fig. 1 Different types of membranes

and different industrial wastewater [15, 16]. Liu et al. (2020) prepared an adsorptive type nanofibrous composite microfiltration membrane by using polyvinyl alcohol (PVA) and polyacrylonitrile (PAN) and applied it for the removal of chromium (VI) and cadmium (II) from wastewater [17]. High selectivity and removal efficiency were observed. A hybrid process was investigated by Keerthi et al. (2013) by combining electro-coagulation and dead-end microfiltration process. Different parameters were varied like pH, initial concentration, current density to observe the removal efficiency of heavy metals from synthetic wastewater. This hybrid process exhibited high removal efficiency (99%) of Zn^{2+}, Ni^{2+} and Cd^{2+} ions from synthetic wastewater [18].

Ultrafiltration: UF is also capable of removing pesticides, viruses and total dissolved solids to some extent. Pore size of UF ranges from 0.1 to 0.01 μm, as a result of which higher feed pressure is required compared to MF. Compared to NF and RO processes, lower energy consumption is noted in UF. All these features of UF are gaining popularity in the treatment of metal-contaminated wastewater. Oka et al. (2017) recently pointed out that fouling control can be avoided by up to 20% of initial permeability [19]. Performance can be further enhanced by relaxing and air sparging. High-quality water can be produced using submerged hollow fiber UF. The hydrophilicity of membrane can be enhanced by incorporating silica resulting in improved flux. Silica-based membrane found effective in the heavy metals separation from contaminated water. The separation is possible by adsorptive type membrane application [20]. UF needs periodic backwashing and chemical cleaning to maintain the flux for operations.

Nanofiltration: For the removal of selective specific dissolved solids, NF is most suitable due to lower pore size below 2 nm with dense active layers. Moreover, NF has shown reduced fouling before reverse osmosis. NF is a pressure-driven technology based on molecular weight cut-offs that can reject solutes like organic and inorganic impurities from water. NF elements are considered with nominal pore size between 0.01 and 0.001 μm. The solvent-resistant principle has helped nanofiltration to explore various applications other than conventional areas [21]. NF is also useful in various pH ranges for non-aqueous applications. The material of construction of the NF plays an important role in its performance. In industries like textile, dairy, chemical and pharmaceutical, food, O&G and refineries, NF has proved its successful performance. NF has also found wide application in the desalination process. NF is incorporated in combination with RO and thermal process for desalination. An increased output of product water is noted from such desalination plants. Negatively charged NF materials help in for high rejections of metals by charge exclusions. NF is also applied for the effective removal of heavy metals from wastewater [22].

Reverse osmosis: RO is more like a synonym for the desalination process. This process is used to eliminate excess total dissolved solid to produce drinking water. The pore size of RO membrane ranges below <0.001 μm. RO process can be of multiple stages depending upon the presence of total dissolve solids. In general, for higher recovery, three-stage processes are practiced. High rejections are observed in heavy metal contaminated wastewater treatment using the membrane bioreactor-reverse osmosis (MBR–RO) combination process. However, a rise in inorganic fouling with heavy metals is noticed [23].

2.3 Polymeric Membrane for Metal Ion Contaminated Wastewater Treatment

Polymeric membranes were widely used for the separation of organic as well as inorganic compounds from wastewater. Different types of polymeric membranes were adopted based on the size impurities. Currently, modifications of membrane materials are reported to make more chemically, thermally and mechanically stable polymeric coating over the support layer. Numerous reports are available on the removal of heavy metals via NF and UF polymeric membranes which were prepared by different routes like grafting, layer-by-layer, phase inversion, sol–gel, etc. methods. Some of the recent results are listed in Table 2. Very recently, simple interfacial polymerization was adopted aimed at enhancing the permeability and selectivity of the membrane by regulating polyamide structure by dendritic amphiphilic hyperbranched polyester acid chloride [24]. Seyed et al. (2020) examined improved hydrophilic surface along with better heavy metal removal capacity via chitosan-coated activated carbon NPs included in PVC membrane [26]. The membrane was prepared layer-by-layer technique. Results showed metal ions removed by the adsorptive nature of the active surface layer of the membrane. Regeneration of fouled membranes was also studied by applying ultrasonic waves thus observing the good performance of metal ion removal. In another study, chitosan and carboxymethyl cellulose polymer were added to polymer enhanced UF membrane and observed to improve Ni (II) removal at certain pH conditions whereas in real effluent it was not so effective [30]. Currently, a trend of incorporation of functionalized or modified nanomaterials in polymeric membrane is observed due to its large surface areas, active surface sites and unique structures. Various nanomaterials were investigated like graphene, zero-valent iron, carbon nanotubes, etc. Sheng et al. (2019) investigated Pb(II) and Cu(II) separation from synthetic wastewater by Poly(ethylenimine) (PEI) modified multiwalled carbon nanotube embedded polyacrylonitrile (PAN) nanofibers membrane [33]. The prepared membrane showed improved mechanical strength and hydrophilicity which exhibits outlet water quality as a drinking water standard.

Table 2 Heavy metal ion contaminated wastewater treatment by polymeric membrane

Sr. no	membrane type	Removal study	Application	References
1	PES (Polyethersulfone) ultrafiltration membrane coated by polydopamine NPs	Pressure—0.1 MPa trans-membrane pressure, pH—5.4, feed—100 ppb (Pb), adsorption capacity: 20.23 mg Pb/g	Pb(II), Cd(II) and Cu(II)	Xiaofeng et al. (2017) [24]
2	Cellulose-supported regenerated membrane grafted with polyelectrolyte	Pressure—1 bar, pH—5–6, multimetal 1 ppm each, metal uptake 194 mg g^{-1}	Cu(II), Pb(II) and Cd(II)	Xiaopeng et al. (2021) [25]
3	Chitosan (CS) coated activated carbon NPs (ACN) inclusive cation exchange polyvinyl chloride (PVC) based membrane prepared by Layer by layer method (LBL)	Feed—10^{-3} M, pH—6.4, Temp—27.8 °C, 98.35% Cu(II), 84.12% Ni(II), 62.42% Pb(II) removal at LBL (CS-0.005ACNs)	Pb(II), Cu(II) and Ni(II)	Seyed et al. (2020) [26]
4	Filler (tetrathioterephthalate) incorporated Polyethersulfone nanofiltration membrane	Feed 20, 50, and 100 ppm, pH—6, Pressure—4 Bar, maximum rejections at 20 ppm: Zn(II) 99.2%; Cu(II) 98.5%; Pb(II) 97.4%	Pb(II), Zn(II) and Cu(II)	Golshan et al. (2020) [27]
5	Polymeric (polysulfone) ultrafiltration membrane prepared by phase inversion method	Transmembrane pressure—1.5 kg cm^{-2}, pH—5.42, feed-50 to 200 ppm, capacity—279.63 mg g^{-1} at 200 ppm feed, rejection: 98.24% Pb(II)	Pb(II), Cu(II), Cd(II) and Fe(II)	Soumen et al. (2020) [28]
6	Composite membrane included metal–organic frameworks/PEI and TMA (polyethyleneimine and trimesic acid)	Pressure: 3–7 bar, rejection: Ni(II) 90.9% at feed—1000 ppm and test pressure—4 bar	Ni(II), Mn(II) and Zn(II)	Gong et al. (2020) [29]
7	Polymer (chitosan and carboxymethyl cellulose) enhanced UF membrane	pH—4.4 to 7.4, feed: Ni-1.7 × 10^{-5} mol L^{-1}, polymer 2 × 10^{-2} mol L^{-1}	Ni(II)	Boukary et al. (2018) [30]
8	Polyamide membrane functionalized by sulfur	Pressure—6 bar, Temperature—25 °C, feed—10 ppm, rejection: 99.99% Hg(II)	Hg(II), Pb(II), Ni(II), Cu(II), Ag(I), and Cd(II)	Hong et al. (2020) [31]

(continued)

Table 2 (continued)

Sr. no	membrane type	Removal study	Application	References
9	Cellulose UF membrane with extracellular polymer cake recovery	Pressure—0.2–2 bar, feed: 0.1–0.8 mM, rejection: Pb(II) 98.4%, Cu(II) 88.9%, Cd(II) 89.2%	Pb(II), Cu(II) and Cd(II)	Cao et al. (2020) [32]
10	MWCNT-PEI nanocomposite PAN nanofiber membrane	Pressure—0.2–0.8 bar, pH—6, feed:, rejection: Pb(II) 97.4% and Cu(II) 93.6% for 3.5 wt% MWCNT-PEI/PAN	Pb(II) and Cu(II)	Sheng et al. (2019) [33]

3 Ceramic Membrane-Based Wastewater Treatment

3.1 Wastewater Management and Ceramic Membrane System

The increasing population and scarcity of freshwater demanded proper wastewater management by controlling its uses, treatments, and collection system. Different industrial activity consumes about 5–20% of total water and generates wastewater in a subsequent manner [34, 35]. Currently, lack of infrastructure and inadequate access to safe water in many areas drives an economic wastewater management scheme. Ceramic membrane for wastewater treatment have provided satisfactory performance for few decades having high temperature and fouling resistance. A search on the topic of ceramic membrane for wastewater treatment for the past 20 years (2003–2023) in Scopus, Elsevier database of abstract and citation provides a total of 1,117 documents on the application of ceramic membranes in different fields like engineering, agricultural, energy, biochemistry, material science, chemical engineering, etc. whereas maximum identified in environmental solutions as about 23% (Fig. 2). Mechanically stable ceramic membrane could be used in wastewater with extreme pH ranges compared to polymeric membranes.

On the other hand, high manufacturing cost of the market-available ceramic membranes, compared to polymeric membrane, propels research on new materials, reduced sintering temperatures for development of low-cost ceramic membranes. Manufacturing of low-cost ceramic membranes is possible by using low-cost membrane materials. Proper pore size distribution, porosity and strength of ceramic supports are required which can be made by extrusion, casting, etc. process. However, integrated ceramic membrane processes like membrane bio reactor (MBR), adsorptive membrane separation recommend an economic approach for the treatment and reuse of wastewater.

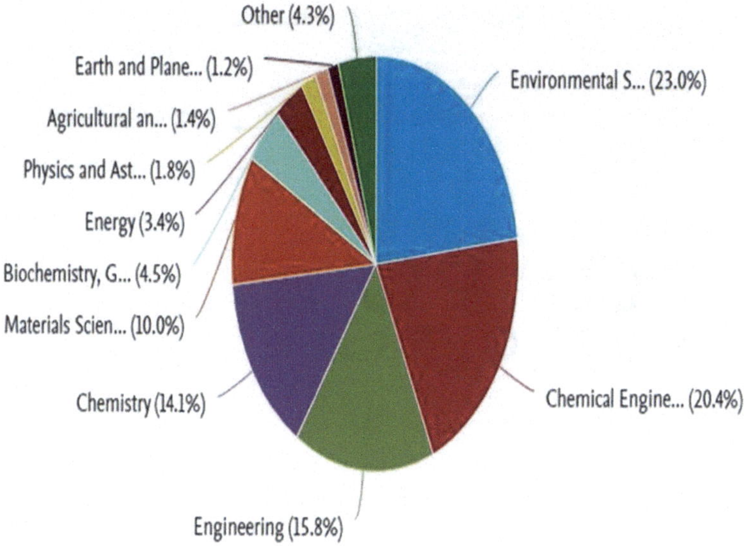

Fig. 2 Published articles by subject area (2003–2023), Scopus by Elsevier

3.2 Ceramic Membrane Structure and Fabrication

Different types of materials are used for the fabrication of ceramic membranes, among which some conventional types are based on alumina, silica, zirconia and titania. Other types of materials now being adopted consist of copper oxide [10], hydroxyapatite [12], fly ash [36], cow bone waste [35], rice husk ash [36], zeolite residue [37], natural kaolinite [38], Zinc oxide [39], solid waste fly ash [40], etc. for water and wastewater application.

The configuration of a ceramic membrane consists of an asymmetric structure, one at the top surface as an active layer responsible for separation performance and another as a support layer that offers physical strength of the membranes (Fig. 3). An intermediate layer between the active and support layers is usually present. Common fabrication methods of support layer are paste processing, colloidal processing and powder pressing methods [41, 42].

Paste processing methods are associated with extrusion and manual paste casting whereas slip, centrifugal and gel casting methods are included as well in colloidal processing. Powder pressing methods are integrated with dry/wet and hot/cold pressing. Several fabrication methods used in the preparation of ceramic membranes are slip casting, dip coating, sol–gel, tape casting, pressing and freeze casting [43]. Depending upon the fabrication method, membrane characteristics vary. Fabrication stages of ceramic membrane commonly consist of coating formulation, structure or surface development, calcinations and modifications. Selective layer is developed over ceramic support based on the application of membrane. This active layer plays an important role in the separation of pollutants in water/wastewater treatment. The

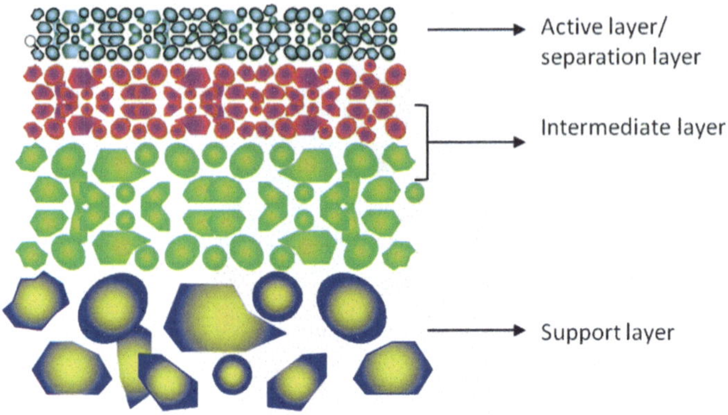

Fig. 3 Multilayer ceramic membrane

demand for ceramic membrane for advance application propels surface modification and structure development in the nanoscale.

3.3 Heavy Metal Ions Removal by Ceramic Membrane

In the reported articles, removal of metal ions by ceramic membrane was mainly conducted by using UF and NF membranes. Ferella et al. (2007) revealed surfactant enhanced ceramic UF for the removal of Pb and As from synthetic industrial wastewater [47]. The performance test was done by using dodecylbenzenesulfonic acid-DSA (anionic surfactant) and dodecylamine (cationic surfactant). The pore size of membrane was about 20 nm. Excellent removal (99%) of Pb (II) ions in the presence of DSA and dodecylamine examined but a lower removal rate (19%) was observed in the case of toxic arsenic ions [As(V)] due to less absorption by microorganisms. However, Pagana et al. (2008) suggested a significant As(V) ion removal process for allowable level in drinking water by combining application of ceramic UF layer deposited over clay based support and adsorption processes using Fe_2O_3 NPs (20–30 nm diameter) as an adsorbent [48]. Ceramic γ-Al_2O_3 UF membrane (with 3 and 4 nm pore size) showed 100% sorbent NPs removal at 3 bar pressure. The same process was applied in parallel for Cr(III) removal. In another study polymer complexing (bovine serum albumin) assisted the ceramic NF system in the rejection of Zn, Cd, Pb, Cu and Fe ions [50]. In this work, studies on the influence of pH showed a high level of rejection of metal ions such as 93% Zn, 99% Cd, 93% Pb, 99% Cu and 85% Fe at pH 9.0 by using ceramic membrane technology. Bhowmick et al., fabricated novel composite NF membranes based on CSP (ceramic-supported-polymeric) concept and embedded copper ions by crosslinking with PEI polymer matrix to modify the surface with positively charge which exhibited promising results regarding rejection of salts (Mg^{2+} and Ca^{2+}) and heavy metal ions(Zn^{2+}, Ni^{2+} and Cd^{2+}) [55]. Agata et al. (2020) investigated on the separation of nitric salt of metals like Cu, Zn, Pb and Fe using 19-channel Al_2O_3/TiO_2 based ceramic NF membrane [57]. (MWCO 450 Da) with a pore radius 0.9 nm and selective layer thickness 0.5 μm to predict the separation performance by mathematical modeling. Many efforts are invested to develop ceramic membrane for the removal of toxic metal ions from wastewater listed below in Table 3.

Permeability (permeation/diffusion rate through membrane barrier) and rate of flux (volumetric liquid flow per unit area per time) are two common parameters to analyze membrane performance. Permeate and retentate are two terminologies used to denote treated water or product and rejected or outlet stream, respectively. The removal mechanism of membrane separation is primarily based on the principle of size separation/Donnan exclusion/diffusive reduction/adsorptive reduction. Metal ion removal process using ceramic UF and NF membranes may involve either one or more of afforesaid mechanisms.

4 Nanoparticle-Incorporated Ceramic Membrane

4.1 Synthesis of Nanoparticle-Enhanced Ceramic Membrane

Characteristics of the materials and processes used in membrane fabrication play an important role in membrane performance. Nanoparticles (NPs) are now being used widely to obtain modified membrane structures with improved separation selectivity and outlet product quality. In many cases, incorporation of NPs into ceramic structure affects the functionality, membrane microstructure or pore sizes considerably. Thus, porestructural modification by inclusion of NPs offers the scope of improving the permeate quality as well as membrane effectiveness. Various fabrication methods are reported like surface coating or entrapment of NPs during the membrane layer formation process. Surface coating with NPs is a common method in which either the substrate is dipped or poured into a calculated amount of coating formulation for a specific time. In situ NPs coating is another technique used in membrane synthesis where the coating is done for a short reaction or co-precipitation method. Other surface coating methods are sol–gel, electroplating, electroless plating, physical or chemical vapor deposition (PVD or CVD) and plasma enhanced CVD. However, the dip coating technique is considered as the most simple and low-cost method compared to others. Many researchers adopted different types of coating material like oxides of titania (TiO_2), alumina (Al_2O_3), silica (SiO_2), copper (CuO), silver (Ag_2O), Zirconia (ZrO_2) iron (Fe_2O_3), zinc oxide NPs. On the other side, selection of NPs depends on the need for desired membrane separation performance. Nanoparticles like TiO_2 and SiO_2 were widely used both in polymeric and ceramic membranes. However, researches are going on to take advantage of the nanoparticle characteristics in membranes as a single element/component or by making nanocomposites. Besides this, silver (Ag) NPs were also largely used in the synthesis of

Table 3 Removal of toxic metal ions from wastewater by ceramic membrane

Membrane type	Removal study	Metal ions removal	References
TiO_2 and $ZnAl_2O_4$ coated ceramic membranes on artificial cordierite support	Pressure 10 bar (TMP), concentration 10^{-3} mol/L, pH 5, rejection rate above 90%	Cr(III), Cd (II) and Pb (II)	Saffaj et al. (2004) [44]
Surfactant-enhanced ceramic UF	Concentration variation 4.4 to 7.6 mg/l, removal of lead ions >99% and arsenate ions 19%	Pb and As	Francesco et al. (2007) [45]
Microporous ceramic UF membrane	As(V) concentration: initial 1 ppm, final 10 ppb; Cr(III) concentration: feed 0.5 ppm, final 10 ppb; Pressure 3 bar(TMP),	As(V) and	Pagana et al. (2008) [46]
Carbosep M5 UF membrane	Cu(II) concentration 125 ppm, time 120 min, pressure 4 bar (TMP), pH 6, fouling behavior studied	Cu(II)	Llanos et al. (2010) [47]
Ceramic nanofiltration membrane	pH variation: 2.0 to 9.0, only metal ion removal 80% at pH > 6.9 and mixture of metal ions and complex at pH 6 Zn, Cd, Pb, Cu removal above 90% and Fe 71%	Zn, Cd, Pb, Cu and Fe complex	Arkadiusz Nędzarek et al. (2015) [48]
Amino-functionalized MOFs combining ceramic UF membrane	Pressure 0.15 MPa (TMP), Temp-35 °C, cross-flow velocity:4.0 $m \cdot s^{-1}$, maximum 61.4% Pb (II) removal	Pb(II)	Yin et al. (2016) [49]
Rice husk ash derived green ceramic hollow fiber membrane	Pressure 2 bar, time: 120 min, pH 5, 99% removal of metal ions	Zn(II), Ni(II), and Pb(II)	Hubadillah et al. (2016) [50]

(continued)

Table 3 (continued)

Membrane type	Removal study	Metal ions removal	References
Integrated direct membrane distillation with microalgae ceramic membrane	Pressure < 0.1 bar (TMP), Temperature: feed 60 °C, permeate 10 °C; flow rate: feed 42 kg/hr, permeate 30 kg/hr, Feed Arsenic concentration 107,500 ppb, final 8 ppb, As rejection 100%	As	Hubadillah et al. (2020) [51]
Fly ash, graphite, and industrial rutile supported ceramic TiO_2 membrane	Pressure: 0.2–1.2 bar, Cu removal rate 90.15%, in photodegradation process Cr(VI) removal 97.09%	Cu(II), Cd(II), Cr(VI), and rhodamine B	Zhang et al. (2021) [52]
Copper ions embedded ceramic polymeric composite membrane	Pure water permeability 8.1 L m^{-2} h^{-1} bar^{-1}, maximum removal Zn(II) 91.73%, Ni(II) 83.15% and Cd(II) 75.50%	Zn(II), Ni(II), Cd(II)	Bhowmick et al. (2022) [53]
Composite of polyamide selective layer on alumina and carbon black interlayer over kaolin substrate	Water flux 2.75 L m^{-2} h^{-1}, removal of Pb(II) 98.44%, and Cd(II) 97.51%	Pb(II) and Cd(II)	Zhou et al. (2022) [54]

ceramic membrane for antibacterial activities in wastewater treatment. Moreover, inclusions of NPs or nanocomposite in ceramic substrates modify/improve its structures enhancing the filtration performance and functionality as well [56]. Though NPs coated ceramic membranes are used for various membrane separation process, UF mode is elaborated in this chapter. Table 4 shows NPs incorporated ceramic UF membrane with a brief description of membrane synthesis and performance analysis application in wastewater treatment.

Three different types of ceramic UF membrane preparation steps and their applications are discussed here [10–12]. Indigenously developed porous silica-alumina ceramic tube was used as membrane support for UF membrane synthesis. The tube pore diameter ranges from 0.5 to 1.5 μm. Ceramic UF membranes were synthesized by using CuO NPs, CuO/hydroxyethyl cellulose (HEC) composite and hydroxyapatite powder (HAp). For the first two types of UF membranes, single channel 150 mm length support tubes of 10 mm diameter, 1.5 mm thickness was used and in case of the last sample, the support tube was made of 19-channel configuration with 500 mm length, 34 mm outer diameter having 4.2 mm channel diameter. The CuO NPs were synthesized (size about 10 nm) by *Catharanthus roseus* leaf extract following the

Table 4 Nanoparticle included ceramic UF membrane for wastewater treatment

Inclusions of nanoparticle	Synthesis of membrane	Performance	References
TiO_2	Flat UF membrane was prepared using bentonite MF support coated by TiO_2 nanoparticle. The process followed making of colloidal suspension with TiO_2 powder, addition of surfactant, peptizing agent and binder, magnetic agitation followed by ultrasonication. Herein, dip coating techniques deposited TiO_2 nanoparticles over the bentonite surface. The results showed excellent adhesion between TiO_2 and support layer which forms about 4 μm thin layer thickness over supports	Dye removal was investigated where 98% Direct Red 80, 85% Acid Orange and 94% Methylene Blue rejection capacities were observed. Membrane pore size and flux were about 10 nm and 16.08 L m^{-2} h^{-1} bar^{-1}, respectively	Bouazizi et al. (2017) [57]
TiO_2	TiO_2 nanoparticle immobilized low-cost ceramic UF membrane was fabricated over clay-alumina-based tubular support. Slurry was prepared by dispersion of TiO_2 nanoparticle, PVA and dispersant agent. Slip casting method was used to coat the support. Finally, calcination was done at 800 °C for the duration of 3 h	Fabricated membrane was applied for removal of Alizarin Red dye. About 50 nm pore diameter and 850 L m^{-2} h^{-1} bar^{-1} permeability were achieved. The membrane exhibited maximum 99% dye removal at 5 bars pressure, pH = 9 and initial dye concentration 150 ppm	Oun et al. (2017) [58]
Titania-Smectite	TiO_2-Smectite mesoporous UF ceramic membrane was prepared over zeolite support. The slip deposition was done by the layer-by-layer method. Deposition was performed by varying coating time followed by room temperature drying and sintering (900 °C). TiO_2-Smectite UF membrane exhibited good antifouling characteristics	Different industrial wastewater was used to analyze its performance. Average pore size of the membrane was 18 nm and the permeability 80 L h^{-1} m^{-2} bar^{-1}. A potential reduction of chemical oxygen demand (COD), turbidity and colors was observed. Regenerated membrane showed relatively the same results in terms of water permeability	Aloulou et al. (2018) [59]

(continued)

Table 4 (continued)

Inclusions of nanoparticle	Synthesis of membrane	Performance	References
Titania-Smectite	Titania-Smectite nanoparticle was used to fabricate UF ceramic membrane over microporous zeolite tubular support. Coating was obtained via a layer-by-layer technique. Sintering was done at 900 °C to get a ceramic composite membrane	Developed membrane showed antifouling properties with a good reduction of turbidity, COD and Chromium (>89%) content. Pore diameter obtained about 3–5 nm while water permeability varied 59–95 L h^{-1} m^{-2} of the coated membrane	Aloulou et al. (2020) [60]
Fe_2O_3	Hybrid of ozonation and ceramic UF membrane incorporated with iron oxide nanoparticles. Fe_2O_3 nanoparticle coating was done over Al_2O_3 ceramic membrane	Treated high molecular weight and aromatic natural organic matter. Reactive ceramic membrane exhibited maximum rejection of 46% of hydroxyl radical probe compound and showed less antifouling tendency due to hydroxyl radicals	Park et al. (2012) [61]
Fe_2O_3	Catalytic membrane was prepared and used in combined ozonation and UF process. Tubular titania-coated commercial ceramic support was coated using Fe_2O_3 nanoparticle by layer-by-layer method. Sintering temperature was 900 °C	Developed membrane was applied in the drinking water system and met water standard limits. The molecular weight cut-off (MWCO) of the membrane was 5 kDa. Moreover, these hybrid membrane-ozonation processes were able to reduce the concentration of total trihalomethanes (TTHMs) by about 90% and halo acetic acids (HAAs) by about 85% potentially	Karnik et al. (2005) [62]
Fe_2O_3	Tight-ultrafiltration membrane was prepared by deposition of Fe_2O_3 nanoparticles over tubular γ-Al_2O_3 ceramic support by a modified layer-by-layer method. The sintering temperature was 450 °C for 30 min	Prepared membrane was applied for the removal of natural organic matter and phenol from wastewater. Coating is not that much effective for phenol rejection but about 72–83% rejections were observed for a natural organic matter where average pore size of the membrane was 4 nm	Harman et al. (2010) [63]
ZrO_2	ZrO_2 deposited UF membrane was synthesized over the mud of hydro cyclone powder content membrane by slip-casting method. The slip was prepared by mixing of ZrO_2 powder, water and polyvinyl alcohol	It was applied for washing cuttlefish effluent treatment where low turbidity, conductivity and COD value were obtained at permeates. The mean pore diameter and water permeability of membrane were 5 nm and 86 L h^{-1} m^{-2} bar^{-1}, respectively	Mouna et al. (2011) [64]

(continued)

Table 4 (continued)

Inclusions of nanoparticle	Synthesis of membrane	Performance	References
ZrO_2	Dip-coating method was followed for the preparation of ZrO_2 nanoparticle-based ceramic UF membrane over silicon carbide substrate. Coated and dried substrate was finally sintered at the temperature range 800–1000 °C for a duration of 45 min	Developed membranes can be applicable for chemically corrosive high-temperature wastewater. Pore size and water flux of membrane were 82 nm (found at 800 °C) and 850 L h^{-1} m^{-2}	Li et al. (2020) [65]
Boehmite (γ-Al_2O_3)	Asymmetric four-layer ceramic membrane was prepared where top most layer signified boehmite nanoparticle coated active surface. This layer was prepared by dip coating method	It was applied for heavy metal removal from contaminated water through a combined adsorption-ultrafiltration process. Membrane pore size and water flux were observed in the range of 3-5 nm and 156 kg h^{-1} m^{-2} (at 3 bar), respectively. Below 10 ppb Cr concentration was achieved	Pagana et al. (2008) [46]
Silica	Modified Al_2O_3 ceramic membrane was prepared by functionalization of tubular ceramic support by tetraethoxysilane and ethylthiocarbamidepropyl triethoxysilane sol. Finally, thiourea functionalized membrane were dried at different temperatures	The functionalized membrane was used to remove heavy metals. Metal ions were adsorbed over the membrane surface and retained as copper (II)-18.6 mg L^{-1}, silver (I)-22 mg L^{-1}, and cadmium (II)-260 mg L^{-1} at 4 bar pressure	Veronika et al. (2019) [66]
SnO_2/ Montmorillonite	Tubular adsorbent type UF ceramic membrane was prepared by dry compaction technique by mixing clay, PEG and glycerin. Sintering temperature maintained at 900–1100 °C for 1–3 h. Further SnO_2/MMT nanocomposite was included in green mixture and prepared membrane	Developed membrane applied for removal of Cu, Zn and Ni aqueous solutions. And observed rejection of Cu(II) 97.88–99.26%, Zn(II) 76.79–92.23%, and Ni(II) 24.97–64.74%. Metal ion concentration varied in the range of 5–50 ppm. About 40% enhancement of rejection was observed by the inclusion of SnO_2/ MMT nanocomposite	Sahar et al. (2022) [67]
CuO	Green-synthesized CuO NPs coated ceramic UF membrane was developed by dip coating method. Porous silica-alumina based ceramic tube (MF range) was used as support tube	Prepared membrane was applied for removal of Cr (VI) from contaminated water which showed excellent water flux. An improved Cr (VI) rejection percentage was observed as 88.08%	Choudhury et al. (2018) [10]

(continued)

Table 4 (continued)

Inclusions of nanoparticle	Synthesis of membrane	Performance	References
Cuo/HEC	Biopolymer-based high pressure ceramic polymeric composite UF membrane was prepared from a porous support tube	The permeability was determined as 34.99 L m^{-2} h^{-1} bar^{-1}. The membrane exhibited significant performances and stability on the rejection of toxic Pb(II) and Cr(VI) ions from spiked solution and simulated wastewater. The operational pressure range was 0–5 bar where the maximum percentage of rejection achieved was 97.14% for Pb(II) and 91.44% for Cr(VI) at 2 bar transmembrane pressure from spiked water	Roy Choud-hury et al. (2018) [11]
Hydroxyapatite	Biocompatible hydroxiapatite ceramic ultrafilatrtion membrane was prepared over porous clay-alumina ceramic support tube. Designing, fabrication and performance evaluation of novel hydroxyapatite ceramic ultrafiltration membrane conducted in bench scale operation	Performances of the developed 500 mm long HAp ceramic membrane was satisfactory, provides high flux (243.5 L m^{-2} h^{-1}) at 3 bar pressure and also it is regenerable. Maximum 99.6% Pb rejection of 5 mg L^{-1} feed solution was achieved	Roy Choudhury et al. (2019) [12]

green synthesis method. The details of the synthesis of HAp powder were described by Choudhury et al. (2018) [10]. Coating slurry was prepared using NPs with binder, dispersant and plasticizer while the slurry casting method was followed to fabricate membrane over cleaned and dried support tubes. After overnight air drying, the coated tubes were sintering at 650 and 750 °C for CuO and HAp coated membrane samples respectively whereas for CuO/HEC membrane, oven drying (100 °C) was used after air drying. Figure 4 represents membrane synthesis steps and Fig. 5 as the firing schedule of HAp-coated membrane.

4.2 Challenges

Inclusion of NPs on ceramic membrane surfaces is a challenging part of membrane fabrication. The binding capacity of NPs with a microstructure of ceramic substrate and its functionality plays a major role in UF membrane formation. Therefore, the stability of NPs enhanced active layer above ceramic support could be the cause of variation in membrane performance in terms of flux, rejections, etc. Chances of penetration of NPs into narrow channels of ceramic support layer are highly possible rather than making a compact top layer. Thus, development NP coated membrane

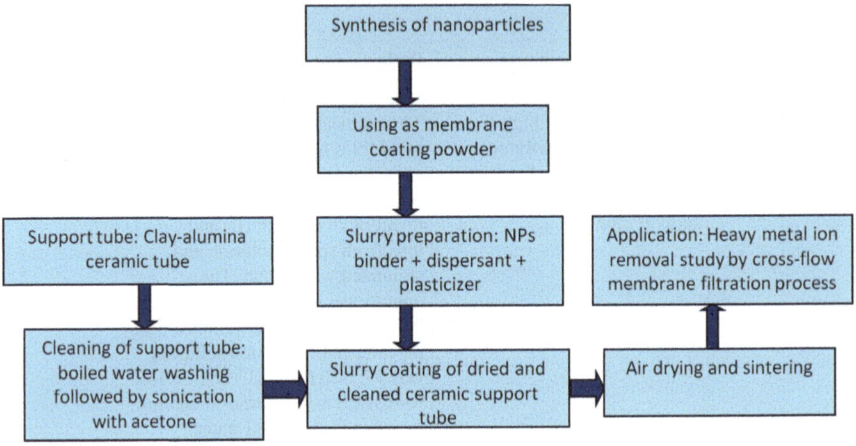

Fig. 4 Schematic of the nanoparticle-enhanced ceramic membrane preparation procedure

Fig. 5 HAp-coated membrane firing schedule (adopted from [12])

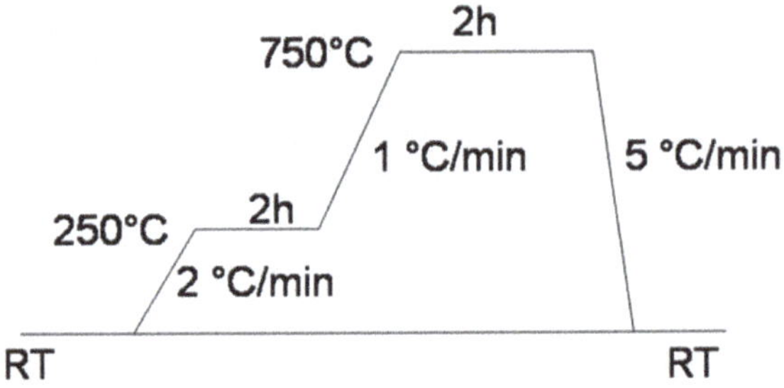

suitable for separation of the toxic metal ions poses a challenge in the formation of stable active layerover porous support layer with higher pore sizes.

Wala et al. (2020) extruded different layers (4–7 layers of smectite/zeolite) based ceramic UF membranes [60]. Decrease in the pore diameter with increase in the number of smectite/zeolite layers was observed up to 6 layers, beyond which the pore size was foud to increase with further increase in the layers.

However, Veronika et al. (2019) demonstrated the possibility of good distribution of NPs in the range of 60–70 nm over the functionalized active layer [66]. They also explored the formation of NPs size depending on the types of catalysts needed for functionalization.

Choudhury et al. (2018) revealed a strong bonding between CuO NPs with porous ceramic support [10]. After calcination, a defect-free homogeneous coating was observed which was confirmed by FESEM images (Fig. 6a, b). To maintain a uniform active layer over the substrate, CuO content was increased from 0.1 wt% to 1wt% in coating slurry composition to restrict the penetration of NPs due to their small sizes into highly porous (microfiltration range) clay-alumina substrate. Thus, a high CuO NPs content active layer was prepared. In another study, Roy

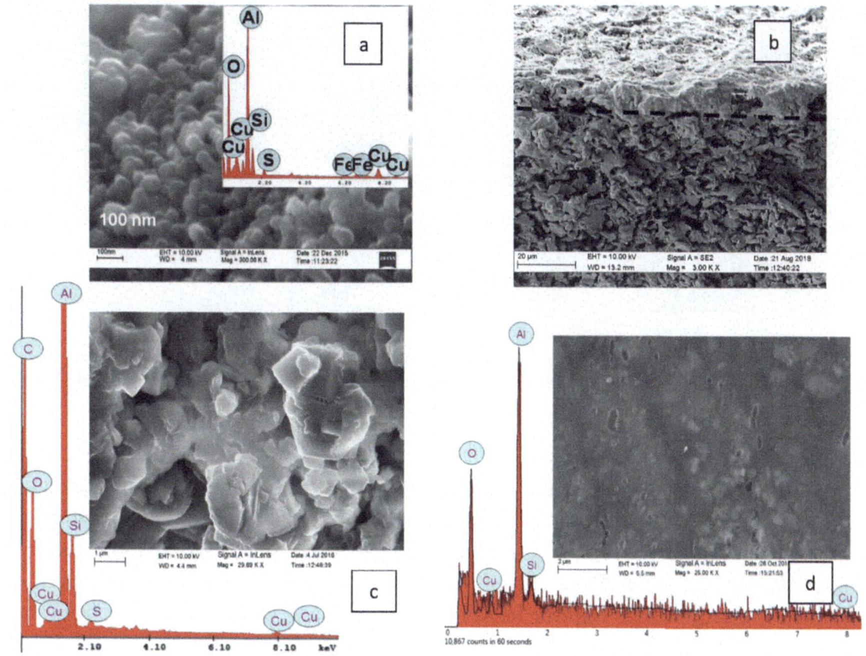

Fig. 6 FESEM images of membrane surface: (**a**) CuO loading 1.0% with EDX in inset and (**b**) cross-sectional membrane thickness); Top surface (**c**) and EDX spectrum (**d**) of CuO/HEC at 40 and 100 °C, respectively (Adopted from [10, 11])

Choudhury et al. (2018) reported the effect of drying temperature (40 and 100 °C) on membrane fabrication [11]. Figure 6c, d depicts the surface morphology of CuO NPs and HEC biopolymer with the corresponding EDX spectrum. It was clearly observed good adhesion between NPs with biopolymer which made a compressed defect-free coating at 100 °C when the thermal shrinkage takes place which controls the morphology as well as membrane thickness. In HAp-coated membrane, a reduced pore structure (2.8 nm from 0.5 to 1 μ range) was formed by dip coating method [12]. Figure 7 depicts a continuous and compact HAp layer formed at top surface of the membrane. Thickness of the membrane layer increases with the coating time although we maintained a fixed coating time of 5 minutes for metal ion separation purpose.

4.3 *Performance Analysis*

Studies on the performance of micellar-enhanced ultrafiltration process (MEUF) was reported for the removal of heavy metal ions from wastewater using commercial ceramic membranes. Francesco et al. (2018) observed affirmative effects

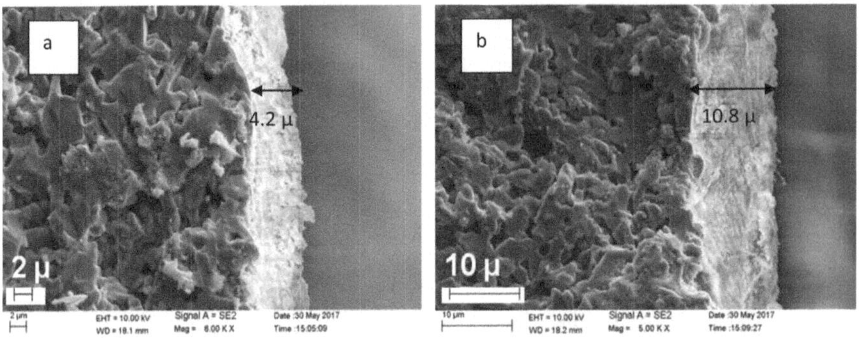

Fig. 7 FESEM images: HAp membrane cross-sectional view, thickness with coating time (**a**) 5 min, (**b**) 10 min (Adopted from [12])

on cobalt removal which could go up to more than 90% using the maximum amount of surfactant (sodium dodecyl sulfate) [68]. Performance analysis also confirmed the feasibility of cost-effectiveness of the MEUF process for metal ion rejection from contaminated water. Pagana et al. (2008) revealed the effectiveness of γ-Al_2O_3 ceramic UF membrane to remove Fe_2O_3 NPs (size: 20–30 nm) completely from the contaminated water [46]. A good rejection (98.1 wt%) of As(V) was found using adsorption-UF technique in series operation along with the regeneration by heating and washing. Also, a satisfactory Cr(III) rejection was achieved by the adsorption-UF process in parallel operation. Na Yin et al. (2016) showed the performance analysis of a ceramic UF system combined with metal–organic framework (MOFs) for adsorptive Pb removal [49]. Sphere-like structure of amino-functionalized MOFs of 100 nm showed good Pb(II) reduction (1795.3 $mg{\cdot}g^{-1}$) from wastewater. In this experiment, a 19-channel commercial zirconia ceramic UF membrane (50 nm pore diameter) was used. Different parameters like pressure, temperature, pH and velocity were varied to observe removal performances as well as regeneration possibilities for potential flux enhancement.

Figure 8 describes the Cr(VI) rejection % by CuO-UF membrane at varied time and pressure range (1 to 3 bar) for a time period of 120 min which showed that cross-flow pressure influences flux and rejection effectively. An appreciable average flux 253 L m^{-2} h^{-1} was achieved at 3 bar TMP which signified the formation of UF range pore structure and established the stability of CuO NPs included in UF membrane as well. The results showed significant Cr(VI) rejection % (88.08%) at 3 bar pressure. Long-term operation and regeneration study will satisfy the membrane performance [10]. It was observed that CuO NPs coated ceramic membrane showed better performance and stability than hydrophilic support tube in terms of flux at varied transmembrane pressure (Fig. 8).

It was observed that the flux was almost stable in the CuO/HEC ceramic composite membrane for each transmembrane pressure [11] and filtration time maintained as 1–5 bar TMP and 120 min. The stable flux also indicated about antifouling

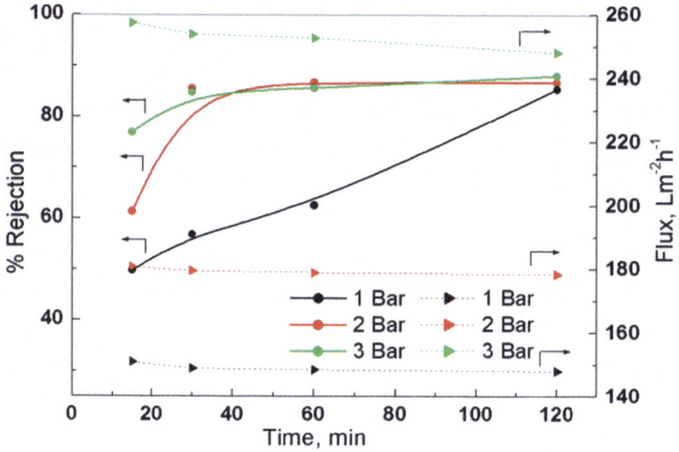

Fig. 8 Cr(VI) rejection and flux of CuO-UF ceramic membrane at varied time and pressure (adopted from [10])

nature of ceramic composite membrane. Further, 5 days-long filtration data reconfirmed the membrane stability. Figure 9 showed highest rejection of 97.14% Pb(II)and 91.44% Cr(VI) in case of composite membrane at 2 bar TMP. Figure 9 depicts that an increase of pressure above 3 bar leads to decrease in the rejection rate due to force penetration of Pb and Cr ions through composite active surface. However, the higher removal rate obtained in this study is attributed to the surface charge of the membrane since the pore size (3 nm) of membrane is much higher than the hydrated radii of Pb^{2+} and CrO_4^{2-} (0.401 nm and 0.375 nm, respectively) which ruled out the scope of separation of the toxic ions based on size exclusion principle. Therefore, it could be assumed that negatively charged CuO membrane surface attract Cr(III) by electrostatic force while repelling anionic Cr(VI).

The applicability of HAp coated membrane was evaluated by studying the separation performance of Pb(II) in spiked water at pH 7.4 (as such) with varied pressure (1–3 bar) and Pb concentration (1, 5 and 10 mg L^{-1}) respectively. Figure 10a depicts an increase in flux with applied transmembrane pressure with decrease in the rejection percentage with increase in the transmembrane pressure but increase in % rejection with increasing time (Fig. 10a, b). Figure 10a revealed a maximum rejection of 99.6% of Pb concentration 5 mg L^{-1} at 1 bar transmembrane pressure.

5 Scope of UF Process

UF process received resurgent importance in wastewater application for a few decades since it ensures a higher removal rate of impurities. Fouling and renovation of UF membrane are another concern which deserves attention to address their

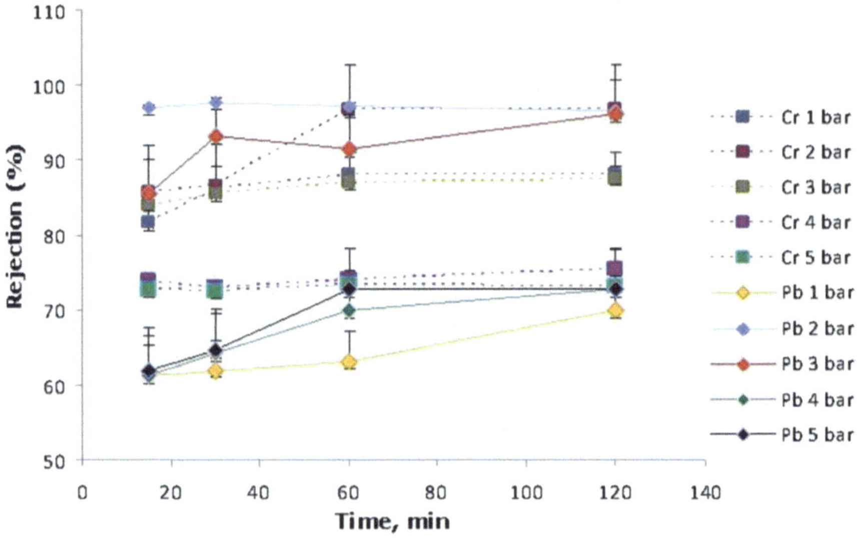

Fig. 9 Cr and Pb ions rejection at varied pressure (adopted from [11])

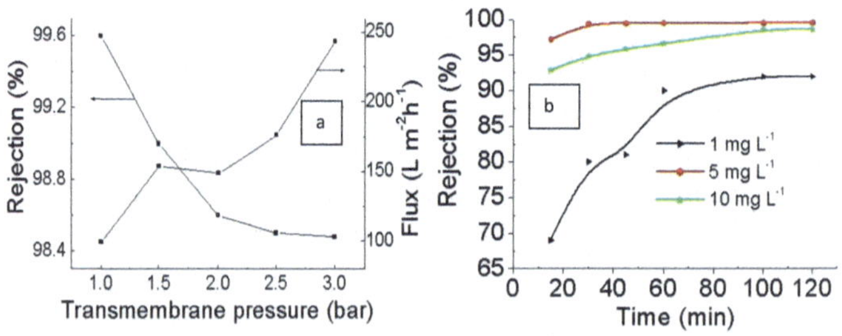

Fig. 10 Rejection % of Pb(II) using HAp coated UF membrane at varied (**a**) pressure and (**b**) operation period (adopted from [12])

feasibility in terms of technical and economic aspects as well. To evaluate the UF process, a systematic analysis of performance data is required to make it more feasible for wastewater treatment [69].

In 2019, the UF market size was estimated at USD 1.13 billion and projected to be USD 3.51 billion by 2027. As per the forecast, the ceramic membrane market is expected to rise to about 15.9% by 2027 [70]. Ceramic membrane's lifespan and antifouling property make it more acceptable in many sectors like food, beverage, chemical and environmental industries. Ceramic membrane is therefore expected to receive more attention, particularly for low-cost treatment process. In Asia Pacific, water and wastewater treatment is receiving resurgent interest with rising economics where ceramic membranes may have a promising market. Acceptance of ceramic UF is also observed with the developing market and incessant membrane manufacturing progress. Herein, UF technology dominates global ceramic membrane market values than other types of membrane [71]. In another study, it was estimated that the global inorganic ceramic UF membrane market could go up significantly by 2026 following steady and sustainable strategies [72].

6 Conclusions

Membrane technology is evolving since 1960 and today it is widely acceptable in water treatment globally. This chapter highlighted the scope of developing ceramic UF membranes for toxic metal ion contaminated wastewater treatment. Different factors limit the membrane performances and enhance energy consumption. Fouling is the major factor which adds energy consumption wherein antifouling ceramic membranes helps to minimize the problem. Incorporation of NPs into membranes' active layers over a ceramic substrate has shown encouraging results. Moreover, the utilization of natural low-cost materials for membrane fabrication helps to brings down the overall membrane cost. Therefore, the removal of toxic heavy metal ions using low-cost ceramic UF membrane demonstrates a new pathway for environment friendly process for wastewater treatment.

Acknowledgements The author (PRC) would like to acknowledge the Department of Rubber and Plastics Technology, Anna University and CSIR-CGCRI, Kolkata, for extending necessary support in conducting the investigations and contributing this book chapter. Also, the author (ARC) would like to thank M/s BGR Energy Ltd., India.

References

1. Özgün H, Sakar H, Ağtaş M (2023) Investigation of pre-treatment techniques to improve membrane performance in real textile wastewater treatment. Int J Environ Sci Technol 20:1539–1550. https://doi.org/10.1007/s13762-022-04034-w
2. Balcik-Canbolat C, Olmez-Hanci T, Sengezer C (2019) A combined treatment approach for dye and sulfate rich textile nanofiltration membrane concentrate. J Water Process Eng 32:100919. https://doi.org/10.1016/j.jwpe.2019.100919
3. Gaikwad MS, Suman SK, Shukla K (2023) A review on recent contributions in the progress of membrane capacitive deionization for desalination and wastewater treatment. Int J Environ Sci Technol. https://doi.org/10.1007/s13762-023-04778-z
4. (2020) Membrane technologies in wastewater treatment: a review. Membranes 10:89. https://doi.org/10.3390/membranes10050089
5. Won Y-J, Lee J, Choi D-C, Chae HR, Kim I, Lee C-H, Kim I-C (2012) Preparation and application of patterned membranes for wastewater treatment. Environ Sci Technol 46:11021–11027. https://doi.org/10.1021/es3020309
6. Padaki M, SuryaMurali R, Abdullah MS, Misdan N, Moslehyani A, Kassim MA, Hilal N, Ismail AF (2015) Membrane technology enhancement in oil-water separation. A review. Desalination 357:197–207. https://doi.org/10.1016/j.desal.2014.11.023

7. Zeming H, Zhiyang L, Qilin G, Lei Z, John W (2019) Ceramic-based membranes for water and wastewater treatment. Colloids Surf A: Physicochem Eng Asp 578:123513. https://doi.org/10.1016/j.colsurfa.2019.05.074
8. Nasim B, Maen MH, Jalel A (2021) Modifying ceramic membranes with in situ grown iron oxide nanoparticles and their use for oily water treatment. J Membr Sci 617:118641. https://doi.org/10.1016/j.memsci.2020.118641
9. Qilin G, How YN, Dan Z, John W (2020) Metal-Organic Frameworks (MOFs)-boosted filtration membrane technology for water sustainability. APL Mater 8:040902. https://doi.org/10.1063/5.0002905
10. Choudhury P, Mondal P, Majumdar S, Saha S, Sahoo GC (2018) Preparation of ceramic ultrafiltration membrane using green synthesized CuO nanoparticles for chromium (VI) removal and optimization by response surface methodology. J Clean Prod 203:511–520. https://doi.org/10.1016/j.jclepro.2018.08.289
11. Roy Choudhury P, Majumdar S, Sahoo GC, Saha S, Mondal P (2018) High pressure ultrafiltration CuO/hydroxyethyl cellulose composite ceramic membrane for separation of Cr(VI) and Pb(II) from contaminated water. Chem Eng J 336:570–578. https://doi.org/10.1016/j.cej.2017.12.062
12. Roy Choudhury P, Majumdar S, Sarkar S, Kundu B, Sahoo GC (2019) Performance investigation of Pb(II) removal by synthesized hydroxyapatite based ceramic ultrafiltration membrane: bench scale study. Chem Eng J 35:510–519. https://doi.org/10.1016/j.cej.2018.07.155
13. Vo TS, Hossain MM, Jeong HM, Kim K (2020) Heavy metal removal applications using adsorptive membranes. Nano Convergence 7:36. https://doi.org/10.1186/s40580-020-00245-4
14. Robert WP, Linda S (1993) Argonne national laboratory. Energy systems division. Publisher Info: Argonne National Lab., IL (United States). Energy Systems Div. Place of Publication: Illinois
15. Shimizu Y, Okuno Y-I, Uryu K, Ohtsubo S, Watanabe A (1996) Filtration characteristics of hollow fiber microfiltration membranes used in membrane bioreactor for domestic wastewater treatment. Water Res 30:2385–2392. https://doi.org/10.1016/0043-1354(96)00153-4
16. Munz G, Gori R, Mori G, Lubello C (2007) Powdered activated carbon and membrane bioreactors (MBRPAC) for tannery wastewater treatment: long term effect on biological and filtration process performances. Desalination 207:349–360. https://doi.org/10.1016/j.desal.2006.08.010
17. Liu X, Jiang B, Yin X, Ma H, Hsiao BS (2020) Highly permeable nanofibrous composite microfiltration membranes for removal of nanoparticles and heavy metal ions. Sep Purif Technol 233:115976. https://doi.org/10.1016/j.seppur.2019.115976
18. Keerthi VV, Balasubramanian N (2013) Removal of heavy metals by hybrid electrocoagulation and microfiltration processes. Environ Technol 34:17–20. https://doi.org/10.1080/09593330.2013.796005
19. Oka PA, Khadem N, Bérubé PR (2017) Operation of passive membrane systems for drinking water treatment. Water Res 115:287–296. https://doi.org/10.1016/j.watres.2017.02.065
20. Othman MHD, Hubadillah SK, Adam MR, Ismail AF, Rahman MA, Jaafar J (2017) Chapter 7—silica-based hollow fiber membrane for water treatment. In: Basile A, Ghasemzadeh K (eds) Current Trends and future developments on (bio-) membranes; silica membranes: preparation, modelling, application, and commercialization, pp 157–180
21. Endre N (2012) Chapter 10—nanofiltration. In: Nagy E (ed) Basic equations of the mass transport through a membrane layer. Elsevier, pp 249–266. https://doi.org/10.1016/B978-0-12-416025-5.00010-7
22. Karki S, Ingole PG (2022) Development of polymer-based new high performance thin-film nanocomposite nanofiltration membranes by vapor phase interfacial polymerization for the removal of heavy metal ions. Chem Eng J 446:137303. https://doi.org/10.1016/j.cej.2022.137303
23. Malamis S, Katsou E, Takopoulos K, Demetriou P, Loizidou M (2012) Assessment of metal removal, biomass activity and RO concentrate treatment in an MBR–RO system. J Hazard Mater 209:1–8. https://doi.org/10.1016/j.jhazmat.2011.10.085

24. Fang X, Li J, Li X, Pan S, Zhang X, Sun X, Shen J, Han W, Wang L (2017) Internal pore decoration with polydopamine nanoparticle on polymeric ultrafiltration membrane for enhanced heavy metal removal. Chem Eng J 314:38–49. https://doi.org/10.1016/j.cej.2016.12.125
25. Pei X, Gan L, Tong Z, Gao H, Meng S, Zhang W, Wang P, Chen Y (2021) Robust cellulose-based composite adsorption membrane for heavy metal removal. J Hazard Mater 406:124746. https://doi.org/10.1016/j.jhazmat.2020.124746
26. Hosseini SM, Alibakhshi H, Jashni E, Parvizian F, Shen JN, Taheri M, Ebrahimi M, Rafiei N (2020) A novel layer-by-layer heterogeneous cation exchange membrane for heavy metal ions removal from water. J Hazard Mater 381:120884. https://doi.org/10.1016/j.jhazmat.2019.120884
27. Moradi G, Zinadini S, Rajabi L (2020) Development of the tetrathioterephthalate filler incorporated PES nanofiltration membrane with efficient heavy metal ions rejection and superior antifouling properties. J Environ Chem Eng 8:104431. https://doi.org/10.1016/j.jece.2020.104431
28. Mondal S, Majumder SK (2020) Fabrication of the polysulfone-based composite ultrafiltration membranes for the adsorptive removal of heavy metal ions from their contaminated aqueous solutions. Chem Eng J 401:126036. https://doi.org/10.1016/j.cej.2020.126036
29. Gong XY, Huang ZH, Zhang H, Liu WL, Ma XH, Xu ZL, Tang CY (2020) Novel high-flux positively charged composite membrane incorporating titanium-based MOFs for heavy metal removal. Chem Eng J 398:125706. https://doi.org/10.1016/j.cej.2020.125706
30. Lam B, Déon S, Crini NM, Crini G, Fievet P (2018) Polymer-enhanced ultrafiltration for heavy metal removal: influence of chitosan and carboxymethyl cellulose on filtration performances. J Clean Prod 171:927–933. https://doi.org/10.1016/j.jclepro.2017.10.090
31. Zhang HL, Cai H, Xia Y, Zhang P, Xiong SW, Gai JG (2020) An l-cystine/l-cysteine impregnated nanofiltration membrane with the superior performance of an anchoring heavy metal in wastewater. RSC Adv 10:3438–3449. https://doi.org/10.1039/C9RA09380J
32. Cao DQ, Wang X, Wang QH, Fang XM, Jin JY, Hao XD, Iritani E, Katagiri N (2020) Removal of heavy metal ions by ultrafiltration with recovery of extracellular polymer substances from excess sludge. J Membr Sci 606:118103. https://doi.org/10.1016/j.memsci.2020.118103
33. Deng S, Liu X, Liao J, Lin H, Liu F (2019) PEI modified multiwalled carbon nanotube as a novel additive in PAN nanofiber membrane for enhanced removal of heavy metal ions. Chem Eng J 375:122086. https://doi.org/10.1016/j.cej.2019.122086
34. Hoekstra AY (2015) Chapter 7—the water footprint of industry. In: Klemeš JJ (ed) Assessing and measuring environmental impact and sustainability. Butterworth-Heinemann, pp 221–254. ISBN 9780127999685. https://doi.org/10.1016/B978-0-12-799968-5.00007-5
35. Hossain MF (2019) Chapter six—water. In: Hossain MF (ed) Sustainable design and build. Butterworth-Heinemann, pp 301–418. ISBN 9780128167229. https://doi.org/10.1016/B978-0-12-816722-9.00006-9
36. Dong Z, Wan F, Jingrui X, Enrico D, Xianfu C, Minghui Q, Yiqun F (2021) One-step engineering of low-cost kaolin/fly ash ceramic membranes for efficient separation of oil-water emulsions. J Membrane Sci 621:118954. https://doi.org/10.1016/j.memsci.2020.118954
37. Hubadillah SK, Othman MHD, Jamalludin MR, Naim R, Kadir SHSA, Puteh MH, Pauzan MAB, Sobri FA (2021) Fabrication and characterisation of superhydrophobic bio-ceramic hollow fibre membranes prepared from cow bone waste. Ceram Int 47:4178–4186. https://doi.org/10.1016/j.ceramint.2020.09.295
38. Ray Chaudhuri A, Behera M (2020) Ceramic membrane modified with rice husk ash for application in microbial fuel cells. Electrochim Acta 363:137261. https://doi.org/10.1016/j.electacta.2020.137261
39. Qian J, Jian Z, Yu M, Surui Y, Ming Z, Zhaoxiang Z, Weihong X (2020) Lower-temperature preparation of SiC ceramic membrane using zeolite residue as sintering aid for oil-in-water separation. J Membr Sci 610:118238. https://doi.org/10.1016/j.memsci.2020.118238
40. Samhari O, Younssi, Saad RB, Murielle L, Patrick M, Achiou MB, Brahim OM (2020) Fabrication of flat ceramic microfiltration membrane from natural kaolinite for seawater pretreatment for desalination and wastewater clarification. Desalin Water Treat 194:59–68. https://doi.org/10.5004/dwt.2020.25859

41. Sheikh M, Pazirofteh M, Dehghani M, Asghari M, Rezakazemi M, Valderrama C, Cortina JL (2020) Application of ZnO nanostructures in ceramic and polymeric membranes for water and wastewater technologies: a review. Chem Eng J 391:123475. https://doi.org/10.1016/j.cej.2019.123475
42. Das D, Nijhuma K, Marsola GA, Daniel GPF, Daniel MIM (2020) Recycling of coal fly ash for fabrication of elongated mullite rod bonded porous SiC ceramic membrane and its application in filtration. J Eur Ceram 40:2163–2172. https://doi.org/10.1016/j.jeurceramsoc.2020.01.034
43. Liu G, Yan C, Jin H (2022) Colloidal processing of complex-shaped ZrB_2-based ultra-high-temperature ceramics: progress and prospects. Materials (Basel) 15:2886. https://doi.org/10.3390/ma15082886
44. Wolfgang MS, Nelson SB, Lennart B (2000) Novel powder-processing methods for advanced ceramics. 83:1557–1574. https://doi.org/10.1111/j.1151-2916.2000.tb01432.x
45. Fane AG, Tang CY, Wang R (2011) Chapter 4.11—membrane technology for water: microfiltration, ultrafiltration, nanofiltration, and reverse osmosis. In: Wilderer P (ed) Treatise on water science. Elsevier, pp 301–335. ISBN 9780444531995. https://doi.org/10.1016/B978-0-444-53199-5.00091-9
46. Saffaj N, Loukili H, Alami Younssi S, Albizane A, Bouhria M, Persin M, Larbot A (2004) Filtration of solution containing heavy metals and dyes by means of ultrafiltration membranes deposited on support made of Moroccan clay. Desalination 168:301–306. https://doi.org/10.1016/j.desal.2004.07.013
47. Ferella F, Prisciandaro M, Michelis ID, Veglio F (2007) Removal of heavy metals by surfactant-enhanced ultrafiltration from wastewaters. Desalination 207:125–133. https://doi.org/10.1016/j.desal.2006.07.007
48. Pagana AE, Sklari SD, Kikkinides ES, Zaspalis VT (2008) Microporous ceramic membrane technology for the removal of arsenic and chromium ions from contaminated water. Micropor Mesopor Mat 110:150–156. https://doi.org/10.1016/j.micromeso.2007.10.013
49. Llanos J, Williams PM, Cheng S, Rogers D, Wright C, Pérez Á, Cañizares P (2010) Characterization of a ceramic ultrafiltration membrane in different operational states after its use in a heavy-metal ion removal process. Water Res 44:3522–3530. https://doi.org/10.1016/j.watres.2010.03.036
50. Arkadiusz N, Arkadiusz D, Filip BH, Agnieszka T (2015) The influence of pH and BSA on the retention of selected heavy metals in the nanofiltration process using ceramic membrane. Desalination 369:62–67. https://doi.org/10.1016/j.desal.2015.04.019
51. Na Y, Ke W, Lizhen W, Zongqun L (2016) Amino-functionalized MOFs combining ceramic membrane ultrafiltration for Pb(II) removal, Author links open overlay panel. Chem Eng J 306:619–628. https://doi.org/10.1016/j.cej.2016.07.064
52. Siti KH, Othman MHD, Harun Z, Ismail AF, Mukhilis AR, Juhana J (2016) A novel green ceramic hollow fiber membrane (CHFM) derived from rice husk ash as combined adsorbent-separator for efficient heavy metals removal. Ceram Int 43:4716–4720. https://doi.org/10.1016/j.ceramint.2016.12.122
53. Hubadillah SK, Othman MHD, Gani P, Sunar NM, Tai ZS, Koo KN, Pauzan MAB, Ismail NJ, Zahari SMSNS (2020) Integrated green membrane distillation-microalgae bioremediation for arsenic removal from Pengorak River Kuantan, Malaysia. Chem Eng Process 153:107996. https://doi.org/10.1016/j.cep.2020.107996
54. Jiayu Z, Mingwei Y, Guangchao S, Kaiqi L (2021) Simultaneous removal of Cu(II), Cd(II), Cr(VI), and rhodamine B in wastewater using TiO_2 nanofibers membrane loaded on porous fly ash ceramic support. Sep Purif Technol 272:118888. https://doi.org/10.1016/j.seppur.2021.118888
55. Bhowmick K, Roy S, Bhowmick S, Jain R, Majumdar S, Sahoo GC, Mondal P (2022) Fabrication of robust ceramic supported polymeric composite nanofiltration membrane for heavy metal contaminated wastewater treatment. J Water Process Eng 50:103287. https://doi.org/10.1016/j.jwpe.2022.103287

56. Zhou W, Wu P, Zhang L, Yao S, Zhu D, Cai Y (2022) Layer-by-layer assembly of nanocomposite interlayers on a kaolin substrate for enhancing membrane performance of Pb(II) and Cd(II) removal. Sci Total Environ 820:153149. https://doi.org/10.1016/j.scitotenv.2022.153149
57. Agata MM, Piotr TM, Arkadiusz NE, Jacek RZ, Waldemar. (2020) Effect of pH on total volume membrane charge density in the nanofiltration of aqueous solutions of nitrate salts of heavy metals. Membranes 10:235. https://doi.org/10.3390/membranes10090235
58. Suresh K, Pugazhenthi G (2017) Cross flow microfiltration of oil-wateremulsions using clay based ceramic membrane support and TiO_2 composite membrane. Egypt J Pet 26:679–694. https://doi.org/10.1016/j.ejpe.2016.10.007
59. Bouazizi A, Breida M, Achiou B, Ouammou M, Calvo JI, Aaddane A, Younssi SA (2017) Removal of dyes by a new nano-TiO_2 ultrafiltration membrane deposited on low-cost support prepared fromnatural Moroccan bentonite. Appl Clay Sci 149:127–135. https://doi.org/10.1016/j.clay.2017.08.019
60. Abdallah O, Nouha T, Samia MC, Benjamin C, Majumdar S, Sarkar S, Sahoo GC, Ben Amar R (2017) Tubular ultrafiltration ceramic membrane based on titania nanoparticles immobilized on macroporous clay-alumina support: elaboration, characterization and application to dye removal. Sep Purif Technol 188:126–133. https://doi.org/10.1016/j.seppur.2017.07.005
61. Wala A, Wiem H, Hajer A, Abdallah O, Sabeur K, Amane J, Chakraborty S, Curcio S, Ben Amar R (2018) Developing of titania-smectite nanocomposites UF membrane over zeolite based ceramic support. Appl Clay Sci 155:20–29. https://doi.org/10.1016/j.clay.2017.12.035
62. Wala A, Hajer A, Mouna K, Joelle D, Michael OD, Ben Amar R (2020) Synthesis and characterization of clay-based ultrafiltration membranes supported on natural zeolite for removal of heavy metals from wastewater. Environ Technol Innov 18:100794. https://doi.org/10.1016/j.eti.2020.100794
63. Hosik P, Yohan K, Byungryul A, Heechul C (2012) Characterization of natural organic matter treated by iron oxide nanoparticle incorporated ceramic membrane-ozonation process. Water Res 46:5861–5870. https://doi.org/10.1016/j.watres.2012.07.039
64. Karnik BS, Davies SH, Baumann MJ, Masten SJ (2005) Fabrication of catalytic membranes for the treatment of drinking water using combined ozonation and ultrafiltration. Environ Sci Technol 39:7656–7766. https://doi.org/10.1021/es05039381
65. Harman BI, Koseoglu HY, N.O, Beyhan M, Kitis M.(2010) The use of iron oxide-coated ceramic membranes in removing natural organic matter and phenol from waters. Desalination 261:27–33. https://doi.org/10.1016/j.desal.2010.05.052
66. Mouna K, Sabeur K, Salwa A, Ben Amar R (2011) Study of ceramic ultrafiltration membrane support based on phosphate industry subproduct: application for the cuttlefish conditioning effluents treatment. Ceram Int 37:3617–3625. https://doi.org/10.1016/j.ceramint.2011.06.020
67. Shuang L, Chuncheng W, Peng W, Peiling G, Lijuan Z, Guangwu W (2020) Zirconia ultrafiltration membranes on silicon carbide substrate: microstructure and water flux. J Eur Ceram 40:4290–4298. https://doi.org/10.1016/j.jeurceramsoc.2020.04.020
68. Veronika VT, Galyna IN, Inna VM (2019) Modification of ceramic membranes by silica nanoparticles with Thiourea functions. J Nanomater 4:1–8. https://doi.org/10.1155/2019/2534934
69. Sahar F, Zerafat MM, Mohammadi Y, Asadnia M (2022) Fabrication of tubular ceramic membranes as low-cost adsorbent using natural clay for heavy metals removal. Cleaner Eng Technol 10:100550. https://doi.org/10.1016/j.clet.2022.100550
70. Francesco T, Valentina I, Ida DM, Francesco V, Giuseppe MC, Marina P (2018). Environ Eng Sci 35:493–500. https://doi.org/10.1089/ees.2017.0211
71. Rehman ZU, Amjad H, Khan SJ, Yasmeen M, Khan AA, Khanzada (2023) N.K. Performance evaluation of UF membranes derived from recycled RO membrane, a step towards circular economy in desalination. Membranes 13:628. https://doi.org/10.3390/membranes13070628
72. Ultrafiltration market size, share and analysis by type (polymeric and ceramic), by module (hollow fiber, plate and frame, tubular and others), and by applications (municipal treatment, industrial treatment, and others), forecasts to 2027, ID: RND_001715, July, 2020. https://www.reportsanddata.com/report-detail/ultrafiltration-market

73. Ceramic membrane market by material (titania, alumina, zirconium oxide), by application (water and wastewater technology, pharmaceuticals, food and beverage, chemical processing, biotechnology), by technology (ultrafiltration, microfiltration, nanofiltration), by region - global trends & forecasts to 2020, ceramic membrane market worth $5.1 billion by 2020. https://www.marketsandmarkets.com/market-reports/ceramic-membranes-market-246576534.html
74. Inorganic ceramic ultrafiltration membrane market (2020–2026), September 2020

Chapter 12
Ecotoxicological and Reusability Aspects of Organic-Pollutant-Bearing Effluents Treated Using Ceramic Membrane Coupled Bioreactors

P. Banerjee, S. Ghosh, and Sibdas Bandyopadhyay

1 Introduction

A dearth of planning and monitoring in the expansion of civilization and industrialization has resulted in the rapid exhaustion of water resources. Therefore, it has become necessary to reclaim water from different types of effluents [14]. Species all over this planet are suffering from the detrimental impacts of being exposed to impure water. Chemical and microbial toxins present in these effluents can be equally threatening for humans and other species exposed to the same [23].

Over the recent decades, the appearance of different emerging organic micropollutants in waterbodies receiving effluent discharge has become an environmental issue of increasing concern [79]. Hence, it has become extremely essential to treat these effluents efficiently before releasing them into the environment or reusing the effluents in domestic and industrial processes. Several studies have reported the reuse of such treated wastewater for tanning [18], dyeing [67], agriculture [80], aquaculture [99], etc.

P. Banerjee (✉)
Department of Environmental Studies, Centre for Distance and Online Education, Rabindra Bharati University, Rabindra Bhavan, EE Block, Sector II, Salt Lake City, Kolkata 700091, India
e-mail: prya_bnrje@yahoo.com

S. Ghosh
Membrane and Separation Technology Division, CSIR-Central Glass and Ceramic Research Institute, 196, Raja S.C. Mullick Road, Kolkata 700032, India

S. Bandyopadhyay
Department of Environmental Studies, Siksha-Bhavana, Visva-Bharati, Santiniketan 731235, India

S. Bandyopadhyay (ed.), *Application of Low Cost Ceramic Membranes in Wastewater Treatment*, Materials Horizons: From Nature to Nanomaterials,
https://doi.org/10.1007/978-981-97-5842-5_12

This chapter discusses the efficiency and applications of ceramic membranes to evaluate their efficiency for municipal and industrial wastewater treatment. The suitability of novel catalytic membranes prepared using ceramic supports for wide-scale application has also been reviewed in this chapter. Moreover, significant emphasis has been laid upon ecotoxicity analysis of untreated and treated effluents.

2 Literature on Micropollutants, Ecotoxicological and Reusability Aspects

Water is a fundamental essential for all living organisms. However, access to decent-quality water is a major challenge faced by developing nations [23]. Many species across the globe are affected by exposure or consumption of contaminated water. Both chemical and microbial pollutants contaminating water bodies exert detrimental impacts on humans and aquatic life alike [23]. The pharmaceuticals and personal care products (PPCPs) most frequently consumed in our daily lives include anionic and non-ionic surfactants, bleaches, dyes, fragrances, soaps, solvents, sunscreen agents, etc. Reactive compounds present in these products pose severe environmental concerns. Triclosan, widely used in PPCPs, reportedly elicits toxicological effects in a few species of algae and aquatic invertebrates even in concentrations lower than 0.4 ppm [35]. Triclosan and surfactants mostly appear in effluents discharged from domestic sectors and PPCP manufacturing industries. These pollutants have been found in different water and soil samples as well as biological samples (like fish tissues, plant tissues, human milk, etc.). Traces of triclosan were also detected in soil samples collected from agricultural fields where biosolids obtained from wastewater treatment plants were used as fertilizer [25].

Hence, sustainable concepts are required to develop efficient strategies for effluent treatment [36]. These novel processes of effluent treatment are expected to be technologically, environmentally and ecologically benign [90, 91]. To uphold the reusability aspects of municipal and industrial wastewater, the toxicity effects of treated wastewater assume great importance and need to be assessed by toxicity studies with a test plant or a biotic species. In a recent review a detailed discussion have been provided on the occurrence, adverse impacts and fate of micropollutants present in aquatic environments [17].

2.1 Wastewater Treatment for Removal of Emerging Micropollutants

Conventional wastewater treatment plants (WWTPs) efficiently reduce the chemical oxygen demand (COD) and biochemical oxygen demand (BOD) of effluents. However, these WWTPs are not efficient for the removal of emerging micropollutants

[5, 56]. In comparison with the technologies like adsorption, ozonation, photocatalytic reactions, etc., pressure driven membrane filtration processes such as nanofiltration (NF) and reverse osmosis (RO) have been reportedly more efficient for treatment of effluents bearing micropollutants [56]. However, some of these processes have their own limitations as well. Adsorption-based processes are limited by the periodic reuse and regeneration of the concerned adsorbents. Although microfiltration (MF) and ultrafiltration (UF) processes are efficient for removal of suspended solids, they are not used for removal soluble impurities. Membranes used for NF and RO are susceptible to fouling and require higher hydraulic pressure for a consistent flux yield [56]. Conventional physicochemical methods like oxidation, ozonation, photocatalysis and adsorption have also been investigated for removal of persistent organic pollutants from effluents [10]. However, these methods have led to partial degradation of the effluents or formation of toxic intermediates like chlorinated phenols, dioxins and trihalomethanes. Biodegradation has been reported by far as the most efficient process for the treatment of such pollutants [10].

Conventional physicochemical methods (like oxidation, ozonation, photocatalysis and adsorption) have also been investigated in previous studies for the removal of persistent organic pollutants from effluents [10]. However, these methods have led to partial degradation of the effluents or formation of toxic intermediates like chlorinated phenols, dioxins and trihalomethanes. Biodegradation has been reported by far as the most efficient process for the treatment of such pollutants [10].

2.2 *Evaluation of Phytotoxic Effect*

Effluents bear significant loads of nutrients and beneficial organic matter. This may improve both crop productivity and soil fertility if utilized properly. Certain countries have specific regulations for utilizing effluents for agricultural purposes. Since effluents may be considered as a potential source for water reclamation and subsequent use for agricultural purposes, several studies have investigated the impact of water pollutants on different plant species at both seed and seedling stages [38]. A recent study investigated the impact of phenolic compounds present in olive mill wastewater on the germination of maize and tomato seeds. Effects exerted by raw and treated wastewater on seed germination were analysed on a comparative scale [38]. Results obtained in this study suggested that integrated effluent treatment processes successfully removed all toxic pollutants and led to a significant reduction of phytotoxicity. According to the authors, this process could be used widely without any long-term detrimental effects on both crop yield and soil quality. In a similar study, Bhattacharya et al. [18] examined the phytotoxicity exerted by Cr^{+6} on germination and enzyme activities of *Vigna radiata* seeds. Biosorbents prepared from the fruit peel of *Trewia nudiflora* were used to treat chromium-bearing effluents collected from a tannery industry. A significant reduction of phytotoxicity ($P < 0.01$) was noted in the case of seeds exposed to biosorbent-treated effluents in comparison to those exposed to untreated ones [18]. Such studies indicate that efficiently treated effluents may serve as a safer source of water for agricultural activities.

2.3 Pressure-Driven Membranes for Wastewater Reclamation

Previous studies have reported the use of pressure-driven membranes for desalination as well as wastewater treatment and reclamation [7]. Membranes can efficiently eliminate suspended impurities and pathogens via processes like adsorption, electrostatic repulsion and size exclusion [4, 6]. In the previous decade, the application of membrane separation processes like forward and reverse osmosis and membrane bioreactors has been widely investigated for water reclamation [74]. Of all these processes, membrane filtration and related advancements have always received more attention for water reclamation and effluent treatment applications [74]. Significant enhancements in module design and cost-effectiveness have resulted in an increased preference for membrane-based treatment of water and wastewater in domestic and industrial sectors. According to a recent study, global returns of ultrafiltration (UF) membrane only reportedly amounted to 2.14 billion USD in 2023 from 0.95 billion USD in 2017 [2].

Polymeric membranes have been used extensively for water and wastewater treatment. Since the 1990s, these membranes have been widely investigated for municipal effluent treatment [7]. Nevertheless, the hydrophobicity of polymeric membranes makes the process unstable in the presence of chemicals and susceptible to fouling. This affects both membrane performance and long-term susceptibility. Owing to these limitations, studies have increasingly focused on ceramic membranes for water treatment in recent decades, as evident from recent publications. In comparison to the polymeric membranes commonly used for effluent treatment, ceramic membranes are reportedly more resistant to chemically, thermally and mechanically induced stresses and therefore considered more suitable for application in extreme environmental circumstances [10, 64, 84]. Ceramic membranes offer elevated permeation flux over longer periods with rigorous regular cleaning on a regular basis [10].

2.4 Separation Process and Ceramic Membrane Characteristics/Configuration

According to the composition and separation application, membranes have been categorised into polymeric, ceramic, liquid and ion-exchange types [84]. Ceramic membranes commonly used for water treatment are usually made up of an asymmetric structure containing several layers including the top separating layer (selective in nature) with one or more intermediate layers coated over a strong porous base that render support [7, 34]. Moreover, on the basis of the molecular weight cut-off of these membranes, the supporting layers may be prepared using one or multiple composite materials [7]. Both types of ceramic membranes have been investigated for water and wastewater treatment [42]. Composite ceramic membranes are prepared through complex processes to achieve desired selectivity and properties [32, 52]. Separation layers of ceramic membranes are usually synthesised using oxides of Al, Si, Ti and Zr [45]. On the basis of the selected support, the ceramic membranes may be cylindrical (hollow-fibre or multichannel tubular) or flat (flat-sheet) having different packing densities [52]. The phase inversion/sintering technique requires heating at only one stage during the fabrication process resulting an assymetric structure due to which it is prefered over other techniques [29, 30, 42] for certain appliactions. The ceramic hollow fiber membranes prepared in the phase inversion technique possess finger like channels with a dense layer (skin layer) responsible for lowering mass transfer resistance through the process of separation [52, 53].

2.5 Low-Cost Ceramic Membranes in Wastewater Treatment

Asymmetric ceramic membranes possess a multi-layered structure having progressively reducing pore dimensions [98]. Generally, thin layers of membrane are deposited over extremely pure alumina supports that are sintered at temperatures higher than 1500 °C [66]. The manufacturing cost of such porous alumina supports are therefore much higher which restricts their wider application. Therefore cost-effective clay-alumina supports have been developed which significantly affect adhesion, behaviour and resistance of the membranes deposited on it [69].

In recent years, several reviews on ceramic membrane [34, 44, 55, 62] have discussed about the methods of membrane synthesis and their efficiency in industrial wastewater treatment. These reviews have also focussed on the development of nanocomposite membranes having enhanced antifouling properties. Nevertheless, ecotoxicity exerted by raw and treated effluents has only been reviewed in a handful of reported studies. Hence, this chapter focuses on reviewing the efficiency and commercial implementation of ceramic membranes to evaluate their scope of application in municipal and industrial wastewater treatment. The prospect of using low-cost ceramic membranes for effluent treatment has been addressed with a significant emphasis on the MBR process along with ecotoxicity analysis of untreated and treated effluents.

3 Membrane Bioreactors (MBR)

Urban wastewater is generally treated using conventional activated sludge (CAS) process (CASPs). This process involves pollutant degradation by heterotrophic bacteria in aerated bioreactors. However, this process is incapable of efficient removal of suspended solids (SSs). An optimal sludge concentration of 5 g L^{-1} requires an aerated bioreactor of large size [24]. MBR is an advancement of this process. In an MBR, the secondary clarifier is replaced by a membrane module for the filtration of biodegraded effluent [47].

In comparison to conventional activated sludge (AS) process reported in previous studies, MBRs having integrated AS-MF processes have been largely accepted as a more efficient process for effluent treatment. MBRs have a lower footprint, greater duration of sludge retention (SRT), lower production of sludge and enhanced quality of effluents. Moreover, MBRs also demonstrate enhanced removal of COD, nutrients (N and P), turbidity, etc. [63]. According to Drioli et al. [33], one of the major advantages of an MBR over CAS is that the former can operate even in the presence of extremely high concentrations of solids in the bioreactor. Owing to limitations on settling, mixed liquor suspended solids (MLSS) concentration in CAS systems are found to be in the range of 2–3 kg L^{-1}, while that in MBR systems may be as high as 8–12 kg L^{-1}. A comparative account of these two processes has been given in Table 1. The cost of operation and maintenance and footprint of both these processes have been compared in Fig. 1 (A and B, respectively).

Furthermore, the integration of MBR with other advanced processes like RO and processes of advanced oxidation reportedly increases the efficiency of the process and renders the treated water suitable for consumption [54, 71]. MBRs may be implemented for the treatment of effluents generated by industries producing automotive, chemicals, cosmetics, dairy and other food products, petrochemicals, pharmaceuticals, pulp and paper, textiles, etc. [63].

The advantages offered by the MBR technology are enlisted as follows [74, 81].

Table 1 Comparative account of CAS and MBR processes for effluent treatment

Key issues (MBR over CAS)	Specific advantages of MBR as compared to CAS systems
Improve water quality	• Meet stringent effluent requirements • Filter out nearly all solids
Allows wastewater reuse	• As part of the treatment scheme, provides water for potable reuse • Reduces wastewater discharge fees and freshwater costs • Provides water for non-potable applications where fresh water is in short supply
Lower capital cost since bacteria concentration is higher	• Clarifier is not needed • Biological step can be scaled down in volume since bacteria concentration is higher
Reduces plant space requirements	• Footprint is up to 50% smaller than conventional plant • Allows for expanded capacity within existing buildings

Reproduced with permission from Attiogbe [8] © Academic Journals

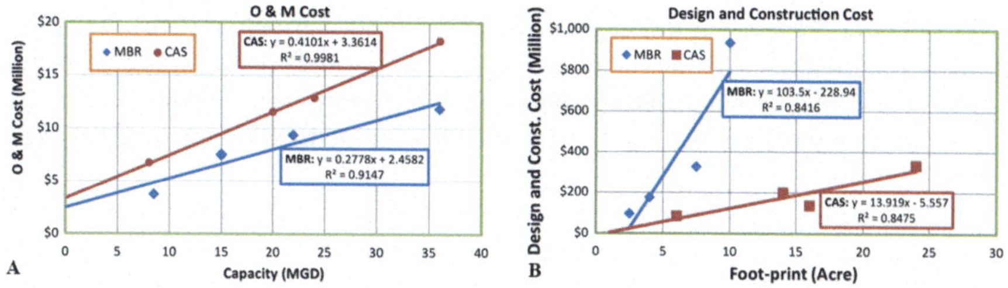

Fig. 1 A preliminary comparative analysis of variations in operation and maintenance (O&M) cost and plant footprint (A and B, respectively) of CAS- and MBR-based wastewater treatment systems [Reproduced with permission from Karim and Mark [48] © 2017 Karim MA, et al.]

- Nearly complete retention of suspended matter and dissolved solids in the bioreactor resulting in superior effluent quality that meets stringent discharge norms and facilitates direct reuse of treated water.
- Nearly complete retention of all microorganisms facilitates increased concentration of sludge and efficient disinfection of effluents as well as subsequent reuse in domestic and industrial sectors.
- Facilitates efficient separation of SRT and hydraulic retention time (HRT). high concentrations of sludge maintained in an MBR ensure effective treatment of high-strength wastewater.
- Removal of the secondary clarifier ensuring a significant reduction in the overall size of the treatment plant.
- Prolonged contact between organic pollutants and activated sludge facilitates the efficient removal of biorecalcitrant pollutants.
- Membranes coupled with bioreactors are able to tolerate extreme temperatures and pH and can be conveniently cleaned in situ for avoiding fouling.

Owing to these advantages, the MBR process is considered an effective option for the removal of a wide range of persistent organic pollutants from different streams of effluents [9].

3.1 Types of Membrane Bioreactors (MBRs)

The term MBR generically refers to aerobic MBRs (AeMBR) which primarily focus on water reclamation instead of recovery of nutrients or energy from effluents [58]. Moreover, AeMBRs are energy intensive as they consume large amounts of electricity for aeration required to sustain the CAS process [50]. Recent initiatives to make the MBR process more energy efficient have inspired the introduction of anaerobic MBRs (AnMBRs) [58]. Anaerobic digestion involved maximizes energy recovery from effluents by converting organic matter present in the effluents in to CH_4-rich biogas through biological processes. Anaerobic digestion also transform the nutrients (nitrogen and phosphorus) in to more chemically available forms which, in turn, facilitate better recovery or direct usage of effluents for agricultural purposes [58]. AnMBR reportedly yields larger quantities of biogas and treated water of better quality than conventional anaerobic digestion [89]. Moreover, this biogas may be used for the generation of electricity with reduction of the footprint of AnMBR even more.

AnMBR has a significantly lower energy requirement than an AeMBR, AnMBRs are also more efficient than AeMBRs in terms of pollutant removal efficiency and volume throughput. AnMBR usually operates under conditions of high biomass concentration, prolonged SRTs and HRTs. It is therefore more appropriate for the treatment of high-strength effluents [58]. A previous study, [41] reported that an AnMBR was more efficient in the removal of sulfamethoxazole and trimethoprim in comparison to an aerobic sequential batch reactor with a subsequent decline in associated antibiotic resistance genes. AnMBR also has lower sludge production in comparison to AeMBR [89].

3.2 MBR Configurations

The two established MBR configurations are the side stream (ssMBR) and the submerged (sbMBR). The former configuration ssMBRs are largely implemented by small-scale industries where the cost per unit time for energy consumption is low [46]. The sbMBR configuration is more widely considered for moderate to large-scale municipal wastewater treatment due to low membrane and operational costs per volume of effluent treated [46]. The air-lift side-stream MBR (A-LssMBR) is at times implemented at the same scale as the conventional ssMBR (Fig. 2).

Both ss and sb MBRs primarily consist of a reactor for the biodegradation of pollutants and a membrane module for the filtration of treated effluents. In the case of sbMBR, the outer skin membranes are located inside the bioreactor (as shown in

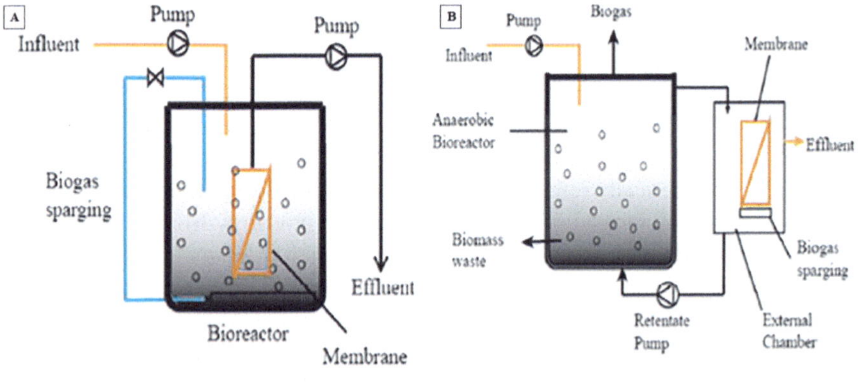

Fig. 2 Schematic representation of MBRs (**a** submerged and **b** side-stream MBRs) [Reproduced with permission from [3]]

Fig. 1a). The driving force across the membrane is generated by the application of pressure on the bioreactor or the creation of a negative pressure on the permeate side [73]. In this configuration, membranes are cleaned by frequent back-washing with permeate and occasional backwashing with chemicals. A diffuser located directly underneath the membrane module facilitates scouring of the membrane surface [100]. Aeration and mixing take place simultaneously in the bioreactor. Anoxic or anaerobic sectors are often included for achieving efficient nutrient removal at the same time [77]. In side-stream MBRs, the mixed liquor is recirculated through a membrane module located outside the bioreactor [57] as shown in Fig. 1b. Both outer-skin and inner-skin membranes may be used in this process [16]. This process is driven by cross-flow velocities formed along the surface of the membranes [68].

3.3 Ceramic Membrane-Based Bioreactors (CMBRs) for Effluent Treatment

3.3.1 Side Stream CMBR for Treatment of Textile Effluent Using Low-Cost Ceramic UF Membrane Over Capillary and Tubular Supports

Sarkar et al. [85] reported the development of novel low-cost ceramic capillary and monochannel supports for the synthesis of UF membranes. The membranes prepared with 3.1/1.9 mm (OD/ID) and 5.5 nm on 8/6 mm (OD/ID) supports having a pore diameter distribution in the UF range (centered at 7.0 nm) were subsequently used in a side stream MBR for the treatment of synthetic and real effluents which showed 98% removal of reactive yellow dye (having an initial concentration of 50 mg L^{-1}) and 95.6% colour removal in case of real effluent as a result of the combined treatment of UF and bioreactor.

In another study by [9], cost-effective single-layer γ-alumina UF membranes (shown in Fig. 3) were used in a pilot-scale cross-flow UF MBR (UMBR) for the treatment of textile effluent. A bacterial consortium composed of *E. coli*, *Bacillus* sp., *Pseudomonas* sp. and *Shigella* sp. was applied in the bioreactor. Results revealed 98.62% colour and 98.66% COD removal from treated effluents. Toxic impacts exerted by both raw and treated wastewater were determined in terms of response to oxidative stress and genotoxic effects in *Heteropneustes fossilis* (catfish). Results indicated that the raw effluent could cause a significant detrimental impact on aquatic life if discharged untreated. However, the treated permeates exerted no significant toxicity and were considered safe for discharge. The treated permeates could also be used for industrial purposes.

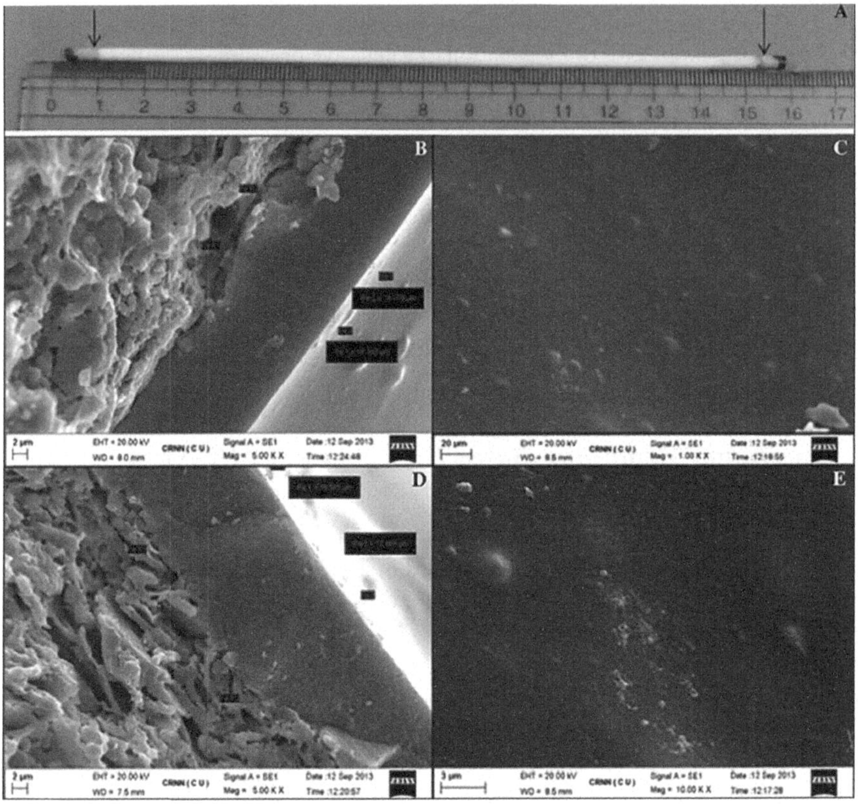

Fig. 3 Features of a UF capillary ceramic membrane over macroporous 3.2/1.87 mm (OD/ID) capillary clay-alumina support (**a**—Picture of a 160 mm-long ceramic capillary membrane with; arrows indicate the area of filtration. **b**, **c**—SEM images of cross-section of the membrane revealing its thickness and *top view* of the membrane before filtration of biotreated effluent respectively. **d**, **e**—SEM images of cross-section of the membrane revealing its thickness and top view of the same, respectively, after filtration of biotreated effluent and a successive backwash) [Reproduced with permission from [9] © 2023 Springer Nature]

3.3.2 Integrated CMBR for Micropollutant Removal

Radjenović et al. [75] compared the effluent treatment efficiency of an integrated CAS-MBR system in terms of removal of 26 micropollutants which demonstrated more efficient removal of 20 of these pollutants in comparison with the CAS system. In a similar study, Suárez et al. [94] reported that SRT was a significant factor in the removal of 16 micropollutants. Moreover, Maeng et al. [59] demonstrated that in case of a few pollutants, biotransformation was positively correlated to the SRT. In a recent study, Li et al. [56] reported the removal of 29 different micropollutants (shown in Fig. 4) from a WWTP-treated effluent using a CMBR operated on a pilot scale over a period of 15 months. The silicon carbide (SiC) ceramic membrane used

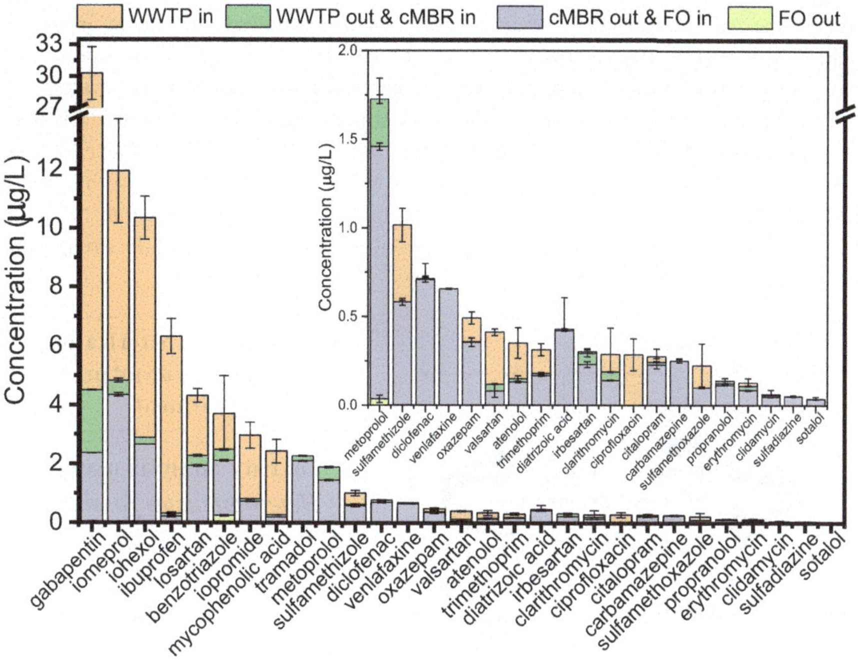

Fig. 4 Micropollutants' concentrations in the WWTP influent, the WWTP effluent (the cMBR inflow), the cMBR outflow (the forward osmosis (FO) inlet water), and the aquaporin-based FO outlet water [Reproduced with permission from [56] © 2021 The Author(s)]

in this study exhibited superior antifouling properties. Membrane fouling severely compromises MBR efficiency and may be controlled via cross-flow mode of operations using ceramic membranes. COD and micropollutant removal recorded in this study established the CMBR coupled with a FO/RO membrane as an efficient process for COD and micropollutant removal from wastewater. This setup also efficiently removed biotransformation products thereby rendering the treated water suitable for direct reuse. However, this integrated system requires both process and energy optimization prior to its implementation on a commercial scale. It is also important to focus on balancing the BOD loading as this will facilitate biomass growth while maintaining biomass activity and selectivity towards recalcitrant compounds [56].

3.3.3 Treatment of PPCP-Rich Wastewater Using Side-Stream and Submerged CMBR Coupled with Low-Cost Ceramic UF Membrane

Owing to their extensive usage, PPCPs appearing in domestic sewage are being recognized as a major environmental concern. These emerging persistent organic pollutants (POPs) are being largely discharged as cosmetic effluents. Therefore, it has become crucial to develop efficient strategies of treatment for the same. Banerjee et al. [10] reported the treatment of cosmetic effluents collected from salons and spas using a bacterial consortium. This consortium was isolated from activated sludge collected from a common effluent treatment plant. The biotreated effluent was filtered using low-cost ceramic UF membranes attached to the bioreactor in both side-stream and submerged modes. The MBR process reported in this study efficiently removed approximately 99% of colour, triclosan and surfactant from the effluent being treated. Comparative accounts of effluent treated by side-stream and

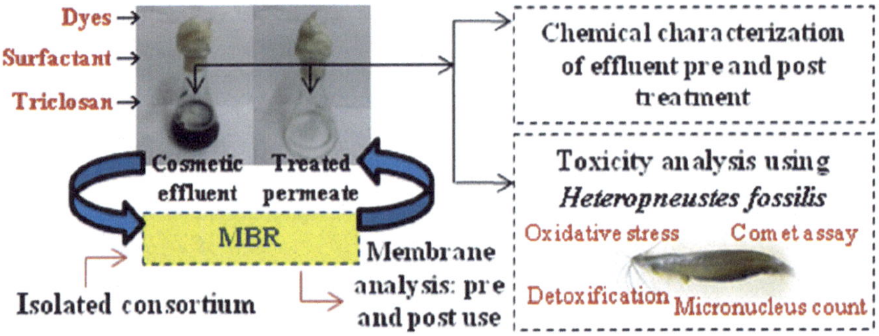

Fig. 5 Schematic representation of treatment of cosmetic effluent in different configurations of ceramic UF membrane-based bioreactor: toxicity evaluation of the untreated and treated wastewater using catfish (*Heteropneustes fossilis*) [Reproduced with permission from [10] © 2023 Elsevier B.V.]

Table 2 Comparative evaluation of biotreated cosmetic effluent filtration using monochannel and capillary ceramic UF membranes associated in side-stream (SS) and submerged (SB) configurations

Performance	Using monochannel membrane		Using capillary membrane	
	Wastewater treated with SS-UMBR	Wastewater treated with SB-UMBR	Wastewater treated with SS-UMBR	Wastewater treated with SB-UMBR
Fresh water flux (Lm^{-2} h^{-1})	80.12	129.81	97.47	142.21
Effluent flux (Lm^{-2} h^{-1})	76.21	127.22	95.11	14.14
Total solid (mg L^{-1})	72.48 ± 1.05	68.54 ± 1.25	60.31 ± 0.07	34.29 ± 0.05
DO (mg L^{-1})	7.14 ± 0.02	7.48 ± 0.05	7.65 ± 0.03	7.85 ± 0.05
COD (mg L^{-1})	56.74 ± 1.12	53.81 ± 2.43	48.42 ± 0.05	42.36 ± 0.05
Residual colour (%)	4.38	3.64	1.24	0.78
Residual triclosan (%)	3.24	2.32	1.86	1.27
Residual surfactant (%)	2.74	2.57	2.18	1.8

Reproduced with permission from [10] © 2023 Elsevier B.V.

submerged UMBRs have been shown in Table 2. A schematic representation of treatment of cosmetic effluent in ceramic UF membrane based bioreactor and toxicity evaluation of the untreated and treated wastewater using catfish is shown in Fig. 5.

4 Integrated MBR Processes

Modern approaches involving the integration of MBR systems with other processes such as advanced oxidation (AOPs), membrane distillation, granular MBR systems, bio-entrapped MBRs, etc. are gaining much attention nowadays. MBR integrated with other wastewater management processes can help improve the quality of permeate and antifouling potential of the membrane. This integrated process can also facilitate more efficient removal of micropollutants. Schematic representations of different integrated MBR setups reported in recent literature have been shown in Fig. 6. Two most widely implemented integrated processes have been discussed as follows.

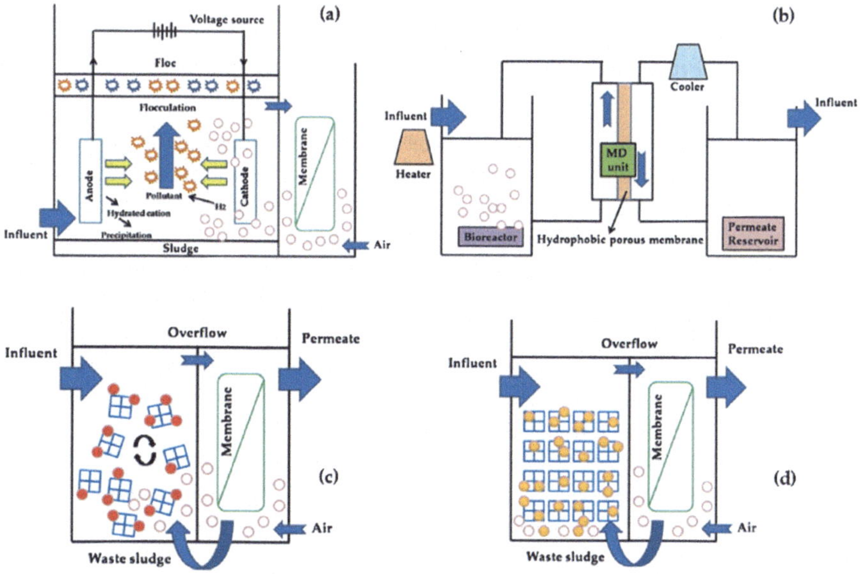

Fig. 6 Schematic diagram of some hybrid MBR systems, **a** electrocoagulation augmented an external side stream membrane bioreactor system, **b** MDBR with an external side stream membrane, **c** Biofilm-Membrane Bioreactor (BF-MBR) and **d** Bio-entrapped-Membrane Bioreactor (BE-MBR) Reproduced/adapted from [40] with permission from Elsevier, © 2018

4.1 Integrated Membrane Filtration and Adsorption

Adsorption is a phenomenon, where the micro-pollutant particles adhere to the solid phases (like activated sludge). Some contaminants are easily adsorbed with the sludge by the processes of adsorption and absorption. Here, adsorption refers to the electrostatic interaction between the negatively charged surface of microorganisms with the positively charged functional groups of micropollutants. Whereas, absorption is the hydrophobic interaction between the cell membrane (lipophilic) of the microbes and the lipid moieties present in the sludge.

The ratio of equilibrium concentration of micropollutants adsorbed and in solution is known as the solid water distribution coefficient (K_d) [83]. Previous studies have reported minimum sorption of hormones and musk fragrances, i.e., compounds having a $K_d < 500$ L kg^{-1} [83]. Moreover, a low value of log K_{ow} is indicative of pollutant hydrophobicity. These pollutants are first taken up by the activated sludge and then subjected to biodegradation [83]. It is mostly observed that the K_d value is comparatively higher for the primary sludge than the activated sludge. This highlights the inhibition occurring in the case of the former.

4.2 Integrated Membrane Filtration and Bioremediation

Bioremediation is an integrated and sustainable approach to employing microorganisms, such as bacteria, fungi, for the treatment of both solid wastes and wastewater that are rich in organic pollutants and generated from various sources [26]. Bioremediation is gaining much appreciation as an innovative tool in the field of environmental science and engineering owing to its eco-friendly nature, cost-effectiveness and comparatively high efficiency. The advantage of this technology is that it allows microorganisms to acclimatize with the toxic environment created by hazardous wastes, resulting in the development of new and toxin-resistant strains naturally. These new strains are hence capable of metabolically transforming a variety of toxic chemical substances into a less harmful form [28, 51, 101].

In the past, several types of biological treatment processes (such as AS, bioaugmentation, biological aerated filters, bioslupring, sequencing batch reactors, rotating biological contactors, and trickling filters) have been reported for effluent treatment [76]. Biological methods are mostly dependent upon the enzymatic and metabolic efficiency of microbes for pollutant degradation [37, 87]. However, bioremediation by indigenous microbial population has proven to be the most eco-friendly, energy-saving and cost-effective process [88, 96]. The most beneficial characteristic of biological processes is that their end products are nontoxic. Mostly harmless gases like nitrogen, carbon, and hydrogen are generated as end products of microbial degradation and can easily get assimilated into the environment [88]. The biodegradation potential of indigenous microorganisms can be further improved by optimizing certain process parameters such as bioavailability, rate of adsorption and mass transfer. However, more inclusive study is essential for the implementation of bioremediation techniques at a large scale and for analysis of their environmental consequences.

Among various different processes reported in the past, biological oxidation gives the best result in terms of complete degradation of pollutants present in the effluent [11]. Evaluation of previous reports implies that bacterial species are more competent in degrading different contaminants over a wide range of environmental factors, as compared to the fungal or algal species. This superior potential of biodegradation displayed by bacterial communities can be credited to their rapid rate of growth and reproduction along with simple culture techniques [86]. Studies have indicated that the MBR technology for wastewater treatment provides some specific advantages like reduced carbon footprint, low sludge production, removal of micropollutants of organic nature with low molecular weights (which are otherwise difficult to remove) along with better separation potential and superior quality of effluent post-treatment [65]. Therefore, the MBR technology is currently being favoured and recommended by many researchers over conventional sludge-based chemical processes. Only one issue that limits wide-scale implementation of MBR technology is the lengthy treatment period.

In the current times, immobilized microbial cell technology is gaining much interest and appreciation as the most suitable process for the treatment of wastewater loaded with reactive azo dyes and their metabolites. Immobilization of cells facilitates increased concentration of biomass, improved stability of the process, amplified hydraulic loading rates, reusability of biocatalysts and enhanced biodegradation of pollutants, as compared to CASP and suspended cultures [78]. Moreover, immobilized cell cultures (ICC) can be efficiently employed for separation of the solid–liquid phase in settling tanks. ICC even eliminates hitches related to bulk amounts unlike suspension cultures [78].

The immobilized cells are therefore often referred to as biocatalysts and have reportedly shown to have a greater efficiency for pollutant degradation, greater resilience to fluctuations in effluent pH and salinity, ambient temperature, varying concentrations of toxic chemicals, etc. [78]. With all these advantages they offer, the ICC or the biocatalysts are regarded as an appropriate technique for large-scale effluent treatment.

Advancement in the field of biotechnology and nanotechnology has empowered the researchers to develop different types of nano-biocatalysts, immobilize cells, enzymes and other biochemical substances into nano environments, manoeuvre the process parameters affecting the macroscale reactions, etc. Studies in the recent past have shown that nano biocatalysts exhibit improved biomolecule loading capacity, appreciably superior efficiency of mass transfer, enhanced biomolecule activity including stability and higher tolerance to fluctuations in experimental parameters. The huge surface area of nanomaterials provides a greater surface-to-volume ratio and hence can sustain a much higher concentration of immobilized cells and biomolecules as compared to other materials with planar surfaces. Some of the advantages offered by nanotechnology for the immobilization of biomolecules include the development of nano biocatalysts with a high percentage of solid content without using any surfactant or toxic chemicals; preparation of well-defined and homogenous supports; convenient modification of nano-particle size as required. Various nanomaterials are being extensively examined and studies are being conducted for concurrent immobilization of several biomolecules on the same substrate for synthetic in vitro applications [40].

5 Ecotoxicological Aspects of Raw and CMBR-Treated Effluents

Ecotoxicity results from various toxicological mechanisms which affect aquatic ecosystems [72]. Most of the processes for effluent treatment are incapable of complete removal of all types of water contaminants [27]. Therefore, wastewater treatment plant discharges are considered as point sources of micropollutants including metals, PPCPs, surfactants, etc. [39, 60]. Only a few studies have addressed the ecotoxicological aspects of CMBR-treated effluents. Assays commonly carried out for determining the toxicity have been discussed in the following sections.

5.1 Toxicity Assays Carried Out with Microorganisms

The impact of untreated and treated effluents on microorganisms may be exerted in a direct or indirect manner [97]. *Direct* methods of toxicity measurement are enlisted as follows [97].

- *OUR-based respiration inhibition* involves incubation of AS with untreated/treated effluents in an air-tight vessel for three hours (or half an hour for hydrolytic or volatile compounds). Then the dissolved oxygen level is monitored for 10 min with an oxygen electrode for determination of the OUR values. This test has been widely utilized for determining the toxic impact of heavy metals, organics and complex wastewater. It is considered as a standard method, for comparison with others.
- The *carbon dioxide evolution test* (modified Sturm test), determines the biodegradability of compounds that are non-volatile in nature. This test requires longer incubation times than the OUR test, and is therefore, more appropriate as a test of biodegradability as well as a chronic toxicity assay.
- The *nitrification (or denitrification) inhibition assay* simultaneously employs an ammonia-oxidizing and a nitrite-oxidizing organism (*Nitrosomonas* sp. and *Nitrobacter* sp., respectively). Both organisms are co-incubated for four hours in effluent samples. Ammonium and nitrite consumed in this assay are obtained colorimetrically.

- The *nitrification inhibition assay* (where the accumulation of nitrous oxide is determined) is used for monitoring nitrification. This assay is also used as an early warning system.
- The *anaerobic toxicity assay* (carried out in terms of methane production) determines the toxicity exerted by a substrate on methanogenesis. This assay is extremely user-friendly and cost-effective. This assay has been widely carried out for determining the toxicity of substrates like phenols (containing halogen groups), metals, nitroaromatics, and plasticizers.
- The *ATP luminescence assay* determines the declining ATP content of organisms exposed to toxic substrates. ATP content is expressed in terms of firefly luciferase activity (emission of light). This assay has been used for determining the toxic effect of effluents bearing imidazolium ionic liquids.
- For the *dehydrogenase activity assay*, concentrations of dyes playing the roles of electron-acceptors (such as triphenyltetrazolium chloride and resazurin) are determined spectrophotometrically. Both in vivo and in situ applications of this assay have been reported previously. This test has also been used commercially for determining cell membrane integrity.
- *Cell growth inhibition assays* are carried out to determine changes in cell growth in terms of optical density. This test is also used for determining toxicity in terms of growth inhibition.
- *Mortality assays* are performed to determine the death rate or viability of organisms exposed to toxic pollutants.
- *Cell membrane integrity assay* is primarily used to determine the viability of bacterial cells exposed to toxicants. Cells are essentially stained using fluorescent dyes such as SYTO9 and propidium iodide (PI) taken as a mixture. The nucleic acids present in bacterial cells emit green fluorescence upon staining with SYTO9. However, PI is able to penetrate the damaged cells. In this case, the emission of red fluorescence is recorded. The integrity of cell membranes is also measured in terms of the quantity of intracellular K^+ released.
- The assay for determining *lactate dehydrogenase (LDH)* activity is also carried out for determining cell membrane integrity. LDH is released from both prokaryotic and eukaryotic cells.

Methos of *indirect* measurement of toxicity has been enlisted as follows [97].

- *Bioluminescent assay* is based on the estimation of luminescence emitted from the bacterial luciferase (lux) gene present in bioluminescent bacteria. Only viable cells are able to emit luminescence. This assay is widely performed for determining the toxicity of pollutants present in air, soil, potable water as well as effluents. Moreover, fungal bioluminescence has also been reported recently for determining the toxicity of contaminated soil samples.
- *Biosensors* produce signals that intensify in parallel to the concentration of pollutants the sensor is exposed to. Biological entities like antibodies, cells, cofactors, enzymes, microbial cells, organelles or tissues may be used for the development of biosensors. Amongst them, enzymes have been mostly used due to their specificity

and sensitivity. However, the biosensors developed using whole cells are conveniently synthesized and implemented. On the basis of the selected transducer, a biosensor may be electrochemical or optical in nature.

5.2 Toxicity Assays Carried Out with Plants

Pollutant uptake in a plant may lead to the production of reactive oxygen species (ROS) which interferes with the growth and metabolism of the plants [12]. Plants have been found to combat this damage by ROS using both enzymatic and non-enzymatic antioxidant systems. Seeds and seedlings of plants like *Vigna radiata* and *Cicer arietinum* have been subjected to untreated and treated effluents [12, 15, 19, 82]. Detoxification of effluents has been established through a comparative analysis of toxic effects exerted by untreated and treated effluents. Toxic effects have been determined in terms of germination, biochemical parameters (protein, carbohydrate and amylase content) as well as oxidative stress build-up (enzyme response) in treated plants. Inferences drawn from results obtained in several such studies have been enlisted as follows [12, 15, 19, 82].

- Alteration in the germination of seeds could be linked up with impairment of certain biochemical processes during seed germination occurring as a result of toxic exposure to pollutants present in untreated effluents. No impact observed in seeds exposed to treated effluent established efficient detoxification of the same.
- Root and shoot development in seedlings developing from seeds exposed to treated effluent were healthier than those exposed to untreated effluents.
- Significant differences recorded in protein, carbohydrate and amylase contents of seeds/tissue exposed separately to raw and treated effluents also suggested the same.
- A significant increase in oxidative stress response activity obtained in terms of enzymes like peroxidase, catalase and superoxide dismutase and parallel decrease in detoxification enzymes like glutathione peroxidase and reduced glutathione in seeds/tissue exposed to raw effluents also suggested the same. This also implied that the stress-combating mechanism of seeds/tissue exposed to raw effluents was severely compromised. However, results obtained in the case of seeds/tissue exposed to treated effluents indicated the detoxification efficiency of the effluent treatment process reported therein.

5.2.1 Case Studies on Phytotoxicity of Wastewater

Phytotoxicity of treated and untreated effluents is determined to understand whether the treated water is suitable for discharge or reuse for agricultural purposes. Phytotoxicity is usually determined in terms of germination parameters or oxidative stress enzymes [19]. In such phytotoxicity assays, seeds or seedlings are exposed to varying concentrations or dilutions of effluent and tested for physiological (root/shoot length,

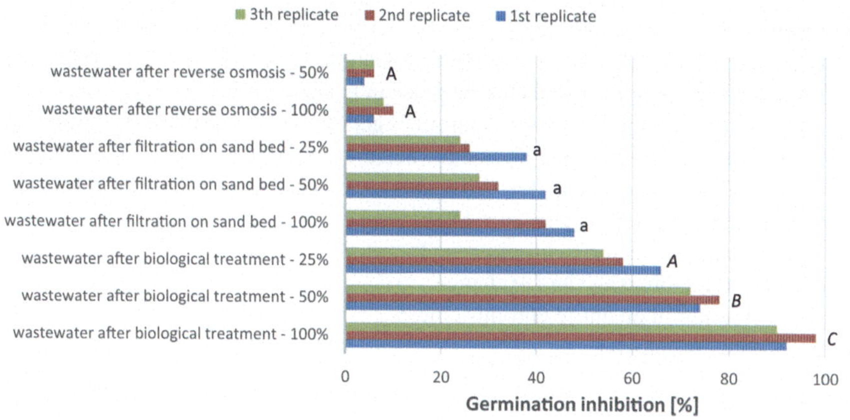

Fig. 7 Germination inhibition of *Vicia faba* (cotton matrix) [Reproduced with permission from [92] Copyright © 2018, The Author(s)]

germination %, etc.), biochemical (amino acid, protein, carbohydrate content, etc.) and enzyme parameters (like oxidative stress and detoxification enzymes) [12]. In a previous study by [95], inhibition of germination (%), was calculated using the following formula where I, A and B represent inhibition % and seed germination in control and test samples, respectively.

$$I = \frac{A - B}{A}. \text{ inhibition} = 100[\%]$$

On the basis of inhibition (%), [1] the degree of toxicity is clssified into three classes 1 (I < 10), 2(10 < I < 50) and 3 (50 < I) denoting non-toxic or slightly toxic, toxic and highly toxic impacts respectively. In a recent study, [92] the toxicity of RO—MBR treated effluents was tested on broad bean (*Vicia faba*). The PAH concentration of the effluent filtered using a sand bed was 45.2 μg L^{-1}. This was further reduced to 5.57 μg L^{-1} post-RO treatment. Phytotoxicity assays indicated efficient detoxification of coke wastewater using integrated RO processes (as shown in Fig. 7.

5.3 *Toxicity Assays Carried Out with Animals*

Detoxification of effluents can also be assayed using animal (fish, molluscs, etc.) models. Widely used assays are described as follows.

- *Enzyme assays* are carried out similar to that of plants.

- *Cytotoxicity assays* (like TUNEL assay) have been developed for detection of apoptotic cells having extensive DNA degradation. This assay is widely conducted for detecting apoptotic events taking place in paraffin-embedded tissue sections.
- The in vitro *micronucleus assay* determines genotoxicity in terms of the formation of one or more small nuclear fragments (i.e. micronuclei) in the interphase cells of various fish and mollusk models. Micronuclei are usually formed by the following four widely accepted mechanisms [43]:
- Degeneration of acentric chromosome fragments;
- Chromosomal breakage and exchange;
- Disintegration of an entire chromosomes;
- Cell death (apoptosis).

Micronuclei formed in blood cells of *Heteropneustes fossilis* are shown in Fig. 8.

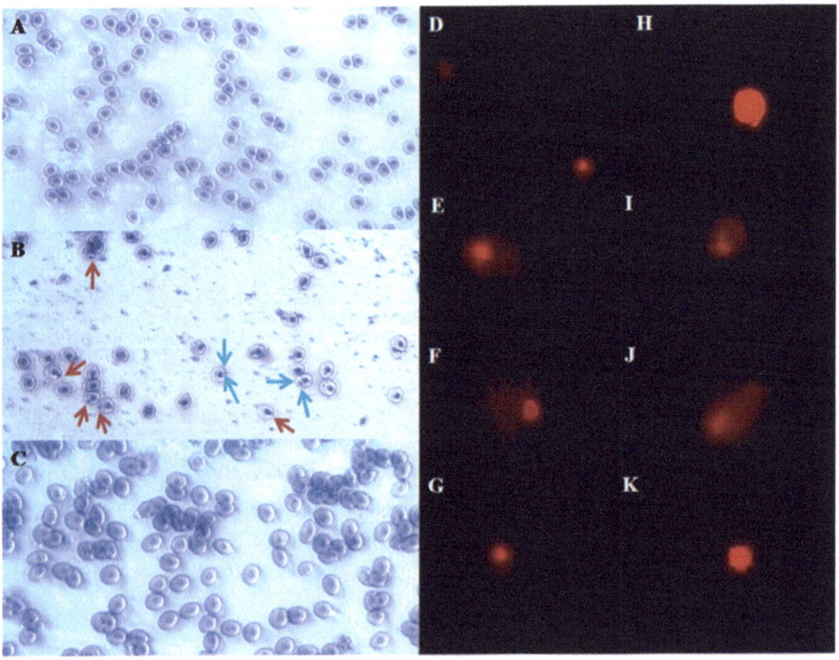

Fig. 8 Showing genotoxicity in *H. fossilis*. Figure **a–c** shows micronuclei formed in peripheral erythrocytes of *H. fossilis*. All data are presented as Mean ± SD (n 1/4 10). A shows normal cells in control individuals, **b** shows micronucleated cells in raw cosmetic effluent treated (25% (v/v)) individuals (red and blue arrows show single and double micronucleus respectively) and **c** shows normal cells in MBR permeate treated individuals. Figures **d–k** show DNA damage (comets formed) in the liver and gill tissues of *H. fossilis*. Figures **d–g** liver cells from control treated, raw cosmetic effluent treated (25% (v/v)) (after 24 h), raw cosmetic effluent treated (25% (v/v)) (after 96 h) and MBR permeate treated individuals respectively. Figure **h–k** shows the same in gill cells respectively Reproduced/adapted from [10] with permission from Elsevier, © 2016

- The *Single Cell Gel Electrophoresis assay* or *Comet Assay* is also indicative of genotoxicity. It is an uncomplicated and sensitive technique for determining DNA damage in fishes and mollusc cells. In this process, cells are encapsulated in a low-melting-point agarose suspension and subjected to lysis in extremely alkaline (pH > 13) conditions and followed by electrophoresis of the suspended lysed cells [9, 10]). On electrophoresis, undamaged DNA stained with ethidium bromide (EtBr) is visible as a compact structure. Damaged DNA is drawn towards the anode to yield a comet-like structure. Comet assay performed on liver and gill cells of *Heteropneustes fossilis* has also been shown in Fig. 5. Similar assays were also conducted using mollusks *Bellamya bengalensis* [13] and *Pila globosa* [21].

6 Reusability of the Treated Effluents

A vital aspect of MBR treatment of effluent is the reusability of the same for domestic or industrial purposes. However, the feasibility of reusing an MBR-treated effluent for agriculture, aquaculture and industrial processes is completely dependent on the nature of effluent [61]. In the previous section, effluent detoxification was established through the use of plants and different aquatic organisms. Results obtained in these studies highlighted the potential of these treated effluents for purposes of agriculture and pisciculture [9, 10, 19, 22]. In another study, CMBR-treated tannery effluents were reportedly reused for tanning of leather [20]. Results obtained in this study indicated that dye uptake and tensile strength of leather processed using treated effluent was higher than that of leather processed using fresh water.

7 Scope for Further Studies

The ceramic membranes (of both tubular and hollow-fibre type) may be considered appropriate for wastewater treatment as they demonstrate enhanced mechanical strength, packing density, as well as greater tolerance of large crossflow velocities in comparison to flat-sheet ones [52, 70]. Flat-sheet membranes demonstrate higher efficiency in submerged configurations owing to convenient cleaning (air sparging) and replacement of the same [55]. However, the ceramic membranes should be optimized in terms of packing density to maintain the environmental footprint of the effluent treatment plants using the same at a bare minimum [52]. This is an urgent requirement for large-scale application of ceramic membranes. Nevertheless, integrated interactions of physicochemical parameters necessitate detailed techno-economic analysis for the selection of the significant parameters guiding the process prior to the application of these membrane-based processes on a commercial scale. Moreover, future studies also need to focus more on improving the stability and reusability of membranes post application of the same for the treatment of real water matrices.

8 Conclusion

Reviews on effluent treatment using membrane bioreactors mostly focus on the membrane characteristics, MBR configuration and efficiency. Studies have seldom addressed the ecotoxicological aspect or reuse potential of treated effluents. This review has discussed in details, the ecotoxicological assays through which the detoxification efficiency of CMBR-based processes has been evaluated and demonstrated the reusability of CMBR-treated effluents with case studies. Similar studies may be undertaken to highlight the efficiency of the treatment techniques and establish whether the treated effluents are actually environmentally benign in nature and suitable for reuse in domestic/ industrial processes.

References

1. Adamcová D, Vaverková MD, Břoušková E (2016) The toxicity of two types of sewage sludge from wastewater treatment plant for plants. J Ecol Eng 17(2):33–37
2. Al Aani S, Mustafa TN, Hilal N (2020) Ultrafiltration membranes for wastewater and water process engineering: a comprehensive statistical review over the past decade. J Water Process Eng 35:101241
3. Al-Asheh S, Marzieh B, Ahmed A (2021) Membrane bioreactor for wastewater treatment: a review. Case Stud Chem Environ Eng 4:100109
4. Ang WL, Mohammad AW, Hilal N, Leo CP (2015) A review on the applicability of integrated/ hybrid membrane processes in water treatment and desalination plants. Desalination 363:2–18
5. Arriaga S, de Jonge N, Nielsen ML, Andersen HR, Borregaard V, Jewel K, Ternes TA, Nielsen JL (2016) Evaluation of a membrane bioreactor system as post-treatment in waste water treatment for better removal of micropollutants. Water Res 107:37–46
6. Asif MB, Ansari AJ, Chen SS, Nghiem LD, Price WE, Hai FI (2019) Understanding the mechanisms of trace organic contaminant removal by high retention membrane bioreactors: a critical review. Environ Sci Pollut Res 26(33):34085–34100
7. Asif MB, Zhang Z (2021) Ceramic membrane technology for water and wastewater treatment: a critical review of performance, full-scale applications, membrane fouling and prospects. Chem Eng J 129481
8. Attiogbe F (2013) Comparison of membrane bioreactor technology and conventional activated sludge system for treating bleached Kraft mill effluent. Afr J Environ Sci Technol 7(5):292–306
9. Banerjee P, Sarkar S, Dey TK, Bakshi M, Swarnakar S, Mukhopadhayay A, Ghosh S (2014) Application of isolated bacterial consortium in UMBR for detoxification of textile effluent: comparative analysis of resultant oxidative stress and genotoxicity in catfish (*Heteropneustes fossilis*) exposed to raw and treated effluents. Ecotoxicology 23(6):1073–1085
10. Banerjee P, Dey TK, Sarkar S, Swarnakar S, Mukhopadhyay A, Ghosh S (2016) Treatment of cosmetic effluent in different configurations of ceramic UF membrane based bioreactor: toxicity evaluation of the untreated and treated wastewater using catfish (*Heteropneustes fossilis*). Chemosphere 146:133–144
11. Banerjee P, Mukhopadhyay A, Das P (2017) Current modifications introduced for improving bioremediation efficiency of polycyclic aromatic hydrocarbons. Bio-remediat: Curr Res Appl 231–251
12. Banerjee P, Barman SR, Sikdar D, Roy U, Mukhopadhyay A, Das P (2017) Enhanced degradation of ternary dye effluent by developed bacterial consortium with RSM optimization, ANN modeling and toxicity evaluation. Desalin Water Treat 72:249–265

13. Banerjee P, Barman SR, Swarnakar S, Mukhopadhyay A, Das P (2018) Treatment of textile effluent using bacteria-immobilized graphene oxide nanocomposites: evaluation of effluent detoxification using *Bellamya bengalensis*. Clean Technol Environ Policy 20(10):2287–2298
14. Banerjee P, Mukhopadhyay A, Das P (2019) Graphene oxide–nanobentonite composite sieves for enhanced desalination and dye removal. Desalination 451:231–240
15. Barman SR, Banerjee P, Mukhopadhayay A, Das P (2017) Biodegradation of acenapthene and naphthalene by *Pseudomonas mendocina*: process optimization, and toxicity evaluation. J Environ Chem Eng 5(5):4803–4812
16. Bhaduri S, Mazumder D (2020) Advanced wastewater treatment using membrane bioreactor system. J Indian Chem Soc 97(12a):2643–2655
17. Bhatt P, Bhandari G, Bilal M (2022) Occurrence, toxicity impacts and mitigation of emerging micropollutants in the aquatic environments: Recent tendencies and perspectives. J Environ Chem Eng 10(3):107598
18. Bhattacharya P, Roy A, Sarkar S, Ghosh S, Majumdar S, Chakraborty S, Mandal S, Mukhopadhyay A, Bandyopadhyay S (2013) Combination technology of ceramic microfiltration and reverse osmosis for tannery wastewater recovery. Water Resour Indus 3:48–62
19. Bhattacharya P, Banerjee P, Mallick K, Ghosh S, Majumdar S, Mukhopadhyay A, Bandyopadhyay S (2013) Potential of biosorbent developed from fruit peel of *Trewia nudiflora* for removal of hexavalent chromium from synthetic and industrial effluent: analyzing phytotoxicity in germinating *Vigna* seeds. J Environ Sci Health Part A 48(7):706–719
20. Bhattacharya P, Ghosh S, Mukhopadhyay A (2013) Combination technology of ceramic microfiltration and biosorbent for treatment and reuse of tannery effluent from different streams: response of defence system in Euphorbia sp. Int J Recycl Organic Waste Agric 2(1):1–11
21. Bhattacharya P, Swarnakar S, Mukhopadhyay A, Ghosh S (2016) Exposure of composite tannery effluent on snail, *Pila globosa*: a comparative assessment of toxic impacts of the untreated and membrane treated effluents. Ecotoxicol Environ Saf 126:45–55
22. Bhattacharya P, Mukherjee D, Deb N, Swarnakar S, Banerjee S (2020) Application of green synthesized ZnO nanoparticle coated ceramic ultrafiltration membrane for remediation of pharmaceutical components from synthetic water: reusability assay of treated water on seed germination. J Environ Chem Eng 8(3):103803
23. Bora LV, Mewada RK (2017) Visible/solar light active photocatalysts for organic effluent treatment: fundamentals, mechanisms and parametric review. Renew Sustain Energy Rev 76:1393–1421
24. Buzatu P, Qiblawey H, Nasser MS, Judd S (2017) Comparative power demand of mechanical and aeration imposed shear in an immersed membrane bioreactor. Water Res 126:208–215
25. Cha J, Cupples AM (2009) Detection of the antimicrobials triclocarban and triclosan in agricultural soils following land application of municipal biosolids. Water Res 43(9):2522–2530
26. Chang YT, Lo T, Chou HL, Laio YF, Lin CC, Chen HT (2016) Anaerobic biodegradation of decabromodiphenyl ether (BDE-209)-contaminated sediment by organic compost. Int Biodeterior Biodegradation 113:228–237
27. Carvalho AR, Pérez-Pereira AI, Couto CMC, Tiritan ME, Ribeiro CMR (2021) Assessment of effluents quality through ecotoxicological assays: evaluation of three wastewater treatment plants with different technologies. Environ Sci Pollut Res 1–14
28. Chen M, Xu P, Zeng G, Yang C, Huang D, Zhang J (2015) Bioremediation of soils contaminated with polycyclic aromatic hydrocarbons, petroleum, pesticides, chlorophenols and heavy metals by composting: applications, microbes and future research needs. Biotechnol Adv 33(6):745–755
29. Chen YR, Chen LH, Chen CH, Ko CC, Huang A, Li CL, Tung KL (2018) Hydrophobic alumina hollow fiber membranes for sucrose concentration by vacuum membrane distillation. J Membr Sci 555:250–257
30. Chi Y, Chong JY, Wang B, Li K (2020) Pristine graphene membranes supported on ceramic hollow fibre prepared via a sacrificial layer assisted CVD approach. J Membr Sci 595:117479

31. Croquesel J, Carry CP, Chaix JM, Bouvard D, Saunier S (2018) Direct microwave sintering of alumina in a single mode cavity: magnesium doping effects. J Eur Ceram Soc 38(4):1841–1845
32. da Silva Biron D, Dos Santos V, Zeni M (2017) Ceramic membranes applied in separation processes. Springer
33. Drioli E, Giorno L (2009) Membrane operations: innovative separations and transformations. Wiley-VCH, Weinheim
34. Drioli E, Macedonio F, Tocci E (2021) Membrane science and membrane engineering for a sustainable industrial development. Sep Purif Technol 275:119196
35. Dussault ÈB, Balakrishnan VK, Sverko ED, Solomon KR, Sibley PK (2008) Toxicity of human pharmaceuticals and personal care products to benthic invertebrates. Environ Toxicol Chem Int J 27(2):425–432
36. Dudziak M, Gryta M (2013) Nanofiltration of fermentation solutions–unfavourable phenomena and membrane cleaning. Ecol Chem Eng A 20(11):1285–1294
37. Elreedy A, Tawfik A, Enitan A, Kumari S, Bux F (2016) Pathways of 3-biofules (hydrogen, ethanol and methane) production from petrochemical industry wastewater via anaerobic packed bed baffled reactor inoculated with mixed culture bacteria. Energy Convers Manage 122:119–130
38. Enaime G, Baçaoui A, Yaacoubi A, Belaqziz M, Wichern M, Lübken M (2020) Phytotoxicity assessment of olive mill wastewater treated by different technologies: effect on seed germination of maize and tomato. Environ Sci Pollut Res 27:8034–8045
39. Franco ME, Burket SR, Sims JL, Lovin LM, Scarlett KR, Stroski K, Lavado R (2020) Multi-approach assessment for the evaluation of spatio-temporal estrogenicity in fish from effluent-dominated surface waters under low instream flow. Environ Pollut 265:115122
40. Goswami L, Kumar RV, Borah SN, Manikandan NA, Pakshirajan K, Pugazhenthi G (2018) Membrane bioreactor and integrated membrane bioreactor systems for micropollutant removal from wastewater: a review. J Water Process Eng 26:314–328
41. Harb M, Wei CH, Wang N, Amy G, Hong PY (2016) Organic micropollutants in aerobic and anaerobic membrane bioreactors: changes in microbial communities and gene expression. Biores Technol 218:882–891
42. He Z, Lyu Z, Gu Q, Zhang L, Wang J (2019) Ceramic-based membranes for water and wastewater treatment. Colloids Surf A 578:123513
43. Heddle JA, Cimino MC, Hayashi M, Romagna F, Shelby MD, Tucker JD, MacGregor JT (1991) Micronuclei as an index of cytogenetic damage: past, present, and future. Environ Mol Mutagen 18(4):277–291
44. Hubadillah SK, Othman MHD, Matsuura T, Ismail AF, Rahman MA, Harun Z, Nomura M (2018) Fabrications and applications of low cost ceramic membrane from kaolin: a comprehensive review. Ceram Int 44(5):4538–4560
45. Jiang LY, Na L (eds) (2017) Membrane-based separations in metallurgy: principles and applications. Elsevier
46. Judd S, Turan F (2018) Sidestream vs immersed membrane bioreactors: a cost analysis. In: Proceedings of the American water works association/American membrane technology association membrane technology conference & exposition; West Palm Beach, Florida, vol 16
47. Kaleem M, Ahmad A, Ahmad W, Nur SM, Tabrez S (2022) Membrane bioreactor as an advanced wastewater treatment technology. In: Development in wastewater treatment research and processes. Elsevier, pp 401–434
48. Karim MA, Mark JL (2017) A preliminary comparative analysis of MBR and CAS wastewater treatment systems. Int J Water Wastewater Treatment 3(2)
49. Kim G, Ahn J, Ahn S (2021) Grain growth and densification of uranium mononitride during spark plasma sintering. Ceram Int 47(5):7258–7262
50. Krzeminski P, van der Graaf JH, van Lier JB (2012) Specific energy consumption of membrane bioreactor (MBR) for sewage treatment. Water Sci Technol 65(2):380–392
51. Kuppusamy S, Thavamani P, Megharaj M, Naidu R (2015) Bioremediation potential of natural polyphenol rich green wastes: a review of current research and recommendations for future directions. Environ Technol Innov 4:17–28

52. Lee M, Wu Z, Li K (2015) Advances in ceramic membranes for water treatment. In: Advances in membrane technologies for water treatment. Woodhead Publishing, pp 43–82
53. Lee M, Wang B, Li K (2016) New designs of ceramic hollow fibres toward broadened applications. J Membr Sci 503:48–58
54. Li T, Zhang L, Ai W, Dong W, Yu Q (2018) A modified MBR system with post advanced purification for domestic water supply system in 180-day CELSS: construction, pollutant removal and water allocation. J Environ Manage 222:37–43
55. Li C, Sun W, Lu Z, Ao X, Li S (2020) Ceramic nanocomposite membranes and membrane fouling: a review. Water Res 175:115674
56. Li R, Kadrispahic H, Jørgensen MK, Berg SB, Thornberg D, Mielczarek AT, Bester K (2022) Removal of micropollutants in a ceramic membrane bioreactor for the post-treatment of municipal wastewater. Chem Eng J 427:131458
57. Lim S, Choi PJ, An AK, Han DS, Phuntsho S, Shon H (2021) Submerged versus side-stream osmotic membrane bioreactors using an outer-selective hollow fiber osmotic membrane for desalination. Desalination 515:115196
58. Liu W, Song X, Huda N, Xie M, Li G, Luo W (2020) Comparison between aerobic and anaerobic membrane bioreactors for trace organic contaminant removal in wastewater treatment. Environ Technol Innov 17:100564
59. Maeng SK, Choi BG, Lee KT, Song KG (2013) Influences of solid retention time, nitrification and microbial activity on the attenuation of pharmaceuticals and estrogens in membrane bioreactors. Water Res 47(9):3151–3162
60. Maia AS, Paíga P, Delerue-Matos C, Castro PM, Tiritan ME (2020) Quantification of fluoroquinolones in wastewaters by liquid chromatography-tandem mass spectrometry. Environ Pollut 259:113927
61. Majumdar S, Sarkar S, Ghosh S, Bhattacharya P, Bandyopadhyay S, Saha A, Roy SN (2018) New trends for wastewater treatment and their reuse using ceramic membrane technology: a case study. In: Water quality management. Springer, Singapore, pp 339–348
62. Mansas C, Atfane-Karfane L, Petit E, Mendret J, Brosillon S, Ayral A (2020) Functionalized ceramic nanofilter for wastewater treatment by coupling membrane separation and catalytic ozonation. J Environ Chem Eng 8(4):104043
63. Marti E, Monclús H, Jofre J, Rodriguez-Roda I, Comas J, Balcázar JL (2011) Removal of microbial indicators from municipal wastewater by a membrane bioreactor (MBR). Biores Technol 102(8):5004–5009
64. Murić A, Petrinić I, Christensen ML (2014) Comparison of ceramic and polymeric ultrafiltration membranes for treating wastewater from metalworking industry. Chem Eng J 255:403–410
65. Neoh CH, Noor ZZ, Mutamim NSA, Lim CK (2016) Green technology in wastewater treatment technologies: integration of membrane bioreactor with various wastewater treatment systems. Chem Eng J 283:582–594
66. Neves TM, Pollo LD, Marcilio NR, Tessaro IC (2021) Alumina supports produced by dry-pressing and sintering at different temperatures for developing carbon molecular sieve membranes. Ceram Int 47(22):32226–32236
67. Núñez J, Yeber M, Cisternas N, Thibaut R, Medina P, Carrasco C (2019) Application of electrocoagulation for the efficient pollutants removal to reuse the treated wastewater in the dyeing process of the textile industry. J Hazard Mater 371:705–711
68. Obotey Ezugbe E, Rathilal S (2020) Membrane technologies in wastewater treatment: a review. Membranes 10(5):89
69. Oun A, Tahri N, Mahouche-Chergui S, Carbonnier B, Majumdar S, Sarkar S, Amar RB (2017) Tubular ultrafiltration ceramic membrane based on titania nanoparticles immobilized on macroporous clay-alumina support: elaboration, characterization and application to dye removal. Sep Purif Technol 188:126–133
70. Pabby AK, Rizvi SS, Requena AMS (2015) Handbook of membrane separations: chemical, pharmaceutical, food, and biotechnological applications, 2nd edn. CRC press

71. Papp K, Moser D, Gerrity D (2020) Viral surrogates in potable reuse applications: evaluation of a membrane bioreactor and full advanced treatment. J Environ Eng 146(2):04019103
72. Pedrazzani R, Cavallotti I, Bollati E, Ferreri M, Bertanza G (2018) The role of bioassays in the evaluation of ecotoxicological aspects within the PEF/OEF protocols: the case of WWTPs. Ecotoxicol Environ Saf 147:742–748
73. Podder A (2020) Application of membrane bioreactors for the modification of microfiltration membrane surface for enhanced antibiofouling capability in wastewater treatment. In: Advances in petroleum technology. Jenny Stanford Publishing, pp 397–413
74. Rabiee H, Vatanpour V, Farahani MHDA, Zarrabi H (2015) Improvement in flux and antifouling properties of PVC ultrafiltration membranes by incorporation of zinc oxide (ZnO) nanoparticles. Sep Purif Technol 156:299–310
75. Radjenović J, Petrović M, Barceló D (2009) Fate and distribution of pharmaceuticals in wastewater and sewage sludge of the conventional activated sludge (CAS) and advanced membrane bioreactor (MBR) treatment. Water Res 43(3):831–841
76. Ramteke LP, Gogate PR (2015) Treatment of toluene, benzene, naphthalene and xylene (BTNXs) containing wastewater using improved biological oxidation with pretreatment using Fenton/ultrasound based processes. J Ind Eng Chem 28:247–260
77. Ratanatamskul C, Katasomboon J (2017) Effect of sludge recirculation pattern on biological nutrient removal by a prototype IT/BF-MBR (inclined tube/biofilm-membrane bioreactor) and microbial population characteristics. Int Biodeterior Biodegradation 124:26–35
78. Rathoure AK (ed) (2015) Toxicity and waste management using bioremediation. IGI Global
79. Reif R, Suárez S, Omil F, Lema JM (2008) Fate of pharmaceuticals and cosmetic ingredients during the operation of a MBR treating sewage. Desalination 221:511–517
80. Reznik A, Feinerman E, Finkelshtain I, Fisher F, Huber-Lee A, Joyce B, Kan I (2017) Economic implications of agricultural reuse of treated wastewater in Israel: a state wide long-term perspective. Ecol Econ 135:222–233
81. Roccaro P, Vagliasindi FG (2020) Techno-economic feasibility of membrane bioreactor (MBR). In: Frontiers in water-energy-nexus—nature-based solutions, advanced technologies and best practices for environmental sustainability. Springer, Cham, pp 269–270
82. Roy U, Sengupta S, Banerjee P, Das P, Bhowal A, Datta S (2018) Assessment on the decolourization of textile dye (reactive Yellow) using *Pseudomonas* sp. immobilized on fly ash: response surface methodology optimization and toxicity evaluation. J Environ Manage 223:185–195
83. Sathe SS, Mahanta C, Mishra P (2018) Simultaneous influence of indigenous microorganism along with abiotic factors controlling arsenic mobilization in Brahmaputra floodplain, India. J Contam Hydrol 213:1–14
84. Samaei SM, Gato-Trinidad S, Altaee A (2018) The application of pressure-driven ceramic membrane technology for the treatment of industrial wastewaters–a review. Sep Purif Technol 200:198–220
85. Sarkar S, Bandyopadhyay S, Larbot A, Cerneaux S (2012) New clay–alumina porous capillary supports for filtration application. J Membr Sci 392:130–136
86. Sharma SK (ed) (2019) Bioremediation: a sustainable approach to preserving earth's water. CRC Press
87. Siddique MNI, Munaim MSA, Zularisam AW (2015) Feasibility analysis of anaerobic co-digestion of activated manure and petrochemical wastewater in Kuantan (Malaysia). J Clean Prod 106:380–388
88. Singh P, Borthakur A (2018) A review on biodegradation and photocatalytic degradation of organic pollutants: a bibliometric and comparative analysis. J Clean Prod 196:1669–1680
89. Smith AL, Stadler LB, Love NG, Skerlos SJ, Raskin L (2012) Perspectives on anaerobic membrane bioreactor treatment of domestic wastewater: a critical review. Biores Technol 122:149–159
90. Smol M, Kulczycka J, Kowlaski Z (2016) Sewage sludge ash (SSA) from small and large incineration plants as a potential source of phosphorus—polish case of study. J Environ Manage 184(3):617–628

91. Smol M, Włodarczyk-Makuła M, Mielczarek K, Bohdziewicz J, Włóka D (2016) The use of reverse osmosis in the removal of PAHs from municipal and fill leachate. Polycyclic Aromat Compd 36(1):20–39
92. Smol M, Włóka D, Włodarczyk-Makuła M (2018) Influence of integrated membrane treatment on the phytotoxicity of wastewater from the coke industry. Water Air Soil Pollut 229:1–14
93. Song X, Luo W, Hai FI, Price WE, Guo W, Ngo HH, Nghiem LD (2018) Resource recovery from wastewater by anaerobic membrane bioreactors: opportunities and challenges. Biores Technol 270:669–677
94. Suárez S, Reif R, Lema JM, Omil F (2012) Mass balance of pharmaceutical and personal care products in a pilot-scale single-sludge system: influence of T, SRT and recirculation ratio. Chemosphere 89(2):164–171
95. Trojanowska-Olichwer A (2013) Assessment of sediment toxicity from the Włocławski reservoir. J Ecol Health 17(3):103–109
96. Wang J, Sandoval K, Ding Y, Stoeckel D, Minard-Smith A, Andersen G, Gardinali P (2016) Biodegradation of dispersed Macondo crude oil by indigenous Gulf of Mexico microbial communities. Sci Total Environ 557:453–468
97. Xiao Y, De Araujo C, Sze CC, Stuckey DC (2015) Toxicity measurement in biological wastewater treatment processes: a review. J Hazard Mater 286:15–29
98. Yu L, Kanezashi M, Nagasawa H, Tsuru T (2020) Phase inversion/sintering-induced porous ceramic microsheet membranes for high-quality separation of oily wastewater. J Membr Sci 595:117477
99. Zaibel I, Arnon S, Zilberg D (2022) Treated municipal wastewater as a water source for sustainable aquaculture: a review. Rev Aquac 14(1):362–377
100. Zhang W, Liang W, Zhang Z, Hao T (2021) Aerobic granular sludge (AGS) scouring to mitigate membrane fouling: performance, hydrodynamic mechanism and contribution quantification model. Water Res 188:116518
101. Zhao HM, Du H, Lin J, Chen XB, Li YW, Li H, Wong MH (2016) Complete degradation of the endocrine disruptor di-(2-ethylhexyl) phthalate by a novel *Agromyces* sp. MT-O strain and its application to bioremediation of contaminated soil. Sci Total Environ 562:170–178

Chapter 13
Biosorption-Assisted Cross-Flow Microfiltration Process for Wastewater Treatment Using Clay-Alumina-Based Ceramic Membranes

Priyankari Bhattacharya, Shritama Mukhopadhyay, Sathi Banerjee, and Sibdas Bandyopadhyay

1 Introduction

Effluents generated from domestic and industrial sectors contain substantial amounts of metals, pathogens, organic and inorganic substances which are considered as the most prominent ecological issues. Profoundly lethal heavy metals like chromium, copper, cadmium are a few examples of cancer-causing pollutants posing serious threats to biological systems. Discharges from domestic sectors are rich in nutrients that cause eutrophication in water bodies and subsequently result in the loss of aquatic life. Traditional treatment of wastewater like coagulation, precipitation, adsorption, ion-exchange, etc. generates a considerable amount of sludge which creates poisonous ooze thus again damaging the environment. An eco-friendly remedy to these issues can be the use of adsorbent materials of natural origin which would be biodegradable. Biosorption includes physico-chemical adsorption and the process is highly selective as well as cost effective. Compared to conventional adsorbents like activated carbon, biosorbents prepared from waste materials are more environments

P. Bhattacharya (✉) · S. Banerjee
Metallurgical and Materials Engineering Department, Jadavpur University, Kolkata 700032, India
e-mail: priyankari2004@yahoo.co.in

S. Mukhopadhyay
Department of Chemical Engineering, Jadavpur University, Kolkata 700032, India

S. Bandyopadhyay
Formerly at Ceramic Membrane Division, CSIR-Central Glass & Ceramic Reseasch Institute, Kolkata 731235, India

S. Bandyopadhyay (ed.), *Application of Low Cost Ceramic Membranes in Wastewater Treatment*, Materials Horizons: From Nature to Nanomaterials,
https://doi.org/10.1007/978-981-97-5842-5_13

friendly. An interesting advantage of the biosorption process is that the sorbed components can be regenerated by desorption. However, biosorption still finds its application limited to the laboratory scale and several factors hinder its real field application for industrial and domestic wastewater. In this chapter, the preparation of biosorbents using different biomaterials and the scope of a hybrid process combining the application of such biosorbents and ceramic microfiltration membrane was discussed and demonstrated aiming at specifying the constraints of biosorption and obliterating their limitation in the field applications. Biosorption-assisted membrane filtration has been discussed as a one-step solution for the treatment and reuse of wastewater.

2 What is Biosorption and Why It is Used?

Biosorption is a type of sorption technique involving a solid phase, i.e., the sorbent derived from the bio-resources and the liquid phase, i.e., solvent containing dissolved unwanted constituents like metals, organics, etc. The process is reversible and continues till equilibrium is reached involving the binding of organic and inorganic ions on the surface of the biosorbent. Biosorption has emerged as an attractive alternative to conventional water treatment processes because of its simple and cost-effective operation, high efficiency, regenerative property, and lesser sludge generation. Moreover, no increase in chemical oxygen demand is observed [1, 2]. The search for new and cost-effective treatment technologies that involve the removal of toxic materials from wastewater has paved the path toward biosorption. The biosorption process involves the application of microorganisms, plant-based materials, agricultural or industrial wastes. Biosorption is capable of removing contaminants even in very dilute concentrations [2–4].

3 Literature Review

Conventional treatment processes for various types of wastewater include primary, secondary and tertiary treatment, disinfection, and solids handling. These types of treatment processes have been primarily designed for domestic wastewater treatment. But effluents emanating from textile, tannery, pharmaceutical and cosmetic industries require some additional treatment apart from the conventional treatment processes. Treatment of wastewater containing dyestuffs was conducted using a biological oxidation process where a sequencing batch reactor was used [5]. In another study magnesium chloride as a coagulant was used for the reduction of COD and BOD from textile effluent. In this study, use of ferrous sulphate as a coagulant was compared which showed that magnesium chloride acted as a better coagulant with about 99.5% and 97.1% colour removal and 63.1% COD removal [6]. Dried mass of *Azolla filiculoides* was used as biosorbent for the adsorption of basic orange dye where 833.33 mg/g biosorption capacity was obtained at pH 7.0, contact time

4 h with a dose of 5 g/L [7]. Wheat bran as a biosorbent was successfully utilized by Sulak and Yatmaz (2012) for the treatment of textile effluent [8]. Removal of remazol red F3B which was used as a model dye was observed using a dose of 0.25 g *Oreganum* stalks where 39.42 mg/g biosorption capacity was obtained. The stalks after surface activation were used for the removal of both acidic and basic dyes [9]. Removal of Cr from tannery effluent was carried out using waste materials like sawdust, rice husk, coir pith, charcoal, etc. Application of coir pith as such and with vermiculite as biosorbent in the column study resulted in about 94% and 93% removal of Cr respectively [10]. Verma et al. (2006) used a biosorbent prepared from tamarind hull for chromium removal [11]. Biosorbents prepared from *Spirogyra condensata* and *Rhizoclonium hieroglyphicum* were used to remove chromium from tannery effluent where maximum chromium uptake of 14 mg Cr (III)/g of algae at pH 5.0 was observed by *Spirogyra condensate* as compared to 11.81 mg of Cr (III)/g of algae at pH 4.0 using obtained for *Rhizoclonium hieroglyphicum* [12]. Neem sawdust in a fixed bed column was applied for removal of chromium from tannery effluent which resulted in an adsorption capacity of 33,009.9 mg g^{-1}[13]. Surface-activated brown seaweed (*Turbinaria* sp.) was carried out using H_2SO_4, $CaCl_2$ and $MgCl_2$. The activated biosorbent was observed for the removal of chromium from tannery wastewater. Maximum chromium uptake of 31 mg was observed at pH 3.5 using 1 g of biosorbent. Reusing the chromium-treated effluent for leather tanning showed comparable results when compared to conventional chrome-tanned leathers [14]. Bisorbents prepared from agricultural wastes like hazelnut shells [15], saw dust, cactus, pine needles, wool, olive cake [16], etc. were used for the removal of Cr (VI). Gupta et al. (1999) studied the removal of Cr (VI) from both real field wastewater and synthetic solution using sugarcane bagasse [17]. *P. putida* 5x-cell was used as a biosorbent combined with SBR process for the removal of Cu^{2+} from municipal wastewater with 87.3 mg × g (−1) adsorption capacity [18].

Carbon-containing waste materials like olive stones, pinewood, rice hulls, palm shell, sawdust, plum kernels, *Moringa oleifera*, seed husks, coconut shells, palm fruit. etc. have been employed for the preparation of biosorbents. Biosorbents prepared from fly ash, brick kiln ash and activated carbon were employed for the removal of organic matter from domestic wastewater resulting in 87.84%, 83.22% and 99.35% removal, respectively [19]. Carbon-containing low-cost biosorbents like coir and saw dust showed promising results in COD removal from domestic wastewater [20]. Molecular carbon-coated material along with activated carbon and CAC-type mixed adsorbent was studied by Devi and Dahiya (2008) for COD and BOD removal from domestic wastewater. They obtained 95.87%, 97.45%, 99.05% and 99.54%, removal, respectively [21]. Highly turbid and organic loaded sulphur black effluent (5912 NTU turbidity and 3910 mg/L COD) was treated with biosorbent prepared out of *Eichhornia crassipes* roots that resulted in 99% removal of dye and 80% removal of COD [22]. *Trewia nudiflora* as biosorbent resulted in the removal of chromium (VI) where 278 mg/g biosorption capacity was obtained [23]. Bioadsorbent prepared from algae *Rhizoclonium riparium* was utilized for the treatment of gray water. A removal efficiency of 96% was observed with 5 g/L biosorbent for the wastewater having COD loading (2400–44,800 mg/L) [24]. A combination of ceramic microfiltration and biosorption was used for the treatment of domestic and industrial effluent. The treatment was also conducted in a single mode using only biosorbent prepared from *Lagerstroemia* sp. The effective result was obtained for the combination process with 88% removal of contaminants for industrial effluent and 89% in case of domestic effluent respectively [25]. Biosorbent from *Trewia nudiflora* resulted in 99% remediation of Cr, Ni and Pb like heavy metals from various streams of tannery effluent [26]. Treatment of textile effluent using a combination process of ceramic microfiltration and biosorbent prepared from sugarcane bagasse resulted in 91% COD removal, 99% dye removal, 90% suspended solids removal and 99% turbidity removal [27]. Combined treatment of ceramic microfiltration and biosorbent developed from *Eichhornia crassipes* was applied for the treatment of pharmaceutical wastewater from ayurvedic medicine manufacturing industries resulting in 88% removal of colour, 88% removal of COD, 99% removal of suspended solids and 99% removal of turbidity respectively using 1 g/L biosorbent [28].

4 Types of Biosorbent

Selection and identification of biosorbents with higher affinity for metal ions or organic matter is a challenging task. Nature offers a variety of materials that can be used as biosorbents. Biosorbents can be derived from plant, microbes, and wastes from industries or agricultural activities. Ideally, biosorbents should have certain characteristics such as; high biosorption capacity, low cost, easily availability, effortless desorption of the material adsorbed, reusability of spent biosorbent [29].

4.1 Biosorbents from Industrial Waste By-Products

Waste by-products from different industries especially food industries are disposed of in large quantities. The use of these materials as biosorbents can reduce pollution loading in terms of zero waste disposal and waste treatment. Examples of food processing and industrial waste by-products are tea waste [30], sugar cane bagasse [27, 31], activated sludge [32, 33], waste green sand [34], fly ash [35], etc.

4.2 Biosorbents from Agricultural Wastes

Agricultural wastes contain a high percentage of cellulose, lignin, lipids, hemicelluloses, simple sugars, starch, hydrocarbons, etc. that have a higher affinity for metal ions. Grape fruit peel [36], bark of *Acacia leucocephala* [37], fruit peel of a *Trewia nudiflora* [26], *Lagerotroemia speciosa* [25], sugarcane bagasse [27, 38], wheat and peanut shell [39–41], vegetable wastes [42], neem bark [43], etc. are some examples of biosorbents prepared from agricultural wastes. Biosorbent prepared from sugarcane bagasse resulted in 91% dye removal and 91% of COD removal for textile effluent having COD loading of 2220 mg/L [27]. Biosorbent prepared from *Trewia nudiflora* resulted in a Cr^{+6} sorption of 278 mg/g capacity at a pH value of 2.0 and 0.75 g/L of biosorbent. [23]. Biosorbents from agricultural wastes are capable of metal uptake and removal of organic matter from effluents.

4.3 Microbial Biosorbents

Several living and dead microorganisms are widely used as biosorbents. Dead biomaterials are ideal as nutrients and monitoring of effluent quality are not required. Dead biomass is effective in metal binding and can treat complex wastewaters having a wide range of metal concentrations. Non-pathogenic microorganisms are used as biosorbents. Several species of macro and micro algae, fungi and bacteria are used as biosorbents. Algae like *Spirogyra* sp. [44], *Cladophora* sp. [45], *Ulva lactuca* [46], *Rhizoclonium riparium* [47], *Spirulina platensis* [48], *Oedogonium* sp. [49], *Laminaria* sp. [50] etc. can be applied as biosorbents. A study with biosorbent from *Rhizoclonium riparium* effectively removed 96% COD with 5 g/L biosorbent [24]. The sorption capacity of Pb and Cu from synthetic solutions using *Spirogyra* sp. and *Cladophora* sp. as biosorbent was found to be 87.2 and 38.2 mg/g respectively. Fungi like *Mucor rouxii* [51], *Rhizopus cohnii* [52], *Saccharomyces cerevisiae* [53], *Pencillium* sp. [54], *Aspergillus* sp. [55] *Pleurotus ostreatus* [56] and bacteria like *Bacillus subtilis* [57], *Streptomyces* sp. [58], *Enterobacter* sp. [59], *Pseudomonas* sp. [60], *Rhizobium* sp. [61] have been reported in the literature. Removal of cadmium from wastewater using *Rhizopus cohnii* as biosorbent resulted in 40.5 mg/g removal capacity. Zinc uptake of about 1699 g/g by yeast *Saccharomyces cerevisiae* was obtained when it was grown in media containing zinc of 1.4372 g/L concentration [62]. Microorganisms as biosorbent are gaining importance as it has high biosorption rate, highly available, more surface area to volume ratio, easy regeneration, etc. Algal cell wall contains cellulose, alginic acid and polysaccharides which participate in metal uptake. Most gram-positive bacteria are negatively charged due to the presence of phosphodiester bonds that help in the biosorption of divalent cations. Although gram-negative bacteria have less Peptidoglycan but outer layer of phospholipid and lipopolysaccharides creates a negative charge thus resulting in metal binding. Polysaccharides and glycoprotein-rich cell walls present in fungi and yeast contain amines, phosphates, carboxyls and hydroxyls that help in metal binding.

4.4 Agricultural or Industrial wastes

Agricultural or Industrial wastes include soybean hulls, cotton seeds hulls, fruit peels like tamarind, banana, sugarcane bagasse, sugarbeet pulp, seeds *of Carica papaya,* etc. [8, 43, 63, 64] These materials contain cellulosic materials like phenolics or carboxylic that help in biosorption process by creating cation exchange between the functional groups with that of metal ions.

The different types of biosorbents and their removal efficiency are shown in Table 1.

5 Factors Affecting Biosorption Process

The following factors govern the process of biosorption:

5.1 Temperature

The biosorption process is generally carried out between 20 and 35 °C as it is not much affected in this temperature range [77, 78]. Some biosorption processes may take place at higher or lower than ambient temperature. Increasing temperature increases the surface activity of biosorbent thereby increasing removal efficiency. The optimum temperature must be determined at which maximum removal is achieved. For the exothermic biosorption process, increasing temperature results in the decrease of removal while for the endothermic process, increasing temperature increases removal efficiency. Tasar et al. observed decreased lead(II) ions removal using peanut shells by increasing temperature [79].

5.2 pH

pH is an important parameter that affects biosorption. While a general pH range of 2.5–6 is best for maximum biosorption. pH for biosorption is very important which affects the biosorption process as well as the mechanism in some cases. Biosorption may increase with increasing pH till an optimum pH is attained after which further increase in pH does not increase biosorption. Low acidic pH favors biosorption of metal ions, i.e., copper, zinc, cobalt, cadmium, etc. [80]. Biosorption capacity increases with increasing pH as negative charges are exhibited by functional groups like carboxyl, hydroxyl, and phosphate groups due to deprotonation [81, 82].

5.3 Contact Time

Contact time might act as a limiting factor in the biosorption process. Under optimum experimental conditions, biosorption capacity increases with increasing contact

Table 1 Sorption capacity of various biosorbents

Water pollutant	Biosorbent	pH	Biosor bent dosage (g L^{-1})	Temperature	Pollutant concen tration (mg L^{-1})	Contact time (min)	Removal percen tage (%)	Maxi mum uptake capacity (mg g^{-1})	References
Methylene Blue Dye	Activated Carbon Prepared from *Citrullus lanatus* Rind	11	0.6	25 °C	100	–	98%	231.48	[65]
Methylene Blue Dye	Watermelon (*Citrullus lanatus*) rinds	5.6	0.6	30 °C	50–400	–	85%	188.68	[66]
Congo red Dye	Titanium oxide (TiO_2) nanoparticles loaded onto activated carbon prepared from bio-waste watermelon rind	4.3	0.8	25 °C	14.7	–	101.45%	17	[67]
Phenol red Dye	Titanium oxide (TiO_2) nanoparticles loaded onto activated carbon prepared from bio-waste watermelon rind	4.3	0.8	25 °C	45.9	–	100%	55.6	[67]
Basic Brown 16 Dye	Durian (*Duriozibethinusmurray*) Shell	8	1.0	–	15	30	74.26%–77.61%	–	[68]
Orange G Dye	Watermelon shell (*Citrullus vulgaris*)	2	25	30 °C	250	–	100%	27	[69]
Basic red 2 Dye	Watermelon shell (Citrullus vulgaris)	4	25	30 °C	250	–	100%	125	[69]

(continued)

Table 1 (continued)

Water pollutant	Biosorbent	pH	Biosor bent dosage (g L^{-1})	Temperature	Pollutant concen tration (mg L^{-1})	Contact time (min)	Removal percen tage (%)	Maxi mum uptake capacity (mg g^{-1})	References
Remazol Brilliant Blue R Dye	*Aspergillus iizukae* 605EAN	6	–	–	571.5	12,312 (8.55 days)	78.34%–80.00%	–	[70]
Crystal Ponceau 6R Dye	Pectin	2	0.25	–	34.32	540	99.20%	–	[71]
Methyl orange Dye	GrewiaVenusta Peel	2	0.5	–	150	–	85.00%	188.68	[71]
Indigo carmine Dye	Moringa Oleifera Seeds	5	0.5	–	150	–	–	31.25	[71]
Methylene Blue Dye	Carbonized watermelon (*Citrullus lanatus*) rind	5.6	0.8	30 °C	100	–	99%	200	[66]
Methyl orange Dye	*Aspergillus iizukae* 605EAN	–	–	–	400	4320 (3 days)	84.90%	–	[70]
Methyl red Dye	*Aspergillus iizukae* 605EAN	–	–	–	400	4320 (3 days)	47.65%	–	[70]
Remazol brilliant blue R Dye	*Aspergillus iizukae* 605EAN	–	–	–	100	12,960 (9 days)	94.78%	–	[70]
Cr(III)	Honeydew peel activated carbon with H_2SO_4 (HDP-ACS)	4.5	1.0	–	1000	40	83.49%	834.94	[72]

(continued)

Table 1 (continued)

Water pollutant	Biosorbent	pH	Biosor bent dosage (g L^{-1})	Temperature	Pollutant concen tration (mg L^{-1})	Contact time (min)	Removal percen tage (%)	Maxi mum uptake capacity (mg g^{-1})	References
Cr(III)	Honeydew peel activated carbon with H_3PO_4 (HDP-ACP)	4.5	1.0	–	1000	40	88.8%	888.85	[72]
Zn(II)	Honeydew peel activated carbon with H_2SO_4 (HDP-ACS)	5.5	1.0	–	400	40	81.55%	326.19	[72]
Zn(II)	Honeydew peel activated carbon with H_2SO_4 (HDP-ACP)	5.5	1.0	–	400	40	84.13%	336.51	[72]
Cd(II)	Watermelon Rind	9.12	1.0	20 °C	500	–	80%	40.16	[73]
Cu(II)	Activated watermelon based biosorbent	5	1.0	40 °C	10	–	88%	31.25	[74]
Cu(II)	Chitin	7	2.0	–	100	360	–	20.50	[75]
Cd(II)	Chitin	7	2.0	–	100	360	–	13.27	[75]
Zn(II)	Chitin	7	2.0	–	100	360	–	5.35	[75]

Source Reprinted from [76], Copyright (2023) with permission from Elseview time. After attaining equilibrium, i.e., when binding sites become fully saturated, increasing contact time would not have any further effect [83, 84].

5.4 Nature of Biosorbent

Surface treatment of biosorbents derived from different sources enhances the biosorption property. Pretreatment of biosorbents including physical and chemical treatments like autoclaving, drying, sonication, acid or alkali improves the biosorption process. It was observed that chemically pretreated olive tree increased the biosorption of lead(II) as compared to the untreated one. The authors observed that chemical treatment resulted in more acidic groups on the surface of the biosorbent making it more negatively charged thereby contributing to higher biosorption capacity [85].

5.5 Surface Area to Volume Ratio

Surface property of biosorbent is essential in the case of microbial biosorption. Metal binding on the surface of microorganisms is generally preferred but intracellular biosorption may also take place [24]. The higher the surface area of the biosorbent, the higher is the biosorption capacity. Shartooh observed that powdered lettuce leaves had more metal removal than dried lettuce leaves [86].

5.6 Concentration of Biomass

Higher biomass concentration results in higher biosorption capacity [24, 87, 88]. More biomass means more available sites for the binding of metal ions which in turn increases biosorption capacity. There exists a strong electrostatic interaction between the cells that results in solid uptake. [89]. But if biomass concentration is very high or metal ions concentration is higher, the interaction between available sites is hampered [90]. At low biomass doses, increased uptake of metal ions has been observed [91]. But, in a contaminated system, there will be competition for available binding sites. So, a lower biosorbent dose will decrease biosorption capacity. Effect of water hyacinth roots on removal of dye and COD showed that 99% dye removal was obtained at 0.5 g/L biosorbent dose. In a study, increasing the biosorbent dose increased COD removal, i.e., 68% at 1 g/L dose but further, increasing biosorbent dose showed no change in removal efficiency [92].

5.7 *Initial Sorbate Concentration*

Initial concentration of the sorbate provides the driving force that causes mass transfer between solvent and solid phase [93]. The biosorption process is triggered with initial metal concentration. Metal ions biosorption on biomass increases at lower initial concentrations. At a given biomass dose, metal removal increases with increasing initial concentration [24]. Biosorption of heavy metals using banana peel showed increasing biosorption capacity with increasing metal ion concentration [94]. Lower initial sorbate concentration may not fully reveal the biosorption efficiency of the biosorbent as available sites may not be completely utilized. When available binding sites are fully saturated, biosorption process is not further affected by increased initial concentration.

6 Desorption

Desorption is the process of reusing the spent biosorbent at the end of the biosorption process by use of specific desorbing agents. The regenerated biosorbent can be reused till the available sites are exhausted [95]. Desorption percent is calculated as follows:

$$\text{Desorption percentage}(\%) = (\text{Amount of Species Desorbed}$$
$$/\text{Amount of Species Loaded}) \times 100\%$$

The desorption percent value determines the reuse of biosorbent. Commonly used desorbing agents are: mineral and organic acids (such as hydrochloric acid, sulphuric acid, nitric acid and acetic acid, citric acid), alkalis (sodium hydroxide, sodium hydrogen carbonate, sodium carbonate and potassium hydroxide) and complexing agents (ethylene diamine tetra acetic acid, thiosulfate) [96]. Acidic desorbing agents are more effective than basic and neutral desorbing agents. The choice of desorbing agent depends on the fact that they should not modify the surface properties of the biosorbent and regain its original structural properties after the desorbing process to make certain that the biosorbent can be successfully reused [97].

7 Mechanism and Kinetics of Biosorption

7.1 *Biosorption Mechanism*

Biosorption of heavy metals occurs through distinct mechanisms, depending on the type and nature of biosorbents, environmental conditions, cellular metabolism as well as the site of metal sorption within the cells [98]. Biosorption mechanism is

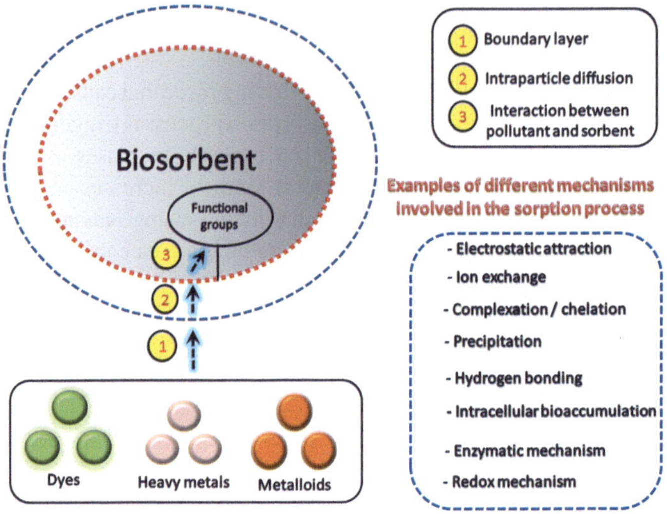

Fig. 1 Mechanism of biosorption (*Source* Reprinted from [76], Copyright (2023) with permission from Elsevier)

depicted in Fig. 1 and details of biosorption capacity and mechanism for different biosorbents are provided in Table 2.

- **Classification of biosorption mechanism based on cellular metabolism**:
- Metabolism dependent biosorption
- Metabolism independent biosorption
- **Classification of biosorption mechanism based on the site of metal sorption within the cells**:
- Extracellular accumulation/precipitation
- Cell surface sorption/precipitation
- Intracellular accumulation

Biosorption using live cells is referred to as a metabolism-dependent and energy-driven process (active mode of metal sequestration requiring ATP) [99]. Mechanisms underlying the process of metabolism-dependent biosorption involve transport across the cell membrane (i.e. intracellular accumulation) and precipitation (i.e. extracellular accumulation). Bioaccumulation of toxic pollutants is often mediated by cell metabolism [100]. However, the uptake of heavy metals by dead cells of biomass is independent of cell metabolic activities and ATP (passive biosorption). This type of biosorption can be attained through physical adsorption, ion exchange, complexation (or coordination) and chelation and precipitation [101]. Cell surface sorption also follows metabolism-independent pathways.

Table 2 Biosorption mechanism for different biosorbents

Biosorbent	Water Pollutant	Biosorption capacity (mg g^{-1})	Biosorption mechanism
Brown seaweed Sargassum muticum	Sb(III)	4 ± 1	Complexation and hydrogen bonding
Magnetized *C. micaceus*	Hg(II)	26.2	Complexation
Sargassum glaucescens	As(III)	116.6	Electrostatic interaction
Phanerochaete chrysosporium	Ni(II)	46.50	Chemical interaction
Renewable durian peels	Zn(II)	36.73	Electrostatic interaction
Sulphur functionalized marine brown algae *Cystoseira indica*	La(III)	185.44	Internal and external diffusion mechanisms
Gliricidia sepium leaf powder	Cr(VI)	35.71	Chelation, Ion exchange, Complexation, Physical adsorption, and chemical adsorption
Magnetized C. micaceus	Co(II)	24.7	Complexation
Yeast *Saccharomyces cerevisiae*	Cu(II)	4.73	Ion exchange
Sargassum glaucescens	As(V)	207.3	Electrostatic interaction
Formaldehyde-modified green tomato husk	Fe(III)	19.83	Precipitation and ion exchange
Bacterium *Bacillus amyloliquefaciens*	U(VI)	179.5	Chemical interaction
Phanerochaete chrysosporium	Cd(II)	71.43	Chemical interaction
Sulphur functionalized marine brown algae *Cystoseira indica*	Ce(III)	172.33	Internal and external diffusion mechanisms
Ion-imprinted Tetraethylenepentamine modified chitosan beads	Pb(II)	259.68	Complexation
Chitin	Au	35	Complexation
Formaldehyde-modified green tomato husk	Mn(II)	15.22	Ion exchange and complexation
Urea calcium alginate beads functionalized with Congo red dye	Cu(II)	441	Electrostatic interaction, ion exchange, and complexation

(continued)

Table 2 (continued)

Biosorbent	Water Pollutant	Biosorption capacity (mg g^{-1})	Biosorption mechanism
Green microalgae *Chlorella pyrenoidosa*	Rhodamine B dye	63.14	Electrostatic interaction
Sepia shells based composite	Methylene blue dye	254.05	Electrostatic interaction
Sepia shells based composite	Reactive black 5 dye	269.18	Electrostatic interaction
Magnetic chitosan Glutaraldehyde composite	Crystal violet dye	105.467	Electrostatic interaction
IL modified AC (ILAC)	Reactive blue dye	364.4	Electrostatic interaction
Ethylenediamine-modified fiber btained from natural Populus tremula	Acid Blue 25 dye	67	Van Der Waals' forces, π–π stacking, and hydrogen bond
Carica papaya wood	Malachite green dye	52.63	Electrostatic interaction
Grape pomace	KROM KGT dye	180.2 ± 3.2	Electrostatic interaction
Chemically modified masau stones	Orange(II) dye	136.8	Hydrogen bonding and electrostatic attraction
Millimeter-sized chitosan/ carboxymethyl cellulose hollow capsule	Methylene blue dye	64.6	Electrostatic attraction, complexation, hydrogen bond, and van der Waals interaction
Millimeter-sized chitosan/ carboxymethyl cellulose hollow capsule	Methyl orange dye	334.8	Electrostatic attraction, complexation, hydrogen bond, and van der Waals interaction
Millimeter-sized chitosan/ carboxymethyl cellulose hollow capsule	Acid blue-113 dye	526.8	Electrostatic attraction, complexation, hydrogen bond, and van der Waals interaction
Bio-sourced composite (activated carbon/TiO_2)	Diclofenac	153.8	Adsorption, photolysis, and photoinduced degradation

(continued)

Table 2 (continued)

Biosorbent	Water Pollutant	Biosorption capacity (mg g^{-1})	Biosorption mechanism
Bio-sourced composite (activated carbon/TiO_2)	Carbamazepine	105.3	Adsorption, photolysis, and photoinduced degradation
Bio-sourced composite (activated carbon/TiO_2)	Sulfamethoxazole	125.0	Adsorption, photolysis, and photoinduced degradation
Biopolymer electrospun nanofibres	Carbamazepine	31.25	Chemisorption
Biopolymer electrospun nanofibres	Ketoprofen	111.11	Chemisorption
Biopolymer electrospun nanofibres	Fenoprofen	333.33	Chemisorption
Biopolymer electrospun nanofibres	Diclofenac	125	Chemisorption
Biopolymer electrospun nanofibres	Ibuprofen	125	Chemisorption
Residual defatted biomass (RDB) of the fungus *Nigrospora* sp.	Procion red H–E7B dye	188.79	Electrostatic interaction
Clay	Phenol	15.11	Electrostatic interaction
Pine fruit shells PFS (BC550)	Phenol	26.738	Dispersive interaction and π–π interaction
Tithonia diversifolia activated carbon	Phenol	50.552	Dispersive interaction and π
Tithonia diversifolia activated carbon	2,4-dinitrophenol	42.607	Dispersive interaction and π
Biochar derived from corn straw	Atrazine	11.566	Electrostatic interaction
Corn cob-derived porous carbons (PCs)	Phenanthrene	692.27	Pore filling, hydrophobic effects, and π–π stacking interactions

Source Reprinted from [76], Copyright (2023) with permission from Elseview

7.1.1 Transport Across Cellular Membranes

Membrane transport of essential ions such as magnesium, sodium and potassium in microorganisms is obligatory for their cell viability and homeostasis. Metal import takes place firstly, by binding to the microbial cell wall (metabolism independent) and secondly, through metabolism-mediated uptake of metal ions into the intracellular compartments [85]. Highly selective heavy metals resemble essential ions with respect to charge and ionic radius and trick their intracellular localization through microbial metal transport systems across biomembranes. Microorganisms including bacteria, fungi, algae and yeasts have evolved different mechanisms of heavy metal removal from effluents with the help of extracellular adsorption and intracellular sequestration: (a) accumulation of metals bound to the cell wall in microbial vacuoles; (b) siderophore (iron chelating agents largely produced by fungi and bacteria with high affinity towards trivalent metals Al, Cr and Ga other than Fe)-assisted sequestration of heavy metals inside the microbial cells; (c) heavy metal detoxification by microbial secretion of metallothioneins (metal binding proteins) and phytochelatins and (d) heavy metal uptake by metal transporters (e.g. membrane proteins) [102–104].

7.1.2 Precipitation

The mechanism of metal ion precipitation may or may not be influenced by cellular metabolism [1]. The precipitation process which is dependent on metabolism is induced by the microbial active defence system as its typical response against heavy metal toxicity. Contrarily, metabolism independent precipitation is activated by physicochemical interactions between the functional groups present on the cell surface and metal ions. Strains of *Bacillus*, *Geobacter*, *Desulfuromonas*, *Pseudomonas*, *Streptomyces* and *Methylobacterium* are reported to alleviate metal stress by reducing them into less toxic forms [103]. Moreover, hydrogen sulfide generated by sulphate-reducing bacteria precipitates metal cations. The role of *Pseudomonas aeruginosa*, *Klebsiella planticola* and *Vibrio harveyi* in cadmium and lead precipitation as insoluble sulphides and lead phosphate salts is demonstrated in many studies [105]. Fungal species like *Aspergillus*, *Mucor*, *Penicillium* and *Trichoderma* also promote metal detoxification by reduction [103].

7.1.3 Physical Adsorption

Adsorption of heavy metals on the surfaces of microbial cell walls via electrostatic interactions like van der Waal's forces (weak intermolecular binding forces) is termed as physical adsorption or physisorption. pH of the solution and surface area of biosorbents strongly control the process of physisorption. Experimental evidences have shown that heavy metal detoxification by the dead cells of bacteria, fungi, algae and yeasts may also occur through non-covalent electrostatic interactions [85, 106].

7.1.4 Ion Exchange

Cell wall polysaccharides of microorganisms assist in their structural formation and proper functioning of the cells. In the presence of a toxic environment, heavy metal ions are interchanged with the counter ions of the polysaccharide chains on the surface of the cell wall, thereby facilitating the biosorption mechanism [105]. Uptake of heavy metals through ionic exchange can be achieved either by cationic (viz. carboxyl and sulfonic acid groups) or anionic exchangers (viz. amino or imidazole groups). pH of the aqueous solution may have a direct impact on the mechanism of ion exchange. In response to heavy metal stress, microbial communities (mainly, species of fungi and algae) release increased levels of Na^{+}, K^{+}, Ca^{2+}, Mg^{2+}, etc. to sorb heavy metals (e.g. Cu^{2+}, Cd^{2+}, Zn^{2+}, Pb^{2+}, Cr^{3+}) through the exchange of ions [85].

7.1.5 Complexation

Complexation is an effective means of heavy metal bioremediation, wherein metal ions are biosorbed through the formation of complexes with the active groups or ligands (i.e. sulfhydryl, carbonyl, thiol, phosphate, hydroxyl, amino and carboxyl groups of microbial polysaccharides, proteins and lipids of cell wall which maintain a uniform negative charge and have high affinity for binding with metal ions) on the cell surfaces by covalent bonding (precisely, coordinate bonding). A wide variety of fungi and bacteria are known to produce organic acids for sequestration of soluble metal ions [107]. Mostly, gluconic, oxalic, acetic, citric, fumaric, lactic and malic acids synthesized by microorganisms act as natural chelating agents that cause the chelation of heavy metals, producing metallo-organic compounds. These organic acids also lead to solubilization of heavy metal compounds, causing eventual permeation within the cell [108–110]. It has been reported that oxalic and citric acids are secreted excessively by the fungus *Beauveria caledonica* in heavy metal stress, thereby neutralizing the ions of Cd, Cu, Pb and Zn [111]. In many cases, hyphae of filamentous mycorrhizal fungi retain oxalate crystals which, in turn, chelate and detoxify heavy metals [112]. Fungal transformation of heavy metals into harmless products with less toxicity, aided by oxalate crystal producing wood rotting fungi *Fomitopsis cf. meliae* and *Ganoderma aff. steyaertanum* has also been observed by many researchers [113]. Additionally, several bacteria of genera *Escherichia, Corynebacterium, Streptomyces, Saccharomyces, Serratia, Pseudomonas* and *Deinococcus* have been found to chelate heavy metals [114].

7.2 *Kinetics of Biosorption*

Biosorption of heavy metals from wastewater is usually evaluated in terms of biosorption capacity and removal efficiency of heavy metals from the solution, which is expressed by the following formulae (1 and 2) [115–117]:

$$q_e = \frac{(C_i - C_e)V}{M} \tag{1}$$

$$R\,(\%) = \frac{(C_i - C_e)}{C_i} \times 100 \tag{2}$$

where q_e is the metal uptake by biomass (mg g^{-1}), C_i is the initial concentration of metals in solution (mg L^{-1}), C_e is the equilibrium concentration of metals (mg L^{-1}), V is the volume of metal-containing solution (L), M is the mass of biosorbent (g) and R is the removal efficiency of metals (%). Multiple parameters, in particular, pH, ionic strength, contact time, temperature, complexation, hydrolysis, ion exchange, stability of the colloidal system, competitive nature of heavy metals for binding sites, physiological activities of living cells and concentrations of adsorbent and adsorbate also modulate the biosorption capacity [118, 119].

Biosorption kinetics defines the rate of binding of heavy metal ions on the surfaces of microbial cells, providing an insight into the biosorption mechanism either through diffusion or chemical interactions. The reaction models of kinetic studies have been developed based on: (a) external diffusion (i.e. diffusion of metals into the exterior surface of biosorbents from the aqueous solution); (b) process of metal transport across plasma membrane; (c) cell surface adsorption of heavy metals and (d) uptake of metal ions bound to the active sites or reactive groups of cell wall of microorganisms [98, 120]. A combination of two or more processes may control the overall adsorption rate for complex systems. Modelling and simulation of kinetic data of biosorption are prerequisites for comparing the efficiencies of biosorbents in treating heavy metal-rich wastewaters under varying operational conditions for industrial applications. The widely used kinetic models are presented in Table 3.

8 Biosorption Isotherms

Biosorption isotherm models are normally applied for illustration of the biosorption mechanism along with the understanding of binding affinity and surface properties of the biosorbents on the basis of experimental data and assumptions of thermodynamic processes [28, 123]. Adsorption isotherms also help optimize the adsorbent dosage for heavy metal removal from the waste streams, affecting the quantities of adsorbed metal ions. The representative isotherm equations are summarized in Table 4.

Langmuir and Freundlich models have most commonly been applied in biosorption equilibrium studies [23, 27, 28]. Many studies have reported that the Sips and Redlich-Peterson models can efficiently predict the biosorption of methylene blue dye and Zn^{2+} using lemon peel and *Rhizopus arrhizus* (zygomycete fungi), respectively [128, 129]. Frumkin model has also been extensively used in determining the characteristic parameters of biosorption of dyes onto activated carbon generated

Table 3 Kinetics of biosorption process [114, 118, 121, 122]

Kinetic model	Model description	Kinetic equation	Kinetic parameters
Zero-order	Rate of adsorption is independent of the concentration of adsorbate or adsorbent	$\frac{dq}{dt} = k_0$ or, $q_t = q_e + k_0t$	q_t = Adsorption capacity (mg g^{-1}) at time t (min); q_e = Adsorption capacity (mg g^{-1}) at the time of equilibrium; k_0 = Zero-order rate constant (mg g^{-1} min^{-1})
Lagergren pseudo-first-order	Biosorption rate is directly proportional to the number of accessible free sites present on the biosorbent surface	$\frac{dq}{dt} = k_1(q - q_e)$ or, $\ln(q_e - q_t) = \ln q_e - k_1 t$	k_1 = First-order rate constant (min^{-1})
Pseudo-second-order	This model predicts the mechanism of ion exchange whereby, the essential metals attached to the microalgal cell surfaces are displaced by heavy metal ions	$\frac{dq}{dt} = k_2(q - q_e)^2$ or, $\frac{t}{q_t} = \frac{t}{q_e} + \frac{1}{k_2 {q_e}^2}$	k_2 = Second-order rate constant (g mg^{-1} min^{-1})
Weber's intraparticle mass transfer diffusion	It determines the mechanism of diffusion in the adsorption and predicts whether the rate-controlling step is diffusive in nature	$q_t = k_{id}t^{1/2} + C$	k_{id} = Intraparticle diffusion rate constant (mg g^{-1} $min^{-1/2}$); C = Intercept (mg g^{-1})
Liquid film diffusion	It leads to the understanding of transport of heavy metal ions at the solid–liquid boundary, i.e., diffusion of ions from the liquid phase to the solid phase boundary during adsorption	$-\ln(1 - F) = k_{fd}t$, $F = \frac{q_t}{q_e}$	k_{fd}= Film diffusion rate constant (min^{-1}); F = Fractional attainment of equilibrium
Elovich	Most useful model which has been introduced to elucidate the kinetics of chemisorption, assuming energetic heterogeneity of the entire adsorption system	$\frac{dq_t}{dt} = \alpha e^{-\beta q_t}$ or, $q_t = \frac{1}{\beta}\ln \alpha \beta + \frac{1}{\beta}\ln t$	α = Initial biosorption rate (mg g^{-1} min^{-1}); β = Degree of surface coverage and activation energy for chemisorption (g mg^{-1})

from apricot wastes [130]. In a separate study conducted by Vijayaraghavan et al. (2006) using marine brown algal species *Sargassum wightii* for Ni^{2+} biosorption, the authors found better model fitting with three-parameter isotherms, namely, Sips, Redlich-Peterson, Khan, Toth and Radke-Prausnitz as compared to two-parameter isotherms such as Langmuir, Freundlich, Dubinin-Radushkevich, Flory–Huggins and Temkin, the best among all of them being the Toth isotherm model [131]. In certain

Table 4 Isotherm models of biosorption process [24, 120, 124–127]

Isotherm model	Model description	Adsorption isotherm equation	Parameters	Remarks
Langmuir	This model explains the chemical sorption process and assumes formation of a uniform monolayer of adsorbate molecules (exhibiting ideal gas behaviour at isothermal conditions) surrounding a homogeneous solid surface having definite adsorption sites (with equivalent energies but no effect of lateral interactions on adsorption)	$\mathrm{Q_e} = \mathrm{Q^e_{th}} \frac{\mathrm{K_{eq} C_e}}{1+\mathrm{K_{eq} C_e}}$ or, $\frac{\mathrm{C_e}}{\mathrm{Q_e}} = \frac{1}{\mathrm{K_{eq} Q^e_{th}}} + \frac{\mathrm{C_e}}{\mathrm{Q^e_{th}}}$	$\mathrm{Q_e}$ = Adsorption capacity by weight at equilibrium (mg $\mathrm{g^{-1}}$) $\mathrm{Q^e_{th}}$ = Theoretical maximum adsorption capacity by weight (mg $\mathrm{g^{-1}}$) $\mathrm{K_{eq}}$ = Adsorption equilibrium constant (L $\mathrm{g^{-1}}$) $\mathrm{C_e}$ = Equilibrium concentration of adsorbate (mg $\mathrm{L^{-1}}$)	
Freundlich	It reflects adsorption onto heterogeneous solid surfaces having different types of adsorption sites with varying binding energies for adsorbate molecules. It is completely an empirical isotherm equation, acceptable for up to a certain concentration, beyond which the equation becomes non-linear	$\mathrm{Q_e} = \mathrm{k_F C_e^{1/n_F}}$ or, $\log Q_e = \frac{1}{n_F} \log C_e + \log k_F$	$\mathrm{k_F}$ = Freundlich model constant (L $\mathrm{g^{-1}}$) n_F = Heterogeneity factor of Freundlich isotherm model (dimensionless)	

(continued)

Table 4 (continued)

Isotherm model	Model description	Adsorption isotherm equation	Parameters	Remarks
Sips	Combination of Freundlich and Langmuir isotherm models, which describes heterogeneous adsorption, avoiding the shortcoming of increased adsorbate concentration in Freundlich isotherm	$Q_e = Q_{th}^{e} \frac{K_{eq}{C_e}^{n_s}}{1+K_{eq}{C_e}^{n_s}}$	n_s = Sips isotherm model exponent	At low concentrations of adsorbed solute, the model equation changes to Freundlich isotherm, while on the contrary, at high adsorbate concentrations, it assumes the phenomenon of monolayer adsorption as in the case with Langmuir isotherm
Koble-Corrigan	Three-parameter hybrid model having combined features of Freundlich and Langmuir isotherms for predicting the equilibrium adsorption data	$Q_e = \frac{a_{KC}C_e^n}{1+b_{KC}C_e^n}$ or, $\frac{1}{Q_e} = \frac{1}{a_{KC}C_e^n} + \frac{b_{KC}}{a_{KC}}$	a_{KC} = Koble-Corrigan isotherm constant (L^n mg^{1-n} g^{-1}) b_{KC} = Koble-Corrigan isotherm constant (L mg^{-1})^n; n = Adsorption intensity	
Redlich-Peterson	Three-parameter adsorption isotherm model, comprising the features of both Freundlich and Langmuir isotherms. It does not obey the ideal characteristics of monolayer adsorption and can be employed for both homogeneous and heterogeneous systems over a broad range of adsorbate concentrations	$Q_e = K_{rp} \frac{C_e}{1+\propto_{rp}{C_e}^{\beta}}$	K_{rp} = Redlich-Peterson isotherm model constant (L g^{-1}) $\propto_{rp}$ = Redlich-Peterson isotherm constant (L mg^{-1}); β = Redlich-Peterson isotherm model exponent ($0 < \beta < 1$)	If $\beta = 1$, the equation modifies to Langmuir isotherm If $\propto_{rp}{C_e}^{\beta} >> 1$, the equation approaches to Freundlich isotherm

(continued)

Table 4 (continued)

Isotherm model	Model description	Adsorption isotherm equation	Parameters	Remarks
Khan	It depicts the process of adsorption of aromatic pollutants on activated carbon with temperature dependency. This model also represents a combined form of Freundlich and Langmuir isotherms	$Q_e = Q_{max}\frac{b_K C_e}{(1+b_K C_e)^{n_K}}$	Q_{max} = Maximum adsorption capacity (mg g^{-1}) b_K and n_K = Khan model constants (dimensionless)	If $n_K = 1$, the equation tends to Langmuir isotherm If $b_K C_e >> 1$, the model equation simplifies to Freundlich isotherm
Toth	The concept of the isotherm is based on the potential theory of adsorption which states that there is no influence of temperature on the adsorption potential and characteristic adsorption curve and hence can be applied for heterogeneous biosorbent surfaces	$Q_e = Q_{max}\frac{b_T C_e}{[1+(b_T C_e)^{1/n_T}]^{n_T}}$	b_T = Toth model constant (dimensionless) n_T = Toth isotherm exponent ($0 < n_T \leq 1$)	If $n_T = 1$, the equation turns into a Langmuir-type isotherm, but unable to explain the Freundlich model
Radke-Prausnitz	It is applicable to adsorption systems with a wide range of adsorbate concentrations. In certain cases, this model can also be used at low concentrations of adsorbate	$Q_e = a_R b_R \frac{C_e^{\propto_R}}{a_R + b_R C_e^{\propto_R - 1}}$ or , $Q_e = a_R \frac{C_e^{\propto_R}}{K_R + C_e^{\propto_R - 1}}$, $K_R = \frac{a_R}{b_R}$	a_R and b_R = Radke-Prausnitz isotherm model constants $\propto_R$ = Radke-Prausnitz model exponent	If $K_R >> C_e^{\propto_R - 1}$, the equation will be converted to Freundlich-type isotherm as follows: $Q_e = \frac{a_R}{K_R} C_e^{\propto_R}$ Now, if $K_R << C_e^{\propto_R - 1}$, the model equation becomes: $Q_e = a_R C_e$

(continued)

Table 4 (continued)

Isotherm model	Model description	Adsorption isotherm equation	Parameters	Remarks
Dubinin-Radushkevich	This model considers the impact of porous structures of adsorbents and can be used for estimating mean adsorption energy that may help in determining the occurrence of either physisorption or chemisorption	$Q_e = Q_{DR}\exp(-K_{DR}\varepsilon^2)$ or, $\ln Q_e =$ $\ln Q_{DR} - K_{DR}R^2T^2\ln^2(1 + \frac{1}{C_e})$, $\varepsilon = RT\ln(1 + \frac{1}{C_e})$	Q_{DR} = Dubinin-Radushkevich constant, signifying theoretical monolayer saturation capacity (mg g^{-1}) K_{DR} = Constant of adsorption energy, directly associated to mean free adsorption energy (E)= $1/\sqrt{2K_{DR}}$ (KJ mol^{-1}) ε = Polanyi adsorption potential (J mol^{-1})	
Frumkin	It takes into consideration the interactions of adsorbed species	$\frac{\theta}{1-\theta}\exp(-f\theta) = K_F C_e$ or, $\ln\left[\left(\frac{Q_e}{Q_F - Q_e}\right)\frac{1}{C_e}\right] = \ln K_F + 2f\frac{Q_e}{Q_F}$	Θ = Extent of coverage of the adsorbent surface ($0 \leq \Theta \leq 1$) f = Interaction coefficient K_F = Equilibrium constant Q_F = Theoretical monolayer saturation capacity (mg g^{-1})	If f = 0, implying no interactions between the adsorbates, the model equation will be reduced to Langmuir isotherm
Flory–Huggins	It deals with the extent of surface coverage of adsorbate on adsorbent surface and is relevant for the molecules that vary in size considerably	$\frac{\theta}{n_{FH}(1-\theta)^{n_{FH}}}\exp(-2n_{FH}\alpha_{FH}\theta) = K_{FH}C_e$	n_{FH}= A constant α_{FH} = A constant representing the interactions of adsorbed molecules K_{FH} = Isotherm equilibrium constant	If $n_{FH} = 1$, the model equation will change to Frumkin isotherm

(continued)

Table 4 (continued)

Isotherm model	Model description	Adsorption isotherm equation	Parameters	Remarks
Brunauer–Emmett–Teller (BET)	It is based on multilayer physisorption process, assuming homogeneous solid surfaces with no lateral interactions	$Q_e = \frac{BC_eQ^0}{(C_S - C_e)[(1+(B-1)C_e)/C_S]}$	Q^0 = Amount of solute adsorbed per unit weight of adsorbent during a complete monolayer formation (mg g^{-1}) B = A constant which is related to the energy of interactions with the surface C_S = Saturation concentration of solute (mg L^{-1})	Provides a basis for the analysis of specific surface area of materials
Temkin	The model suggests a linear decline in heat of adsorption or adsorption energy with the increase in surface coverage of a specific adsorbate molecule on a given surface	$Q_e = \frac{RT}{b_t}\ln(a_tC_e)$	R = Universal gas constant (8.314 J K^{-1} mol^{-1}) T = Absolute temperature (Kelvin) b_t = A constant depicting the heat of adsorption a_t = Temkin isotherm constant	
Hill	It describes adsorption as a homotropic cooperative process, where binding of a ligand at one site of a molecule facilitates the binding of other ligands to the different binding sites of the same molecule	$Q_e = \frac{Q_{S_H}C_e^{n_H}}{K_D + C_e^{n_H}}$ or, $\log\left(\frac{Q_e}{Q_{S_H} - Q_e}\right) = n_H \log C_e - \log K_D$	Q_{S_H} = Hill isotherm maximum uptake saturation (mg L^{-1}); K_D = Hill constant n_H = Hill cooperativity coefficient of the binding interaction	

cases of organic load reduction from gray water using algae as biosorbents, it has been described that Koble-Corrigan isotherm outperformed Redlich-Peterson, Langmuir and Freundlich models [24]. Nevertheless, investigations have also indicated that BET, Hill and Freundlich isotherms well fitted the data generated experimentally during biosorption of ochratoxin A (mycotoxins) onto yeast industry byproducts [132]. Furthermore, Dubinin-Radushkevich, Flory–Huggins and Temkin models have been evaluated for analysing the equilibrium experimental data of Al^{3+}, Ag^{+} and Co^{2+} biosorption by agro-wastes over a broad range of initial concentrations of metal ions [133].

9 Biosorption Thermodynamics

Thermodynamic parameters such as changes in enthalpy (ΔH^0—liberation or absorption of heat during adsorbate-adsorbent interactions at standard state in kJ mol^{-1}) and entropy (ΔS^0—binding affinity of the adsorbates at standard state in kJ K^{-1} mol^{-1}) relate to the standard Gibbs free energy change (ΔG^0 in kJ mol^{-1}) representing the spontaneity of an adsorption process and is given by the following expressions (Eqs. 3 and 4) [23, 24, 121]:

$$\Delta G^0 = \Delta H^0 - T\Delta S^0 \tag{3}$$

$$\Delta G^0 = -RT\ln K_{eq} \tag{4}$$

By equating Eqs. 3 and 4, the resultant expression (Eq. 5) becomes:

$$\ln K_{eq} = \frac{\Delta S^0}{R} - \frac{\Delta H^0}{R}\frac{1}{T} \tag{5}$$

Equation 5 is also known as van't Hoff equation.

Biosorption equilibrium constant K_{eq} (L g^{-1}) is expressed as follows (Eq. 6):

$$K_{eq} = \frac{Q_e}{C_e} \tag{6}$$

where T is the absolute temperature (K), R is the gas constant (8.314 J K^{-1} mol^{-1}), Q_e is the adsorption capacity by weight at equilibrium (mg g^{-1}) and C_e is the concentration at equilibrium of adsorbate (mg L^{-1}).

The spontaneity of biosorption depends on the value of ΔG. If $\Delta G < 0$ (negative), the process is said to be spontaneous, but if $\Delta G > 0$ (positive), it is non-spontaneous [119]. It is evident that both ΔH^0 and ΔS^0 of biosorption can be estimated through the linear plot of $\ln K_{eq}$ vs 1/T. It is important to analyse the parameters thermodynamically to explain the viability and nature of the temperature-dependent biosorption process for eradicating heavy metals together with dyes, pesticides and phenolic compounds [134]. Most biosorption researches in the last decade focused on the investigation of thermodynamics for elucidating the overall adsorption and desorption processes via the interactions between biosorbates and biosorbents having distinct structural forms, composition and binding affinities [135].

10 Applications of Biosorption Process

During the past few years, the application of biosorption has acquired appreciable recognition as an eco-friendly and green technology employing bio-based materials as adsorbents for sustainable treatment of wastewater or effluents charged with dyes, heavy metals and other organic pollutants. Toxicity, abundance, recalcitrance and persistence of the contaminants in the environment affect both humans and other biotic communities equally badly due to their accumulation with time. High efficiency, cost-effectiveness and reduction of sludge (chemical or biological) volume generated make the biosorption techniques more applicable for large-scale commercial exploitation.

10.1 Wastewater Treatment

Numerous reports have revealed the contributions of potential biosorbents prepared from the roots of aquatic macrophytes *Eichhornia crassipes* (water hyacinth), biomass of green algae *Rhizoclonium riparium*, fruit peels of *Lagerstroemia speciosa* and *Trewia nudiflora* and sugarcane bagasse in remediation of wastewaters discharged from households and industries including tannery, textile or garment processing and pharmaceutical as well as composite wastewaters loaded with both domestic and industrial contaminants [22–28, 115, 136]. The authors proved the efficacy of an integrated wastewater treatment approach using biosorbents and ceramic microfiltration process with a substantial decrease in the obnoxious constituents, viz. biochemical and chemical oxygen demands, turbidity, total suspended and dissolved solids, colour and hexavalent chromium. Therefore, hybrid systems, viz. biosorbent-assisted membrane filtration for wastewater treatment have made remarkable progress in the reduction of organic and inorganic pollutant load.

10.2 Heavy Metal Removal

For addressing the consequences of heavy metal (Pb, Cu, Co, Li, Cd, As, Cr (trivalent and hexavalent), Zn, Mn, Ni, dichromate, etc.) contamination, several prospective biosorbents such as live and dead biomass of *Bacillus xiamensis* (marine bacterium), modified rice husk, nanoparticles of wild herbs, leaves of *Thevetia peruviana*, soya bean seeds, olive stone, leaf biomass of *Jatropha curcas*, groundnut husk, extracellular polymeric substances (EPS) of *Agrobacterium tumefaciens*, green filamentous macroalgae and brown seaweeds have also been studied [45, 116, 119, 120, 134, 137–142]. In recent times, the heavy metal biosorption performances of genetically engineered microorganisms with extracellular recombinant metal-binding proteins and peptides have been tested owing to their enhanced capacity and specificity after genetic modifications [65]. However, the engineered microbes may likely to be susceptible to the variability in pH and ionic strength of wastewaters [114].

In addition, the process of heavy metal biosorption has been implemented in continuous dynamic systems, for example, column reactors (fluidised-bed reactor column, fixed-bed column, packed-bed column and flow-through column) other than batch stirred tank reactors. The dynamic removal of heavy metals requires critical examination of the operating conditions such as column length, interstitial velocity, time of operation, feed solution concentration, metal-sorbent affinity and mass transfer coefficient for their optimization for maximum removal efficiency [101, 143–147]. The dynamic equilibrium data of biosorption are usually approximated with the help of a breakthrough curve which is a graphical representation of the time required for biosorption against the relative adsorbate concentration in the effluent stream.

10.3 Removal of Dyes and Other Organic Pollutants

Removal of dyes from industrial effluents has been a challenging task for the scientific communities because of the carcinogenicity and mutagenicity of dyes hampering the entire ecosystem as a result of biomagnification. They may be classified as azoic, reactive, direct, acidic, basic, disperse and sulphur dyes, of which some are highly soluble in water and others are water-insoluble. In this context, the practical applicability of biosorption in dye removal has also been acknowledged. Widespread usage of agricultural wastes, microbial biomass, leaf sheath, fermentation waste biomass, seaweeds and fruit pomace (skin, pulp, stem and seeds of fruits) in biosorption of toxic dyes is well-documented in multiple investigations. Diverse groups of dyes, for instance, vinyl sulphone reactive dyes, rhodamine-B, reactive blue 5G, congo red, basic red 2, orange G, bromocresol green, brilliant green, bromophenol blue, basic violet, coomassie brilliant blue, methylene blue, methyl orange, etc. in wastewaters have been treated effectively utilizing biosorbents with high adsorption capacities [69, 148–154].

Besides, biosorption with live and dead microbial biomass also turns out to be a propitious alternative for the removal of phenolics, pesticides, petroleum pollutants, antibiotics, endocrine disruption compounds, pharmaceuticals and personal care products, dioxins, dibenzofurans and polychlorinated biphenyls for detoxification of pollutant-bearing effluents [155–162]. The authors suggested that different pollutants undergo different biosorption patterns, depending on the specific strain, pH, incubation time, temperature and concentration of pollutants present and hence proved the profound impact of microbial cell surface structures on biosorption.

10.4 Biosorption in the Production of Feed Additives and Recovery of Valuable Metals and Pharmaceuticals

Biosorption technology can also be used for the production of dietary feed supplements (enriched with macro and microelements) for animals and fertilizers for plants [101]. Other new aspects of the biosorption process deal with the recovery of gold, silver, platinum and palladium as precious metals and proteins, peptides and antibiotics as pharmaceuticals from aqueous solutions/industrial wastes [163, 164].

11 Functional Characteristics of Different Biosorbents

Biosorbents obtained from microbial biomass (bacteria, fungi, yeasts, microalgae, cyanobacteria, archaea), macroalgae, agricultural wastes, natural plant residues, tree bark, industrial wastes (sludge, fermentation and food wastes, etc.), sawdust, peat moss and some other materials including cellulose, chitosan, etc. are currently under consideration for their practical use in biosorption [164]. Adsorption capacities of biosorbents are controlled by the experimental conditions and process of pretreatment or chemical modifications by oxidizing agents, organic compounds, organic acids, bases, alkali, etc. Chemically modified biosorbents display higher adsorption capacities than untreated forms [165]. Analytical techniques that have been explored for studying the characteristics of different kinds of biosorbents constitute scanning or transmission electron microscopy (SEM or TEM) with energy-dispersive X-ray microanalysis (for examining biosorbent morphology and elemental composition), potentiometric titration (for quantitative analysis of the biosorbent active sites), Fourier-transform infrared (FTIR), electron spin resonance (ESR) and nuclear magnetic resonance (NMR) spectroscopies (for determination of active centres), thermogravimetric analysis (TGA) and differential scanning calorimetry (DSC) (for checking the thermal stability of biosorbents) and XPS (in order to analyse the oxidation state and constitution of functional groups of biosorbent surface) [98].

Chemistry of the functional groups of microbial cell surfaces contributes to metal sorption to a great extent. Peptidoglycans are the principal structural units of the cell wall of bacteria and generally contain cross-linked glycan chains with periodic N-acetylglucosamine (NAG) and N-acetylmuramic acid (NAM) residues [166]. Carboxyl groups of thick and dense peptidoglycan layer of Gram-positive bacterial cell wall serve as the prime sites of metal cation binding in comparison with the Gram-negative bacteria where biosorption is mainly aided by the phosphate groups of cell wall [81]. Protein or glycoprotein surface layer (S-layer) and sheaths found in the majority of bacterial species can also adsorb metal ions [98]. Cell walls of cyanobacteria possess a thick peptidoglycan layer and an outer membrane consisting of proteins, lipids and carotenoids. Thus, cyanobacterial sheath, peptidoglycans and mucilage polysaccharides (EPS) are proven to be active sorption centres [98]. Archaeal cell wall structures are species-specific and have some unique features which lack peptidoglycans, but contain pseudomurein (resembling bacterial peptidoglycans), S-layer, protein sheaths, halomucin, glutaminylglycan, heteropolysaccharides, methanochondroitin and lypoglycan that provide sites for binding of metal ions [98, 167]. The morphological diversity of the algal cell wall constituents makes them a good choice as biosorbents. Polysaccharides and proteoglycans are the major cell wall components, cellulose being the most common in almost all the algal species. Fibrillar, matrix and crystalline polymers present in the algal cell walls interact with various metal ions through phosphate, sulphate, hydroxyl, amino, amine and imidazole groups [49, 81].

Filamentous fungi are also among the most candidate biosorbents due to the presence of their vegetative parts mycelia produced from hyphal branching [168]. Fungal biomass can easily be grown in relatively low-cost culture media at a large scale with high yield. Their abundance in waste products of fermentation during industrial processing renders them more advantageous over other microbes [169]. The rigid cell wall of fungi is composed of glycoproteins, polysaccharides and pigments e.g. chitins, chitosan (deacetylated derivative of chitin), mannoproteins, glucans and melanin which contain active functional groups (carboxyl, alcoholic, carbonyl, phenolic, methoxyl and hydroxyl) for heavy metal biosorption [170]. Accumulation of metal ions in fungal mycelia and spores occurs through intracellular and extracellular processes. It is well-known that fungal chitin and chitosan increase the metal uptake and binding capacities [170].

Zhou et al. (2016) assessed the adsorption characteristics of biosorbents derived from waste-activated sludge for eradicating Pb^{2+} and Zn^{2+} ions from aqueous metal solutions and specified the combined impact of electrostatic interactions, complexation of metal ions with –CN and $-NH_2$ groups and ionic exchange between –COOH groups and metal ions [171]. Biosorption with immobilized microbial biomass has also displayed greater removal of toxic metals, dyes, fluorides, phthalates and pharmaceuticals from industrial effluent streams with more than 98% efficiency [172]. Immobilization results in highly porous surfaces of the biosorbents with a huge increase in specific surface area and high adsorption of heavy metals forming plaque-like hard crystals on the surface and within the walls of the pores. Bind et al. (2018) evaluated the biosorption potential of aquatic macrophytes *Azolla filiculoides* and *Hydrilla verticillata* for remediation of heavy metals and indicated the role of hydroxyl, ester, amine, alcohol, carboxyl, aldehyde and alkene functional groups in sorption. However, the authors noticed no significant changes in the morphology of biosorbents due to biosorption [173]. Bhattacharya et al. (2014a) explored the biomass of algae *Rhizoclonium riparium* as a biosorbent for reducing organic load from kitchen wastewater and observed that the pores of the biosorbent were covered with the organic and inorganic components, creating larger agglomerations with apparent smooth surface and decrease in the number of pores. *Lagerostroemia* sp.

was used as a biosorbent for the remediation of wastewater that is composite in nature having a high organic load. Treatment study was conducted in a single mode and combined with a ceramic membrane. FESEM analysis depicted porous particles of biosorbent arranged irregularly whereas after biosorption, embedded particles in pore suggest attachment of organic and inorganic materials on surface (Fig. 2a–c) [24, 25]. Fungal variants, namely *Rhizopus*, *Aspergillus*, *Saccharomyces*, *Streptoverticillum*, *Trichoderma*, *Geotrichum*, *Fusarium*, *Alternaria* and *Penicillium* have gained importance for heavy metal pollution remediation on account of the presence of their cell wall materials at a very high percentage, thereby increasing the diversity of functional groups and hence the metal sequestration ability [169, 174].

Biosorbents of plant origin (predominantly leaves and other plant residues) exhibit uneven (or rough) and heterogeneous surfaces with various cavities, clefts, perforations, voids, breaches, crevices and convolutions which enhance surface sorption [175]. Biosorbent surface heterogeneity increases its surface area for greater metal ion sorption. The pores on the surface of biosorbents get clogged after biosorption, making the surfaces smoother with reduced unevenness and convolutions [176]. Bhattacharya et al. (2011) studied the efficacy of *Eichhornia crassipes* roots (ECR) as biosorbent for domestic wastewater treatment and found organic particles adhered to the porous biosorbent surfaces in close contact and major involvement of –OH group, –C–C– group and C–O stretching in biosorption [115]. FTIR analysis of

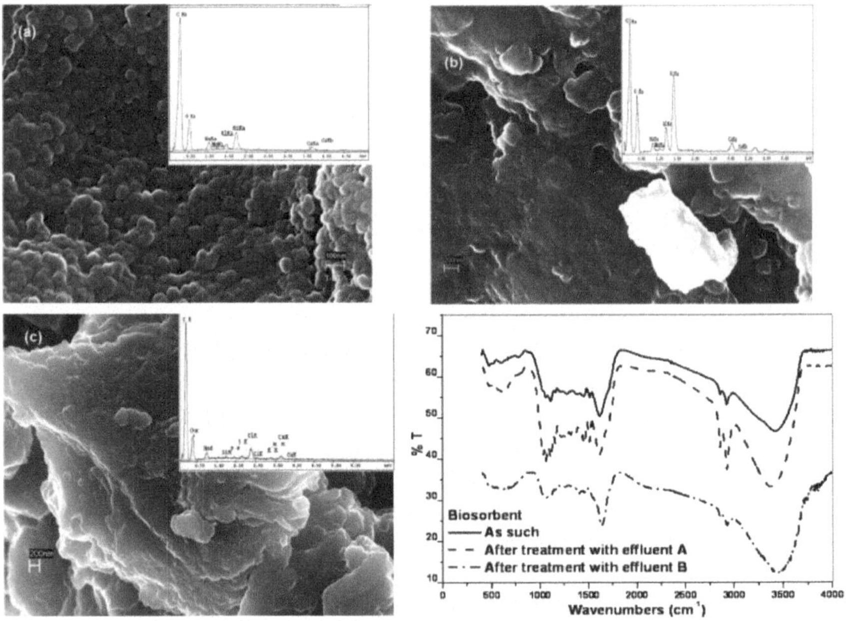

Fig. 2 FESEM and EDX analysis of Lagerostroemia sp. **a** Unspent, **b** spent biosorbent with effluent type A, **c** spent biosorbent with effluent type B, **d** FTIR analysis of Lagerostroemia sp. (*Source* Reprinted from [25], Copyright (2023) with permission from Elsevier)

Lagerostroemia sp. showed the involvement of various functional groups in adsorption process. The involvement of cellulose as well as hemicelluloses of biosorbent in the biosorption process was observed (Fig. 2d) [25]. Gupta et al. (2019) discussed the viability of Na_2CO_3-modified *Aloe barbadensis* Miller (*Aloe vera*) leaf (MABL) powder in removing Ni^{2+} ions from synthetic aqueous solution. The porous and non-uniform surface of the biosorbent shrunk after adsorption owing to the cross-linkages between Ni^{2+} ions [117]. Biosorbent functionality depended on hydroxyl, carboxyl and amide groups. Thus, for further developments in bioremediation research, biosorbent properties should be thoroughly reviewed for a better understanding of the complex biosorption mechanisms using novel materials.

12 Performance Comparison of Biosorbents with Activated Carbon

Carbonaceous materials with high carbon and low inorganic contents can be utilized for the production of activated carbon. Though, it has been proven to be very promising for heavy metal adsorption because of its complete porous structure having high surface area and mechanical strength, activated carbon derived from coal, peat, bone char, petroleum coke, lignite and wood raw materials is too costly which compelled the researchers to make serious attempts for producing much cheaper natural adsorbents from biomaterials [177]. Economic feasibility of the biosorption process is the foremost criterion determined by the adsorbent costs. More attention has been paid to the production of activated carbon from readily available cheap agricultural and forest wastes [177]. Raw materials and preparation methods govern the adsorption capacities of activated carbons. Chemical activation methods consume lower energy and give higher productivity compared to physical techniques [177].

Lloyd-Jones et al. (2001) investigated wood-based commercial granular activated carbon (unmodified and acid-oxidized) and leaves of *Azolla filiculoides* for cadmium adsorption. The authors found the adsorption capacities of the sorbents as 0.08, 0.33 and 1.40 mmol g^{-1} for unmodified activated carbon, biosorbent and acid-oxidized activated carbon respectively. Acid oxidation increased the quantity of functional groups containing oxygen of activated carbon by 3.3 times, making it more efficient [178]. Cochrane et al. (2006) studied and compared biosorbents, such as peat, crab carapace and macroalgae *Fucus vesiculosus* with commercially available AC and resin for removing Cu from aqua solution [179]. Macroalgae and crab carapace, which exhibited the highest capacities of adsorption—114.9 and 79.4 mg g^{-1} with more than 95% of copper removal efficiency, were found to be much more effective than other adsorbents used. Hydari et al. (2012) examined the performances of commercialized activated carbon, chitosan (biosorbent) and composite adsorbent of chitosan-activated carbon for cadmium removal from dilute solutions [180]. For all the three adsorbents, 100% cadmium removal was achieved under optimum operating conditions (adsorbent dosage = 6 g L^{-1}; pH = 6; adsorbent particle size =

0.425 mm) after 2 h. The authors reported maximum adsorption capacities of 10.3, 10 and 52.63 mg g^{-1} for activated carbon, chitosan and composite adsorbent. The adsorption abilities of activated carbon and chitosan were quite comparable, whereas the composite adsorbent gave the best yield. Sorption behaviours of low-cost natural biosorbents (shells of edible crab species *Cancer pagurus*, soil enriched with iron, brown seaweed *Fucus vesiculosus* and wood chips) along with commercial biochar and activated carbon were analysed by Richards et al. (2019) for treating aqueous solutions of zinc and copper [181]. Among the biosorbents, untreated, chemically and thermally treated macroalgae or seaweeds displayed the highest uptake capacities (mg g^{-1}) of 72.37 ± 0.37, 66.77 ± 0.19 and 63.06 ± 0.82 for Cu^{2+} and 52.40 ± 0.80, 48.83 ± 2.01 and 39.57 ± 0.80 for Zn^{2+} respectively than that of commercial activated carbon (36.71 ± 2.20 for Cu^{2+} and 20.78 ± 1.63 for Zn^{2+}) and biochar (18.07 ± 1.09 for Zn^{2+}). Though all the biosorbents indicated the ability to entirely remove heavy metals at all concentrations, macroalgae manifested greater affinity and metal removal capacity with respect to activated carbon in the presence of high levels of metal ions. Hence, natural biomaterials seem to be the most economical and ecologically sound biosorbents for metal ion removal. *Moringa oleifera* pods (with an adsorption capacity of 60.81 mg g^{-1}) have also been proven to be potent replacements of conventional adsorbents in wastewater treatment for diclofenac removal [182]. Raw wastes (silk and husk) of *Zea mays* variants and commercial activated carbon were tested by Lakshmi et al. (2021) for biosorption of oxybenzene from synthetic solution [183]. Marginal difference in the uptake of oxybenzene was observed for commercial AC (around 2.8 mg g^{-1}) and baby corn silk (2.6 mg g^{-1}). Therefore, the authors proposed that the biosorbents prepared from *Zea mays* can be utilized in lieu of commercial AC for wastewater treatment polluted with chemical ingredients of body care products.

13 Wastewater Treatment Using Biosorbent and Ceramic Membrane Based Integrated Hybrid System

13.1 Treatment of Domestic Wastewater

Biosorbent prepared from roots of *Eichhornia crassipes* was utilized in biosorptive pretreatment of domestic wastewater collected from the kitchen-sink using a ceramic microfiltration membrane in both single and multichannel configuration. The membrane process in combination with pretreatment by biosorption provided about 97.5% BOD removal and 99% COD removal along with considerable removal of suspended solids, turbidity and other waste materials [136]. Figure 3 depicts a schematic diagram of the combined process. The membrane module consists of multichannel ceramic elements (outer diameter 35 mm, channel diameter 4 mm and 200 mm length having 19 no of channels of average pore size 1 μm).

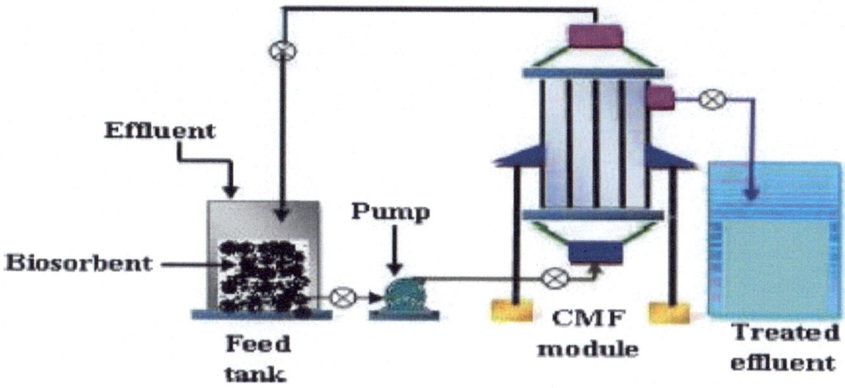

Fig. 3 Representation of schematic depicting the combined process of biosorption and ceramic membrane (*Source* Reprinted from [26], Copyright (2023) with permission from Springer open)

The combination of biosorption and ceramic membrane based microfiltration process was also used for the treatment of domestic sewage water. *Eichhornia crassipes* roots were used for preparation of biosorbent. Low-cost ceramic MF membrane (1 μm average pore size) was prepared using a clay-alumina mixture and the microfiltration (MF) process was carried out in crossflow mode at a pressure of 1 bar. For an initial COD loading of 800 mg/L and 660 mg/L, about 77.5% and 86% removal was obtained using a biosorbent dose of 0.25 g/L. In this study, it was observed that an increase in temperature from 10 to 30 °C favoured the biosorption process. For different temperatures, the equilibrium data fitted well with Langmuir isotherm. Kinetic data fitted well with the pseudo-second-order kinetic model. The results when compared with direct microfiltration without any biosorptive pretreatment provided only 70% COD and 73% BOD removal whereas the combination process resulted in 93% and 94% COD and BOD removal respectively along with effective separation of suspended solids and turbidity. [115].

Biosorbent prepared from fruit peels of *Lagerstroemia* sp. was applied for the treatment of composite wastewater having a high initial organic load. Two types of wastewater, i.e., wastewater A and wastewater B were used in the combination process of biosorption and membrane filtration. Wastewater A was collected from a sewage treatment plant for three consecutive days and mixed. Wastewater B was a mixed system containing 5, 10 and 15 % textile wastewater (sulphur black effluent) with domestic wastewater (kitchen-sink wastewater) COD removal was about 88% for all three sample of wastewater B and 89% for wastewater A (Fig. 4) [25].

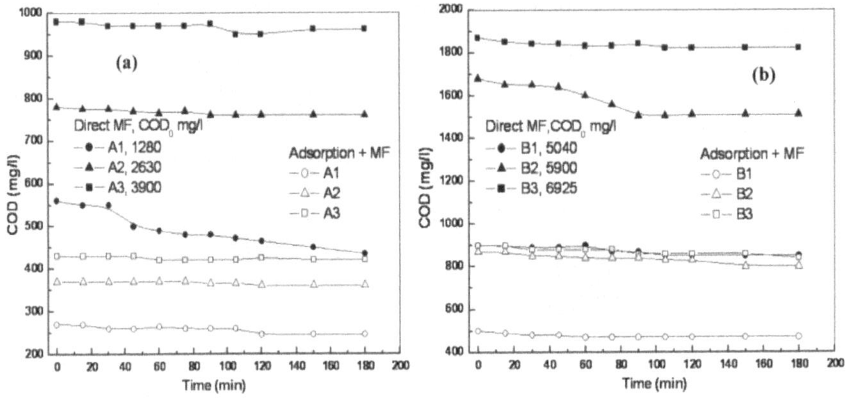

Fig. 4 COD content in the treated wastewater in membrane based hybrid process using biosorbent (prepared from fruit peels of *Lagerstroemia* sp.) and clay-alumina based ceramic microfiltration membrane (*Source* Reprinted from [25], Copyright (2023) with permission from Elsevier)

13.2 Treatment of Tannery Effluent Using Ceramic MF and Biosorbent

The efficiency of combined technology involving ceramic microfiltration and biosorbent was studied for treatment of tannery effluent from different streams, viz. composite effluent, effluent from primary clarifier and secondary clarifier. The study was conducted on three different loadings of wastewater, viz. composite wastewater (effluent 1), effluent from primary clarifier (effluent 2) and secondary clarifier (effluent 3). Biosorption study was conducted at varying pH (1 to 9) and biosorbent dose (0.25 to 5 g/L) for three different effluents to optimize the process parameters. A novel biosorbent prepared from fruit peels of *Trewia nudiflora* was used. From batch equilibrium study conducted to optimize the dose of biosorbent and pH of sample solution, it was observed that for effluent 1, the optimum dose was 3 g/L, while for effluent 2, the dose was 2 g/L and for effluent 3 the dose was 0.5 g/L. It was observed that COD reduction for effluents 1, 2 and 3 for different days were about 96% to 97%, 92% to 94% and 90% to 93%, respectively. TKN reduction for effluents 1, 2 and 3 was about 86%, 86.3% and 94.2%, respectively. Since COD loading of effluent 1 was higher compared to effluents 2 and 3, a the flux was found to be somewhat lower—about 9 to 15 $L/m^2/h$ (LMH). For effluent 2, the flux was about 17.9 to 21 LMH while for effluent 3 with lowest COD, it was the highest—about 23.5 to 27.5 LMH. About 99% reduction of heavy metals like Cr, Ni and Pb was achieved in the combined process (Figs. 5, 6) [26].

Biosorbent prepared from *Trewia nudiflora* was explored for the treatment of tannery wastewater from various streams having different organic loading. A combination treatment process using ceramic microfiltration (MF) membrane process along with biosorption was used. The study was carried out at 1 bar transmembrane pressure and CFV of 2 m/sec using ceramic membrane in 19-channel configuration having

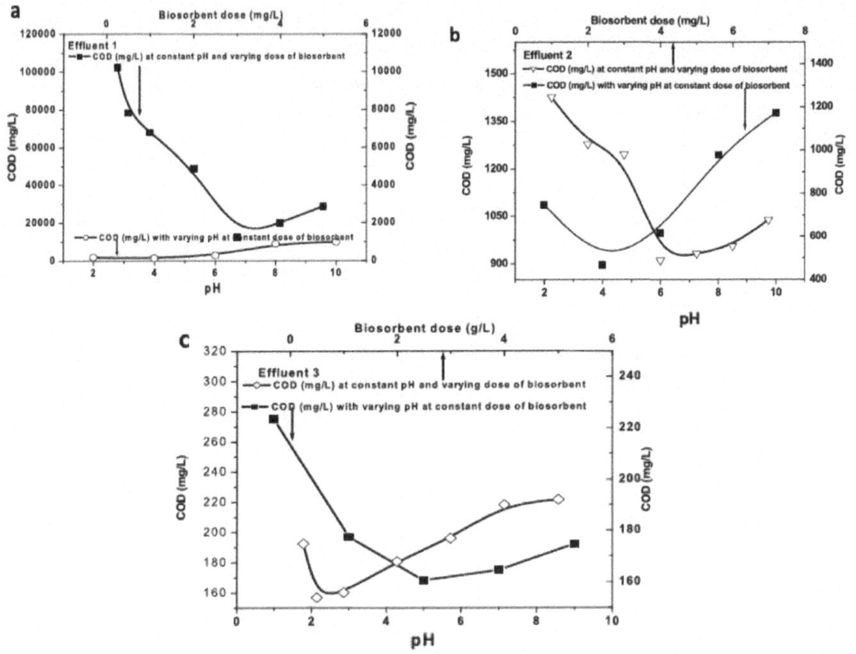

Fig. 5 Effect of pH and biosorbent dose on COD content after treatment for effluent 1 (**a**), effluent 2 (**b**), effluent 3 (**c**) (*Source* Reprinted from [26], Copyright (2023) with permission from Springer open)

200 mm length with outer diameter 35 mm, channel diameter 4 mm with apparent porosity of 36% and average pore size 1 μm. Tannery wastewater was collected from different chambers, i.e., initial composite raw wastewater, wastewater from primary clarifier and wastewater from the secondary clarifier each denoted as wastewater 1, 2 and 3. Each wastewater was collected on different days and COD removal obtained for wastewater1 was 96–97% using 3 g/L biosorbent. While COD removal of 92–94% was for wastewater 2 using 2 g/L biosorbent and for wastewater 3, about 90–93% removal was observed using 0.5 g/L biosorbent. Removal of total Kjeldahl nitrogen for all three wastewaters was 86%, 86.3% and 94.2%, respectively. Biosorbent-assisted microfiltration process resulted in removal of 99% heavy metals like chromium, nickel and lead [26].

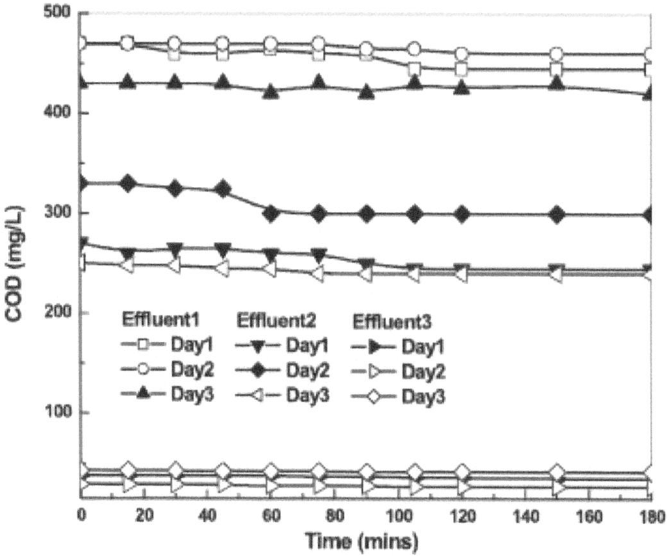

Fig. 6 COD content in the permeate with time for different effluents using biosorbent (prepared from fruit peels of *Trewia nudiflora*) and clay-alumina based ceramic MF membrane (*Source* Reprinted from [26], Copyright (2023) with permission from Springer open)

13.3 Ceramic Microfiltration Membrane Based Process for Treatment of Dye Effluent

13.3.1 Ceramic MF Membrane in Combination with Biosorbent for Treatment of Textile Effluent

An attempt was undertaken for the treatment of highly concentrated dyebath effluent collected from a cotton fabric dyeing unit. Ceramic microfiltration membrane (19 channels of 200 mm length with outer diameter 35 mm, channel diameter 4 mm with apparent porosity of 36% and average pore size 1 μm) prepared from a cost-effective composition of alumina and clay was used in combination with an adsorptive treatment. Biosorbent was prepared from sugar cane bagasse with a surface area of 705 m^2/g. Filtration performance showed that COD reduction increased with increasing time i.e. at the onset of experiment removal was about 88% which increased to 91.4% within 90 min of operation and remained constant till 180 min. Similarly, colour removal also increased to about 99.7% with increasing time after 90 min of operation. In addition, there was about 93% reduction in total organic carbon and a substantial reduction of heavy metals like chromium and lead [27].

13.3.2 Combined Process for Treatment of Sulphur Black Effluent

Crossflow microfiltration (CMF) using a ceramic membrane (19 channel of 200 mm length with outer diameter 35 mm, channel diameter 4 mm with apparent porosity of 36% and average pore size 1 μm) was used alone, and in combination with biosorption process for treatment of wastewater collected from the sulphur dyeing process of a garment processing industry. Roots of *E. crassipes* were used as biosorptive pretreatment of sulphur black wastewater using clay-alumina-based ceramic MF membrane in crossflow mode. This wastewater was very rich in organic loading i.e. COD of 3910 mg/L and turbidity of 5912 NTU. Using 1 g/L biosorbent, the combination process resulted in 99% removal of dye within 15 min of operation while 80% removal was achieved within 1 hour with complete removal of turbidity. Membrane based separation process without biosorptive pretreatment resulted in only 51% dye removal and 24% COD removal [22].

13.4 *Ceramic Microfiltration and Biosorption Process for Treatment of Herbal Pharmaceutical Wastewater*

Herbal pharmaceutical wastewater (HPWW) is generated during the manufacturing of different ayurvedic medicines including the finishing process of honey filtration, purification, packing, etc. The COD/BOD ratio of the wastewater was >6 indicating that the effluent was not amenable to biological treatment. The wastewater collected from a herbal pharmaceutical company in India was used for treatment with the biosorbent (ECR) prepared from the roots of water hyacinth (*Eichhornia crassipes*). The potential of chemically stable low-cost multichannel ceramic MF membrane (19 channel with 4mm, channel diameter, apparent porosity of 36% and average pore size 1 μm) was explored for treatment of HPWW in combination with ECR in crossflow microfiltration (CMF) mode. Biosorption batch study showed effective removal of colour and COD where in the equilibrium data fitted well in the Langmuir isotherm equation and the kinetic data followed closely to the pseudo-second-order model. Encouraging results were obtained in CFM of the biosorbent-treated wastewater with respect to colour, turbidity, TSS and COD reduction. Turbidity of all the permeate samples was below 1 NTU with >99% removal. Treatment of HPWW with 1 g/L adsorbent loading led to about 87% colour removal with 99% TSS and 88% COD reduction [28].

14 Case Studies on Biosorbent and Membrane Based Hybrid Process

The performance of ceramic membrane based hybrid biosorption process for treatment of domestic wastewater was compared with the conventioal activated sludge treatment using 19 channel ceramic membrane (apparent porosity of 36% and average pore size 1 μm) at 1 bar transmembrane pressure and a flow rate of 2 ml/min. Domestic wastewater collected from the kitchen sink of the office canteen was subjected to activated sludge treatment using 0.1–5 ml/L sludge per 500 mL of wastewater under optimized concentration of nutrients with proper aeration and temperature. The hybrid process resulted in 97% and 95% BOD and COD removal respectively. Biosorptive pre-treatment using Eichhornia crassipes roots followed by ceramic microfiltration of domestic wastewater that resulted in 97.5% and 99% removal of BOD and COD, respectively. Bioadsorption-based hybrid process was found to be more effective for removal of organic matter than that of activated sludge treatment involving use of nutrients and aeration for sludge handling [136].

Hybrid process consisting of biosorption and clay-alumina based ceramic MF membrane was also used for treatment of domestic and municipal wastewater having different organic composition at 1 bar transmembrane pressure and flow rate of 2 ml/min using biosorbent from *Eichhornia crassipes* roots. Domestic wastewater sample was draw from a residential building complex consisting of 354 houses, while the municipal wastewater was sample was collected from sewage pumping station discharging 4.6 million gallons/day of wastewater. The hybrid process resulted in 93% and 94% removal of COD and BOD respectively with significant removal of turbidity and suspended solids [25].

A hybrid process using ceramic MF membrane and biosorption as well as coagulation followed by microfiltration was employed for the treatment of sulphur black wastewater. Sulphur black is made of very high molecular weight compounds containing amines and phenols. Biosorbent was prepared from roots of *E. crassipes* and various inorganic coagulants were used. The wastewater had a high organic loading of 3910 mg/L COD and turbidity of 5912 NTU. Biosorptive pretreatment resulted in >99% of dye and 80% removal of COD respectively as compared to coagulant pretreated effluent. On the other hand coagulant pretreated effluent showed higher removal of suspended solids, i.e., ~99% as compared to biosorptive pretreatment i.e. 94–98% respectively. Turbidity removal for both hybrid processes was >99% [22].

Hybrid combined treatment process was applied for herbal pharmaceutical wastewater having >6 COD/BOD ratio, using 1 g/L biosorbent resulted in 88% removal of COD, 99% removal of suspended solids, 87% removal of colour and >99% removal of turbidity [28].

Biosorbent prepared from *Lagerstroemia speciosa* was used in the hybrid treatment process for two wastewater samples having a high organic load. Two types of wastewater were used where wastewater (W1) was made of composite wastewater collected over three consecutive days from sewage treatment plant while the other wastewater (W2) consisted of mixed effluent from the textile industry effluent and kitchen-sink water [184, 185]. The hybrid treatment process resulted in 89% and 88% COD removal for W1 and W2 respectively [186].

Treatment of tannery effluent with the hybrid process of biosorbent and microfiltration was also explored using biosorbent prepared from *Trewia nudiflora* fruit peels. Ceramic microfiltration was followed after the biosorption process. Tannery effluent was collected from three different discharge points viz. composite raw effluent, effluent discharged from the primary clarifier and effluent discharged from the secondary clarifier. Significant removal of organic loads, suspended solids, organic carbon and nitrogen was observed. Heavy metals like chromium, nickel, lead were reduced to about 99% [26].

Removal of colour from textile effluent was also attained through the hybrid process of biosorption and ceramic MF membrane based crossflow microfiltration. Waste material like sugar cane bagasse was used for the preparation of biosorbent having a surface area of 705 m^2/g [187]. The hybrid process demonstrated about 99.2% colour reduction with 91.4% removal of COD and a total organic carbon reduction of 93%. Additionally, the hybrid process reduced heavy metals like chromium and lead to significant levels [27].

15 Future Study

In the biosorption process, enhancements of performance and selectivity are of prime concern. Simulation of complex biosorption systems should be emphasized for a detailed interpretation of the plausible mechanism of metal sequestration. Microorganisms sequester heavy metals intracellularly through their uptake via membrane proteins. Very few studies have employed genetically engineered microorganisms (GEMs) with increased capacity of metal uptake from the periplasm into the cytoplasm for the purpose of wastewater treatment. Therefore in order to meet the increasing demands of wastewater treatment, GEM-mediated biosorption is recommended. Immobilized biomass is more useful for metal removal than the free cells due to high cell wall permeability, increasing the metal accumulation manifold. Critical assessment of the preparation techniques of biosorbents by immobilization technology is therefore imperative in the context of effluent treatment. Biosorbents with multipurpose applications (such as purification or extraction of pharmaceuticals and simultaneous removal of a vast array of pollutants) should be given additional focus to compensate for the overall cost of biosorbent pretreatments. It is always preferable to develop new pretreatment techniques that are both cheap and effective. Hu et al. (2018) used corncob biomass for both extraction of polysaccharides as healthcare products and removal of aquatic pollutant malachite green [188]. As of now, a large spectrum of biosorbents has already been 2 established, yet further research may be carried out for enhancement of adsorption capacities, optimization of biosorbent preparation methods, biosorbent regeneration for reuse and commercial viability.

Due emphasis should also be given to the development membrane bioreactor (MBR) for treating wastewater with high loading of organics (domestic, industrial, etc.) to utilize their high separation performance. Moreover, the MBR process generates high-quality effluents after treatment. Consumption of a considerable amount of energy to maintain high circulation velocity over the membrane surface for fouling control to obtain high filtration flux is the major drawback in the conventional MBR process. Such limitations can be overcome through the development of new MBR configurations which need further investigations. Ceramic membranes in wastewater treatment demand close attention for their thermal and chemical stabilities, long lifespan and lower susceptibility to fouling under harsh treatment conditions. Ceramic MBRs are preferred because of their higher efficiency in the separation of pollutant to achieve higher flux. The only disadvantage of the ceramic membrane is the higher production cost. The thrust should be given to the synthesis of cost-effective ceramic membranes, investigation of the biofouling aspects effects for mitigating membrane fouling.

16 Conclusion

Wastewater or effluent treatment deploying membranes has been widely recognized to be a breakthrough innovation. Combined biosorptive wastewater treatment with membrane separation provides an added advantage with respect to separation efficiency and cost competitiveness. Membrane-based separation techniques involving inorganic and composite membranes are becoming increasingly important in view of their biocompatibility and versatility in controlling surface chemistry. Technological advancements in manufacturing of low-cost ceramic membranes must satisfy the targeted properties for achieving selectivity and flux for specified applications. For making wastewater treatment more cost-efficient, fabrication of low-cost ceramic membranes with cheaper raw materials is necessary for their enhanced application in the separation of emerging pollutants. In reference to the aforementioned case studies, it is evident that the integrated system (biosorbents combined with cost-effective ceramic MF membranes) offers a unique opportunity to enhance the separation efficiency and permeate quality for reusing the treated wastewater. The hybrid process consisting of biosorption and crossflow microfiltration opens up a new vista for pilot-scale demonstration using real effluents for its implementation in the domestic and industrial sectors.

References

1. Kanamarlapudi LRK, Chintalpudi VK, Muddada S (2018) Application of biosorption for removal of heavy metals from wastewater. In: Derco J, Vrana B (eds) Biosorption, pp 70–115
2. Vieira RH, Volesky B (2000) Biosorption: a solution to pollution? Int Microbiol 3:17–24
3. Chojnacka K (2010) Biosorption and bioaccumulation–the prospects for practical applications. Environ Int 36:299–307

4. Kuyucak N, Volesky B (1988) Biosorbents for recovery of metals from industrial solutions. Biotech Lett 10:137–142
5. Ghoreishi M, Haghighi R (2003) Chemical catalytic reaction and biological oxidation for treatment of non-biodegradable textile effluent. Chem Eng J 95:163–169
6. Verma AK, Bhunia P, Dash RR (2012) Decolorization and COD reduction efficiency of magnesium over iron based a salt for the treatment of textile wastewater containing diazo and anthraquinone dyes. Int J Chem Biol Eng 6:116–123
7. Tan C-Y, Li M, Lin Y-M, Lu X-Q, Chen Z-L (2011) Biosorption of basic orange from aqueous solution onto dried *A. filiculoides* biomass: equilibrium kinetics and FTIR studies. Desalination 266:56–62
8. Sulak MT, Cengiz Yatmaz H (2012) Removal of textile dyes from aqueous solutions with eco-friendly biosorbent. Desalin Water Treat 37:169–177
9. Ayan EM, Secim P, Karakaya S, Yanik J (2012) Oreganum stalks as a new biosorbent to remove textile dyes from aqueous solutions. Clean: Soil, Air, Water 40:856–863
10. Sumathi KMS, Mahimairaja S, Naidu R (2005) Use of low-cost biological wastes and vermiculite for removal of chromium from tannery effluent. Biores Technol 96:309–316
11. Verma A, Chakraborty S, Basu JK (2006) Adsorption study of Hexavalent chromium using tamarind hull-based adsorbents. Sep Purif Technol 50:336–341
12. Onyancha D, Mavura W, Ngila JC, Ongoma P, Chacha J (2008) Studies of chromium removal from tannery wastewaters by algae biosorbents, *Spirogyra condensate* and *Rhizocolonium hieroglyphicum*. J Hazard Mater 158:605–614
13. Vinodhini V, Das N (2010) Packed bed column studies on Cr(VI) removal from tannery wastewater by neem sawdust. Desalination 264:9–14
14. Aravindhan R, Madhan B, Rao JR, Nair UN (2004) Recovery and reuse of chromium from tannery wastewaters using *Turbinaria ornata* seaweed. J Chem Technol Biotechnol 79:1251–1258
15. Cimino G, Passerini A, Toscano G (2000) Removal of toxic cations and Cr(VI) from aqueous solution by hazelnut shell. Water Res 34:2955–2962
16. Dakiky M, Khamis M, Manassra A, Mereb M (2002) Selective adsorption of Cr(VI) in industrial wastewater using low cost abundantly available adsorbents. Adv Environ Res 6:533–540
17. Gupta VK, Morhan D, Sharma S, Park KT (1999) Removal of Cr (VI) from electroplating industry wastewater using bagasse fly ash-a sugar industry waste material. Environmentalist 19:129–136
18. Wu XL, Wang L, Gao Y, Su YD, Li FT (2005) Combined bio-process for treatment of municipal wastewater containing Cu^{2+}. J Environ Sci Health, Part A. Toxic/Hazard Substances Environ Eng 26:131–136
19. Devi R, Dahiya RP (2006) Chemical oxygen demand (COD) reduction in domestic wastewater treatment by fly ash and brick kiln ash. Int J Water, Air Soil Pollut 174:33–46
20. Devi R, Dahiya RP, Gadgil K (2002) Investigation of coconut coir carbon and sawdust based adsorbents for combined removal of COD and BOD from domestic wastewater. Water Environ Manage Ser Int Water Assoc 2002:1209–1218
21. Devi R, Dahiya RP (2008) COD and BOD removal from domestic wastewater generated in decentralised sectors. Biores Technol 99:344–349
22. Bhattacharya P, Dutta S, Ghosh S, Vedajnananda S, Bandyopadhyay S (2010) Crossflow microfiltration using ceramic membrane for treatment of sulfur black effluent from garment processing industry. Desalination 261:67–72
23. Bhattacharya P, Banerjee P, Mallick K, Ghosh S, Majumdar S, Mukhopadhyay A, Bandyopadhyay (2013) Potential of biosorbent developed from fruit peel of *Trewia nudiflora* for removal of hexavalent chromium from synthetic and industrial effluent: analyzing phytotoxicity in germinating Vigna seed. J Environ Sci Health, Part A: Toxic/Hazard Substances Environ Eng (Taylor and Francis) 48:706–719
24. Bhattacharya P, Mallick K, Ghosh S, Banerjee P, Mukhopadhyay A, Bandyopadhyay S (2014) Algal biomass as potential for reduction of organic load in gray water and subsequent reuse: effect on seed germination and enzyme activity. Bioremediat J 18:56–70

25. Bhattacharya P, Ghosh S, Mukhopadhyay A (2013) Efficiency of combined ceramic microfiltration and biosorbent based treatment of high organic loading composite wastewater: an approach for agricultural reuse. J Environ Chem Eng 1:38–49
26. Bhattacharya P, Ghosh S, Mukhopadhyay A (2013) Combination technology of ceramic microfiltration and biosorbent for treatment and reuse of tannery effluent from different streams: response of defense system in Euphorbia sp. Int J Recycl Organic Waste Agric 2:1–11
27. Bhattacharya P, Ghosh S, Swarnakar S, Mukhopadhyay A (2014) Reuse of textile effluent for dyeing using combined technology of ceramic microfiltration and surface treated sugarcane bagasse: toxicity evaluation using Channa punctatus as model. Desalin Water Treat 54:715–735
28. Bhattacharya P, Ghosh S, Majumdar S, Dasgupta S, Bandyopadhyay S (2011) Biosorbent-assisted ceramic microfiltration process for treatment of herbal pharmaceutical wastewater with high organic loading. Int J Environ Technol Manage 14:132–146
29. De Gisi S, Lofrano G, Grassi M, Notarnicola M (2016) Characteristics and adsorption capacities of low-cost sorbents for wastewater treatment: a review. Sustain Mater Technol 9:10–40
30. Malkoc E, Nuhoglu Y (2007) Potential of tea factory waste for chromium(VI) removal from aqueous solutions: thermodynamic and kinetic studies. Sep Purif Technol 54:291–298
31. Kumar A, Sahu O (2013) Sugar industry waste as removal of toxic metals from waste water. World J Chem Educ 1:17–20
32. Suryan S, Ahluwalia S (2012) Biosorption of heavy metals by paper mill waste from aqueous solution. Int J Environ Sci 2:1331–1343
33. Jana A, Bhattacharya P, Sarkar S, Majumdar S, Ghosh S (2016) An ecofriendly approach towards remediation of high lead containing toxic industrial effluent by a combined biosorption and microfiltration process: a total reuse prospect. Desalin Water Treat 57:5498–5513
34. Lee T, Park JW, Lee JH (2004) Waste green sands as reactive media for the removal of zinc from water. Chemosphere 56:571–581
35. Alinnor I (2007) Adsorption of heavy metal ions from aqueous solution by fly ash. Fuel 86:853–857
36. Torab-Mostaedi M, Asadollahzadeh M, Hemmati A, Khosravi A (2013) Equilibrium, kinetic, and thermodynamic studies for biosorption of cadmium and nickel on grapefruit peel. J Taiwan Inst Chem Eng 44:295–302
37. Munagapati VS, Yarramuthi V, Nadavala SK, Alla SR, Abburi K (2010) Biosorption of Cu(II), Cd(II) and Pb(II) by *Acacia leucocephala* bark powder: kinetics, equilibrium and thermodynamics. Chem Eng J 157:357–365
38. Alomá I, Martín-Lara M, Rodríguez I, Blázquez G, Calero M (2012) Removal of nickel (ii) ions from aqueous solutions by biosorption on sugarcane bagasse. J Taiwan Inst Chem Eng 4:275–281
39. Bansal M, Singh D, Garg V, Rose P (2009) Use of agricultural waste for the removal of nickel ions from aqueous solutions: equilibrium and kinetics studies. Int J Civil Environ Eng 418:763–776
40. Basci N, Kocadagistan E, Kocadagistan B (2004) Biosorption of copper(II) from aqueous solutions by wheat shell. Desalination 164:135–140
41. Johnson P, Watson M, Brown J, Jefcoat I (2002) Peanut hull pellets as a single use sorbent for the capture of Cu(II) from wastewater. Waste Manage 22:471–480
42. Patel S (2012) Potential of fruit and vegetable wastes as novel biosorbents: summarizing the recent studies. Rev Environ Sci Bio/Technol 11:365–380
43. Naiya T, Bhattacharya A, Das S (2008) Adsorption of Pb(II) by sawdust and neem bark from aqueous solutions. Environ Prog Sustain Energy 27:313–328
44. Gupta V, Rastogi A (2008) Biosorption of lead from aqueous solutions by green algae Spirogyra species: kinetics and equilibrium studies. J Hazard Mater 152:407–414
45. Lee YC, Chang SP (2011) The biosorption of heavy metals from aqueous solution by Spirogyra and *Cladophora filamentous* macroalgae. Biores Technol 102:5297–5304

46. Lupea M, Bulgariu L, Macoveanu M (2012) Biosorption of Cd(II) from aqueous solution on marine green algae biomass. Environ Eng Manag J 11:607–615
47. Al-Homaidan AA, Al-Houri HJ, Al-Hazzani AA, Elgaaly G, Moubayed NM (2014) Biosorption of copper ions from aqueous solutions by Spirulina platensis biomass. Arab J Chem 7:57–62
48. Gupta VK, Rastogi A, Nayak A (2010) Biosorption of nickel onto treated alga (*Oedogonium hatei*): application of isotherm and kinetic models. J Colloid Interface Sci 342:533–539
49. Davis TA, Volesky B, Mucci A (2003) A review of the biochemistry of heavy metal biosorption by brown algae. Water Res 37:4311–4330
50. Yan G, Viraraghavan T (2000) Effect of pretreatment on the bioadsorption of heavy metals on *Mucor rouxii*. Water SA-Pretoria 26:119–124
51. Luo JM, Xiao X (2010) Biosorption of cadmium(II) from aqueous solutions by industrial fungus *Rhizopus cohnii*. Trans Nonferrous Metals Soc China 20:1104–1111
52. Shet AR, Patil LR, Hombalimath VS, Yaraguppi DA, Udapudi BB (2011) Enrichment of Saccharomyces cerevisiae with zinc and their impact on cell growth. Biotechnol Bioinform Bioeng 1:523–527
53. Say R, Yilmaz N, Denizli A (2003) Removal of heavy metal ions using the fungus Penicillium canescens. Adsorpt Sci Technol 21:643–650
54. Dursun A, Uslu G, Tepe O, Cuci Y, Ekiz H (2003) A comparative investigation on the bioaccumulation of heavy metal ions by growing *Rhizopus arrhizus* and *Aspergillus niger*. Biochem Eng J 15:87–92
55. Arbanah M, Miradatul Najwa M, Ku Halim K (2013) Utilization of Pleurotus ostreatus in the removal of Cr(VI) from chemical laboratory waste. Int Refreed J Eng Sci 2:29–39
56. Joo JH, Hassan SH, Oh SE (2010) Comparative study of biosorption of Zn^{2+} by *Pseudomonas aeruginosa* and *Bacillus cereus*. Int Biodeterior Biodegrad 64:734–741
57. Selatnia A, Boukazoula A, Kechid N, Bakhti M, Chergui A, Kerchich Y (2004) Biosorption of lead(II) from aqueous solution by a bacterial dead *Streptomyces rimosus* biomass. Biochem Eng J 1:127–135
58. Lu WB, Shi JJ, Wang CH, Chang JS (2006) Biosorption of lead, copper and cadmium by an indigenous isolate Enterobacter sp. J1 possessing high heavy-metal resistance. J Hazard Mater 134:80–86
59. Green-Ruiz C, Rodriguez-Tirado V, Gomez-Gil B (2008) Cadmium and zinc removal from aqueous solutions by *Bacillus jeotgali*: Ph, salinity and temperature effects. Biores Technol 99:3864–3870
60. Abd-Alla MH, Morsy FM, El-Enany AWE, Ohyama T (2012) Isolation and characterization of a heavy-metal-resistant isolate of *Rhizobium leguminosarum* bv. Viciae potentially applicable for biosorption of Cd^{2+} and CO^{2+}. Int Biodeterior Biodegrad 67:48–55
61. Pahlavanzadeh H, Keshtkar A, Safdari J, Abadi Z (2010) Biosorption of nickel(II) from aqueous solution by brown algae: equilibrium, dynamic and thermodynamic studies. J Hazard Mater 175:304–310
62. Aksu Z, Isoglu IA (2005) Removal of copper(II) ions from aqueous solution by biosorption onto agricultural waste sugar beet pulp. Process Biochem 40:3031–3044
63. Anirudhan TS, Radhakrishnan PG (2010) Adsorptive performance of an amine-functionalized poly (hydroxyethylmethacrylate)-grafted tamarind fruit shell for vanadium (V) removal from aqueous solutions. Chem Eng J 165:142–150
64. Anwar J, Shafique U, Waheed-uz-Zaman DA, Anwar S (2010). Removal of Pb(II) and Cd(II) from water by adsorption on peels of banana. Bioresource Technol 101:1752–1755
65. Ueda M (2016) Establishment of cell surface engineering and its development. Biosci Biotechnol Biochem 80:1243–1253
66. Jawad AH, Ngoh YS, Radzun KA (2018) Utilization of watermelon (*Citrullus lanatus*) rinds as a natural low-cost biosorbent for adsorption of methylene blue: kinetic, equilibrium and thermodynamic studies. J Taibah Uni Sci 12:371–381
67. Masoudian N, Rajabi M, Ghaedi M (2019) Titanium oxide nanoparticles loaded onto activated carbon prepared from bio-waste watermelon rind for the efficient ultrasonic-assisted adsorption of congo red and phenol red dyes from wastewaters. Polyhedron 173:114105

68. Gopalakrishnan Y, Al-Gheethi A, Abdul Malek M, Marisa Azlan M, Al-Sahari M, Mohamed RMSR, Alkhadher S, Noman E (2020) Removal of basic brown 16 from aqueous solution using durian shell adsorbent, optimisation and techno-economic analysis. Sustainability 12:1–22
69. Chigbundu EC, Adebowale KO (2017) Equilibrium and fractal-like kinetic studies of the sorption of acid and basic dyes onto watermelon shell (*Citrullus vulgaris*). Cellulose 24:4701–4714
70. Noman E, Al-Gheethi AA, Talip B, Mohamed R, Kassim AH (2020) Mycoremediation of Remazol Brilliant Blue R in greywater by a novel local strain of *Aspergillus iizukae* 605EAN: optimisation and mechanism study. Int J Environ Anal Chem 100:1650–1668
71. Ishak SA, Murshed MF, Md Akil H, Ismail N, Md Rasib SZ, Saeed Al-Gheethi AA (2020) The application of modified natural polymers in toxicant dye compounds wastewater: a review. Water 12:1–32
72. Yunus ZM, Al-Gheethi A, Othman N, Hamdan R, Ruslan NN (2020) Removal of heavy metals from mining effluents in tile and electroplating industries using honeydew peel activated carbon: a microstructure and technoeconomic analysis. J Clean Prod 251:119738
73. Husein DZ, Aazam E, Battia M (2017) Adsorption of cadmium (II) onto watermelon rind under microwave radiation and application into surface water from Jeddah, Saudi Arabia. Arab J Sci Eng 42:2403–2415
74. Gupta H, Gogate PR (2016) Intensified removal of copper from waste water using activated watermelon based biosorbent in the presence of ultrasound. Ultrason Sonochem 30:113–122
75. Sofiane B, Sofia KS (2015) Biosorption of heavy metals by chitin and the chitosan. Dev Pharma Chem 7:54–63
76. Elgarahy AM, Elwakeel KZ, Mohammad SH, Elshoubaky GA (2021) A critical review of biosorption of dyes, heavy metals and metalloids from wastewater as an efficient and green process. Clean Eng Technol 4:100209
77. Gilbert UA, Emmanuel IU, Adebanjo AA, Olalere GA (2011) Biosorptive removal of Pb^{2+} and Cd^{2+} onto novel biosorbent: defatted *Carica papaya* seeds. Biomass Bioenergy 35:2517–2525
78. Abbas SH, Ismail IM, Mostafa TM, Sulaymon AH (2014) Biosorption of heavy metals: a review. J Chem Sci Technol 3:74–102
79. Taşar Ş, Kaya F, Özer A (2014) Biosorption of lead(II) ions from aqueous solution by peanut shells: equilibrium, thermodynamic and kinetic studies. J Environ Chem Eng 2:1018–1026
80. Sao K (2014) A review on heavy metals uptake by plants through biosorption. Int Proc Econ Develop Res 75:78–83
81. Gadd GM (2009) Biosorption: critical review of scientific rationale, environmental importance and significance for pollution treatment. J Chem Technol Biotechnol 84:13–28
82. Bilal M, Rasheed T, Sosa Hernandez J, Raza A, Nabeel F, Iqbal H (2018) Biosorption: biosorption: an interplay between marine algae and potentially toxic elements—a review. Mar Drugs 16:1–16
83. Abdi O, Kazemi M (2015) A review study of biosorption of heavy metals and comparison between different biosorbents. J Mater Environ Sci 6:1386–1399
84. Hajahmadi Z, Younesi H, Bahramifar N, Khakpour H, Pirzadeh K (2015) Multicomponent isotherm for biosorption of Zn(II), CO(II) and Cd(II) from ternary mixture onto pretreated dried *Aspergillus niger* biomass. Water Resources Indus 11:71–80
85. Redha AA (2020) Removal of heavy metals from aqueous media by biosorption. Arab J Basic Appl Sci 27:183–193
86. Calero M, Perez A, Blazquez G, Ronda A, Martin-Lara M (2013) Characterization of chemically modified biosorbents from olive tree pruning for the biosorption of lead. Ecol Eng 58:344–354
87. Shartooh SM (2013) Lettuce leaves as biosorbent material to remove heavy metal ions from industrial wastewater. J: Baghdad Sci 11:1164–1170
88. Gadd GM, White C (1985) Copper uptake by *Penicillium ochro-chloron*: Influence of pH on toxicity and demonstration of energy dependent copper influx using protoplasts. J Gen Microbiol 131:1875–1879

89. Modak JM, Natarajan KA (1995) Biosorption of metals using nonliving biomass-a review. Miner Metall Process 12:189–196
90. Gourdon R, Bhande S, Rus E, Sofer SS (1990) Comparison of cadmium biosorption by grampositive and gram-negative bacteria from activated sludge. Biotech Lett 12:839–842
91. Nuhoglu Y, Malkoc E (2005) Investigations of nickel (II) removal from aqueous solutions using tea factory waste. J Hazard Mater 127:120–128
92. Perpetuo EA, Souza CB, Nascimento CAO (2011) Engineering bacteria for bioremediation. Prog Mol Environ Bioeng 605–632
93. Zouboulis AL, Matis KA, Hancock IC (1997) Biosorption of metals from dilute aqueous solutions. Sep Purif Methods 26:255–295
94. Ashraf M (2011) Low cost biosorbent banana peel (*Musa sapientum*) for the removal of heavy metals. Sci Res Essays 6:4055–4064
95. Benila Smily JRM, Sumithra PA (2017) Optimization of chromium biosorption by fungal adsorbent, Trichoderma sp. BSCR02 and its desorption studies. HAYATI J Biosci 24:65–71
96. Lata S, Singh PK, Samadder SR (2015) Regeneration of adsorbents and recovery of heavy metals: a review. Int J Environ Sci Technol 12:1461–1478
97. Ahemad M, Kibret M (2013) Recent trends in microbial biosorption of heavy metals: a review. Biochem Mol Biol 1:19–26
98. Fomina M, Gadd GM (2014) Biosorption: current perspectives on concept, definition and application. Biores Technol 160:3–14
99. Taty-Costodes VC, Fauduet H, Porte C, Ho YS (2005) Removal of lead(II) ions from synthetic and real effluents using immobilized *Pinus sylvestris* sawdust: adsorption on a fixed-bed column. J Hazard Mater 123:135–144
100. Kulkarni RM, Vidya Shetty K, Srinikethan G (2020) Biosorption of Nickel(II) and Cadmium(II). In: Inamuddin AMI, Lichtfouse E, Asiri AM (eds) Methods for bioremediation of water and wastewater pollution. Environmental chemistry for a sustainable world, vol 51. Springer, Cham, pp 373–391
101. Michalak I, Chojnacka K, Witek-Krowiak A (2013) State of the art for the biosorption process—a review. Appl Biochem Biotechnol 170:1389–1416
102. Mosa KA, Saadoun I, Kumar K, Helmy M, Dhankher OP (2016) Potential biotechnological strategies for the cleanup of heavy metals and metalloids. Front Plant Sci 7:1–14
103. Tiwari S, Lata C (2018) Heavy metal stress, signaling, and tolerance due to plant-associated microbes: an overview. Front Plant Sci 9:1–12
104. Li N, Qin L, Jin M, Zhang L, Geng W, Xiao X (2021) Extracellular adsorption, intracellular accumulation and tolerance mechanisms of *Cyclotella* sp. to Cr(VI) stress. Chemosphere 270:128662
105. Shamim S (2018) Biosorption of heavy metals. Intech Open Publishing, pp 21–49
106. Derco J, Vrana B (2018) Biosorption. Intech Open Publishing, pp 1–19
107. Seneviratne M, Seneviratne G, Madawala H, Vithanage M (2017) Role of rhizospheric microbes in heavy metal uptake by plants. In: Singh JS, Seneviratne G (eds) Agro-environmental sustainability: managing environmental pollution, vol 2. Springer International Publishing, Cham, pp 147–163
108. Ullah A, Heng S, Munis MFH, Fahad S, Yang X (2015) Phytoremediation of heavy metals assisted by plant growth promoting (PGP) bacteria: a review. Environ Exp Bot 117:28–40
109. Gube M (2016) Fungal molecular response to heavy metal stress. In: D. Hoffmeister (ed) Biochemistry and molecular biology. Springer International Publishing, Cham, pp 47–68
110. Mishra J, Singh R, Arora NK (2017) Alleviation of heavy metal stress in plants and remediation of soil by rhizosphere microorganisms. Front Microbiol 8:1–7
111. Fomina M, Hillier S, Charnock JM, Melville K, Alexander IJ, Gadd GM (2005) Role of oxalic acid overexcretion in transformations of toxic metal minerals by *Beauveria caledonica*. Appl Environ Microbiol 71:371–381
112. Gadd GM, Bahri-Esfahani J, Li Q, Rhee YJ, Wei Z, Fomina M et al (2014) Oxalate production by fungi: significance in geomycology, biodeterioration and bioremediation. Fungal Biol Rev 28:36–55

113. Kaewdoung B, Sutjaritvorakul T, Gadd GM, Whalley AJS, Sihanonth P (2016) Heavy metal tolerance and biotransformation of toxic metal compounds by new isolates of wood-rotting fungi from Thailand. Geomicrobiol J 33:283–288
114. Diep P, Mahadevan R, Yakunin AF (2018) Heavy metal removal by bioaccumulation using genetically engineered microorganisms. Front Bioeng Biotechnol 6:1–20
115. Bhattacharya P, Ghosh S, Sarkar S, Majumdar S, Bandyopadhyay S (2011) Effectiveness of biosorption-assisted microfiltration process for treatment of domestic wastewater. Bioremediat J 15:206–217
116. Rawat AP, Giri K, Rai JPN (2014) Biosorption kinetics of heavy metals by leaf biomass of *Jatrophacurcas* in single and multi-metal system. Environ Monit Assess 186:1679–1687
117. Gupta S, Sharma SK, Kumar A (2019) Biosorption of Ni(II) ions from aqueous solution using modified *Aloe barbadensis* Miller leaf powder. Water Sci Eng 12:27–36
118. Rasool K, Lee DS (2015) Characteristics, kinetics and thermodynamics of Congo Red biosorption by activated sulfidogenic sludge from an aqueous solution. Int J Environ Sci Technol 12:571–580
119. Medhi H, Chowdhury PR, Pulakananda D, Baruah PD, Bhattacharyya KG (2020) Kinetics of aqueous Cu(II) biosorption onto *Thevetia peruviana* leaf powder. ACS Omega 5:13489–13502
120. Tatah VS, Ibrahim KLC, Ezeonu CS, Otitoju O (2017) Biosorption kinetics of heavy metals from fertilizer industrial waste water using groundnut husk powder as an adsorbent. J Appl Biotechnol Bioeng 2:221–228
121. Liu Y, Liu YJ (2008) Biosorption isotherms, kinetics and thermodynamics. Sep Purif Technol 61:229–242
122. Riahi K, Chaabane S, Thayer BB (2017) A kinetic modeling study of phosphate adsorption onto *Phoenix dactylifera* L. date palm fibers in batch mode. J Saudi Chem Soc 21:S143–S152
123. Yadav M, Singh NK (2017) Isotherm investigation for the sorption of fluoride onto Bio-F: comparison of linear and non-linear regression method. Appl Water Sci 7:4793–4800
124. Dang VBH, Doan HD, Dang-Vu T, Lohi A (2009) Equilibrium and kinetics of biosorption of cadmium(II) and copper(II) ions by wheat straw. Biores Technol 100:211–219
125. Foo KY, Hameed BH (2010) Insights into the modeling of adsorption isotherm systems. Chem Eng J 156:2–10
126. Gunay A, Ersoy B, Dikmen S, Evcin A (2013) Investigation of equilibrium, kinetic, thermodynamic and mechanism of Basic Blue 16 adsorption by montmorillonitic clay. Adsorption 19:757–768
127. Khan TA, Mukhlif AA, Khan EA, Sharma DK (2016) Isotherm and kinetics modelingof Pb(II) and Cd(II) adsorptive uptake from aqueous solution by chemically modified green algal biomass. Model Earth Syst Environ 2:1–13
128. Preetha B, Viruthagiri T (2005) Biosorption of zinc (II) by *Rhizopusarrhizus*: equilibrium and kinetic modelling. Afr J Biotech 4:506–508
129. Kumar KV, Porkodi K (2006) Relation between some two- and three-parameter isotherm models for the sorption of methylene blue onto lemon peel. J Hazard Mater 138:633–635
130. Basar CA (2006) Applicability of the various adsorption models of three dyes adsorption onto activated carbon prepared waste apricot. J Hazard Mater 135:232–241
131. Vijayaraghavan K, Padmesh TVN, Palanivelu K, Velan M (2006) Biosorption of nickel(II) ions onto *Sargassum wightii*: application of two-parameter and three-parameter isotherm models. J Hazard Mater 133:304–308
132. Ringot D, Lerzy B, Chaplain K, Bonhoure JP, Auclair E, Larondelle Y (2007) In vitro biosorption of ochratoxin A on the yeast industry by-products: comparison of isotherm models. Biores Technol 98:1812–1821
133. Horsfall JM, Spiff AI (2005) Equilibrium sorption study of Al^{3+}, Co^{2+} and Ag^{+} in aqueous solutions by fluted pumpkin (*Telfairia occidentalis* HOOK f) waste biomass. Acta Chimica Slovenica 52:174–181
134. Gaur N, Kukreja A, Yadav M, Tiwari A (2018) Adsorptive removal of lead and arsenic from aqueous solution using soya bean as a novel biosorbent: equilibrium isotherm and thermal stability studies. Appl Water Sci 8:1–12

135. Doke KM, Khan EM (2013) Adsorption thermodynamics to clean up wastewater; critical review. Rev Environ Sci Bio/Technol 12:25–44
136. Ghosh S, Bhattacharya P, Majumdar S, Dasgupta S, Bandyopadhyay S (2010) Comparative study on treatment of kitchen-sink wastewater using single and multichannel ceramic membrane. Int J Environ Technol Manage 13:336–347
137. Calero M, Hernainz F, Blazquez G, Martin-Lara MA, Tenorio G (2009) Biosorption kinetics of Cd(II), Cr(III) and Pb(II) in aqueous solutions by olive stone. Braz J Chem Eng 26:265–273
138. Lee SH, Park CH (2012) Biosorption of heavy metal ions by brown seaweeds from southern coast of Korea. Biotechnol Bioprocess Eng 17:853–861
139. Al-Senani GM, Al-Fawzan FF (2018) Adsorption study of heavy metal ions from aqueous solution by nanoparticle of wild herbs. Egypt J Aquat Res 44:187–194
140. Kapsalaki K, Psychogiou M, Mitrogiannis D, Sgoubopoulou A (2019) Adsorption isotherm study of copper removal from aqueous solutions onto agricultural by-product. In: 16th international conference on environmental science and technology (CEST 2019), Rhodes, Greece, pp 1–2
141. Mohapatra RK, Parhi PK, Pandey S, Bindhani BK, Thatoi H, Panda CR (2019) Active and passive biosorption of Pb(II) using live and dead biomass of marine bacterium *Bacillus xiamenensis* PbRPSD202: kinetics and isotherm studies. J Environ Manage 247:121–134
142. Cui D, Tan C, Deng H, Gu X, Pi S, Chen T, Zhou L, Li A (2020) Biosorption mechanism of aqueous Pb^{2+}, Cd^{2+}, and Ni^{2+} ions on extracellular polymeric substances (EPS). Archaea, pp 1–9
143. Figueira MM, Volesky B, Azarian K, Ciminelli VST (2000) Biosorption column performance with a metal mixture. Environ Sci Technol 34:4320–4326
144. Lee HS, Suh JH (2000) Continuous biosorption of heavy metal ions by Ca-loaded *Laminaria japonica* in fixed bed column. Korean J Chem Eng 17:477–479
145. Pakshirajan K, Swaminathan T (2007) Continuous biosorption of Pb, Cu, and Cd by *Phanerochaete chrysosporium* in a packed column reactor. Soil Sediment Contam 15:187–197
146. Naja G, Volesky B (2008) Optimization of a biosorption column performance. Environ Sci Technol 42(15):5622–5629
147. Mishra A, Tripathi BD, Rai AK (2016) Packed-bed column biosorption of chromium(VI) and nickel(II) onto Fenton modified *Hydrilla verticillata* dried biomass. Ecotoxicol Environ Saf 132:420–428
148. Aksu Z, Tezer S (2005) Biosorption of reactive dyes on the green alga *Chlorella vulgaris*. Process Biochem 40:1347–1361
149. Ju DJ, Byun IG, Park JJ, Lee CH, Ahn GH, Park TJ (2008) Biosorption of a reactive dye (Rhodamine-B) from an aqueous solution using dried biomass of activated sludge. Biores Technol 99:7971–7975
150. Fiorentin LD, Trigueros DEG, Módenes AN, Espinoza- Quinones FR, Pereira NC, Barros STD, Santos OAA (2010) Biosorption of reactive blue 5 G dye onto drying orange bagasse in batch system: kinetic and equilibrium modeling. Chem Eng J 163:68–77
151. Ibrahim MB, Sani S (2014) Comparative isotherms studies on adsorptive removal of Congo red from wastewater by watermelon rinds and neem-tree leaves. Open J Phys Chem 4:139–146
152. Ibrahim MB, Ahmed A (2017) Adsorption of dyes onto activated watermelon (*Citrullus lanatus*) rinds (AWR). Bayero J Pure Appl Sci 10(1):606–614
153. Sahu S, Pahi S, Sahu JK, Sahu UK, Patel RK (2020) Kendu (*Diospyros melanoxylon* Roxb) fruit peel activated carbon—an efficient bioadsorbent for methylene blue dye: equilibrium, kinetic, and thermodynamic study. Environ Sci Pollut Res 27:1–14
154. Sintakindi A, Ankamwar B (2021) Fungal biosorption as an alternative for the treatment of dyes in waste waters: a review. Environ Technol Rev 10:26–43
155. Hong HB, Hwang SH, Chang YS (2000) Biosorption of 1,2,3,4-tetrachlorodibenzo-p-dioxin and polychlorinated dibenzofurans by *Bacillus pumilus*. Water Res 34:349–353
156. Juhasz AL, Smith E, Smith J, Naidu R (2002) Biosorption of organochlorine pesticides using fungal biomass. J Ind Microbiol Biotechnol 29:163–169

157. Bayramoglu G, Gursel I, Tunali Y, Arica MY (2009) Biosorption of phenol and 2-chlorophenol by *Funaliatrogii* pellets. Biores Technol 100:2685–2691
158. Xu L, Chen X, Li H, Hu F, Liang M (2016) Characterization of the biosorption and biodegradation properties of *Ensifer adhaerens*: a potential agent to remove polychlorinated biphenyls from contaminated water. J Hazard Mater 302:314–322
159. Al-Gheethi AA, Efaq AN, Mohamed RM, Norli I, Kadir MO (2017) Potential of bacterial consortium for removal of cephalexin from aqueous solution. J Assoc Arab Univ Basic Appl Sci 24:141–148
160. Wang L, Xiao H, He N, Sun D, Duan S (2019) Biosorption and biodegradation of the environmental hormone nonylphenol by four marine microalgae. Sci Rep 9:1–11
161. Bozorginia S, Jaafari J, Taghavi J, Taghavi K, Ashrafi SD, Roohbakhsh E, Naghipour D (2021) Biosorption of ceftriaxone antibiotic by *Pseudomonas putida* from aqueous solutions. Int J Environ Anal Chem 190:1–15
162. Vahabisani A, An C (2021) Use of biomass-derived adsorbents for the removal of petroleum pollutants from water: a mini-review. Environ Syst Res 10:1–10
163. Kaushik P, Malik A (2009) Fungal dye decolourization: recent advances and future potential. Environ Int 35:127–141
164. Park D, Yun Y-S, Park J-M (2010) The past, present, and future trends of biosorption. Biotechnol Bioprocess Eng 15:86–102
165. Ngah WSW, Hanafiah MAKM (2008) Removal of heavy metal ions from wastewater by chemically modified plant wastes as adsorbents: a review. Biores Technol 99:3935–3948
166. Van Heijenoort J (2001) Formation of the glycan chains in the synthesis of bacterial peptidoglycan. Glycobiology 11:25R-36R
167. Albers S, Eichler J, Aebi M (2017) Archaea. In: Varki A, Cummings RD, Esko JD et al (eds) Essentials of glycobiology [Internet], 3rd edn. Cold Spring Harbor Laboratory Press, Cold Spring Harbor (NY), pp 2015–2017
168. Shah SS, Palmieri MC, Sponchiado SRP, Bevilaqua D (2020) Enhanced bio-recovery of aluminum from low-grade bauxite using adapted fungal strains. Braz J Microbiol 51:1909–1918
169. Dhankhar R, Hooda A (2011) Fungal biosorption—an alternative to meet the challenges of heavy metal pollution in aqueous solutions. Environ Technol 32:467–491
170. Naveena J, Latha L (2018) Fungal cell walls as protective barriers for toxic metals. Adv Med Biol 53:180–198
171. Zhou Y, Zhang Z, Zhang J, Xia S (2016) New insight into adsorption characteristics and mechanisms of the biosorbent from waste activated sludge for heavy metals. J Environ Sci 45:248–256
172. Velkova Z, Kirova G, Stoytcheva M, Kostadinova S, Todorova K, Gochev V (2018) Immobilized microbial biosorbents for heavy metals removal. Eng Life Sci 18:871–881
173. Bind A, Goswami L, Prakash V (2018) Comparative analysis of floating and submerged macrophytes for heavy metal (copper, chromium, arsenic and lead) removal: sorbent preparation, characterization, regeneration and cost estimation. Geol Ecol Landscapes 2:61–72
174. Dusengemungu L, Kasali G, Gwanama C, Ouma KO (2020) Recent advances in biosorption of copper and cobalt by filamentous fungi. Front Microbiol 11:1–16
175. Ighalo JO, Adeniyi AG (2020) A mini-review of the morphological properties of biosorbents derived from plant leaves. SN Appl Sci 2:1–16
176. Peng SH, Wang R, Yang LZ, He L, He X, Liu X (2018) Biosorption of copper, zinc, cadmium and chromium ions from aqueous solution by natural foxtail millet shell. Ecotoxicol Environ Saf 165:61–69
177. Anirudhan TS, Sreekumari SS (2011) Adsorptive removal of heavy metal ions from industrial effluents using activated carbon derived from waste coconut buttons. J Environ Sci 23:1989–1998
178. Lloyd-Jones PJ, Rangel-Mendez JR, Streat M (2001) Sorption of cadmium using a natural biosorbent and activated carbon. Department of Chemical Engineering, Loughborough University, Leicestershire, UK. Symposium Series No. 148, IChemE, pp 847–866

179. Cochrane EL, Lu S, Gibb SW, Villaescusa I (2006) A comparison of low-cost biosorbents and commercial sorbents for the removal of copper from aqueous media. J Hazard Mater 137:198–206
180. Hydari S, Sharififard H, Nabavinia M, Reza Parvizi M (2012) A comparative investigation on removal performances of commercial activated carbon, chitosan biosorbent and chitosan/activated carbon composite for cadmium. Chem Eng J 193–194, 276–282
181. Richards S, Dawson J, Stutter M (2019) The potential use of natural vs commercial biosorbent material to remediate stream waters by removing heavy metal contaminants. J Environ Manage 231:275–281
182. Viotti PV, Moreira WM, dosSantos OAA, Bergamasco R, Vieria AMS, Vieira MF (2019) Diclofenac removal from water by adsorption on *Moringa oleifera* pods and activated carbon: mechanism, kinetic and equilibrium study. J Clean Prod 219:809–817
183. Lakshmi S, Baker S, Shivamallu C, Prasad A (2021) Biosorption of oxybenzene using biosorbent prepared by raw wastes of *Zea mays* and comparative study by using commercially available activated carbon. Saudi J Biol Sci 28:3469–3476
184. Venckatesh R, Amudha T, Sivaraj R, Chandramohan M, Jambulingam M (2010) Kinetics and equilibrium studies of adsorption of direct red-28 onto *Punica granatum* carbon. Int J Eng Sci Technol 2:2040–2050
185. Ljupkovic RB, Mitrovic J, Radovic M, Kostic M, Bojic D, Mitic-Stojanovic D-L, Bojic A-L (2011) Removal Cu(II) ions from water using sulphuric acid treated *Lagenaria vulgaris* Shell (Cucurbitaceae). Biologica Nyssana 2:85–89
186. Bhattacharya P, Sarkar S, Ghosh S, Majumdar S, Mukhopadhyay A, Bandyopadhyay S (2013) Potential of ceramic microfiltration and ultrafiltration membrane for treatment of grey water for an effective reuse. Desalin Water Treat 51:1–10
187. Girgis BS, Khalil LB, Tawfik TAM (1994) Activated carbon from sugar cane bagasse by carbonization in the presence of inorganic acids. J Chem Technol Biotechnol 61:87–92
188. Hu H, Liang W, Zhang Y, Wu S, Yang Q, Wang Y, Zhnag M, Liu Q (2018) Multipurpose use of a corncob biomass for the production of polysaccharides and the fabrication of a biosorbent. ACS Sustain Chem Eng 6:3830–3839

Chapter 14
Techno-commercial Comparison of Membrane Filtration Solutions for Separating Hemicelluloses from Highly Alkaline Process Waste Streams

Saurabh C. Singh, Gorakshnath Takalkar, and Z. V. P. Murthy

1 Brief Overview of Pulp, Paper, and Cellulosic Fibre Industries

Pulp and derived products like cellulosic fibre is of utmost significance considering the rising world population and increasing consumer demands. Cellulose is a renewable biopolymer and one of the sustainable resources for manufacturing fibre extensively used in textile industries. Cellulose-based products have a wide range of applications such as medical, pharmacy, biofuel, paper, acetate including rayon for textile and nitrocellulose for paints and explosives along with the commercial products like membranes, fibres, non-woven materials, nanoparticles and sponges [1]. Cellulose, an organic polymer, is the most prevalent and sustainable substance available on Earth. It has the potential to serve as a substitute for other synthetic polymeric substances [1, 2]. Cellulose-derived viscose staple fibres (VSF) or artificial cotton are natural, biodegradable and derived from dissolving grade pulp (DGP), mainly consist of a high percentage of cellulose (more than 90%) and fewer impurities like lignin and hemicellulose. The VSF is the most important man-made cellulose

S. C. Singh · Z. V. P. Murthy (✉)
Department of Chemical Engineering, Sardar Vallabhbhai National Institute of Technology, Surat, Gujarat 395007, India
e-mail: zvpm@ched.svnit.ac.in

S. C. Singh
e-mail: saurabh.c.singh@adityabirla.com

S. C. Singh · G. Takalkar
Fibre Research Centre, Birla Cellulosic, Kharach, Bharuch, Gujarat 394120, India
e-mail: gorakshanth.takalkar@adityabira.com

S. Bandyopadhyay (ed.), *Application of Low Cost Ceramic Membranes in Wastewater Treatment*, Materials Horizons: From Nature to Nanomaterials,
https://doi.org/10.1007/978-981-97-5842-5_14

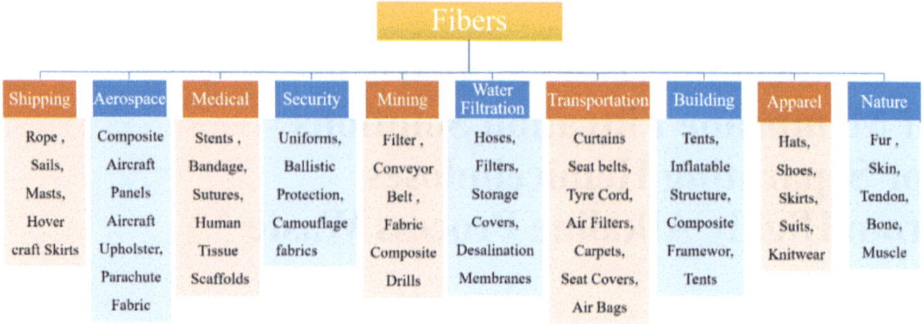

Fig. 1 Various applications of fibres (adapted from [4])

fibre, with a market share of around 79% of all man-made cellulosic fibres. The VSF is a highly adaptable and absorbent material that can be easily mixed with other fibers. It is commonly used in various textile applications, particularly in the production of clothing, dress materials, furnishings, home textiles, and woven and knitted garments. The increasing need for textiles is anticipated to fuel the demand for VSF in these specific applications Woven textiles are displaying good growth due to the rising demand in Asian countries like India, China, etc. Cellulosic man-made fibres are pertinent to make yarns and fabrics required by textile industries and find a wide range of applications in aerospace, shipping, packaging, security, domestic and medical applications [3] as shown in Fig. 1. For example, fibres for bulletproof jackets or more commonly known as Kevlar need to have exceptional impact strength. On the other hand, for applications related to apparel, dyeability, extensibility, and dimensional stability are necessary [4].

Cellulose-derived fibres are primarily made from regenerated cellulose. This can be achieved by two methods: treating native cellulose with highly alkaline solutions, which are then converted to viscose and then regenerated, or by directly precipitating cellulose from solutions. These fibres are commonly identified by a fibrillary structural model that can be comprehended by various microscopic techniques [2]. Cellulose fibres are a type of man-made regenerated cellulosic fibre that has been in existence for over a century. They possess various desirable characteristics, including strong hydrophilicity, easy dyeability, abundant availability, and the ability to naturally degrade. The production of cellulose fibre from high-purity dissolving grade pulp (DGP) entails a series of procedures, including steeping, pressing, depolymerization, cellulose xanthogenate formation, dissolution, spinning, and drying, as depicted in Fig. 2.

The first and a very key process step in viscose rayon manufacture is steeping where high-purity pulp is preserved with the concentrated caustic (aqueous NaOH) solution for a quantified duration and leads to the formation of alkali cellulose (also called alk-cell). Dissolving grade pulp used for viscose production primarily contains α-cellulose (>90%) and hemicelluloses in the range of 5–8%. Pulp cellulose is activated through swelling in the steeping process by using a high percentage of aqueous

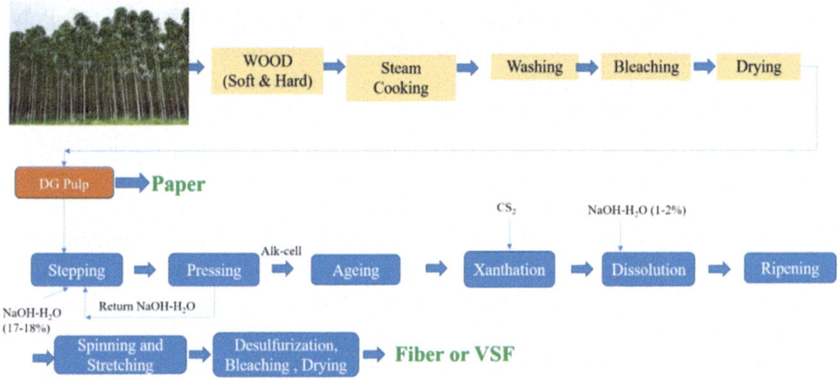

Fig. 2 Regenerated cellulose fibre via viscose process

caustic and rupture of intra and intermolecular bonds of the cellulose chain and this improves the xanthation reaction via enhanced CS_2 accessibility [5].

During the fibre synthesis process, a portion of the cellulose degrades, leading to the generation of low molecular weight polymers and branching substituted structures. Having a high-quality xanthation and ensuring the complete dissolving of cellulose into a weak alkali solution is crucial. The xanthation reaction is generally advantageous for celluloses and hemicelluloses with low molecular weight. However, the presence of these substances negatively impacts the overall reactivity of the pulp. Moreover, these hemicelluloses might have a negative effect on the efficiency of specific equipment, as well as the quality of the fibres, particularly in terms of tensile strength and dyeability. Therefore, it is crucial to extract hemicellulose at the outset of the fibre production process [6, 7].

In recent years, there has been extensive research on membrane separation due to its economic advantages in various industrial sectors such as wastewater treatment, chemicals, pharmaceuticals, and food production. Ultrafiltration is a very efficient method that requires minimal energy to remove hemicelluloses from industrial wastewater. [8]. However, literature work has been limited to polymeric-type membranes [9, 11]. Polymeric membranes, despite their compactness, incur substantial replacement costs due to their susceptibility to corrosion when exposed to 18% caustic substances. The practical application of ceramic membranes in industries such as food, chemical, and pharmaceutical is owing to their advantageous characteristics, including high flux, chemical resistance, superior strength, and high thermal and chemical stability [12, 13]. Cellulosic fibres morphological features can be modified using hemicelluloses [14].

The existing literature on the use of ceramic membranes as a substitute for polymeric ultrafiltration membranes, particularly in situations where chemical compatibility is challenging, is insufficient [15]. A study was conducted earlier, in which the performance of different easily accessible membranes was tested for separating hemicelluloses from the alkaline process stream of the viscose process [16]. Polymeric nanofiltration membranes showed good hemicelluloses separation efficiency along with sustenance of performance. Ceramic membranes also presented a good opportunity for the referred application as they have been already tested in an alkaline medium such as treatment of bleaching effluent from sulphite pulp production [17] and also for recovery of lignin [18].

The primary aim of the current book chapter is to understand the role of hemicelluloses in the pulp-to-viscose-making process and its separation using membranes. Accordingly, this study is divided into three major parts: Benefits of hemicellulose conversion into value-added products like biofuels, biomaterials, chemicals, etc. Characteristics of effluent generated after hemicellulose separation, Use of membranes for hemicellulose separation from concentrated effluent generated in pulp industries during the stepping process.

2 Benefits of Removing Hemicellulose

Lignocellulose biomass consists mainly of hemicellulose, cellulose, and lignin wherein the hemicellulose percentage is as much as 30–40%. The extraction and retrieval of hemicelluloses from the pulp industry provide a chance to utilize them as primary resources for additional value-enhancing industrial procedures, as illustrated in Fig. 3. The extraction of hemicellulose from the pulp and fibre production process can yield various valuable products or chemicals, including organic acids such as acetic acid, methane, monosaccharide, fuel ethanol, xylose, xylitol, alternative solvents to those derived from petroleum, plasticizers, hydrogels, and dyes [19, 22]. Ethanol as biofuel is obtained via the fermentation of hemicellulose sugars can become one additive to petroleum-beaded fuel [23, 25]. The process of conversion of hemicellulose is needed to be efficient and economical [26]. Hemicelluloses can be transformed into sugars and then utilized to synthesize bulk chemicals like C5 and C6 sugars. Hemicellulose like hexoses and pentose can be transformed into high value-added chemicals like 5-hydroxymethylfurfural and furfural.

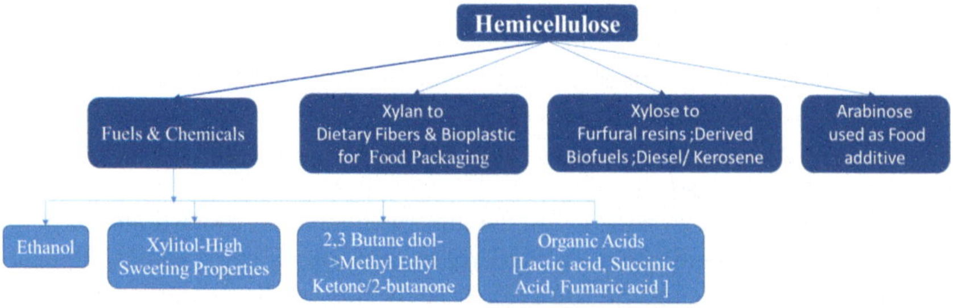

Fig. 3 Hemicellulose into value-added products like biofuel, biomaterials, chemicals (adapted from [26])

3 Characteristics of Effluents Generated in Pulp and Paper Industry and Their Impact

Lignin, cellulose, and hemicellulose are present in the primary and secondary cell walls of biomass. The percentage of lignocellulose components in various types of wood is listed in Table 1. Hemicellulose is a short-chain and branched polymer containing different monomers like hexoses, pentose attached by different functional groups; typically, represents 25–30% of the biomass. Characteristically glucomannans and xylans are monomers of pentose and hexose and linked by short or branched chains. Hemicelluloses have many free hydroxyl groups in their molecular structure and are present in amorphous form only [5]. Many new functional groups are generated by oxidation, hydrolysis, reduction, etherification, esterification, and cross-linking reactions. High usage of biomass is governed by the efficient process for hemicellulose separation.

In the VSF production process, hemicellulose within the DGP deteriorates the quality of fibre produced after spinning. Therefore, hemicellulose removal is necessary without alkali losses [11]. The large percentage of hemicelluloses in the cellulose comes from re-circulated aqueous caustic solution after the pressing of alkali-treated dissolving grade pulp slurry. Hemicellulose is a water-soluble hetero-polysaccharide comprising several different monomers such as glucose, mannose, xylose, and arabinose [20, 28, 29]. Hemicellulose has a low molecular weight, is branched, and has a degree of polymerization (DP) much lower than α-cellulose. While α-cellulose is insoluble in the caustic solution, hemicelluloses are soluble species. Thus, the steeping process helps to eliminate the hemicelluloses present in the DG pulp. The hemicellulose-rich caustic stream is usually recycled in the process. Efficient removal of hemicelluloses in the initial stages of viscose manufacture is extremely critical else it creates several issues related to pulp swelling, reactivity, slurry pressing, aging, equipment efficiency, reactant consumption, and fibre properties. Therefore, it is crucial to extract hemicellulose during the earliest stage of the fiber production

Table 1 Various biomass and its lignocellulose composition [3, 27]

Biomass source	Cellulose	Hemicellulose	Lignin
Hardwood	43–47	25–35	16–24
Softwood	40–44	25–29	25–31
Bamboo	45–47	26–28	23–24
Pinecome	42–46	27	20–23
Coconut	31–32	25–26	33–37
Cotton stalk	48–52	25–27	24–26
Sugarcane baggage	45	39	20–22
Pineapple leaf	34–40	21–25	25–29
Wheat straw	37–43	31.37	18–22

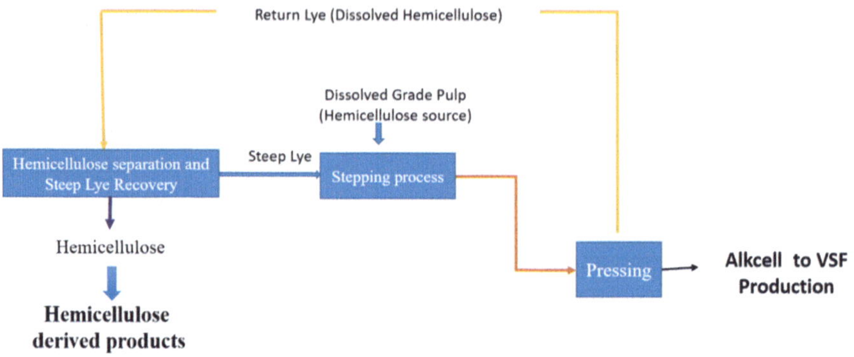

Fig. 4 Hemicellulose separation in VSF production process

process. Given that a significant portion of hemicelluloses in alkali, cellulose originates from the excess steeping lye that is reused after pressing, the initial stage of lowering hemicelluloses involves regulating the concentration of hemicelluloses that are dissolved in the steeping lye, as illustrated in Fig. 4.

The removal of hemicelluloses from the pulp is a primary goal of pulp steeping. Initial hemicelluloses in steeping lye greatly reduce pulp hemicellulose removal efficiency. If the steeping lye does not contain any hemicelluloses, then nearly all of the hemicellulose in the pulp is dissolved. The inclusion of hemicelluloses in alkali-cellulose will affect the reactivity of pulp. Hemicelluloses, which have a lower molecular weight, are more reactive and tend to consume CS_2 preferentially. This limits the amount of CS_2 available for reaction with cellulose molecules, leading to a higher amount of unreacted and undissolved cellulose in the viscose solution. The pulp steeping process can be described as a series of steps. First, NaOH and water are absorbed by the pulp, causing it to swell. This leads to the reaction of NaOH with the pulp, resulting in the formation of alkali-cellulose. The excess NaOH, water, and dissolved hemicelluloses are then removed from the pulp using either a centrifuge or by squeezing/pressing. The presence of hemicelluloses in the early steeping lye leads to a notable rise in solution viscosity due to the polymeric nature of hemicelluloses. The high viscosity of the pulp hinders the extraction of hemicelluloses during the desorption process using pressing or centrifugation. High resistance to NaOH can be useful to handle concentrated alkali water solution generated after the stepping process to separate hemicellulose [7]. Further, the author presented that the pulp reactivity reduces with an increase in hemicellulose composition beyond 20 g/L in the recirculated steeping lye. The presence of a significant amount of hemicellulose in the recirculated lye leads to reduced swelling, a less effective alkalization reaction, and lesser extraction of hemicelluloses. These factors combined result in poor pulp reactivity. Therefore, it is necessary to separate and eliminate hemicelluloses from the steeping lye-employing membranes.

4 Current Strategies to Handle the Hemicelluloses-Containing Effluent

The presence of higher amounts of hemicelluloses in the recycled caustic stream can lead to operational challenges in the production process of pulp to fiber. This includes an increase in the viscosity of the caustic solution, which can impede the operation of the slurry press. Additionally, hemicelluloses, being smaller and more reactive molecules, can selectively react with other substances, thereby affecting the strength of the fibers and their ability to be dyed. Simultaneously, it is crucial to recover and recycle the caustic stream that contains hemicelluloses in order to minimize chemical usage in the process. Previously, semi-permeable dialysis membranes were employed to retrieve caustic substances from the hemicellulose-rich alkali solution that had been used. The dialysis process is marked by significant spatial demands, suboptimal efficacy, and the need to dilute the stream for separation. These aspects contribute to the high complexity and costliness of the operation. The dialysis procedure has typically been employed due to the overall advantages of hemicellulose separation, which ultimately surpass the associated expenses. Traditionally, ultrafiltration (UF) membranes used in industrial applications were only suitable for treating wastewater that may contain specific pollutants. This limitation was due to the fact that the conventional spiral wound UF membranes were not able to withstand rigorous workihg conditions.

Recently, there have been multiple documented research on the utilization of membranes to separate hemicelluloses from alkaline streams, as outlined in Table 2. Hemicelluloses are composed of polymer chains with different molecular lengths. Table 2 seeks to select an appropriate membrane for the separation of hemicelluloses from a caustic solution. The study found that when hydrophobic membranes were employed for treating process water from thermo-mechanical pulping of spruce, there was observed fouling and loss in flux. Additionally, the study also investigated the impact of temperature and pressure on the performance of the membranes [16]. The study examined the efficacy of five alkali-resistant polymeric NF and UF membranes in the separation of hemicelluloses from process liquors generated during viscose fibre manufacture [11]. It has been discovered that the operational parameters of actual operations might result in significant variations in the optimal performance of the membrane. The study examined the extraction of hemicelluloses from wheat bran and the subsequent recovery of the alkaline solvent from the residual stream. It was discovered that a combination of ultrafiltration and nanofiltration systems in series can be utilized cost-effectively [9]. The researchers investigated the use of alkali-resistant ultrafiltration (UF) membranes to separate hemicelluloses from a process stream containing caustic in the production of viscose rayon fibre and in pulp mills [30].

Table 2 Hemicellulose separation and recovery technique

References	Hemicellulose extraction techniques	Study objective or application	Feed	Study details	Membranes details	Remarks
[10]	Used three-stage membrane and enzyme treatment to separate high molecular weight hemicellulose	Renewable food packaging barrier film	Chemithermo-mechanical pulp process water	Concentration of hemicellulose 1 to 54 gpl	Polymeric: UF (5 and 10 kDa)	–
[27]	Freeze–thaw assisted alkali (7%) treatment to recover hemicellulose	Separation of hemicellulose	Bamboo	64.7% hemicellulose extraction using 7% alkali at high temperature	NA	Obtained higher purity, higher molecular weight, and lower polydispersity Index
[9]	Membrane—UF and NF	Oxygen barrier films for food packaging	Alkaline solvent from a residual stream after the extraction of hemicelluloses in wheat bran		10 kDa ceramic membrane for UF Four flat-sheet polymeric and two hollow-fibre polymeric NF membranes with nominal cut-offs from 0.2 to 1.4 KDa	Hemicellulose recovery using UF NaOH recovery process using NF – Pay-back period <2 months

(continued)

Table 2 (continued)

References	Hemicellulose extraction techniques	Study objective or application	Feed	Study details	Membranes details	Remarks
[11]	NF + UF	Performance of five alkali-resistant polymeric NF and UF membranes	Alkali liquid (200 gpl), 17%	Studied the effect and Temperature and pressure	Polymeric NF and tight UF membranes	Hemicellulose retention above 1,000 g/mole achieved using polyethersulfone-type membranes MWCOs are found on the upper side due to higher temperature of 49 °C, swelling and enlarging effective pore size
[7]	Ultrafiltration ceramic membranes	Long-term performance	Highly alkaline stream of viscose process containing 17–18 wt% sodium hydroxide	Studied the effect of feed composition, temperature, and pressure	Tubular Ceramic Membranes with MWCO 3, 5 and 15 kDa	Pulp swelling, reactivity, porosity, absorption capacity was noted to decrease with a rise in hemicelluloses of steeping lye. Drop-in permeate flux was found to reduced ceramic membranes than polymeric membrane (UF, 3 kDa)
[31]	Ultrafiltration membranes	Lab- and pilot-scale experiment	Spent sulphite liquor of the sulphite pulping process	An alkaline cleaning step (pH 11, 1 h) is required to control constant flux and retention	–	The observed difference in retention and flux in lab and pilot-scale experiment

5 Membrane-Based Processes for Separating Hemicelluloses from Effluent Streams

The membrane-based separation and chemical recovery process have been widely applied for desalination and wastewater treatment due to the selective removal of impurities based on molecular size. Two types of membranes namely, polymeric and ceramic membranes are widely employed mostly for various wastewater treatment applications [32, 33]. Polymeric membranes primarily govern waste-water treatment due to their ease of fabrication and low cost in full-scale applications. However, degradation, instability, and incompatibility with chemicals, high fouling reduce their efficiency which can also impact the sustainability of the long-term separation process. This generates more interest in ceramic-type membranes [17, 30, 32]. In comparison to the polymeric membrane, the ceramic membrane is mechanically, thermally, and chemically more stable with long-term performance and is considered better for the treatment of wastewater and alkaline solutions generated in pulp and paper industries [17].

The major difference between the two membranes as described in Table 3 shows that ceramic membranes have demonstrated enhanced removal efficiency over a long period, better permeate flux, and fouling performance than polymeric membranes. However, the major drawback of ceramic membranes is the high cost of materials used for fabrication which limits commercial use for wastewater treatment.

Table 3 Difference between polymeric and ceramic membranes [13, 32]

Parameters	Polymeric membrane	Ceramic membrane
Fouling	High due hydrophobicity	Less due to hydrophilic nature
Life	Short	Long
Effect of alkali	High	Low
Operation time	High due to enhanced flux	Lower
Strength	Low	Mechanically, thermally and chemically more stable
Cost	Low	High
Technology	Matured	Under development and commercialization
Polluted water handling capacity	Low concentration only	Concentrated solution
Scale-up and commercialization	Well-known and commercialized at an industrial scale	Under development
Oxidation with membrane materials	Degrade membrane	No degradation so can be integrated with an oxidation process
Cleaning efficiency	Low	Better

Ceramic membranes are classified into integral and composite based on the materials utilized for membrane fabrication. The fabrication of composite membranes is based on the development of multi-layered structures over mechanically stronger supports of hollow-fibre, multi-channel tabular, and flat geometry of various densities. In comparison to flat geometry, hollow fibres and multichannel configurations have been widely considered for wastewater applications because of enhanced packing density, stability and strength.

The filtration layers within the ceramic membrane are generally fabricated using metal oxides like Al_2O_3, ZrO_2, TiO_2 and SiO_2. Worldwide commercial ceramic membranes manufacturers are listed in Table 4. This indicates that ceramic membranes can work over a wide range of process operating conditions like pressure (up to 40 bar), temperature (300 °C), and pH (0–12). As shown in Fig. 5a–d, ceramic membranes consist of three layers; namely a thin selective layer, intermediate layer(s), and a supporting layer fabricated into three main geometry tubular, flat, and hollow fibre which provide good strength, stability and separation selectivity. The cost of ceramic membranes mainly depends on the type of material used for the fabrication. As revealed in Fig. 5e relative comparison of ceramic membrane materials like titanium, zirconia, alumina, and silicon carbide, wherein SiC is the most expensive but has better permeability than the widely used Al_2O_3. The tubular multichannel ceramic membrane has been used widely in wastewater treatment due to tailored modules and customization according to filtration requirements.

The techno-economic feasibility of polymeric and ceramic membranes for large wastewater treatment in terms of CAPEX, OPEX, and total life cycle cost is evaluated in Table 5. In this, the initial cost or CAPEX of ceramic membranes is 5 to 10 times higher than polymeric membranes however its replacement cost is null due to long-period performance (20 years). Therefore, the life cycle costs of both membranes are nearly the same over the period of 20 years. In the upcoming years, the rising demand for ceramic membranes can reduce the price and life cycle cost.

6 Separation of Hemicelluloses from Highly Alkaline Industrial Process Stream

In a recent review study, Lipnizki et al. [35] have highlighted the significant global importance of the pulp and paper business. They emphasize that this industry has immense potential to serve as a foundation for the future bio-economy, namely in the development of lignocellulosic bio-refineries. The membrane processes were emphasized for their potential in three applications: kraft/sulphate, sulphite and thermomechanical pulping. Furthermore, this text delves into the issues surrounding membrane fouling and cleaning in the context of the pulping business. It also provides insights into the potential advancements of membranes in biorefineries.

Table 4 Commercial ceramic membrane used for wastewater treatment [32]

Manufacturer	Geometry	Membrane size	Materials	Design and operating range
ALSYS, USA TAMI	Tubular	MF: 100–200 nm UF: 5–50 nm	Al_2O_3, SiO_2, TiO_2, SiC	Up to 10 bar, pH 0–14, up to 130 °C
Pall, USA		MF: 100–5000 nm UF: 20, 50 nm	Al_2O_3, ZrO_2, TiO_2	8–10 bar, upto 95 °C
Fraunhofer IKTS, Germany		MF: 70–800 nm UF: 5, 10, 30 nm	Al_2O_3, ZrO_2, TiO_2, Al_2O_3	Up to 40 bar, up to 100 °C
Jiuwu Hi-tech, China Jiangsu		MF: 200, 500, 1200 nm UF: 20, 50 nm	Al_2O_3/ZrO_2	Up to 10 bar, pH 0–14, up to 150 °C
Jiangsu Kaimi, China		MF: 100, 200, 800 nm UF: 20, 50 nm	Al_2O_3/TiO_2	2–10 bar, pH 0–14, 0–150 °C
TAMI, France Deltapore		MF: 100–1400 nm UF: 1 to 150 kDa	ZrO_2, TiO_2	Up to 4 bar, pH 2–12, up to 350 °C
LiqTech, Denmark		MF: 100, 1000 nm UF: 40 nm	SiC	Up to 10 bar, pH 0–14, up to 800 °C
Meidensha, Japan	Flat sheet	MF: 100 nm MF	Al_2O_3	Up to −1 bar, pH 2–12, −20 to 80 °C
Ceraflo, Singapore		MF: 500 nm MF	Al_2O_3	−0.2 to −0.5 bar, pH 2–12, up to 60 °C
ItN-Nanovation, Germany CFM		MF: 200 nm MF	Al_2O_3	Up to −0.5 bar, pH 2–12, up to 130 °C
TAMI, France	Disc flat sheet filter	MF: 100–1400 nm UF: 1 to 150 kDa	ZrO_2, TiO_2	Up to 4 bar, pH 2–12, up to 350 °C
Deltapore, Netherland	Hollow fibre	Symmetric MF: 100–200 nm Asymmetric UF: 20, 40 nm	Al_2O_3	pH 0–12, up to 70 °C
Nanostone, USA		UF: 30 nm	α-Al_2O_3	Up to 7 bar, pH 1–13, up to 45 °C
Media and Process Tech., USA		MF: 100, 200 and 40–500 A∘	α-Al_2O_3	–

Al-Rudainy et al. [31] have reported the recovery of hemicellulose by ultrafiltration from lab-scale to on-site pilot trials, which included experimental and theoretical evaluations.

Singh et al. (2018) have demonstrated the impact of dissolved hemicelluloses on pulp swelling and reactivity with ceramic membranes and found that they are more effective for hemicelluloses separation from lye. They observed that the alkali and hemicellulose concentration of feed strongly impacts membrane performance. Singh

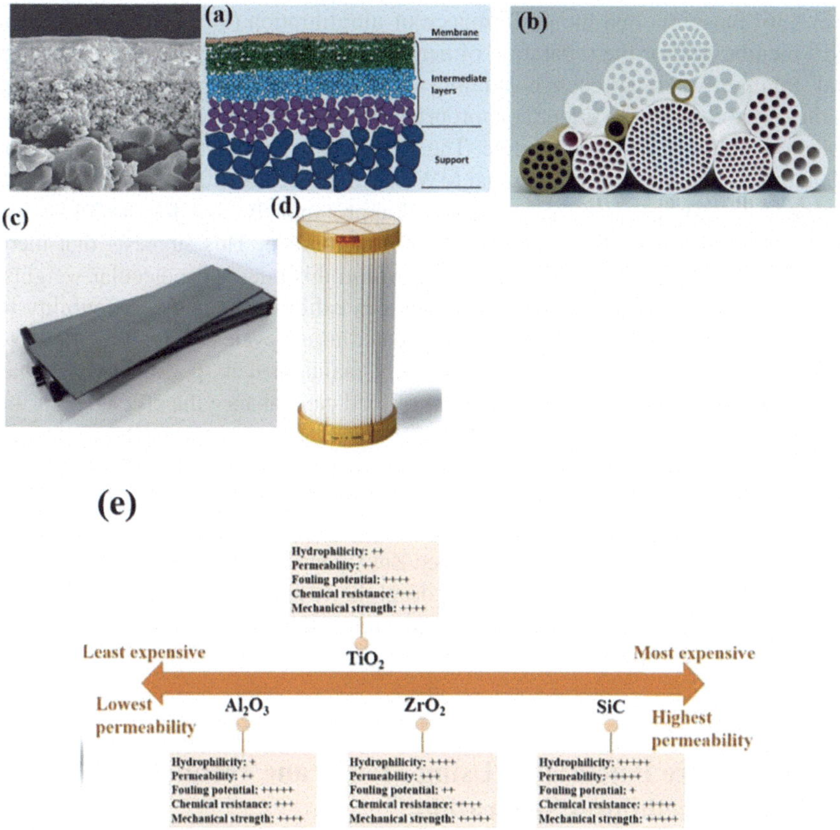

Fig. 5 Ceramic membranes. **a** Cross-section with three layers **b** tubular multi-channel membranes **c** flat membrane **d** hollo-fibre **e** material used for ceramic membranes fabrication and relative cost and permeability (reproduced with permission from Elsevier Inc. [32])

Table 5 Economic analysis of membranes based on wastewater treatment in terms of CAPEX, OPEX, and total life cycle cost [32, 34]

Membrane and waste water treatment parameter	Polysulphonate	Alumina	Polymeric	Ceramic
Processing capacity (m^3/d)	18,925		30,000	
Flux (L/m^3h)	100	125	41.7	95.8
Membrane cost	50	300–500		600
Life (year)	3–5	20		20
Membrane replacement cost	–	–	0.031	Nil
CAPEX (USD/m^3)	0.047	0.05–0.059		
OPEX (USD/m^3)	0.07–0.09	0.06–0.08	0.05	0.02
Lifecycle cost USD/m^3)	0.12–0.138	0.12–0.14	0.274	0.28

et al. [36] have reported the performance of ultrafiltration (UF) and nanofiltration (NF) membranes for the separation of hemicellulose from a highly alkaline industrial stream of viscose process containing 17–18 wt% sodium hydroxide. Screening studies were conducted to identify appropriate membranes, which were subsequently tested on a pilot-scale spiral module. The screening experiments demonstrated that UF and NF membranes, with nominal MWCO values of 3 and 0.5 kDa respectively, exhibited an average retention efficiency of approximately 70% for the 3 kDa UF membrane and around 80% for the 0.5 kDa membrane. This suggests that there is no significant presence of hemicellulose within this range of molecular weights. According to their findings, the UF membranes exhibited greater susceptibility to performance degradation over time and repeated usage cycles during the pilot size testing, while the NF membrane consistently maintained its performance. It has been observed that increasing the amount of feed can enhance the efficiency of the membrane in recovering sodium hydroxide. The viscosity of the feed was significantly reduced by dilution, leading to a 50% increase in flow after accounting for concentration. The addition of nanostructured zinc oxide as filler into both polymeric and ceramic membranes reduces the fouling, improves permeability and roughness. Sheikh et al. [37] reviewed nanosized ZnO embedded membranes for wastewater treatments and it gives enhanced hydrophilicity and in turn permeate rate and rejection. The authors concluded further research is required to commercialize with ZnO-inserted membranes.

7 The Future Potential of Using Membrane Technology for Hemicelluloses Separation

Viscose staple fibers are regenerated cellulose fibers that are made by treating purified dissolving grade wood pulp with concentrated alkaline solutions. This process results in the formation of a soluble sodium xanthate ester, which is then converted into an alkaline cellulose solution called viscose. The viscose is subsequently regenerated into cellulose by reacting it with an acidic solution. The dissolving grade pulp, which is utilized as a raw material for cellulose in the fiber production process, also contains minor amounts of branching short-chain hemicelluloses as impurities. It is necessary to separate hemicelluloses at the beginning of the fiber production process to avoid negative effects on the efficiency of specific equipment and the qualities of the fibers, such as strength and ability to absorb colour. During the initial stage of the viscose process, the cellulosic pulp is enlarged and immersed in an abundant amount of aqueous sodium hydroxide to isolate soluble hemicelluloses. The presence of elevated levels of hemicelluloses in the recycled caustic stream following slurry pressing might lead to operational challenges in the process, such as the caustic solution becoming highly viscous, hence impeding slurry pressing. Due to their smaller size and higher reactivity, hemicelluloses can selectively react with CS_2 during the xanthation process. Hemicelluloses have a significant impact on the physical qualities of fibers, such as their shape and size, as well as on the ability of pulp to absorb water and the chemical properties that determine its reactivity. Simultaneously, it is crucial to recover and recycle the caustic stream that contains hemicelluloses in order to minimize chemical usage in the process and transform it into valuable compounds. Malmali has reported the application of membrane processes for the concentration and separation of sugar streams in biofuel production [38]. Abels et al. [39] reviewed the application of membrane processes in biorefinery. Zabed et al. [25] reviewed the prospects of lignocellulosic bioethanol production prospects of emerging membrane technologies to improve the process. Figoli et al. [40] compiled the articles related to the application of membrane technologies for Biorefining.

It is possible to use membrane filtration for the separation of hemicelluloses in the recycled aqueous alkaline solution but polymeric membranes were found to be plagued by issues of clogging and deformation, especially at the time of use increased. Ceramic membranes based on their characteristics like using cheaper material of construction in membrane development, suitable geometry and appropriate molecular weight cut-off, can be selected to control the level of hemicellulose separation.

8 Summary and Conclusion

Ceramic membranes, in comparison to polymeric membranes, possess greater potential for effectively separating hemicellulose from the concentrated caustic solution. The literature documents a significant correlation between flow and the characteristics of the feed, such as the concentration of alkali and the percentage of hemicelluloses. Additionally, the molecular weight cut-off and feed pressure also play a role in this dependence. When comparing ceramic membranes to polymeric ultrafiltration membranes, it is evident that ceramic membranes provide a stronger and more durable solution in extremely alkaline feed conditions, and they maintain their performance over a longer length of time. Optimum membrane flux and permeate concentration is required to be achieved by regulating process conditions of hemicelluloses separation from recycled caustic stream.

Amongst limited options available for the separation of such dissolved solid from extremely corrosive alkaline media, the scope of using ultrafiltration and nanofiltration was addressed owing to their increasing dominance in industrial applications. The effect of different process parameters including feed characteristics such as hemicellulose concentration, alkalinity of the lye, membrane operating pressure is very important to improve the overall performance of membranes. The viscosity and hemicellulose concentration are the most critical parameter that governs on permeate flux performance of the membranes. Detailed study of membranes at different molecular weight cut-offs indicated that ultrafiltration membranes tend to foul and clog early as compared to nanofiltration membranes. Further, as expected smaller pore size of nanofiltration membranes enables better separation and higher retention of hemicellulose as compared to ultra-filtration membranes. Fouling was predominant in polymeric membranes as compared to ceramic membranes.

In the future, membrane separation technology for efficient separation of hemicellulose from the highly alkaline stream of viscose needs to be explored on an industrial scale in order to evaluate their performance and feasibility. Techno-economic comparison study of wastewater treatment plants showed the price attractiveness towards ceramic membrane-based plants. The rising demand for ceramic membranes for wider applications primarily relies on a reduction in production cost facilitating their growing application on a larger scale. Commercialization with reduced-cost ceramic membranes appears to be a prospective area for hemicellulose separation from steep lye which demands detailed investigation with an aim to validate their long-term performance and economic feasibility. This chapter specified the requirements for an efficient and robust process for hemicellulose separation from the recycled alkaline stream. This book elaborated on the pathway to address such future challenges by highlighting the prospect of using low-cost ceramic membranes in different industries.

References

1. Klemm D, Heublein B, Fink H-P, Bohn A (2005) Cellulose: fascinating biopolymer and sustainable raw material. Angew Chem Int Ed 44:3358–3393. https://doi.org/10.1002/anie.200460587
2. Moon RJ, Martini A, Nairn J, Simonsen J, Youngblood J (2011) Cellulose nanomaterials review: structure, properties and nanocomposites. Chem Soc Rev 40:3941. https://doi.org/10.1039/c0cs00108b
3. Sharma A, Thakur M, Bhattacharya M, Mandal T, Goswami S (2019) Commercial application of cellulose nano-composites—a review. Biotechnol Rep 21:e00316. https://doi.org/10.1016/j.btre.2019.e00316
4. Johnson N, Keith C, Xungai W, Peter K, Lachlan C (2008) Australian science, Australia's future. https://www.aph.gov.au/DocumentStore.ashx?id=d0fc8a53-3606-4d3b-b11f-3038eebe4ba1
5. Strunk P, Lindgren Å, Agnemo R, Eliasson B (2012) Properties of cellulose pulps and their influence on the production of a cellulose ether. Nord Pulp Pap Res J 27:24–34. https://doi.org/10.3183/npprj-2012-27-01-p024-034
6. Singh SC, Khare RA, Murthy ZVP (2021) Separation of hemicelluloses from an industrial stream by nanofiltration and ultrafiltration processes. Cellul Chem Technol 55:989–1000. https://doi.org/10.35812/CelluloseChemTechnol.2021.55.85
7. Singh SC, Khare RA, Murthy ZVP (2018) Effect of hemicelluloses on pulp characteristics and use of ceramic membranes in the separation of hemicelluloses from highly alkaline industrial process stream. Cellulose 25:2577–2588. https://doi.org/10.1007/s10570-018-1716-4
8. Amore A, Ciesielski PN, Lin C-Y, Salvachúa D, i Nogué VS (2016) Development of lignocellulosic biorefinery technologies: recent advances and current challenges. Aust J Chem 69:1201. https://doi.org/10.1071/CH16022
9. Arkell A, Krawczyk H, Thuvander J, Jönsson AS (2013) Evaluation of membrane performance and cost estimates during recovery of sodium hydroxide in a hemicellulose extraction process by nanofiltration. Sep Purif Technol 118:387–393. https://doi.org/10.1016/j.seppur.2013.07.015
10. Krawczyk H, Oinonen P, Jönsson A-S (2013) Combined membrane filtration and enzymatic treatment for recovery of high molecular mass hemicelluloses from chemithermomechanical pulp process water. Chem Eng J 225:292–299. https://doi.org/10.1016/j.cej.2013.03.089

11. Schlesinger R, Götzinger G, Sixta H, Friedl A, Harasek M (2006) Evaluation of alkali resistant nanofiltration membranes for the separation of hemicellulose from concentrated alkaline process liquors. Desalination 192:303–314. https://doi.org/10.1016/j.desal.2005.05.031
12. Fan P, Zhen K, Zan Z, Chao Z, Jian Z, Yun J (2016) Preparation and development of porous ceramic membrane supports fabricated by extrusion technique. Chem Eng Trans 55:277–282. https://doi.org/10.3303/CET1655047
13. Arumugham T, Kaleekkal NJ, Gopal S, Nambikkattu J, Rambabu K, Aboulella AM, Ranil Wickramasinghe S, Banat F (2021) Recent developments in porous ceramic membranes for wastewater treatment and desalination: a review. J Environ Manag 293:112925. https://doi.org/10.1016/j.jenvman.2021.112925
14. Singh SC, Murthy ZVP (2017) Study of cellulosic fibres morphological features and their modifications using hemicelluloses. Cellulose 24:3119–3130. https://doi.org/10.1007/s10570-017-1353-3
15. Toledano A, García A, Mondragon I, Labidi J (2010) Lignin separation and fractionation by ultrafiltration. Sep Purif Technol 71:38–43. https://doi.org/10.1016/j.seppur.2009.10.024
16. Persson T, Jönsson AS (2010) Isolation of hemicelluloses by ultrafiltration of thermomechanical pulp mill process water-Influence of operating conditions. Chem Eng Res Des 88:1548–1554. https://doi.org/10.1016/j.cherd.2010.04.002
17. Ebrahimi M, Busse N, Kerker S, Schmitz O, Hilpert M, Czermak P (2015) Treatment of the bleaching effluent from sulfite pulp production by ceramic membrane filtration. Membranes 6:1–15. https://doi.org/10.3390/membranes6010007
18. Humpert D, Ebrahimi M, Czermak P (2016) Membrane technology for the recovery of lignin: a review. Membranes 6:1–13. https://doi.org/10.3390/membranes6030042
19. Gírio FM, Fonseca C, Carvalheiro F, Duarte LC, Marques S, Bogel-Łukasik R (2010) Hemicelluloses for fuel ethanol: a review. Bioresour Technol 101:4775–4800. https://doi.org/10.1016/j.biortech.2010.01.088
20. Ji X-J, Huang H, Nie Z-K, Qu L, Xu Q, Tsao GT (2011) Fuels and chemicals from hemicellulose sugars. Adv Biochem Eng Biotechnol 128:199–224. https://doi.org/10.1007/10_2011_124
21. Patel A, Shah AR (2021) Integrated lignocellulosic biorefinery: Gateway for production of second generation ethanol and value added products. J Bioresour Bioprod 6:108–128. https://doi.org/10.1016/j.jobab.2021.02.001
22. Sedlmeyer FB (2011) Xylan as by-product of biorefineries: characteristics and potential use for food applications. Food Hydrocoll 25:1891–1898. https://doi.org/10.1016/j.foodhyd.2011.04.005
23. Chaturvedi V, Verma P (2013) An overview of key pretreatment processes employed for bioconversion of lignocellulosic biomass into biofuels and value added products. 3 Biotech. 3:415–431. https://doi.org/10.1007/s13205-013-0167-8
24. Robak K, Balcerek M (2020) Current state-of-the-art in ethanol production from lignocellulosic feedstocks. Microbiol Res 240:126534. https://doi.org/10.1016/j.micres.2020.126534
25. Zabed H, Sahu JN, Boyce AN, Faruq G (2016) Fuel ethanol production from lignocellulosic biomass: an overview on feedstocks and technological approaches. Renew Sustain Energy Rev 66:751–774. https://doi.org/10.1016/j.rser.2016.08.038
26. Qaseem MF, Shaheen H, Wu AM (2021) Cell wall hemicellulose for sustainable industrial utilization. Renew Sustain Energy Rev 144:110996. https://doi.org/10.1016/j.rser.2021.110996
27. Li J, Liu Z, Feng C, Liu X, Qin F, Liang C, Bian H, Qin C, Yao S (2021) Green, efficient extraction of bamboo hemicellulose using freeze-thaw assisted alkali treatment. Bioresour Technol 333:125107. https://doi.org/10.1016/j.biortech.2021.125107
28. M. Kihlman (2012) Dissolution of cellulose for textile fibre applications, pp 1–61
29. Sjöman E, Mänttäri M, Nyström M, Koivikko H, Heikkilä H (2008) Xylose recovery by textile from different hemicellulose hydrolyzate feeds. J Membr Sci 310:268–277. https://doi.org/10.1016/j.memsci.2007.11.001
30. Jönsson AS (2016) Membranes for lignin and hemicellulose recovery in pulp mills. Membr Technol Biorefining 105–133. https://doi.org/10.1016/B978-0-08-100451-7.00005-0

31. Al-Rudainy B, Galbe M, Wallberg O (2020) From lab-scale to on-site pilot trials for the recovery of hemicellulose by ultrafiltration: experimental and theoretical evaluations. Sep Purif Technol 250:117187. https://doi.org/10.1016/j.seppur.2020.117187
32. Asif MB, Zhang Z (2021) Ceramic membrane technology for water and wastewater treatment: a critical review of performance, full-scale applications, membrane fouling and prospects. Chem Eng J 418:129481. https://doi.org/10.1016/j.cej.2021.129481
33. Thuvander J, Lipnizki F, Jönsson AS (2019) On-site recovery of hemicelluloses from thermo-mechanical pulp mill process water by microfiltration and ultrafiltration. J Wood Chem Technol 39:214–223. https://doi.org/10.1080/02773813.2019.1565865
34. Guerra K, Pellegrino J (2013) Development of a techno-economic model to compare ceramic and polymeric membranes. Sep Sci Technol 48:51–65. https://doi.org/10.1080/01496395.2012.690808
35. Lipnizki F, Rudolph G, Thuvander J, Al-Rudainy B, Vives MB (2021) Application potential of membrane processes in the concept of lignocellulose biorefineries. Chem-Ing-Tech 93:1369–1374. https://doi.org/10.1002/cite.202100018
36. Singh SC, Murthy ZVP (2017) Hemicelluloses separation from caustic-containing process stream by ultrafiltration. Sep Sci Technol 52:2252–2261. https://doi.org/10.1080/01496395.2016.1273952
37. Sheikh M, Pazirofteh M, Dehghani M, Asghari M, Rezakazemi M, Valderrama C, Cortina JL (2020) Application of ZnO nanostructures in ceramic and polymeric membranes for water and wastewater technologies: a review. Chem Eng J 391:123475. https://doi.org/10.1016/j.cej.2019.123475
38. Malmali M (2014) Application of membrane processes for concentration and separation of sugar streams in biofuel production. University of Arkansas, Fayetteville. https://scholarworks.uark.edu/etd/2171
39. Abels C, Carstensen F, Wessling M (2013) Membrane processes in biorefinery applications. J Membr Sci 444:285–317. https://doi.org/10.1016/j.memsci.2013.05.030
40. Figoli A, Cassano A, Basile A (2016) Membrane technologies for biorefining. Elsevier (Amsterdam)

The manufacturer's authorised representative in the EU is Springer Nature Customer Service Centre GmbH, Europaplatz 3, 69115 Heidelberg, Germany. If you have any concerns regarding our products, please contact ProductSafety@springernature.com

Printed and bound by CPI Group (UK) Ltd, Croydon, CR0 4YY
01/10/2026
02225949-0016